QM Library

23 1200053 5

TM 113 CHI (OWL)

DO NOT PASS OVER
CHECK UNIT

Physical Ceramics

Principles for Ceramic Science and Engineering

MIT Series in Materials Science & Engineering Series Statement

In response to the growing economic and technological importance of polymers, ceramics, advance metals, composites, and electronic materials, many departments concerned with materials are changing and expanding their curricula. The advent of new courses calls for the development of new textbooks that teach the principles of materials science and engineering as they apply to all classes of materials.

The MIT Series in Materials Science and Engineering is designed to fill the needs of this changing curriculum.

Based on the curriculum of the Department of Materials Science and Engineering at the Massachusetts Institute of Technology, the series will include textbooks for the undergraduate core sequence of courses on Thermodynamics, Physical Chemistry, Chemical Physics, Structures, Mechanics, and Transport Phenomena as they apply to the study of materials. More advanced texts based on this core will cover the principles and technologies of different materials classes, such as ceramics, metals, polymers, and electronic materials.

The series will define the modern curriculum in materials science and engineering as the discipline changes with the demands of the future.

The MIT Series Committee

Samuel M. Allen
Yet-Ming Chiang
Merton C. Flemings
David V. Ragone
Julian Szekely
Edwin L. Thomas

Physical Ceramics
Principles for Ceramic Science and Engineering

Yet-Ming Chiang
Massachusetts Institute of Technology
Cambridge, Massachusetts

Dunbar P. Birnie, III
University of Arizona
Tucson, Arizona

W. David Kingery
University of Arizona
Tucson, Arizona

John Wiley & Sons, Inc.
New York • Chichester • Toronto • Brisbane • Singapore

Acquisitions Editor   Cliff Robichaud
Production Editor   Ken Santor
Designer   Kevin Murphy
Manufacturing Manager Dorothy Sinclair
Illustration Coordinator   Jaime Perea

This book was set in 10.5/12.5 Times Roman by John Wiley & Sons.

Recognizing the importance of preserving what has been written, it is a policy of John Wiley & Sons, Inc. to have books of enduring value published in the United States printed on acid-free paper, and we exert our best efforts to that end.

The paper on this book was manufactured by a mill whose forest management programs include sustained yield harvesting of its timberlands. Sustained yield harvesting principles ensure that the number of trees cut each year does not exceed the amount of new growth.

Copyright © 1997, by John Wiley & Sons, Inc.

All rights reserved. Published simultaneously in Canada.

Reproduction or translation of any part of
this work beyond that permitted by Sections
107 and 108 of the 1976 United States Copyright
Act without the permission of the copyright
owner is unlawful. Requests for permission
or further information should be addressed to
the Permissions Department, John Wiley & Sons, Inc.

Library of Congress Cataloging in Publication Data:
Chiang, Yet-ming
     Physical ceramics / Yet-ming Chiang, Dunbar P. Birnie III, W. David Kingery.
     p. cm. --(MIT series in materials science and engineering)
     Includes bibliographical references.
     ISBN 0-471-59873-9 (cloth : alk. paper)
     1. Ceramic materials.   I. Birnie, Dunbar P.   II. Kingery, W.D.   III. Title
 IV. Series.
    TA455.C43C53   1997
     620.1'4--dc20   95-32997
     CIP

Printed in the United States of America

10 9 8 7 6 5 4

To Chia Chi and Hsu Sha Chiang, for good instruction early on,
and
To April Ruby Rose Birnie, whose boundless energy is always an inspiration.

Preface

It is a generally accepted paradigm of Materials Science and Engineering that materials selection, synthesis and processing give rise to a product's internal structure that determines properties and thus performance in a desired function and use. The core of this model is **structure**. The nature and origin of structure and its influence on properties is the central theme of *Physical Ceramics*. Analysis and practice have shown that this focus on structure is intellectually satisfying and empirically successful.

Of the principle classes of engineering materials, ceramics are in many ways the most interesting and challenging. These inorganic nonmetallic crystals and glasses have an enormous range of structures, properties and applications. They include materials that are weak and strong; friable and tough; opaque and transparent; insulators, conductors, and superconductors; low melting and high melting; diamagnetic, paramagnetic and ferromagnetic; linear and non linear dielectrics; single crystal, polycrystalline and composite; crystalline and glassy; porous and dense, ... they possess many properties or combinations of properties not achievable in other classes of materials. Experience has shown that this wonderful variety and complexity can be ordered, appreciated and learned by concentrating on structure.

Our focus as scientists and engineers is mostly on those materials, which are at the heart of existing technologies and the cutting edge of new technologies. However, it is worth remembering that the use and manufacture of ceramics began about 7000 BC; by 6500 BC, almost all of the techniques for working clay had been invented except for the potter's wheel (circa 3500-4000 BC). Ceramics have the capacity to be formed into an infinite variety of shapes with an enormous range of color, transparency, reflectivity, translucency—forms and visual effects that provide a wonderful medium for aesthetic creation. The central role of structure on properties makes this book equally applicable to understanding and interpreting the creation and performance of these objects.

Any complete study of ceramics requires investigation of many functions and uses - social and ideological as well as utilitarian (During the 1970's high tech ceramics served as a symbol for advanced technology in Japan.) Performance of ceramics in these roles influences, even determines, materials selection, product design and methods of production. We have to admit that selecting *Physical Ceramics* alone for concentrated study is much too narrow at a time when engineering is coming to be widely recognized as a socio-technical activity. Our justification is that this textbook serves as an introduction to the most critical core of the ceramic technological system, and it provides a coherent unit that can be fit into one or two terms of study for well-prepared students. The proposed method of learning allows students to effectively continue into more varied and complex topics.

This book is first and foremost intended to be a teaching text. Each of the authors has had experience teaching introductory courses in ceramics (at the Massachusetts Institute of Technology and at the University of Arizona). The primary intended audience is juniors and seniors in materials science and engineering with a background in inorganic chemistry, chemical thermodynamics and basic crystallography. We have also found the level of presentation suitable for beginning graduate students who have had little prior experience with materials science or ceramics. The material covered builds on prior courses in a satisfying way, including other texts in *The MIT Series in Materials Science and Engineering*, and it provides students with the core understanding necessary to pursue the subject of ceramics as it now exists and to be prepared for new surprises likely to emerge.

Not only for ceramics, but for all materials, we provide an effective framework for learning how to learn. Key concepts are developed in a sequence that builds on firm foundations in a cumulative way, always using the material learned in such a way that its significance is continuously reinforced.

In the first chapter we analyze how atoms and ions combine to form three-dimensional crystals and glasses. Because ceramics consist of atoms and ions of many different sizes and charge and orbital configuration, there is a rich but sometimes intimidating variety of structures. We see how these are constructed from variations on a very few themes. Difficult structures such as the cuprate superconductors, hydrated aluminosilicate clays and complex glasses can become rational; order can emerge from chaos.

As a basis for getting at real processes and real properties, Chapter II introduces and discusses the nature of defects which intrude upon the perfect geometry of ideal crystal structures. Point defects—missing or misplaced, atoms, ions or electrons—are introduced. Ideas we already have in hand about ion size and charge constitute a basis for discussing defect interactions. Defect structures and defect equilibria are seen as a consequence of requirements for mass and charge balance. It is a natural step to extend these structures to linear and planar defects as a basis for understanding dislocation structure and grain boundary structure. Structure remains the key.

A major consequence of point, line, and planar defects is that atoms and ions and electrons are able to move from one place to another. This allows for the migration of matter and of charge, discussed in Chapter III. Because of the wide range of ceramic crystal and glass structures and the diversity of defect mass and charge, these phenomena are much more complex than for metals or semiconductors. Rates of material and electrical transport vary by some twenty orders of magnitude. Examining these processes allows us to investigate how matter transport can effect changes in structure and how both matter and charge transport can lead to many interesting and useful properties. In this process we begin to connect structure and properties, an essential learning paradigm for materials science and engineering.

In Chapter IV we step back and look at the chemical equilibria affecting relationships between the crystalline, glass, and defect structures we have encountered in the first two chapters. Combined with the kinetic understanding developed in Chapter III, phase equilibria examines the driving forces that determine particular crystalline and glass structures that will coexist as distinguishable condensed phases in the complex world of real ceramics.

Finally, in Chapter V we bring all these strands together into a tapestry corresponding to the complex microstructure exhibited by almost all widely used ceramics. Capillary forces direct changes of grain size, grain shape, and pore elimination during firing at a rate determined by the defect mobilities discussed in Chapter III. Phase transformations in glasses lead to more complex and useful structures. Composites are formulated to achieve particular useful microstructures. Microstructures affect all properties; control of strength and toughness provides a practical illustration of what we can accomplish by the rational selection of crystal and glass structures and their arrangement in a microstructure designed for optimal properties.

In maximizing the utility of the method of this book the visualization of three dimensional structures is an essential component of the learning process. We have observed an enormous variability in the ease and confidence with which students transform two-dimensional pictures into three-dimensional images. In our discussion of crystal structures we recommend that students make three-dimensional models with ping pong balls or styrofoam balls. As a further aid toward effective visualization, we are providing a computer disk written by Stephen A. Newcomb and Dunbar P. Birnie III which runs on IBM-compatible computers. It is more visually effective when used with a color monitor, but also works with monochrome displays. It is menu driven and self-explanatory. From a prompt (>) type "keramos" to run the program.

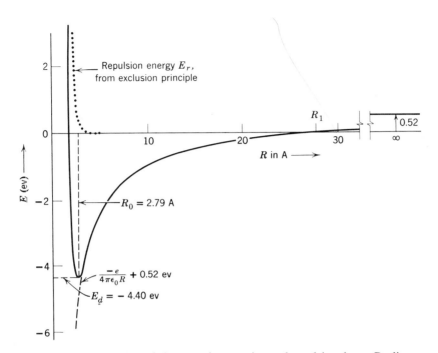

Fig. 1.10 Contribution of electrostatic attraction and repulsion due to Pauli exclusion to the ionic bond energy, where R is the ion separation. The energy zero is referenced to neutral atoms. At infinite separation, the energy is that for forming individual ions from neutral atoms. In this example for KCL, the ionization energy to form K⁺ and the electron affinity for Cl⁻ together total +0.52 eV.

tween ions of like charge for which the energy is repulsive as well as those of opposite charge for which it is attractive, and for a crystal of N molecules is:

$$E_C = N \sum_j \left[\frac{Z_i Z_j e^2}{4\pi\varepsilon_0 R_{ij}} + \frac{B_{ij}}{R_{ij}^n} \right] \quad (1.2)$$

Eq. 1.2 can be rewritten in a form that allows the summation terms to be separated. For a compound of formula "MX" in which the cation valence is Z_c and the anion valence is Z_a, upon writing the separation between ions as $R_{ij} = x_{ij} R_0$, where R_0 is the minimum possible separation (the sum of the ionic radii, $R_o = R_A + R_c$), we have

$$E_C = N \left[\alpha \frac{Z_C Z_A e^2}{4\pi\varepsilon_0 R_0} + C \right] \quad (1.3)$$

In this relation α is the summation of the electrostatic interactions, given by:

Chapter 1 / Structure of Ceramics

$$\alpha = -\sum_j \frac{(Z_i/|Z_i|)(Z_j/|Z_j|)}{X_{ij}} \quad (1.4)$$

α is termed the *Madelung constant*, and represents the electrostatic energy of the crystal relative to the energy of the same number of isolated molecules.[3] The sum of the short-range repulsive interactions in Eq. 1.2 is given by the term C:

$$C = -\sum_j \frac{B_{ij}}{X_{ij}{}^n} \quad (1.5)$$

and will be evaluated later.

The Madelung constant is a measure of the magnitude of the electrostatic stabilization, and for stable crystals has a value greater than unity. It can be evaluated exactly for a particular structure type. Table 1.1 lists the values of the Madelung constant for a few ionic crystal structures. We see that the electrostatic energy of crystals can be substantially lower than that of the corresponding single pairs of ions. It can also be seen that the differences between some structures are relatively small. In some cases, such as the zincblende and wurtzite structures in Table 1.1 (both of which are described later) the difference in electrostatic energy is truly minor (~0.2%). When the energy difference between different structure types of the same stoichiometry is small, we often have *polymorphism*, in which a single compound can take on more than one structure.

For ionic crystals the majority of the interaction energy lies in the electrostatic term, with the short-range repulsion accounting for only about 10% of the interac-

Table 1.1 Madelung Constants for Some Common Ionic Crystal Structures

Structure Type	α
Rocksalt	1.748
Cesium Chloride	1.763
Zincblende	1.638
Wurtzite	1.641
Fluorite	2.519
Corundum	4.040

---

[3]This statement is strictly true only for compunds "MO" of 1:1 stoichiometry, such as NaCl, MgO, ZnO. For other stoichiometries, such as M_2O, MO_2, and M_2O_3, the electrostatic energy of an isolated molecule is no longer given exactly by the term $Z_c Z_a e^2/4\pi\varepsilon_0 R_o$. We may still use this term as an energy reference and use Eq. 1.3 to define the Madelung constant, α, as the energy of the crystal relative to this value. However, α is no longer given by the summation in Eq. 1.4. Some other conventions are also used to define the Madelung constant, as discussed in greater detail in D. Quane, *J. Chem. Education*, **47**[5], 396 (1970) and the original references contained therein.

1.2 Stability of Ionic Crystal Structures

tion. We show this to be the case by first noting that the energy of interaction is a minimum at an ion separation of R_0. Then, upon differentiating Eq. 1.2 with respect to R and setting the result equal to zero at $R = R_0$, the constant C is

$$C = -\alpha \frac{Z_C Z_A e^2}{4\pi\varepsilon_0} R_0^{n-1} \tag{1.6}$$

Substituting this expression back into Eq. 1.3 yields a total energy of interaction of

$$E_C = N\alpha \frac{Z_C Z_A e^2}{4\pi\varepsilon_0 R_0}\left[1-\frac{1}{n}\right] \tag{1.7}$$

where the second term in the brackets is ~10% of the total (i.e., n ~ 10).

Pauling's Rules

The Madelung constant gives the electrostatic energy of a particular crystal structure type relative to isolated molecules or other ionic structures, but does not by itself allow us to predict structures. After all, even though the first four structure types in Table 1.1 are formed by compounds of MO stoichiometry, not all such compounds form in the structure of the largest α, which is cesium chloride. The differences lie in the coordination of anions around cations, and vice versa, in these respective structures. In any structure with a large fraction of ionic bonding character, minimum electrostatic energy is achieved when cation–anion attractions are maximized and like-ion electrostatic repulsion is minimized. That is, cations prefer to be surrounded by the maximum number of anions as *first nearest neighbors*, and vice versa. At the same time, however, cations prefer to maintain maximum separation from the other cations that are their *second nearest neighbors*. Another way of saying this is that ions of like charge prefer to be electrostatically "shielded" from one another as much as possible by ions of the opposite charge. Often, the larger of the ions will form an FCC or HCP array, the interstices of which are occupied by the oppositely charged ion in an orderly way. However, any such arrangement must conform to the need for local charge neutrality, which when extended over the entire crystal maintains the *stoichiometry* or cation/anion ratio of the compound.

Pauling's rules are a set of five general statements, given as follows in approximate order of decreasing influence, which allow us to understand how known ionic structures satisfy the preceding requirements. Conversely, we may use them to predict the structure in which a compound is likely to crystallize. Pauling's rules are based on the geometric stability of packing for ions of different sizes, combined with simple electrostatic stability arguments. These geometric arguments treat ions as hard spheres, which is clearly an oversimplification. However, while ionic radii (as defined by interatomic spacings) do vary from compound to compound, they tend to vary most strongly with the valence state of the ion and the number of nearest-neighbor ions of the opposite charge. Thus for present pur-

14   Chapter 1 / Structure of Ceramics

poses we will consider an ionic radius to be constant for a particular valence state and a nearest-neighbor coordination number. This is a useful approximation supported by empirical observations of the interatomic spacings (determined by the sum of cation and anion radii) in a wide variety of oxides and halides. In general, ion size increases as valence decreases (as electrons are added) and also increases as the number of nearest neighbors increases. These trends can be seen in Table 1.2, which lists ionic radii from the 1976 tabulation of R. D. Shannon. Data for octahedral (CN = 6) and tetrahedral (C N = 4) coordination, and some radii for CN = 8 and 12 are included.

Rule 1   Pauling's first rule states that each cation will be coordinated by a polyhedron of anions, the number of ions in which is determined by the relative sizes of the cation and anion. When anions form a regular polyhedron, there is a single characteristic size for the interstice if the anions (assumed to be spherical) are in contact. For instance, the largest sphere that can fit in the tetrahedral and octahedral interstices of the FCC and HCP arrays when all atoms are touching can be calculated to be 0.225 and 0.414 times the radius of the close-packed atom. The filling of an interstice by a cation smaller than this characteristic size (and which therefore can rattle about) tends to be unstable. A stable configuration is obtained when the cation is as large or slightly larger than this characteristic dimension, as depicted in Fig. 1.11. (Recall that in Figs. 1.1 and 1.2 we showed the close-packed anion arrays as slightly separated to account for the intervening cations in the interstices.) Beyond a certain cation size, however, a larger polyhedron becomes necessary.

We can therefore determine from the cation/anion *radius ratio*, r_C/r_A, the largest polyhedron for which the cation can completely fill the interstice. This is then the local structural unit most likely to form. These polyhedra and the corresponding limiting radius ratios are shown in Fig. 1.12. When the radius ratio is less than this geometrically determined critical value, the next lower coordination becomes preferred. (Remember, however, that a different effective radius applies when we change coordination number.) Although the above is concerned with cations that are smaller than the anion, some important compounds exist for which the reverse is true. In those instances, we can apply the same principle using a cation coordi-

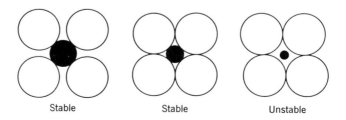

Fig. 1.11 Stable and unstable coordination configurations.

Table 1.2 Ionic Crystal Radii (nm)

Coordination Number = 6

Ion	r	Ion	r	Ion	r	Ion	r	Ion	r	Ion	r	Ion	r	Ion	r	Ion	r	Ion	r	Ion	r	Ion	r	Ion	r
Ag^{1+}	0.115	Al^{3+}	0.054	As^{5+}	0.046	Au^{1+}	0.137	B^{3+}	0.027	Ba^{2+}	0.135	Be^{2+}	0.045	Bi^{3+}	0.103	Bi^{5+}	0.076	Br^{1-}	0.196	C^{4+}	0.016	Ca^{2+}	0.100	Cd^{2+}	0.095
Ce^{4+}	0.087	Cl^{-}	0.181	Co^{2+}	0.075	Co^{3+}	0.055	Cr^{2+}	0.080	Cr^{3+}	0.062	Cr^{4+}	0.055	Cs^{1+}	0.167	Cu^{1+}	0.077	Cu^{2+}	0.073	Cu^{3+}	0.054	Dy^{3+}	0.091	Er^{3+}	0.089
Eu^{3+}	0.095	F^{-}	0.133	Fe^{2+}	0.078	Fe^{3+}	0.065	Ga^{3+}	0.062	Gd^{3+}	0.094	Ge^{4+}	0.053	Hf^{4+}	0.071	Hg^{2+}	0.102	Ho^{3+}	0.090	I^{-}	0.220	In^{3+}	0.080	K^{1+}	0.138
La^{3+}	0.103	Li^{1+}	0.076	Mg^{2+}	0.072	Mn^{2+}	0.083	Mn^{4+}	0.053	Mo^{3+}	0.069	Mo^{4+}	0.065	Mo^{6+}	0.059	N^{5+}	0.013	Na^{1+}	0.102	Nb^{5+}	0.064	Nd^{3+}	0.098	Ni^{2+}	0.069
Ni^{3+}	0.056	O^{2-}	0.140	OH^{-}	0.137	P^{5+}	0.038	Pb^{2+}	0.119	Pb^{4+}	0.078	Rb^{1+}	0.152	Ru^{4+}	0.062	S^{2-}	0.184	S^{6+}	0.029	Sb^{3+}	0.076	Sb^{5+}	0.060	Sc^{3+}	0.075
Se^{2-}	0.198	Se^{6+}	0.042	Si^{4+}	0.040	Sm^{3+}	0.096	Sn^{4+}	0.069	Sr^{2+}	0.118	Ta^{5+}	0.064	Te^{2-}	0.221	Te^{6+}	0.056	Th^{4+}	0.094	Ti^{2+}	0.086	Ti^{3+}	0.067	Ti^{4+}	0.061
Tl^{1+}	0.150	Tl^{3+}	0.089	U^{4+}	0.089	U^{5+}	0.076	U^{6+}	0.073	V^{2+}	0.079	V^{5+}	0.054	W^{4+}	0.066	W^{6+}	0.060	Y^{3+}	0.090	Yb^{3+}	0.087	Zn^{2+}	0.074	Zr^{4+}	0.072

Continued

Coordination Number = 4

Ion	Radius	Ion	Radius	Ion	Radius	Ion	Radius	Ion	Radius	Ion	Radius	Ion	Radius
Ag^{1+}	0.100	Al^{3+}	0.039	As^{5+}	0.034	B^{3+}	0.011	Be^{2+}	0.027	C^{4+}	0.015	Cd^{2+}	0.078
Co^{2+}	0.058	Cr^{4+}	0.041	Cu^{1+}	0.060	Cu^{2+}	0.057	F^{1-}	0.131	Fe^{2+}	0.063		
Fe^{3+}	0.049	Ga^{3+}	0.047	Ge^{4+}	0.039	Hg^{2+}	0.096	In^{3+}	0.062	Li^{1+}	0.059	Mg^{2+}	0.057
Mn^{2+}	0.066	Mn^{4+}	0.039	Na^{1+}	0.099	Nb^{5+}	0.048	Ni^{2+}	0.055	O^{2-}	0.138		
OH^-	0.135	P^{5+}	0.017	Pb^{2+}	0.098	S^{6+}	0.012	Se^{6+}	0.028	Sn^{4+}	0.055	Si^{4+}	0.026
Ti^{4+}	0.042	V^{5+}	0.036	W^{6+}	0.042	Zn^{2+}	0.060						

Coordination Number = 8

Ion	Radius	Ion	Radius	Ion	Radius	Ion	Radius	Ion	Radius	Ion	Radius	Ion	Radius
Bi^{3+}	0.117	Ce^{4+}	0.097	Ca^{2+}	0.112	Ba^{2+}	0.142	Dy^{3+}	0.103	Gd^{3+}	0.105	Hf^{4+}	0.083
Ho^{3+}	0.102	In^{3+}	0.092	Na^{1+}	0.118	Nd^{3+}	0.111	O^{2-}	0.142	Pb^{2+}	0.129	Rb^{1+}	0.161
Sr^{2+}	0.126	Th^{4+}	0.105	U^{4+}	0.100	Y^{3+}	0.102	Zr^{4+}	0.084				

Coordination Number = 12

Ion	Radius	Ion	Radius	Ion	Radius	Ion	Radius
Ba^{2+}	0.161	Ca^{2+}	0.134	La^{3+}	0.136	Pb^{2+}	0.149
Sr^{2+}	0.144						

Source: R. D. Shannon, *Acta Crystallographica*, A32, 751 (1976).

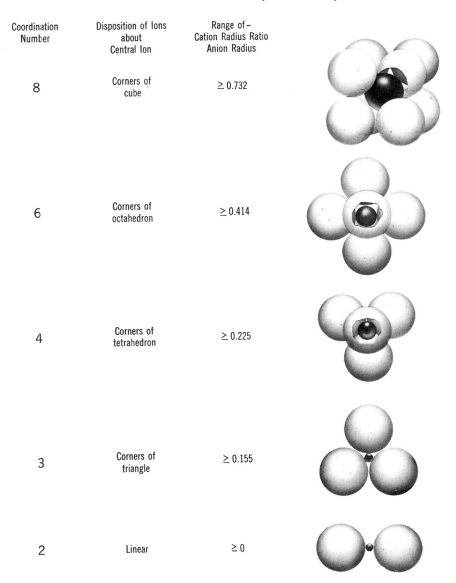

Coordination Number	Disposition of Ions about Central Ion	Range of – Cation Radius Ratio Anion Radius
8	Corners of cube	≥ 0.732
6	Corners of octahedron	≥ 0.414
4	Corners of tetrahedron	≥ 0.225
3	Corners of triangle	≥ 0.155
2	Linear	≥ 0

Fig. 1.12 Critical radius ratios for various coordination numbers. The most stable structure is usually the one with the maximum coordination number allowed by the radius ratio.

nation polyhedron as a structural unit, and use the anion/cation radius ratio to determine the likely coordination number for cations around anions. The fluorite structure discussed later is an example of of such structures.

Exceptions to the exact statement of Pauling's first rule are not difficult to find and occur because ions are not really hard spheres as assumed, but are somewhat deformable. Local electrical neutrality requirements as discussed for the following rule can result in a different coordination number than the first rule predicts, particularly for high atomic number elements. Covalent and metallic bonding character also tend to shorten interatomic distances compared to highly ionic bonds.

Rule 2 This rule ensures that the basic coordination polyhedra are arranged in three dimensions in a way that preserves local charge neutrality. A cation–anion "bond strength" is defined as *the valence of the ion divided by its coordination number*. Taking the compound MgO as an example, octahedrally coordinated Mg^{2+} will have a bond strength of 2/6. This is qualitatively a measure of the relative fraction of the 2+ charge that is being allocated to or shared with each of the coordinating anions. Next, the oxygen anion must be coordinated by sufficient cations to satisfy its 2– valence. This is achieved when *the sum of the bond strengths reaching the ion equals its valence*. This means that in MgO each oxygen atom must be coordinated by six Mg^{2+} ions: $6 \times (2/6) = 2$. In multicomponent compounds an anion may be coordinated by more than one type of cation, in which case each cation may have a different bond strength. Still, the sum of all bond strengths to the anion must equal its valence. This calculation of the contribution of bond strength to local charge neutrality can be performed around either the cation or the anion, and must be satisfied for both. In applying the second rule it thus becomes important to understand the coordination of *cations around anions* as well as those of *anions around cations*.

Rule 3 Coordination polyhedra prefer linkages where they share corners rather than edges, and edges rather than faces. This rule is simply based on the fact that cations prefer to maximize their distance from other cations in order to minimize electrostatic repulsion. (With model polyhedra, one can visualize the decreasing distance between cations upon proceeding from corner-sharing to edge-sharing to face-sharing.) Notice that this preference must be balanced by the fact that corner sharing does not always permit local charge neutrality since it restricts the number of cations that can coordinate an anion (see the second rule).

Rule 4 This rule states that rule 3 becomes more important when the coordination number is small or the cation valence high. A prime example is SiO_2, in which SiO_4^{4-} tetrahedra are linked at corners. This also is based on electrostatics; the repulsive energy between a pair of cations is proportional to the charge squared and inversely proportional to their separation.

Rule 5 Simple structures are usually preferred over more complicated arrangements. For example, when several cations of similar size and identical valence are incorporated into a lattice, they frequently occupy the same type of site but are distributed at random, forming a solid solution or "alloy." As the cations become increasingly dissimilar (and as temperature is lowered), a tendency to form an ordered arrangement or "superlattice" may occur. Finally, when the cations are sufficiently different, they may each take on entirely different coordinations increasing the complexity of the structure.

In summary, then, how can we apply Pauling's rules to deduce the unknown structure of an ionic compound? If the cation is smaller than the anion, as is often the case, we can deduce that FCC or HCP close-packing of the anions will occur. The cation/anion radius ratio helps us to decide which interstitial sites will be occupied; and by far the most common are octahedral and tetrahedral sites. Then, keeping in mind the ratio of interstitial sites to atoms in the FCC and HCP structures, we examine the stoichiometry of the compound. For example, a binary metal oxide MO has a 1:1 cation to anion ratio, so if octahedral coordination (CN = 6) is preferable, we can deduce that all of the octahedral sites will be filled since the ratio of octahedral sites to atoms is also 1:1. If tetrahedral coordination (CN = 4) is preferred, then only one-half of the tetrahedral sites need be filled since the ratio of tetrahedral sites to atoms is 2:1. These sites will tend to be filled in a way that maximizes the cation separation, according to Pauling's third and fourth rules. What follows is a straightforward analysis that helps us understand the more common ceramic structures.

1.3 CERAMIC CRYSTAL STRUCTURES

The majority of ceramic crystal structures are based on either FCC or HCP close-packing of one type of ion, with the other ion(s) occupying a specific set of interstitial sites. We will discuss some of the most important structures, beginning with those based on FCC anion close-packing with the cations occupying octahedral sites. Then, structures that are based on occupation of tetrahedral sites will be considered. Finally, we will discuss some structures of mixed occupancy. Table 1.3 lists a number of representative compounds for each crystal structure type.

For example, the *rocksalt* structure, which is the simplest and perhaps most familiar of the FCC-based structures, has all of the octahedral sites filled with cations. A corresponding FCC structure with all tetrahedral sites occupied is *antifluorite*. On the other hand, the *zincblende* structure has only one-half of the tetrahedral sites filled. It is closely related to the diamond cubic structure, which is that of the diamond phase of carbon as well as the elemental semiconductors silicon and germanium. The *perovskite* structure, which is quite important for understanding ferroelectric materials and ceramic superconductors, is loosely based on an FCC-based structure with octahedral site filling, but here the FCC array includes both anions and cations.

Perhaps the most important example of an HCP-based oxide with octahedral site filling is *corundum*, the structure of the widely used ceramic, aluminum oxide (Al_2O_3). It is based on the HCP stacking of oxygen with $2/3$ of the octahedral sites filled. We will examine this and some closely related derivative structures: $LiNbO_3$ and $FeTiO_3$. (Derivative structures are more complex structures that can be derived, and are most easily understood, using a simpler structure.) A tetrahedrally

Table 1.3 Some Ceramic Crystal Structures

Structure	Stoichiometry	Anion Packing	Coordination Number around M and X	Examples	Derivative Structures
Rocksalt	MX	FCC	6, 6	NaCl, KCl, LiF, KBr, MgO, CaO, SrO, BaO, NiO, CoO, MnO, FeO, TiN, ZrN	
Zincblende	MX	FCC	4, 4	ZnS, SiC (3C), BN, GaAs, CdS, InSb	Diamond cubic (Si, Ge, C)
Wurtzite	MX	HCP	4, 4	ZnO, ZnS, AlN, SiC (2H), BeO	
Nickel arsenide	MX	HCP	6, 6	NiAs, FeS, FeSe, CoSe	
Anti-fluorite	M_2X	FCC	4, 8	Li_2O, Na_2O, K_2O, Rb_2O	
Fluorite	MX_2	Primitive Cubic	8, 4	CaF_2, ZrO_2, UO_2, ThO_2, CeO_2	Pyrochlore ($A_2B_2O_7$), e.g., $Pb_2Ru_2O_7$, $Gd_2Ti_2O_7$, $Gd_2Zr_2O_7$ Bixbyite (e.g. Y_2O_3)
Cesium Chloride	MX	Primitive Cubic	8, 8	CsCl, CsBr, CsI	
Rutile	MX_2	Distorted HCP	6, 3	TiO_2, SnO_2, GeO_2, MnO_2, VO_2, NbO_2, RuO_2, PbO_2, PbO_2	Defective "Magneli phases" of formula Ti_nO_{2n-1}

Table 1.3 continued

Name	Formula	Structure	Coordination	Examples	Notes
Corundum	M_2X_3	HCP	6, 4	Al_2O_3, Cr_2O_3	Ilmenite ($FeTiO_3$) Lithium Niobate ($LiNbO_3$, $LiTaO_3$)
Perovskite	ABX_3	AO sublattice forms FCC	12, 6, 6	$CaTiO_3$, $SrTiO_3$ $BaTiO_3$, $PbTiO_3$ $LaGaO_3$, $LaAlO_3$ $BaZrO_3$, $PbZrO_3$ $Ba(Pb_{1-x}Bi_x)O_3$, $Ba_{1-x}K_xO_3$ $Pb(Zr, Ti)O_3$ $(Pb, La)(Zr, Ti)O_3$	Ordered solid solutions, e.g., $Pb(Mg_{1/3}Nb_{2/3})O_3$, $Pb(Sc_{1/2}Ta_{1/2})O_3$
Spinel	AB_2X_4	FCC	4, 6, 4	$MgAl_2O_4$, $FeAl_2O_4$ $ZnAl_2O_4$, $ZnFe_2O_4$ $MnFe_2O_4$, $LiTi_2O_4$	Many solid solutions are possible.
Inverse spinel	$B(AB)X_4$	FCC	4, 6, 4	Fe_3O_4, $CoFe_2O_4$ $NiFe_2O_4$, $MgFe_2O_4$	Many solid solutions.
"K_2NiF_4"	A_2BX_4	Alternating perovskite and rocksalt type layers	9, 6, 6	K_2NiF_4, La_2CuO_4 $La_{2-x}Sr_xCuO_4$, La_2NiO_4	"Ruddlesden-Popper" phases $AO \cdot nABO_3$, e.g., Sr_2TiO_4, $Sr_3Ti_2O_7$; high temperature oxide superconductors of formula $mAO \cdot nABO_3$ such as $Bi_2Sr_2CaCu_2O_8$, $Tl_2Ba_2Ca_2Cu_3O_{10}$.

Table 1.3 *continued*

"$Y Ba_2Cu_3O_7$"	$AB_2C_3X_7$	Perovskite-like with missing oxygens	8, 10, 5 or 4, 6	$YBa_2Cu_3O_7$, $MBa_2Cu_3O_7$, where M = Eu, Dy, Ho, Er, Yb	$Y_2Ba_4Cu_7O_x$, $Y_2Ba_4Cu_8O_x$
Silicates (quartz, tridymite crystobalite)	AX_2	Corner-shared SiO_4 tetrahedra	4, 2	SiO_2, GeO_2	β-eucryptite ($LiAlSiO_4$) is a quartz derivative; many other crystalline silicates of similar coordination; network glasses.
Silicon nitride	A_3X_4	Corner shared SiN_4 tetrahedra	4, 3	α-Si_3N_4, β-Si_3N_4	Silicon oxynitrides solid solutions (sialons) with valence-compensating cation and anion substitutions.

Contents

Chapter 1.   Structure of Ceramics   1

Assumed Knowledge   2
1.1   Close-Packed Lattices   3
       FCC and HCP Lattices   3
       Location and Density of Interstitial Sites   7
             Sites Between Two Close-Packed Layers   7
             Three-Dimensional Arrangements of Interstitial Sites   8
1.2   Stability of Ionic Crystals   9
       The Madelung Constant   9
       Pauling's Rules   13
1.3   Ceramic Crystal Structures   19
       FCC Based Structures   23
             Rocksalt   23
             Anti-Fluorite and Fluorite   24
             Zincblende   27
             Special Topic: Polymorphs and Polytypes   29
       HCP Based Structures   31
             Wurtzite   31
             Corundum   32
             Ilmenite and Lithium Niobate   34
             Rutile   37

x   Table of Contents

   Perovskite   38
     Special Topic: Ferroelectrics and Piezoelectrics   42
   Spinel   49
     Special Topic: Magnetic Ceramics   52
   Perovskite/Rocksalt Derivatives: Cuprate Superconductors   59
     Special Topic: Structure, Conductivity, and Superconductivity   65
   Covalent Ceramics   68
    Silicon Nitride   68
    Oxynitrides: Charge Compensating Solid Solutions   69
1.4 Crystalline Silicates   72
   Oxygen/Silicon Ratio   73
   Clay Minerals   73
1.5 Glass Structure   80
   Glass Formation   80
   Continuous Random Networks   83
   Random Close-Packing
   Radial Distribution Function   85
   Oxide Glasses   87
   Borates and Borosilicates   91

Additional Reading   93
Problems   95

Chapter 2 Defects In Ceramics   101

2.1 Point Defects   102
   Intrinsic Ionic Disorder   104
    Concentration of Intrinsic Defects   105
    Intrinsic vs. Extrinsic Behavior   107
    Units for Defect Concentration   110
   Special Topic: Kröger-Vink Notation   110
   Defect Chemical Reactions   111
    Solute Incorporation   113
    Electrons, Holes, and Defect Ionization   115
    Oxidation and Reduction Reactions   116
   Extent of Nonstoichiometry   117
   Electronic Disorder   118
    Bandgaps   119
    Concentration of Intrinsic Electrons and Holes   119
    Example: Intrinsic Electronic and Ionic
     Defect Concentrations in MgO and NaCl   125
    Donors and Acceptors   126
    Electronic vs. Ionic Compensation of Solutes   129
   Special Topic: Point Defects and Crystalline Density in ZrO_2   131
   Special Topic: Color and Color Centers   133

2.2 Simultaneous Defect Equilibria:
     The Brouwer Diagram   136
     *Special Topic: Some Simple Procedures for Constructing a
               Brouwer Diagram*   141
     Special Topic: Oxygen Sensors based on Nonstoichiometric TiO_2   142
2.3 Defect Association and Precipitation   146
         Point Defect Association   146
         Precipitation   148
         Debye-Huckel Corrections   152
     *Special Topic: Cation Nonstoichiometry, Disorder, and Defect
               Energetics in Lithium Niobate*   152
2.4 Interactions Between Point Defects and Interfaces   155
         Ionic Space Charge Potential   157
         Intrinsic Potential   157
         Extrinsic Potential   158
2.5 Line and Planare Defects   165
     Dislocations   166
     Grain Boundaries   171
         Special Boundaries   172
         General Boundaries   176
         Boundary Films   176
Additional Reading   182
Problems   183

Chapter 3   Mass and Electrical Transport   185

3.1 Continuum Diffusion Kinetics   186
3.2 Atomistic Diffusion Processes   191
     Random Walk Diffusion   192
     Diffusion as a Thermally Activated Process   193
     Types of Diffusion Coefficients   195
     Diffusion in Lightly-Doped NaCl   201
     Diffusion in a Highly Stoichiometric Oxide: MgO   202
     Diffusion in Cation-Deficient Oxides: the Transition Metal Monoxides   205
     Diffusion in a Highly Defective Oxide: Cubic Stabilized ZrO_2 208
3.3 Electrical Conductivity   211
     Relationship Between Mobility and Diffusivity   212
     Ionic and Electronic Conductivity 216
     Cobalt Oxide: a *p*-Type Electronic Conductor   217
     Mixed Electronic-Ionic Conduction in MgO   219
     Ionic Conduction in Cubic ZrO_2 221
     Conductivity in $SrTiO_3$ 222
     *Special Topic: Nonlinear Conducting Ceramics: Varistors
               and Thermistors*   225

xii Table of Contents

3.4 The Electrochemical Potention   233
 The Nernst Equation and Application to Ionic Conductors
 Ambipolar Diffusion   236
  Equilibration of Defect Structures   238
  Ambipolar Diffusion in Sintering   242
 Special Topic: Diffusional Creep as an Example of
  Ambipolar Diffusion   245
 Special Topic: Kinetic Demixing   251
Additional Reading   256
Problems   256

Chapter 4 Phase Equilibria   263

4.1 Thermodynamic Equilibrium   263
 Special Topic: Metastability in Carbon: Diamond and
  Diamond-like Materials   265
4.2 The Gibbs Phase Rule   267
4.3 Binary Phase Diagrams   271
 Complete Solid Solution   271
 Limited Solid Solution   273
  Binary Eutectic System   273
  Intermediate Compounds
 Special Topic: Free Energy Curves and the Common
  Tangent Construction   277
 Peritectic Diagrams and Incongruent Melting   283
 Subsolidus Phase Equilibria   285
 Solidus and Liquidus Temperatures   287
 Variable Valence Systems: Example in Fe-O   288
 Binary Lever Rule   290
 Special Topic: Crystal Growth and Phase Equilibria   291
4.4 Features of Ternary Phase Diagrams   293
 Reading Compositions   294
  Example: SiO_2-Al_2O_3-"FeO"   295
 Primary Phase Fields   297
 Congruently and Incongruently Melting Compounds   297
 Boundary Curves and Temperature Contours   299
 Ternary Invariant Points   300
 Compatibility Triangles   301
 Solidus Temperatures   301
 Liquid-Liquid Immiscibility   302
 Special Topic: Reciprocal Salt Diagrams   302
4.5 Operations Using Ternary Diagrams   307
 Constructing Binary Diagrams from Ternary Diagrams   307
  Example: "FeO"-SiO_2 Binary Diagram   308

Table of Contents   xiii

       Constructing Isothermal Sections   310
       The Ternary Lever Rule   315
4.6   Reactions Upon Heating and Cooling   318
       Ternary Eutectic Reaction   318
       Ternary Peritectic Reaction   319
       Reactions Upon Heating   321
            Heating Through a Ternary Eutectic   322
            Heating Through a Ternary Peritectic   327
       Equilibrium Crystallization Paths   332
            Eutectic Solidification   332
            Crystallization with Partial Resorption   335
       Nonequilibrium Crystallization   340
       Special Topic: Porcelain   342
Additional Reading   345
Problems   345

Chapter 5   Microstructure   351

5.1   Capillarity   351
       Pressure Due to Curved Surfaces   354
       Chemical Potential Changes at Curved Surfaces   356
       Wetting and Dihedral Angles   360
       Special Topic: Rayleigh Instability and Microstructures   368
5.2   Grain Growth and Coarsening   371
       Grain Boundary Migration and Grain Growth   372
       Particle Coarsening (Ostwald Ripening)   388
5.3   Single Phase Sintering   392
       Viscous Sintering   392
       Crystalline Ceramics   398
       Later Stage Sintering   404
       Packing, Agglomeration, and Pore Growth   409
       Special Topic: Magnesia-doped Alumina   413
5.4   Reactive Additive Sintering   421
5.5   Hot-Pressing   429
5.6   Glasses and Glass-Ceramics   430
       Crystallization and Glass Formation   431
       Controlled Crystallization in Glass-Ceramics   430
       Phase Separation   449
       Multiple Phase Separation   460
       Special Topic: Thermal Shock Resistance   464
5.7   Composite Properties   466
       Rules of Mixing   468
       Percolation   474

xiv   Table of Contents

5.8   Strength and Toughness   477
      Brittle Fracture   478
      Stress Intensity Factor and Fracture Toughness   481
      Variability in Strength   485
      Surface Flaws   486
      Microstructural Toughening   487
      Transformation Toughened Zirconia   488
      Other Flaw-Tolerant Microstructures   492
      Silicon Nitride   494
      Fiber- and Whisker-Reinforced Ceramics   496
Additional Reading   500
Problems   501

Index   515

Chapter 1

Structure of Ceramics

Atomic-level structure is a foundation for understanding the physical properties of materials. In this chapter we will discuss the idealized structure of crystalline and glassy ceramics, as the first level in a hierarchy of structures to be explored in subsequent chapters. The structure of perfect crystals is the basis for understanding defects in crystals and their many related physical properties, as discussed in Chapter 2. In Chapter 3, crystalline and defect structures are then used as the foundation for understanding mass and electrical transport in ceramics. Phase equilibria and microstructure-property relationships in single and polyphase ceramics, treated in Chapters 4 and 5, respectively, also cannot be meaningfully discussed without an understanding of the atomic-level structure of the phases involved.

Our principal objective in the following discussion will be to develop a systematic understanding of why particular crystal structures form, and how they may be predicted. Well-known ionic structures and their important derivatives are described, with an emphasis on viewing even complex structures from the basis of a few simple structure types. A second objective is to illustrate important relationships between the atomic arrangements of crystalline and glassy structures and the physical properties of ceramics, using examples with particular technological relevance.

Assumed Knowledge

As background, we will assume familiarity with general chemistry principles including the periodic table of the elements, the Bohr model of the atom, and the fundamentals of interatomic bonding. In the great majority of ceramic materials, it is ionic and covalent bonding that is of interest; ceramics with predominantly hydrogen bonding or metallic character exist but are relatively uncommon. With respect to the following discussion of crystal structures, a feature of particular importance is the *nondirectional* nature of the ionic bond compared to the *directional* nature of the covalent bond. This simply means that in the case of the ionic bond, electrostatic attraction is equally favorable in all directions and does not by itself promote a certain local bonding geometry. In contrast, the allowed geometries of the covalent bond are very much constrained by the electron orbital configuration. For example, the most commonly encountered type of covalent bonding in ceramic materials occurs where atoms have sp^3 hybridized orbitals, which may be visualized as extending in four lobes toward the vertices of a tetrahedron, with a symmetric interbond angle of 109.5°. This causes many covalently bonded ceramics to adopt a tetrahedral local coordination.

We will assume an elementary understanding of crystallography and of the system of Miller indices.[1] If not familiar, this information is available in a number of texts to which the reader may wish to refer, some listed at the end of the chapter. We will not emphasize the formal, crystallographic description of structures. A complete and quantitative description of crystals requires an identification of the periodicity (unit cell) of the structure, the three-dimensional symmetry, and the detailed positions of all atoms. There exists a widely accepted convention for describing these essential characteristics of crystals, embodied in an important reference known as the *International Tables for Crystallography*. The possible atomic arrangements of crystals are quantitatively described therein. However, while an understanding of this system is important background for the practicing materials scientist and engineer, it is a language that may not be familiar to many readers. In the interest of simplicity, we will only refer to it in passing. Sources for additional reading on the subject are given at the end of this chapter.

The theory of crystallography is extremely powerful in another respect; it permits the prediction of anisotropy in physical properties from crystal symmetry. We will discuss a number of properties that are related to crystal symmetry. Here also, a detailed discussion is not possible without a more extensive background in crystallography. The reader is referred in particular to the text by J. F. Nye, *Physical Properties of Crystals* (Clarendon Press, Oxford, 1957), for further reading on this topic.

---

[1] Whereby single crystallographic directions are referred to by the notation [hkl] and a family of directions by <hkl>, and a single crystallographic plane by (hkl) and a family of planes by {hkl}.

1.1 CLOSE-PACKED LATTICES

A majority of ceramic compounds crystallize in structures based on close packing of at least one of the atomic constituents. From the two simple close-packed lattice face-centered cubic (FCC) and hexagonal close packed (HCP)—most ionic crystals can be derived by the substitution of atoms into available *interstices*, the geometrically regular voids between atoms. It is generally the larger of the ions (usually the anion) that forms the close-packed structure, with the smaller ion or ions occupying the interstices. In the following discussion, we will often take the anion to be oxygen since so many important ceramics are oxides, but this anion may just as easily be, for example, a halogen ion (F, Cl, Br, I), rogen, or sulfur. In the case of high atomic number cations, for example zirconium and uranium, the cations may be larger than oxygen and the structure better viewed as a close-packed arrangement of cations, with oxygen inserted in the interstices. This is the basis of the fluorite structure, which we will use extensively as an example in this text. And, in some structures such as perovskite, the cations and anions are close in size and together form a close-packed lattice that is a convenient basis for viewing the crystal structure.

FCC and HCP Lattices

We begin by examining the characteristics of the FCC and HCP close-packed lattices, both of which consist of a sequential stacking of planar layers of close-packed atoms. These structures may be quite familiar; we will especially emphasize the location and number density of interstitial sites available to cations. Figure 1.1 shows one of the close-packed planes used to form an FCC or HCP lattice. Let's assume this to be an oxygen plane in a metal oxide crystal. Within the layer, each oxygen has six *nearest-neighbor* oxygen ions. Notice that in this hard-sphere representation, the oxygens are not in contact; this illustrates the situation in a real crystal where the anions are slightly separated from each other by the intervening cations. Thus the crystallographic unit cell size is determined by the size of the cation as well as the anion. We will arbitrarily call this an "A" layer, defining all positions that are directly above the centers of the oxygen atoms as "A" positions, whether they are occupied or not.

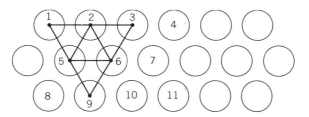

Fig. 1.1 Hexagonal close-packed layer of atoms.

4   Chapter 1 / Structure of Ceramics

We arrive at the FCC and HCP lattices by stacking like layers on top of this first layer, in the most densely packed arrangement possible. This is achieved by nesting the oxygen atoms of the next layer in the depressions between three oxygen atoms in the A layer, Fig. 1.1. The spacing of atoms in the next close-packed layer "B" requires that every other triangular array of atoms in layer A be occupied. Such positions are defined by atoms 1, 2, 5 and 2, 3, 6 in Fig. 1.1. Figure 1.2 shows how this B layer of close-packed oxygen atoms is positioned with respect to the A layer. The two are identical except for a lateral translation; the B-layer atoms are denoted in Fig. 1.2 by lowercase letters and do not lie directly above any of the A-layer atoms.

Continued stacking of close-packed layers on top of the B-layer generates the FCC and HCP lattices. The FCC lattice is formed when a third layer is stacked with its atoms nested in the triangular arrays of atoms in the B layer that have their apex downward in the orientation of Fig. 1.3. It turns out that these positions do not lie directly over atoms in either the A or B layer, and so we denote it as a "C" layer (shown shaded). The FCC lattice repeats when a fourth layer is added over C with its atoms directly above those in A. (Here again the new layer occupies triangles in the C layer that have their apex pointing downward on the page.) The FCC stacking sequence (ABCA) is then repeated indefinitely to form the lattice: ABCABCABC.... Even though this lattice is made by a stacking sequence of hexagonal planar layers, in three dimensions the unit cell is cubic. A perspective showing the cubic FCC unit cell appears in Fig. 1.4, where the {111} planes of atoms are the original A, B, C, and A layers of close-packed oxygen ions. If the unit cell in Fig. 1.4 is oriented so that the body diagonal (a $\langle 111 \rangle$ direction) is vertical, we then obtain the side-view perspective in Fig. 1.6 showing clearly the ABCA.... stacking sequence.

The HCP lattice is formed by stacking another A layer directly above the A–B layer sequence shown in Fig. 1.2. Figure 1.5 shows this arrangement, where the second A layer is now directly above the initial one. Notice that this results from the second A layer being nested in triangular arrays of atoms in the B layer which

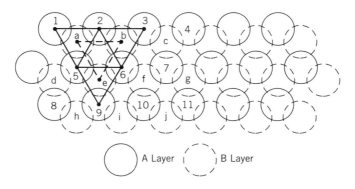

Fig. 1.2 Close-packing of two atomic layers.

1.1 Close-Packed Lattices 5

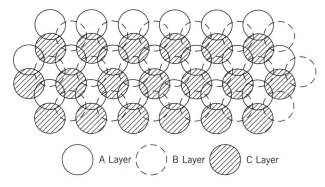

Fig. 1.3 A-B-C close-packing of the FCC structure.

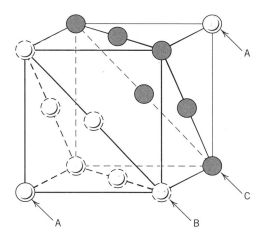

Fig. 1.4 A-B-C-A stacking of close-packed layers in the FCC structure. Each plane is normal to the body diagonal (a <111> type direction).

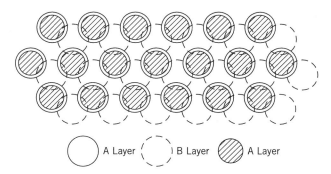

Fig. 1.5 A-B-A close-packing of the HCP structure.

6 Chapter 1 / Structure of Ceramics

have their apex upwards, rather than downwards as in the FCC structure. This sequence alternates as successive layers are added as A and B layers. The final infinitely-repeating stacking sequence is ABABABAB, which we observe from a side-view perspective in Fig. 1.6.

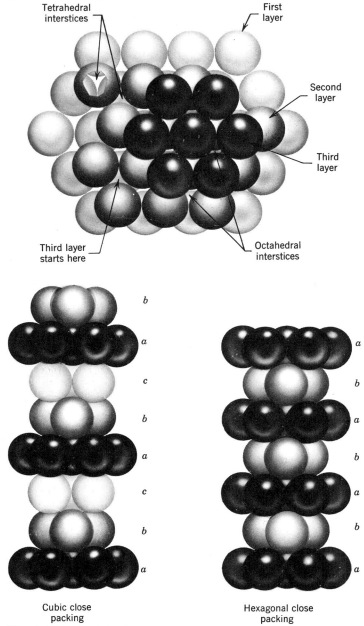

Fig. 1.6 Perspective of FCC and HCP structures viewed parallel to close-packed layers.

1.1 Close-Packed Lattices   7

Location and Density of Interstitial Sites

Sites Between Close-Packed Layers. We now examine the interstices, defined as the polyhedral cavities formed between packed atoms, which exist in close-packed structures.[2] Two principle types of interstitial sites, *tetrahedral* and *octahedral*, exist between layers of close-packed ions. These are the most common locations for cations in ceramic structures. Each site is defined by a local *coordination polyhedron* formed between any two adjoining close-packed layers and which does not depend on the configuration of the third layer. The nearest-neighbor configuration of oxygen atoms around octahedral and tetrahedral cations is thus independent of whether the basic structure is derived from FCC or HCP. From this fact we deduce directly that *FCC and HCP lattices have the same number density of octahedral and tetrahedral sites*.

The location of interstitial sites between an A and B layer are illustrated in Fig. 1.7, where the atoms are labeled the same way as in Figs. 1.1 and 1.2. A single octahedral site is defined by the six oxygen atoms labeled 3, 6, 7, and b, c, f; i.e., three of these atoms are in the A layer and three are in the B layer. (Recall that a regular octahedron has eight sides and six vertices; therefore these six atoms define the vertices of the octahedron.) The octahedral site is centered between these six atoms, equidistant from the two oxygen close-

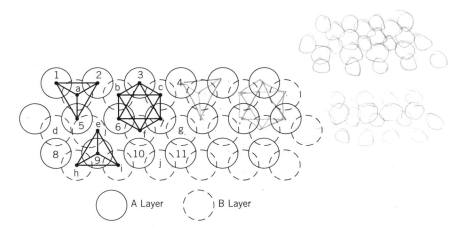

Fig. 1.7 Octahedral (3-6-7-b-c-f) and tetrahedral (1-2-5-a and e-h-i-9) interstices between two close-packed layers of atoms.

---

[2]Physical models are invaluable in helping to visualize the spatial relationships between atoms and interstitials. It is highly recommended that models be obtained or constructed (glued-together ping-pong balls or styrofoam balls are effective) for better visualization.

packed layers. This site is neither directly above nor directly below the atoms in the oxygen layers (A and B). It is co-linear with third layer C atom if the three-dimensional arrangement is FCC). Upon identifying the other octahedral sites between the A and B atomic layers, we find that the octahedral sites together form a hexagonal array, centered halfway between the adjacent close-packed layers and with the same periodicity as the close-packed atoms.

Turning now to the tetrahedral sites, we find that there are two different types defined by any two adjacent close-packed oxygen layers. One group of tetrahedra are defined by having three A-layer atoms and one B-layer atom (see the atoms labeled 1, 2, 5, and a). These tetrahedra can be visualized as having an apex pointing out of the plane of the paper. The other group of tetrahedra are defined by three B-layer atoms and one A-layer atom (see the atoms labeled e, h, i, and 9), and have an apex pointing into the plane of the plane of the paper. Upon identifying all sites of each type, it can be seen that a hexagonal array of each exists as well.

Note that the geometric centers of the tetrahedral sites, unlike those of the octahedral sites, are not exactly halfway between the adjacent oxygen planes but are slightly closer to the plane that forms the base of the tetrahedron. For all tetrahedral sites the center of the tetrahedron is either directly above or directly below an atom in one of the adjacent oxygen layers.

Three-Dimensional Arrangements of Interstitial Sites. Let's now examine the position of these sites with respect to the conventional depiction of the FCC and HCP unit cells. Figure 1.8*a* shows the location of the octahedral sites in the FCC structure: There is one at the cube center, and one halfway along each edge at symmetrically equivalent positions. The FCC unit cell contains at total of four atoms (each face atom contributes one-half atom to the cell, each edge atom one-quarter, and each corner atom one-eighth). If we total the number of octahedral sites, there are also four. *The ratio of octahedral sites to atoms is therefore 1:1.*

Figure 1.8*b* shows the location of the tetrahedral sites in the FCC structure. There is one inside each corner, coordinated by the corner atom and the three closest face atoms. As there are eight of these sites in each unit cell containing four atoms, *the ratio of tetrahedral sites to atoms is 2:1.*

We already deduced that the density of octahedral and tetrahedral sites is the same in FCC and HCP. In Fig. 1.9, the locations of each type of site in the HCP unit cell are shown. It is somewhat more difficult to visualize than in the FCC structure; we find two tetrahedral sites contained entirely within the cell above and below the body atom, and two on each vertical edge of the cell. The total number is therefore four tetrahedral sites per unit cell. Since this cell contains just two atoms, the ratio of tetrahedral sites to atoms is 2:1. There are two octahedral sites per unit cell, situated within the unit cell as shown in Fig. 1.9, for the expected site to atom ratio of 1:1. Notice that the octahedral sites form continuous rows oriented normal to the close-packed atom planes of the HCP structure.

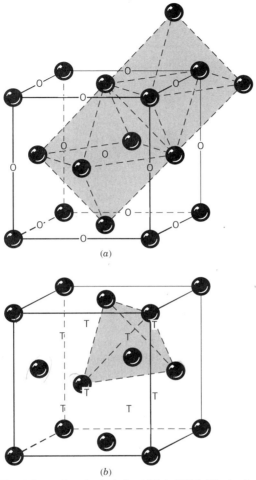

Fig. 1.8 (*a*) Octahedral sites (O) in FCC. The body-centered site and one edge site are outlined. (*b*) Tetrahedral sites (T) in FCC. There is one within each corner, and together they form a simple cubic array.

1.2 STABILITY OF IONIC CRYSTAL STRUCTURES

The Madelung Constant

With this understanding of the atom and interstitial positions in the FCC and HCP structures, we can now develop a geometric approach to ceramic crystal structures through the systematic filling of interstices. First, however, let us consider the question of energetic stability. That is, why is an ionic crystal preferred over the same number of isolated molecules? The *Madelung constant* is a precise definition of the energy of a particular crystal structure relative to the same number of isolated molecules.

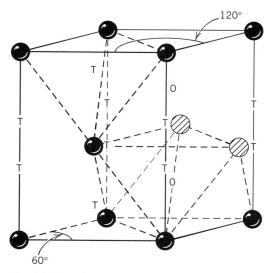

Fig. 1.9 Location of tetrahedral sites (T) and octahedral sites (O) in HCP. Dotted lines show one of each type.

Beginning with cations and anions rather than neutral atoms, we have already expended the *ionization energy* for the cation and gained back the *electron affinity* for the anion. The energy of an ionic bond between a cation and anion pair can then be described by two terms: one is the *coulombic attraction* that is the basis for the bond in the first place, and the second is a *repulsion* due to the Pauli exclusion principle that becomes strong at very close separations. Between two ions of charge $Z_1 e$ and $Z_2 e$, respectively, where e is the electron charge, this interaction energy has the form:

$$E = \frac{Z_i Z_j e^2}{4\pi\varepsilon_0 R_{ij}} + \frac{B_{ij}}{R_{ij}^n} \qquad (1.1)$$

where ε_0 is the permittivity of free space, B_{ij} is an empirical constant, R_{ij} is the interatomic separation, and the exponent n has a value of ~10. The contributions of these two terms to the ionic bond are shown in Fig. 1.10 for the example of KCl. For attractive bonds there exists an equilibrium separation R_0 given by the sum of the cation and anion radii at which the total energy is a minimum, which we will denote as E_0.

Consider now a crystal composed of N such molecules. As separate pairs, the energy of the whole would be NE_0. In order for the crystal to form and be stable, its energy must be less than NE_0. The interaction energy of the crystal is obtained by summing the interaction of each of the $2N$ ions in the crystal, using Eq. 1.1, with every other ion in the crystal (and dividing by two to avoid counting interactions between any ions i and j twice). This summation includes interactions be-filled HCP structure of importance is *wurtzite*, in which one half of the tetrahedral sites are filled. (ZnO is one example.) A final HCP-based structure we will examine is *rutile*, the mineral name of the compound TiO_2, in which one-half of the octahedral sites are also filled, but in such a way that the resulting symmetry of the crystal is actually tetragonal.

The structures of ceramics with a large degree of covalent bonding are not determined by Pauling's rules, but by the directional bonding. These are discussed later. Coincidentally, however, some of the simpler ionic structures also satisfy this requirement. Two such structures are zincblende and wurtzite, in which all atoms have a tetrahedral nearest-neighbor coordination satisfying the tetrahedral sp^3 hybridized orbital orientation.

FCC-Based Structures

Rocksalt. This structure consists of an FCC anion lattice in which all octahedral sites are filled with cations. The radius ratio r_C/r_A should be between 0.414 and 0.732 according to Pauling's first rule, and since the ratio of octahedral sites to atoms in FCC is 1:1, compounds of this structure have an ideal stoichiometry of MX. (There are cases such as $Fe_{1-x}O$ where a deficiency exists on one sublattice, discussed later when we consider lattice defects.) Examples of ceramic materials that form this structure include NaCl, KCl, LiF, MgO, CaO, SrO, NiO, CoO, MnO, and PbO. For all of these, the anion is the larger and forms the FCC lattice. Figure 1.13a shows one unit cell of the rocksalt structure. The lattice constant of the cubic unit cell is "a" and each unit cell contains four formula units.

Let's examine the structure with respect to Pauling's second rule, for the cases of NaCl and MgO. For NaCl the bond strength contributed by a cation is 1/6. The coordination number of anions around cations is the same as that of cations around anions: six. Therefore the sum of bond strengths reaching a chlorine ion is the valence of the chlorine ion ($6 \times (1/6) = 1$) as predicted by the second rule. For MgO, the bond strength is 2/6 and the sum of bond strengths reaching the oxygen ion is $6 \times (2/6) = 2$, the valence of the oxygen ion. Upon examining the structure closely, we find that the octahedra share edges. If they were to share corners, we would have a coordination of two cations per anion and the second rule would not be satisfied.

In order to help visualize atom positions, it is useful to examine specific crystallographic planes. Figure 1.14 shows the (110) plane of the rocksalt structure, in which the row of cations corresponds to a row of octahedral sites at one-half the unit cell height. In this plane also lies unoccupied tetrahedral sites at heights of 1/4 and $3/4a_o$, which may be visualized by referring back to Figure 1.8. Note that simultaneous occupation of the octahedral and tetrahedral sites would bring two cations into close proximity with no shielding anions in between; this is an electrostatically unfavorable arrangement that is avoided. In structures that do have both octahedral and tetrahedral sites occupied, such as the spinel structure discussed below, partial occupancy of each type of site allows cations to be arranged in a way that avoids this like-neighbor repulsion. In Fig. 1.14, the close-packed

24 Chapter 1 / Structure of Ceramics

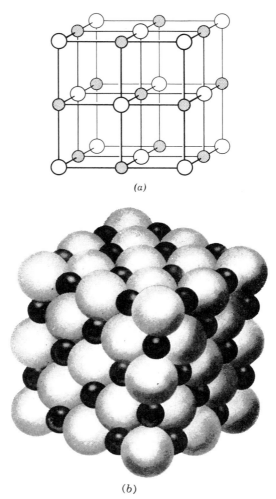

Fig. 1.13 Crystal structure of sodium chloride.

direction shown is the shortest distance between anions and lies in the close-packed plane, which for FCC is the (111) plane (shown earlier in Figs. 1.1 and 1.3). Note that neither the octahedral nor tetrahedral sites lie exactly in this plane since interstitial sites are formed *between* close-packed planes. Parallel to the close-packed anion planes exist alternating layers of similarly packed cations fully occupying the octahedral sites. We thus observe that between any two close-packed layers of anions in the rocksalt structure, there lies a hexagonal array of cations with an identical periodicity.

Antifluorite and Fluorite. What kinds of compounds can be based on FCC anion packing with all of the tetrahedral sites filled? Since there are twice as many

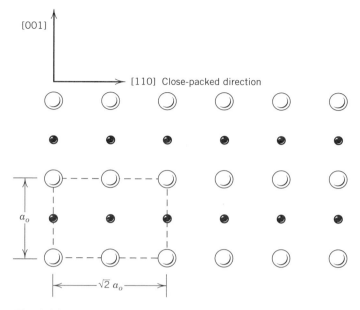

Fig. 1.14 (110) plane of the rock salt structure.

tetrahedral sites as atoms, the stoichiometry must be M_2X. Examples include the alkaline oxides such as Li_2O, Na_2O, and K_2O. They crystallize in the *antifluorite* structure, which is the structural counterpart to the fluorite structure but with the cation and anion positions reversed and a stoichiometry of MX_2. Figure 1.15a shows one unit cell of the antifluorite structure compound Li_2O, in which all tetrahedral sites are occupied by the alkali ion.

The Li ions are somewhat larger than we would expect to find in the tetrahedral sites according to Pauling's first rule, but there are too many for the available octahedral sites, so occupying the tetrahedral sites is the next best thing. The bond strength is 1/4, so according to the second rule the coordination of Li^{1+} ions around each oxygen must be eight. If we examine Fig. 1.15a, this coordination can be verified. The connectivity of the tetrahedra is completely edge sharing; each tetrahedron shares two of its oxygen ions with a neighboring tetrahedron.

The fluorite structure is named for the mineral fluorite, CaF_2. Oxides of the fluorite structure, particularly ZrO_2, UO_2, and CeO_2, have great technological importance in structural, electrical, and chemical applications. It is a common structure for binary compounds in which the cation is large enough that eight-fold coordination is preferred. The structure is based on FCC close-packing of the cations with all tetrahedral interstices filled by anions. We can envision it simply by reversing the cation and anion positions in Fig. 1.15a. The coordination of cations around anions is 4. Using ZrO_2 as an example, if we apply Pauling's second rule to the anion polyhedra as above, the bond strength contributed by each cation is 4/8. Since there are four cations surrounding each anion, the sum of bond strengths

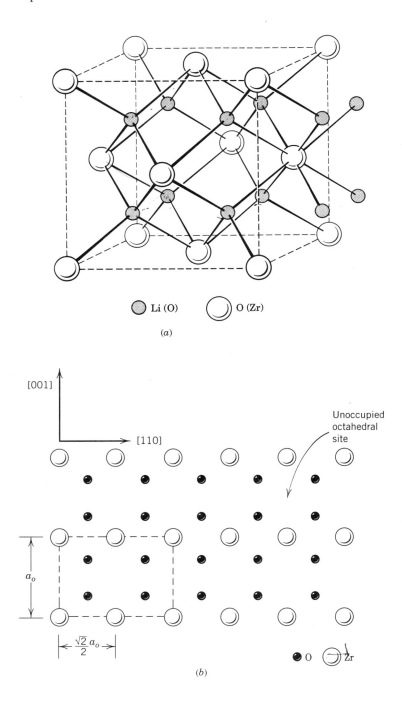

Fig. 1.15 (a) Antifluorite (fluorite) structure, typified by compounds Li_2O and ZrO_2. (b) (110) plane of the fluorite structure compound ZrO_2.

is 4 × (4/8) = 2, which is the valence of the anion. Applying the second rule to the cation coordination polyhedra, the bond strength of an anion is 2/4, and each cation is coordinated by eight anions so that the sum is 8 × (2/4) = 4, which is the cation valence. If we examine Fig. 1.16 closely we find that the anion polyhedra and cation polyhedra each share edges only. Note also that there exists a large unoccupied octahedral site in the body-center and along each edge of the fluorite unit cell. (UO_2 is a useful basis for nuclear fuels in part because this large interstice is able to accommodate various fission products.) A (110) plane of the fluorite structure is shown in Fig. 1.15b, in which the octahedral site at $1/2\ a_0$ and the anions at 1/4 and $3/4\ a_0$ along [001] can be seen.

Cubic bixbyite, the structure in which Y_2O_3 crystallizes, and pyrochlore, the structure of compounds such as $Pb_2Ru_2O_7$, $Gd_2Ti_2O_7$, $Gd_2Zr_2O_7$, are derivatives of the fluorite structure. They are oxygen deficient with respect to the stoichiometry of fluorite; in bixbyite one out of every four oxygen ions is missing, while in pyrochlore one out of every eight is missing. The ideal structures have an ordered arrangement of oxygen vacancies leading to a cubic unit cell composed of eight fluorite type cells, with approximately twice the lattice parameter of flourite. The intrinsically high concentration of oxygen vacancies yields particularly good oxygen ion conduction in some pyrochlores. The FCC cation sublattice is also ordered in the ideal pyrochlore structure. Disorder on both the cation and anion sublattices is possible at elevated temperatures, and further increases the ionic conductivity.

Zincblende. Oxides and sulfides (such as ZnO, ZnS, BeO) with smaller cations that prefer tetrahedral coordination tend to form this structure, as well as covalent compounds such as SiC, BN, and GaAs (for reasons discussed as follows). It is named after the mineral zincblende, ZnS. In contrast to the rocksalt structure in which all octahedral sites were filled, here we need only fill one-half of the tetrahedral sites with the divalent cation in order to satisfy the MX stoichiometry. Recalling from Fig. 1.8b that the tetrahedral sites in FCC form a primitive cubic array, we can fill half of them on the opposing corners of the cube in order to achieve maximum cation separation. The resulting structure is shown in Fig. 1.16. With four anions coordinating each cation the bond strength is 2/4. We also see that each anion is coordinated by four cations; hence, the sum of bond strengths to the anion is 2 and satisfies the second rule. The anion coordination tetrahedra share only corners in this structure.

A number of covalent compounds crystallize in the zincblende structure, but for quite different reasons than the electrostatic arguments embodied in Pauling's rules. Zincblende is one structure type of SiC (others are discussed in the special topic *Polymorphs and Polytypes*), because the covalent bond between Si and C results from the overlap of sp^3 hybridized orbitals with tetragonal directionality. Since atoms of both types are tetrahedrally coordinated in the zincblende structure, this need is mutually satisfied. This tendency is also shared by III-V semiconductor compounds such as GaAs, CdS, GaP, and InSb. For these covalently

28 Chapter 1 / Structure of Ceramics

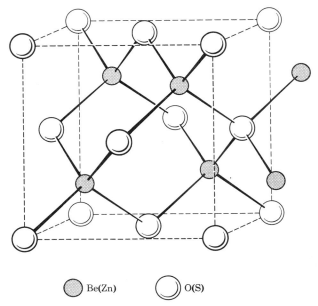

Fig. 1.16 Zincblende (ZnS) structure.

bonded binary compounds the zincblende structure can be viewed as a derivative of the simpler diamond cubic structure. If all atoms in the zincblende structure are identical, we obtain the diamond cubic structure (Fig. 1.17), which is the crystalline form of this well-known phase of carbon as well as that of silicon and germanium. All three are covalent elemental semiconductors with sp^3 hybridized orbital covalent bonding.

Many compounds of zincblende structure can also crystallize in the *wurtzite* structure, discussed in the next section, which shares identical coordination num-

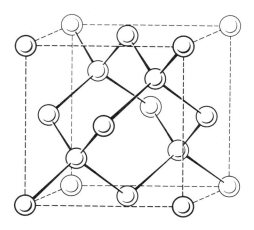

Fig. 1.17 Crystal structure of diamond.

1.3 Ceramic Crystal Structures

bers. For ionic compounds, the electrostatic energy difference (Madelung constant, Table 1.1) between these two structures is very small; hence, there is little preference to form one over the other. Since the two differ only in the stacking sequence of a two-dimensional layer, they may also be considered polytypes of one another.

SPECIAL TOPIC 1.1

POLYMORPHS AND POLYTYPES

Although we discussed ZrO_2 as a *cubic* fluorite structure, in reality there are three distinct crystalline forms or *polymorphs* of this one compound. These are the cubic, tetragonal, and monoclinic phases (Fig. ST1), in which the crystal symmetry differs, as do the detailed interatomic spacings, but the numerical coordination of ions does not. Transformations between the different phases in this system can occur by simple atom displacements, and are known as *displacive* transformations. Other phase transformations require bond breaking and rearrangement, and are known as *reconstructive* transformations.

Cubic ZrO_2 is the highest temperature phase, only stable in the pure form at temperatures between ~2370°C and the melting point at ~2680°C, as shown in the phase diagram in Fig. ST1. However, upon the addition of a few percent of "stabilizers," such as CaO, MgO, or Y_2O_3, this cubic phase can be preserved to lower temperatures (Fig ST1). While the region of thermodynamic stability (equilibrium single phase field) does not reach room temperature for cubic ZrO_2-MgO or other stabilizers (see the ZrO_2-CaO diagram in Fig. 4.8), it can be retained upon cooling as a metastable phase. The cubic forms of ZrO_2-CaO and ZrO_2-Y_2O_3 are technologically important as

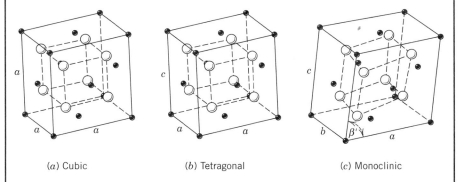

(a) Cubic   (b) Tetragonal   (c) Monoclinic

Fig. ST1 Three polymorphs of ZrO_2: (*a*) c phase, (*b*) t phase (c/a ~1.02), (*c*) m phase. (From A. H. Heuer and M. Rühle, "Phase Transformations in ZrO_2-Containing Ceramics: I, The Instability of c-ZrO_2 and the Resulting Diffusion-Controlled Reactions," *Advances in Ceramics*, Vol. 12, *Science and Technology of Zirconia II*, Nils Claussen, M. Rühle, and A. H. Heuer, Eds., The American Ceramic Society, 1984.)

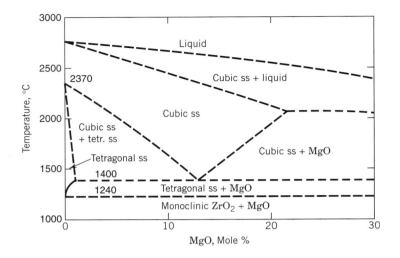

Fig. ST2 Phase diagram of the ZrO_2-rich region in the MgO-ZrO_2 binary system. (From S. C. Farmer, T. E. Mitchell, and A. H. Heuer, "Diffusional Decomposition of c-ZrO_2 in Mg-PSZ," *Advances in Ceramics*, Vol. 12, *Science and Technology of Zirconia II*, N. Claussen, M. Rühle, and A. H. Heuer, Eds., The American Ceramic Society, 1984.)

oxygen ion conductors used in sensors and fuel cells, and as diamond substitutes in the gem trade. The tetragonal-to-monoclinic phase transformation is important for structural applications since it is the basis for *transformation toughening*. While Fig. ST2 shows this transformation occuring at ~1240°C in pure zirconia, in lightly doped ZrO_2 the tetragonal phase can often be retained to room temperature where it is metastable. The transformation to the monoclinic phase involves not only a change in symmetry but also a volume expansion of about 4.7%. When this phase change is stimulated by the local stress field at the end of a sharp crack, the associated energy absorption and volume expansion can be used to impede the propagation of the crack, leading to toughening and strengthening of an otherwise brittle ceramic body, as discussed further in Chapter 5.

Polytypism refers to a special type of polymorphism in which the different crystalline forms of the compound are related by different stacking sequences of a two-dimensional layer. One example is the zincblende and wurtzite structures, which differ only in the stacking sequence of tetrahedrally filled close-packed anion layers. Note that the transformation between the two is necessarily reconstructive, and requires atom diffusion unlike a displacive transformation such as that in zirconia. Another example of a polytypic material is silicon carbide, which can form in not only the zincblende and wurtzite structures but also in a wide variety of stacking sequences with larger periodicity (more than 74 different polytypes have been identified). One reasonably stable polytype is called "6H" (in what is known as Ramsdell notation) because it requires six close-packed layers to repeat the stacking sequence, and the resulting symmetry is hexagonal. According to this notation, cubic or "β" SiC is called "3C," signifying a three-layer stacking sequence and cubic symmetry, while "2H" is the polytype corresponding to wurtzite. "15R" is another type, which repeats after 15 layers and is rhombohedral in symmetry. By convention, all SiC polytypes other than 3C are generally considered to be "α" silicon carbide.

HCP-Based Structures

Wurtzite. The wurtzite structure is based on the HCP close-packing of anions, with one-half of the tetrahedral sites occupied by the cations. The coordination number of each ion is 4. Let's examine the spatial distribution of tetrahedral sites in HCP to determine how they can be one-half filled, with maximum separation between cations. We earlier showed that there are two orientations of tetrahedral sites between close-packed anion layers: those with the apex "up" and those with the apex "down" (Figs. 1.7 and 1.9). Examining Fig. 1.7 closely, it can be seen that there are equal numbers of each, and that each forms a hexagonal array with the same spacing as the anions. One can therefore fill one-half the total tetrahedral sites with maximum separation by filling only tetrahedra of one orientation. This is shown in the unit cell for the wurtzite structure in Fig. 1.18, where all of the filled tetrahedral sites have their apex upward. The coordination of anions around

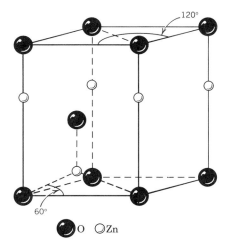

Fig. 1.18 Unit cell of the wurtzite structure.

32 Chapter 1 / Structure of Ceramics

cations and of cations around anions is four, as in the zincblende structure, thus we can be satisfied that the wurtzite structure also meets the requirements of Pauling's second rule. The unit cell for wurtzite is smaller than that of zincblende (it contains only two MX formula units rather than four), so it is more difficult to visualize the interconnections between tetrahedra; they are also linked by corners. We mentioned previously that the Madelung constants of the zincblende and wurtzite structures are very similar (1.638 and 1.641, respectively). Since each has one-half of the tetrahedral sites filled as well it is reasonable that compounds of one structure often have a polytype of the other. Furthermore, since both atoms are tetrahedrally coordinated in wurtzite, it, like zincblende, is the preferred structure of a number of covalent compounds, including AlN and the α phases of SiC.

Corundum. The corundum structure is named for the compound Al_2O_3, also commonly referred to as aluminum oxide, α-alumina (several other polymorphs exist), or sapphire (which commonly refers to the single crystalline form, as it is the basis of ruby, blue sapphires, and star sapphires). Other compounds of corundum structure include Fe_2O_3 and Cr_2O_3. The ilmenite ($FeTiO_3$) and lithium niobate ($LiNbO_3$) structures discussed below are derivatives of corundum. Based on the 2:3 cation:anion stoichiometry of these compounds, cations that take on octahedral coordination must fill two-thirds of the available sites. To see how this occurs in an orderly way with maximum cation separation, the location of octahedral sites between two layers of close-packed oxygen ions have been drawn in Fig. 1.19 (only one oxygen layer is shown). These octahedral sites form a hexagonal array with the same spacing as the oxygen layers. We show the two-thirds that are occupied by aluminum ions in corundum with filled circles, and the one-third remaining empty octahedral sites with x's. The next cation layer has the same honeycomb configuration but is shifted by one atomic spacing, in the direction of the vector labeled "1" in Fig. 1.19. After another close-packed oxygen layer, a third cation layer is placed, now shifted by the vector labeled "2." If we take a vertical

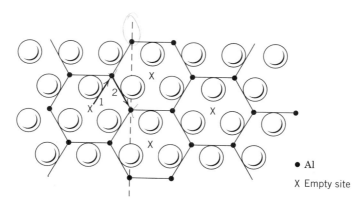

Fig. 1.19 Filling of 2/3 of the octahedral sites in the basal plane of corundum. Only one close-packed anion plane is shown.

1.3 Ceramic Crystal Structures   33

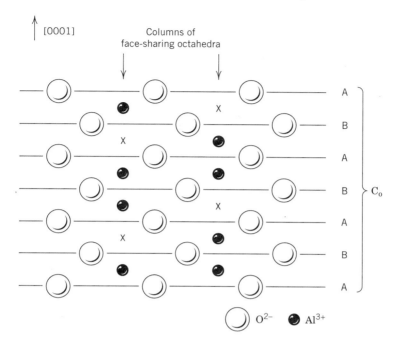

Fig. 1.20 Plane shown by dashed line in Fig. 1.19. Two-thirds occupancy of the columns of octahedral sites is shown.

slice (normal to the plane of the paper) as indicated by the dashed line in Figure 1.19, we have a plane of $\{10\bar{1}0\}$ type in which the arrangements of ions are as shown in Figure 1.20. The columns of octahedral sites perpendicular to the (0001) or basal plane of corundum alternate in having every two sites occupied and one empty. Adjacent columns are staggered in their site occupancy. Fig. 1.21 shows the cation sublattice alone, which repeats after three layers. Taking into account the periodic spacing of both the cation and anion layers, the structure repeats itself after six layers, giving rise to a 12.99 angstrom c_0 unit cell dimension as shown in Fig. 1.20 and 1.21.

The octahedra in corundum share faces, and since two out of every three are occupied (Fig. 1.20), the coulombic repulsion between Al^{3+} ions causes each to move slightly toward the adjacent unoccupied octahedral site. As a result, in real corundum the cations form a slightly puckered layer in the basal plane rather than the ideal structures in Figs. 1.20 and 1.21. The oxygen ions also shift slightly from idealized hexagonal close-packed positions when this occurs.

Applying Pauling's second rule to this structure, we find a bond strength for the octahedrally coordinated cations of 3/6. In Fig. 1.19 we can see that each oxygen ion has three octahedral sites above it (and also three below). Since only two of the three are occupied, the total number of cations around oxygen is four. This satisfies the second rule since the sum is 4(3/6) = 2.

34 Chapter 1 / Structure of Ceramics

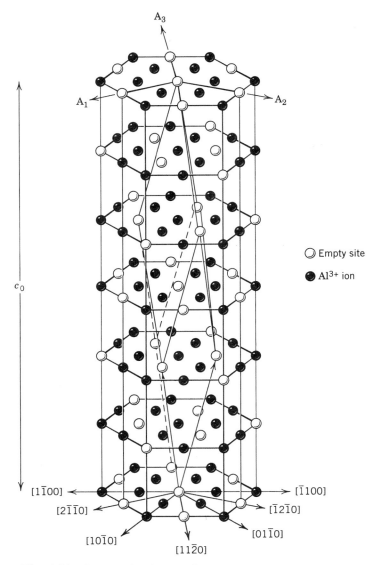

Fig. 1.21 Structural unit cell of corundum (Al_2O_3), showing only the cation sublattice. A_i's are the hexagonal base vectors.

Ilmenite and Lithium Niobate. Two structures that directly follow from the corundum structure are ilmenite and lithium niobate. Ilmenite is the compound $FeTiO_3$; other ABO_3 compounds in which A and B both prefer octahedral coordination can also take on this structure. As in corundum, two-thirds of the octahedral sites are filled, in this case by an ordered substitution of Fe and Ti for Al. The two-thirds-filled arrangement shown in Fig. 1.20 is preserved, but with alternat-

1.3 Ceramic Crystal Structures 35

ing cation layers occupied by Fe and Ti alone. This gives rise to the vertical stacking arrangement shown in Fig. 1.22, which again requires a six cation layer sequence before the structure is repeated. Applying Pauling's second rule, we find that the bond strength of Fe is 2/6 and that of Ti is 4/6. Instead of each oxygen being coordinated by four Al as in the case of corundum, here it is coordinated by two Fe ions and two Ti ions. The net bond strength 2(2/6) + 2(4/6) = 2 is still equal to the valence of oxygen, satisfying Pauling's second rule.

$LiNbO_3$ is another interesting derivative of corundum, in which there is again an ordered substitution of Li and Nb for Al. However, unlike ilmenite, here each basal layer has the two-thirds-filled octahedral sites occupied by equal numbers of Li and Nb ions, as shown in Figure 1.23. The Li and Nb alternate in-plane so that like cations do not occupy immediately adjacent sites. In the vertical (c-axis) direction, we also find a different arrangement than for ilmenite. Figure 1.24 shows a cross-section parallel to the c-axis, taken at a different orientation ($\{11\bar{2}0\}$ plane) than that shown in Figs. 1.20 and 1.21 in order to emphasize first nearest-neighbor octahedra in the basal plane. (From Fig. 1.19 we can see that those in the $\{10\bar{1}0\}$ planes in Figs. 1.20 and 1.21 are actually second-nearest-neighbor octahedra.) In every pair of adjacent occupied octahedral sites there is one Li and one Nb ion. The bond strengths of the Li and Nb ions are 1/6 and 5/6, respectively, and each oxygen ion is coordinated by two of each type, satisfying the second rule.

Notice that each of the Li-Nb pairs in Figure 1.24 is oriented in the same direction, unlike ilmenite in which the Fe-Ti pairs alternate direction. Due to the charge

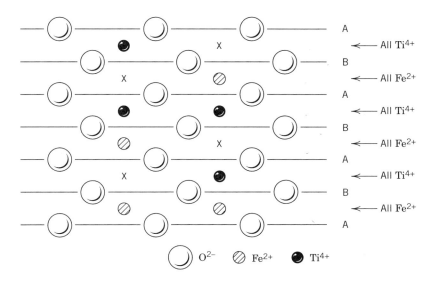

Fig. 1.22 Ilmenite structure in same projection as Fig. 1.20.

36   Chapter 1 / Structure of Ceramics

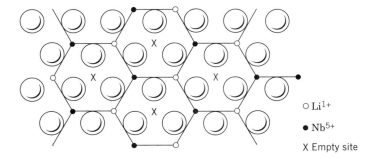

Fig. 1.23 Basal plane of LiNbO$_3$ structure, showing mixed Li, Nb occupancy.

distribution between the Li-Nb pair, a net electric dipole exists for each of the pairs in the structure. This yields a ferroelectric crystal with a permanent net polarization. The anisotropy in structure also yields a highly anisotropic refractive index (birefringence), which is furthermore responsive to applied electric field (high electro-optical coefficients). Because of these properties, lithium niobate has been a leading candidate material for electro-optic devices.

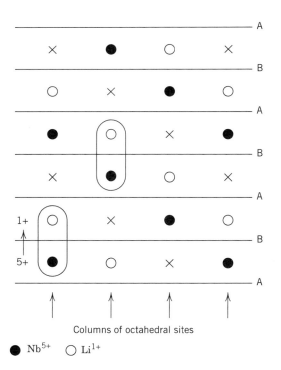

Fig. 1.24 Cation arrangement in LiNbO$_3$, showing orientation of dipole between Li^{1+} and Nb^{5+}.

1.3 Ceramic Crystal Structures 37

Rutile. Rutile is one polymorph of the mineral TiO_2 (anatase and brookite are the others) and has a structure based on quasi-HCP packing of oxygen atoms. Although the cations fill one-half of the available octahedral sites in HCP, the resulting unit cell is tetragonal. We will show how this unit cell arises.

Beginning again with a close-packed oxygen plane and the octahedral sites that exist between it and the next oxygen layer, we can fill one-half of the octahedral sites by completely filling alternating diagonal rows, as shown in Fig. 1.25. Any horizontal row then obviously has one-half of the octahedral sites filled. The next layer of octahedral sites lies directly above the octahedral sites shown here (since the stacking is HCP). In this layer, the Ti atoms fill sites that are directly above the empty ones shown here, and the empty sites are directly above filled ones. The third cation layer is directly above the first one, so that the ABA stacking sequence is simply repeated. The resulting unit cell is tetragonal. However, the atoms do relax from their positions in the perfect HCP lattice due to the partial occupancy of sites by the highly charged Ti^{4+} ions. The dotted line in Fig. 1.25 outlines one side of the unit cell, shown in its usual tetragonal perspective in Fig. 1.26. The octahedron of oxygen ions in Fig. 1.26 comes from two close-packed layers of the type in Fig. 1.25. Comparing these two figures, it is possible to see that the tetragonal structure then has columns of filled octahedral sites along the c-axis direction, and a similar parallel column of empty octahedral sites. This site arrangement results in very anisotropic diffusion properties for some cations in rutile structure oxides. Small cations in particular are able to diffuse interstitially through the empty octahedral sites, and therefore diffuse much faster along the c-axis direction than in the a-axis direction.

Rutile TiO_2 has a large and highly anisotropic refractive index (high birefringence), and has great scattering power as a fine powder of a few tenths of a micron diameter. The largest application of rutile (and its polymorph anatase) is as an opacifying pigment for paints, paper, and fabric. A great advantage is that it is nontoxic, unlike its predecessor in paint applications, PbO.

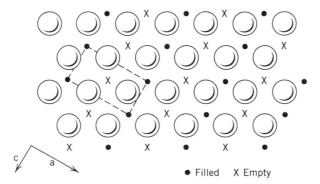

Fig. 1.25 One-half filling of octahedral sites above a close-packed plane (rutile structure).

38 Chapter 1 / Structure of Ceramics

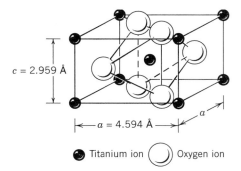

Fig. 1.26 Unit cell of rutile structure.

Perovskite

Many ternary compounds of formula ABO_3 for which the A and B cations differ considerably in size crystallize in the perovskite structure. While no one sublattice is actually close-packed, the structure can be considered an FCC-derivative structure in which the larger A cation and oxygen together form an FCC lattice. The smaller B cation occupies the octahedral sites in this FCC array and has only oxygen as its nearest neighbors. The perovskite family includes many titanates used in electroceramic applications, such as $BaTiO_3$, $CaTiO_3$ (which is the mineral perovskite), $SrTiO_3$, and $PbTiO_3$, zirconates such as $PbZrO_3$ and $BaZrO_3$, and a number of other compounds including $LaGaO_3$, $LaAlO_3$, and $KNbO_3$. It is also a partial structural unit in some more complex structures, such as the superconducting cuprates. For the present discussion, we will use the high-temperature cubic form of $BaTiO_3$ as our model perovskite.

The radius ratio expected for Ba^{2+} to O^{2-} is large, ranging from 0.96 to 1.15 depending on the coordination number we choose (Table 1.2). Although this ratio would appear to be too large for octahedral coordination, the binary compound BaO nonetheless forms in the rocksalt structure since the stoichiometry requires a 1:1 site-to-atom ratio. In $BaTiO_3$, however, a more favorable 12-fold coordination of Ba^{2+} can be adopted. This is accomplished when one barium and three oxygen ions together form an FCC lattice in which the barium ions occupy corner positions (forming a primitive cubic Ba^{2+} sublattice). Charge neutrality is accomplished by filling the body-centered octahedral site with the Ti^{4+} ion. Figure 1.27 shows the unit cell of cubic $BaTiO_3$. Notice that the Ti and Ba ions are shielded from one another by the oxygen ions. The bond strength of Ba and Ti are 2/12 and 4/6, respectively, in this coordination. Each oxygen ion (in face positions on the unit cell) is coordinated by four Ba (at the corners of the unit cell) and two Ti ions. Pauling's second rule is satisfied by the bond sums $4(2/12) + 2(4/6) = 2$. The titanium-occupied oxygen octahedra share only corners with each other, but they unavoidably share faces with the 12-fold Ba polyhedra (dodecahedra). It is difficult to imagine a simpler way of accommodating two cations that are very different in size. Perovskites can be of $A^{2+}B^{4+}O_3$ stoichiometry, as in $BaTiO_3$ and $PbZrO_3$, or

1.3 Ceramic Crystal Structures

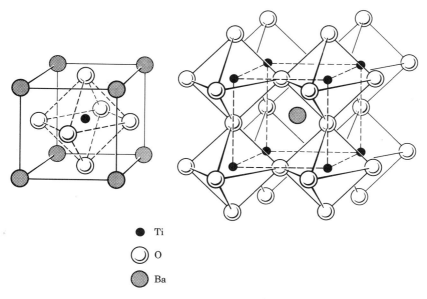

Fig. 1.27 Ion positions in ideal cubic perovskite structure.

$A^{3+}B^{3+}O_3$ stoichiometry, as in $LaGaO_3$ and $LaAlO_3$. Mixed $A(B^{2+}_{1/3}B^{5+}_{2/3})O_3$ or $A^{2+}(B^{3+}_{1/2}B^{5+}_{1/2})O_3$ compositions are also possible, as in $Pb(Mg_{1/2}Nb_{2/3})O_3$ and $Pb(Sc_{1/2}Ta_{1/2})O_3$. In each instance the A-site cations are the larger. An enormous range of perovskite compositions and solid solutions have been developed for technical applications.

Like ZrO_2, many perovskite compounds exist in polymorphs of different crystal symmetry related by simple displacive transformations. The most important of these is the cubic-to-tetragonal transformation. In $BaTiO_3$, the oxygen octahedron coordinating Ti is larger than necessary, being expanded by the large Ba nearest neighbors. This causes the Ti ion to be somewhat unstable in the sense described by Pauling's first rule. The "rattling titanium" ion can be easily displaced from the body-centered position, causing a change in crystal symmetry. The room-temperature structure of $BaTiO_3$ is tetragonal, with the center Ti^{4+} occupying a minimum energy position that is displaced by ~0.12 angstroms from the center toward one face of the unit cell, yielding a structure that is noncentrosymmetric (Fig. 1.28). A most important feature of this spontaneous transformation is that it results in a permanent electrical dipole. A cooperative alignment of adjacent dipoles then occurs, leading to a net polarization that extends over many unit cells. The temperature of the cubic to tetragonal transformation is known as the Curie point and occurs at 130°C for pure $BaTiO_3$. There are also lower temperature transformations to orthorhombic and rhombohedral phases; the transitions between the different phases of $BaTiO_3$ are depicted in Fig. 1.29. The phase transformation temperatures vary widely among perovskites. In $PbTiO_3$, which has a larger Pb^{2+} ion in place of Ba^{2+} and a still more unstable octahedral environment for Ti^{4+}, the

40 Chapter 1 / Structure of Ceramics

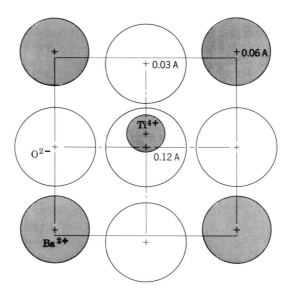

Fig. 1.28 Ion positions in tetragonal BaTiO$_3$. [From G. Shirane, F. Jona, and R. Pepinsky, *Proc. I.R.E.*, 42, 1738 (1955).]

cubic to tetragonal transformation occurs at a higher temperature (490°C) than in BaTiO$_3$. Conversely, SrTiO$_3$ has its Curie point at a lower temperature of -55°C. This can be rationalized from the point of view that the smaller ionic size of Sr^{2+} results in a smaller oxygen octahedron than in BaTiO$_3$, which stabilizes the Ti ion in its centralized position. In practice the Curie point can be continuously varied over a wide range by forming solid solutions between BaTiO$_3$ and PbTiO$_3$ (to raise T$_c$) or SrTiO$_3$ (to lower T$_c$).

Perovskite-based compounds are the basis for a large fraction of the existing electronic, electro-optical, and electromechanical applications of ceramics. In almost every instance, the property of interest is derived from the spontaneous or field-assisted electrical polarization of the noncubic forms. The existence of a permanent electrical dipole allows coupling between an applied electric field and the dielectric, mechanical, or optical properties of the crystal. During the cubic-to-tetragonal transformation, the unstable central titanium ion is easily displaced in one of the six [001] directions by an external electric field; the material can be "poled." In some capacitor applications of perovskites, the orientation of an impending polarization yields unusually high effective dielectric constants. Also accompanying the transformation is a change in crystal dimensions. The coupling between crystal dimensions and applied field is the origin of the *piezoelectric* effect, useful in electromechanical applications where electrical energy is converted to mechanical energy, or vice versa. Lead titanate–lead zirconate solid solutions (given the acronym PZT) have been developed as piezoelectrics for such applications as speakers and microphones, sonar transducers, ultrasonic cleaners, and actuators for high-preci-

1.3 Ceramic Crystal Structures 41

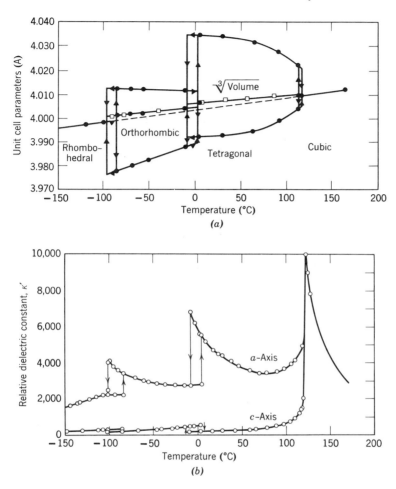

Fig. 1.29 (a) Dimensions of pseudocubic unit cell of $BaTiO_3$. [From H. F. Kay and P. Vousdan, *Phil. Mag.*, 7, 40, 1019 (1949).] (b) Temperature dependence of dielectric constant. [From W. J. Merz, *Phys. Rev.*, 76, 1221 (1949).]

sion positioning. Anisotropy in the optical refractive index (birefringence) also accompanies the electric polarization. The switchable birefringence of ferroelectric perovskites such as La-substituted $PbTiO_3$-$PbZrO_3$ (PLZT) enables applications such as optical shutters, displays, and optical memory. Another application that utilizes the ferroelectric transition is the $BaTiO_3$ positive-temperature-coefficient (PTC) resistor, in which the spontaneous polarization upon cooling below the Curie temperature causes the removal of grain boundary barriers to conduction, resulting in enormous changes in resistivity. PTC resistors are used as temperature sensors and as self-regulating heating elements, discussed further in Chapter 3.

SPECIAL TOPIC 1.2

FERROELECTRICS AND PIEZOELECTRICS

Most materials are *paraelectric*. This means that polarization can be inducd by an applied electric field even though no permanent electric dipole exists. The polarization increases linearly with applied field over some range, and the proportionality between applied field (E, in V/m) and polarization (P, in C/m^2) is the dielectric susceptibility:

$$\chi = \frac{P}{\varepsilon_0 E}$$

The dielectric constant κ and the susceptibility are related by: $\kappa = 1 + \chi$. The mechanisms giving rise to paraelectric behavior are the distortion of individual ions (displacement of the electron cloud from the nucleus) and the polarization of molecules or combinations of ions or defects. The paraelectric susceptibility is completely reversible; removal of the field returns the material to zero polarization.

Ferroelectric crystals are polar: They contain a permanent electrical dipole at the unit cell level as a result of the local atomic arrangement, and this electrical dipole spontaneously aligns with those in adjacent unit cells to yield a net polarization over many unit cell dimensions. Ferroelectricity can only occur in *noncentrosymmetric* crystal structures.[*] The defining characteristic of a ferroelectric is that the direction of polarization is switchable; it can be changed in direction (within a limited set determined by the crystal symmetry) through the application of a sufficiently high electric field. Many crystals are polar; but unless the polarization is reversible with field, the crystal is not considered ferroelectric.

---

[*] From the viewpoint of crystal symmetry, ferroelectricity is quite a restrictive condition. The hierarchy of symmetry-limited properties is as follows. There exist 32 crystal classes, of which 21 are noncentrosymmetric. Of these 21, 20 are piezoelectric: stress induces a polarization or change in polarization. However, only 10 of the 21 are polar in the absence of applied stress. The 10 polar classes are also pyroelectric: changes in temperature cause changes in polarization due to thermal expansion. Finally, if a crystal is not only polar but the direction of polarization is switchable, then it is ferroelectric. This means that ferroelectric crystals are also pyroelectric and piezoelectric. Electro-optical ceramics are broadly defined as those with electrically controllable birefringence. The component of refractive index that varies linearly with electric field is known as the Pockels effect, and that which varies with the square of the electric field is known as the Kerr effect. The magnitude of the response (among other practical factors) defines whether a material is a useful electro-optic, and not surprisingly, is large in a number of polar and ferroelectric materials. Symmetry-dependent crystal properties such as these constitute a very rich area of science and technology. They cannot be discussed in detail without the use of tensor concepts, for which a classic reference is the text by Nye. For further study, see references listed at the end of this chapter.

FERROELECTRIC DOMAINS

The spontaneous alignment of dipoles over many unit cells results in formation of a microstructural entity known as a *ferroelectric domain*. For instance, when the cubic paraelectric phase of $BaTiO_3$ transforms to the tetragonal phase upon cooling below the Curie point, the displacement of the Ti^{4+} ion (Fig. 1.28) can occur along one of the six <100> directions of the cubic phase. A cooperative alignment among neighboring unit cells results in the formation of a ferroelectric domain which may be oriented in any one of these six directions. If only one orientation were to form throughout a single crystal, then we have a single domain, and the opposing surfaces of the crystal where the polarization terminates would be oppositely charged. This kind of long-range charge separation is energetically unfavorable, and instead a number of domain orientations tend to form in all but the smallest crystallites, resulting in a net zero macroscopic polarization, as illlustrated in Fig. ST3. (*Antiferroelectricity* refers to the spontaneous anti-parallel alignment of unit cell dipoles, resulting in zero net polarization, and is distinct from both paraelectric behavior and the random domain orientation that gives zero polarization in a ferroelectric. $PbZrO_3$ is antiferroelectric.)

In tetragonal $BaTiO_3$, adjacent domains can have their polarization vectors in antiparallel directions or at right angles to one another. The boundaries between these domains are, correspondingly, known as 180° or 90° domain walls. *Domain walls* are about one unit cell in thickness and have a small positive energy (2–4 ergs/cm² for 90° walls, 7–10 ergs/cm² for 180° walls). The formation of ferroelectric domains to avoid long-range charge separation comes at the expense of this interfacial energy. In $BaTiO_3$ the ferroelectric domains form as lamellae that are typically a few tenths of a micron to a micron in thickness (Fig. ST4).

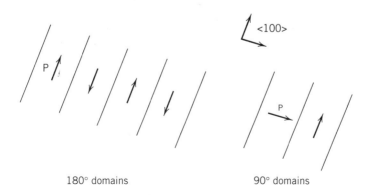

Fig. ST3 Polarization vectors in ferroelectric domains of 180° and 90° misorientation.

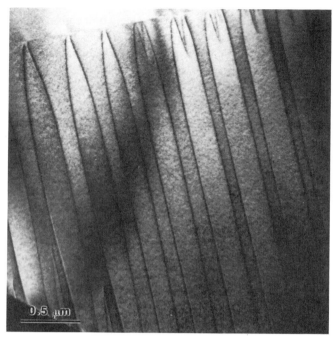

Fig. ST4 Transmission electron microscope image of ferroelectric domain structure in $BaTiO_3$. (Courtesy of H. M. Chan, Lehigh University.)

Upon application of a sufficiently high electric field, ferroelectric domains can be oriented or "poled." While domains cannot be perfectly aligned with the field except when the grain or crystal is coincidentally oriented with its c- or a-axis in the field direction, their polarization vectors can be aligned to maximize the component resolved in the field direction. The process of poling requires the motion of domain walls. In principle, a randomly oriented polycrystal of a tetragonal ferroelectric can achieve a maximum polarization that is 83% of the single crystal value. In practice much lower values are frequently observed due to incomplete domain orientation (domain wall *pinning*) wall motion. Domains misoriented by 180° tend to switch more easily than 90° domain walls since no net physical deformation is required; domains misoriented by 90° are inhibited from switching by the strain which accommodates the switching of c- and a-axes.

HYSTERESIS

A consequence of resistance to domain switching is that polarization in a ferroelectric is hysteretic; it is not precisely reversible with field. This behavior is illustrated in Fig. ST5. Starting with a material in which domains are randomly oriented and yield zero macroscopic polarization, the initial

1.3 Ceramic Crystal Structures 45

shallow slope of P vs. E is that of the paraelectric component of susceptibility. At some higher field, on the order of several kV/cm for $BaTiO_3$ and the PZT/PLZT family of ferroelectrics, domain orientation begins to take place. This results in a sharply rising P with increasing field E; that is, a high dielectric susceptibility. The polarization cannot increase without limit, however, and reaches a *saturation polarization* P_s, which corresponds to the maximum degree of domain orientation possible for that material. Upon decreasing the field back to zero, some of the polarization will be lost (the paraelectric component as well as some ferroelectric contributions) but a *remnant polarization* P_r is retained. Upon reversing the direction of the field E, the polarization is removed until at some *coercive field* E_c a net zero polarization is again obtained. Further increases in the reverse field lead to saturation of P in the opposite direction.

The other physical properties of a ferroelectric which originate from polarization are hysteric as well. Optimizing the size and shape of the hysteresis loop for properties such as the piezoelectric strain, optical birefringence, and dielectric constant is an important aspect of the engineering of ferroelectric materials. The desired characteristics are very much application specific. For instance, an optical memory ferroelectric requires well-defined switching (see the analogous "square loop" magnetic hysteresis in Special Topic 1.3). If domain switching is too difficult, applications can be limited. Even though $LiNbO_3$ is ferroelectric, its Curie point is so high (1210°C) and the fields required to pole it at room temperature are so great that it is not a useful ferroelectric, although it is a useful electro-optic material. Compositions with large hysteresis are also "lossy" in alternating-current applications (the area swept by the loop corresponds to the energy consumed in one cycle), and the crystal distortion accompanying polarization can cause me-

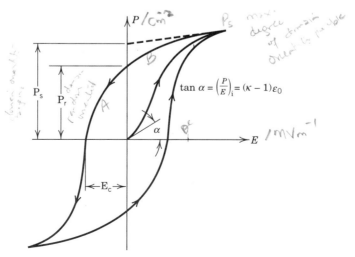

Fig. ST5 A typical ferroelectric hysteresis loop.

chanical fatigue over time. The size and shape of hysteresis loops depends on many factors, chief among which is composition. In perovskites in particular, the extremely broad range of solid solutions which are possible has permitted extensive compositional tailoring of properties for technological applications. An additional important factor is microstructure; polycrystals generally have a smaller and less square loop than the corresponding single crystal, and the grain size and porosity play major roles in determining specific behavior. Finally, the hysteresis loop is inherently temperature-dependent, being larger at lower temperatures and closing completely at the Curie point. The engineering of ferroelectrics for technical applications takes into account all of these factors.

Figure ST6 displays the piezoelectric strain which accompanies polarization of a ferroelectric. Piezoelectric materials are those in which an applied electric field induces a change in lattice constant, or conversely, an applied stress results in the generation of a voltage across the crystal. While piezoelectricity also occurs in many crystals which are not ferroelectric (e.g., ZnO and quartz), the magnitude of the piezoelectric effect is particularly large in compositionally engineered ferroelectrics such as PZT and PLZT. The electromechanical coupling factor is a measure of the efficiency with which electrical energy is converted to mechanical energy in a piezoelectric, and is given by:

$$k^2 = \text{mechanical energy output/electrical energy input}$$

Values of k in useful piezoelectrics range from 0.1 for quartz to 0.4 for barium titanate to 0.72 in a well-poled PLZT containing 7% La substituted for Pb, and a 60/40 ratio of Zr to Ti. The piezoelectric strain coefficient, d, defined as the longitudinal strain obtained per unit applied electric field, is high for well-poled PLZT, reaching values of 6.8×10^{-8} cm/V as compared to 1.9×10^{-8} cm/V for $BaTiO_3$. Note that on an absolute scale, piezoelectric

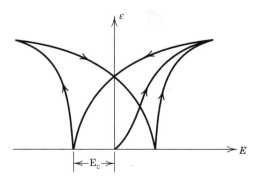

Fig. ST6 Piezoelectric strain accompanying ferroelectric polarization.

strains remain small at reasonable operating fields (below that of dielectric breakdown 10^5-10^6 V/cm). For example, at 10^4 V/cm the maximum strain is only about a tenth of a percent. However, response times are fast; the displacement is precise, and forces generated can be large. Actuators with mechanical leverage can be designed when larger displacements are needed.

DIELECTRIC BEHAVIOR

Many perovskites are also the basis for ceramic capacitors, particularly where circuit miniaturization requires a maximum capacitance per unit volume. As seen in Fig. 1.29, the dielectric constant of $BaTiO_3$ reaches a maximum value just above the Curie point. This behavior results from the impending instability of the paraelectric phase, which makes it easy to polarize the material with just a little applied field. Below the Curie point, the dielectric constant drops substantially from this peak value, and within the tetragonal phase is much higher along the a-axis than along the c-axis. This indicates that once the structure has polarized, the ions are less easily displaced in the direction of polarization than normal to it. The highly peaked behavior of $BaTiO_3$ is problematic in practical applications which require a stable capacitance over a broad temperature range. The ideal capacitor material would show a high but temperature-independent dielectric constant, with low ac loss. A number of approaches have been taken in pursuit of this ideal.

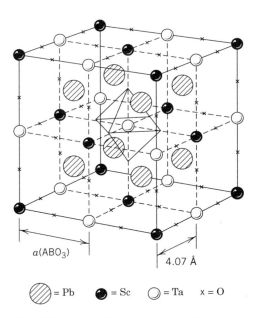

Fig. ST7 Ordered B-site cation sublattice in relaxor perovskite structure [From J. Chen, H. M. Chan, and M. P. Harmer, *J. Am. Ceram. Soc.*, 72[4], 593 (1989).]

48 Chapter 1 / Structure of Ceramics

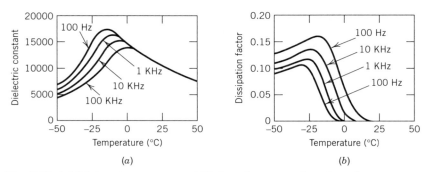

Fig. ST8 (*a*) Dielectric constant and (*b*) loss factor as a function of temperature and frequency for lead magnesium niobate (PMN) relaxor ferroelectric. [From S.L. Swartz, T. R. Shrout, W. A. Schulze, and L. E. Cross, *J. Am. Ceram. Soc.*, 67[5] 311 (1984).]

Relaxor ferroelectrics (sometimes *relaxator*), of which $Pb(Mg_{1/3}Nb_{2/3})O_3$ is an example, are also based on the perovskite structure but with an ordering of the B-site cations that requires eight perovskite units to form a unit cell (Fig. ST7). This ordering occurs over a broad range of temperature as a *diffuse phase transition* occurring in microdomains of the crystal, rather than as a single sharp phase transition at the Curie point. The resulting dielectric constant shows a broad peak with values as high as 25,000 ε_o over

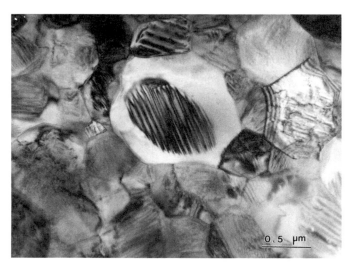

Fig. ST9 Transmission electron microscope view of core-shell structure in a $BaTiO_3$ X7R type capacitor material. (Courtesy of Masayuki Fujimoto, Taiyo Yuden Co., Ltd., Gunma, Japan.)

the temperature range where B-site cation ordering takes place (Fig. ST8). Hysteresis is small in the absence of conventional ferroelectric domains.

Another means of achieving temperature stability of capacitance in a ferroelectric is to incorporate a range of compositions and therefore a range of Curie points in a single material. This is the principle behind so-called "X7R" dielectrics (named after an technical specification, not a material!) which require that the dielectric constant vary by no more than +15% over the temperature range of -55°C to +125°C. These are based on polycrystalline $BaTiO_3$ into which dopants such as Bi and Nb have been diffused along grain boundaries to create a "shell" in each grain of lower Curie point and a lesser-doped "core" of higher Curie point. Figure ST9 shows a room-temperature transmission electron microscope image of this core-shell structure, where the undoped center has undergone its ferroelectric transition and shows a domain structure. This process in effect creates a composite dielectric structure and broadens the Curie transition; the dielectric constant is typically 3000-5000 ε_o. Barrier layer capacitors are another type of dielectric based on polycrystalline $SrTiO_3$ or $BaTiO_3$ in which the grains are conductive but the grain boundaries insulating. These materials utilize space-charge polarization within each grain in addition to an intrinsically high dielectric constant to achieve dielectric constants that can exceed 20,000 ε_o.

Spinel

The spinel structure is based on an FCC close-packed oxygen sublattice in which a fraction of the octahedral and tetrahedral sites are filled. Compounds of stoichiometry AB_2O_4 in which the cations A and B are divalent and trivalent, respectively, ($AO \cdot B_2O_3$) often form as spinels. The unit cell of spinel contains eight FCC oxygen subcells in a cubic array. One-half of the octahedral sites and one-eighth of the tetrahedral sites are occupied. In a *normal* spinel, such as $MgAl_2O_4$ (the mineral spinel), the B^{3+} cations occupy one-half of the octahedral sites and the A^{2+} cations one-eighth of the tetrahedral sites. The A^{2+} bond strength is therefore 2/4 and the B^{3+} bond strength 3/6. In order to satisfy the second rule each oxygen must be coordinated by three octahedral cations and one tetrahedral cation. This structure is shown in Fig. 1.31 as sequential atomic layers parallel to the (100) plane of the unit cell, at height increments of 1/8 a_o. Figure 1.30 shows a three-dimensional depiction. An *inverse* spinel $B(AB)O_4$ has a slightly different occupancy, in which one half of the B^{3+} cations occupy the one-eighth-filled tetrahedral sites, and the A^{2+} ions and the remaining half of the B^{3+} ions occupy octahedral positions. The occupied sites remain those shown in Fig. 1.30, but the cations are exchanged. In reality, most spinels whether they are normal or inverse are disordered to some degree by exchange of the A^{2+} and B^{3+} cations. They may be thought of

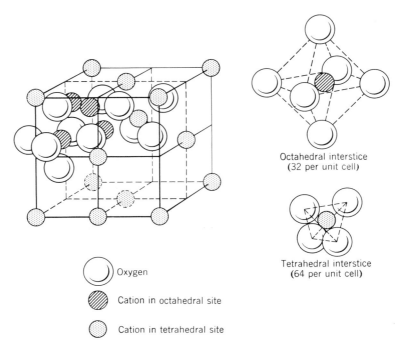

Fig. 1.30 Atomic layers parallel to the (001) plane in spinel.

as a solid solution between an ideal end member (normal or inverse) and a disordered spinel in which the site occupation is completely random. $MgAl_2O_4$ forms the highly ordered normal spinel structure only in its naturally occurring mineral form, in which slow crystallization over geologic times has allowed a high degree of ordering to take place. Synthetic $MgAl_2O_4$ that is grown at elevated temperatures and then cooled at laboratory rates is always disordered to some degree. Magnesium aluminate spinel is a highly refractory compound of some utility for structural and optically transmitting applications. The most commercially important spinel structure ceramics by far are soft magnetic ferrites, used in a broad range of applications such as inductors, transformer cores, and read/write heads for magnetic storage media.

1.3 Ceramic Crystal Structures 51

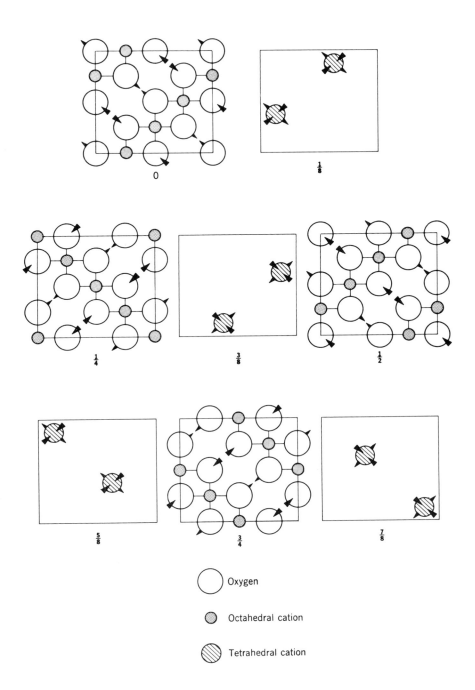

Fig. 1.31 Spinel structure. (From A. R. von Hippel, *Dielectrics and Waves*, John Wiley and Sons, New York, 1954.)

SPECIAL TOPIC 1.3

MAGNETIC CERAMICS

Magnetism in ceramics originates from two principal sources. One, common to all materials, is *diamagnetism*, the induced magnetic moment that results from the effect of an applied magnetic field on the angular velocity of electrons orbiting the nucleus of an atom. Materials that we do not think of as "magnetic" are in fact diamagnetic. According to Lenz's law, the change in velocity of a circulating current in a magnetic field always results in a magnetic moment opposite to the applied field. Thus, in diamagnetism the induced magnetic moment, M, opposes the applied magnetic field and causes a net reduction in the magnetic flux density, B. Diamagnetism is normally very weak and of little practical utility (the susceptibility, or ratio of induced magnetization to applied field, $\chi = M/H$, is only about one part in 10^5 or 10^6). A superconductor, however, can be a *perfect diamagnet*.[*] This results from the fact that being a perfect electrical conductor, the induced magnetization M from surface currents are able to exactly oppose an applied field H, and no magnetic field penetrates through the material.

The second and more broadly useful type of magnetism results when *paramagnetic* ions are incorporated into a crystal structure, resulting in *ferromagnetic* and *ferrimagnetic* properties. Each electron in an atom contributes a quantized amount of magnetic moment from its orbital angular momentum as well as its spin angular momentum. When the sum of the two contributions is nonzero the atom or ion is paramagnetic; this occurs for transition metal, rare-earth, and actinide ions. The contributed magnetic moment and the susceptibility χ are positive. For the transition metal ions used in most ferrites the contribution from the orbital angular momentum is negligible, and the magnetic moment of an ion is determined by the number of unpaired electron spins. Each unpaired electron contributes a moment of one *Bohr magneton*, μ_B (= 9.27 x 10^{-24} Å m²/electron = 9.27 x 10^{-21} erg/gauss). (For rare-earth containing magnetic ceramics such as the garnets there is an additional contribution from the orbital momentum.) Hund's rule requires that electrons fill the orbital energy levels with aligned spins whenever possible. However, the Pauli exclusion principal allows only two electrons in any energy level and requires that their spins be opposite. Thus, in the transition metal series where $3d$-shells are partially filled, we can determine the magnetic moment of an ion by filling the $3d$-shell with aligned spins up to a maximum of five electrons, beyond which the spins must be in

---

[*] This ideal condition is only achieved below a certain *critical field*, H_c. In *Type I* superconductors the material is no longer superconducting above H_c. In *Type II* superconductors there is a *lower critical field* H_{c1} above which magnetic flux is able to penetrate into the material, and an *upper critical field* H_{c2} above which the material is no longer superconducting. See additional references at the end of this chapter.

antiparallel alignment to the first five. The moment is determined by the net number of unpaired spins. The resulting magnetic moments for the transition metal series are shown in Table ST1.

The magnetic properties of the crystal are ultimately determined by the way in which moments of individual magnetic ions interact. *Ferro*magnetism refers to the cooperative alignment of magnetic moments throughout the crystal, in much the same way that ferroelectricity refers to cooperative alignment of electrical dipoles, and leads to a net magnetic moment even in the absence of an applied field. *Antiferro*magnetism is the exact cancellation of magnetic moment by anti-parallel interactions between cations on adjacent sites. This is a highly ordered state, distinct from the zero net moment of a paramagnetic substance in the absence of applied field. *Ferri*magnetism refers to the condition where the moments of ions on one type of site are partially offset by antiparallel interactions with ions of another site, but there remains a net magnetization.

Whether a metal oxide containing magnetic ions is ferromagnetic, antiferromagnetic, or ferrimagnetic depends on the magnitude of the individual moments, the type and number of sites that are occupied, and the nature of the interaction between sites. The *exchange interaction* between any two cations is mediated by the intervening oxygen ions, and is known as a *superexchange* interaction, to contrast with the *direct exchange* in metals where atoms are on immediately adjacent sites. The superexchange interaction involves the temporary transfer of an electron from one of the oxygen ion's dumbbell-shaped $2p$ orbitals to one of the adjacent cations, leaving behind an unpaired $2p$ elec-

Table ST1 Outer-Shell Electron Configuration and Number of Unpaired Electrons for Several Spinel-Forming Ions

Ion	Electron Configuration	Number of Unpaired Electrons
Mg^{2+}	$2p^6$	0
Al^{3+}	$2p^6$	0
O^{2-}	$2p^6$	0
Sc^{3+}	$3p^6$	0
$Ti^{4+}(Ti^{3+})$	$3p^6(3d^1)$	0
$V^{3+}(V^{5+})$	$3d^2(3p^6)$	2(0)
$Cr^{3+}(Cr^{2+})$	$3d^3(3d^4)$	3(4)
$Mn^{2+}(Mn^{3+})(Mn^{4+})$	$3d^5(3d^4)(3d^3)$	5(4)(3)
Fe^{2+}	$3d^6$	4
Fe^{3+}	$3d^5$	5
$Co^{2+}(Co^{3+})$	$3d^7(3d^6)$	3(4)
Ni^{2+}	$3d^8$	2
$Cu^{2+}(Cu^+)$	$3d^9(3d^{10})$	1(0)
Zn^{2+}	$3d^{10}$	0
Cd^{2+}	$4d^{10}$	0

tron to interact with the opposing cation. For cations with more than half-filled d levels, such as the transition metals that appear in ferrites, this interaction generally results in antiparallel spin between the cations. The interaction is stronger for more closely separated cations and for metal-oxygen-metal angles closer to 180°. If we examine the spinel structure in Figs. 1.30 and 1.31, it is reasonable that the a-b interaction is the strongest, the b-b interaction less strong, and the a-a interaction the weakest.

Consider for instance the transition metal monoxides, MnO, FeO, CoO and NiO. In each the cation has at least a half-filled d-level, and since there are no a cations we expect an antiparallel b-b interaction to dominate. These oxides are thus antiferromagnetic, ordered such that cations within a (111) plane have parallel spins and adjacent (111) planes have antiparallel spins. $ZnFe_2O_4$ is a normal spinel in which the magnetic ions (Fe^{2+}) are all on b sites. It likewise is antiferromagnetic. Ferrimagnetic spinels are generally those with some degree of inverse structure. Magnetite, Fe_3O_4, is an inverse spinel with the cation ordering

$$\begin{array}{ccc} Fe^{3+} & (Fe^{2+} \quad Fe^{3+}) & O_4 \\ \uparrow\uparrow\uparrow\uparrow\uparrow & (\downarrow\downarrow\downarrow\downarrow \ \downarrow\downarrow\downarrow\downarrow\downarrow) & \\ a & b & \end{array}$$

The antiparallel a-b interaction dominates, illustrated in Fig. ST10, and the net magnetization or *saturation magnetization* per formula unit is $4\mu_B$.

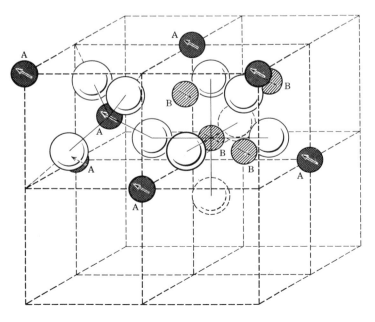

Fig. ST10 Antiparallel alignment of magnetic moments between A-site and B-site cations in ferrimagnetic spinel.

1.3 Ceramic Crystal Structures

Inverse spinel ferrites of formula $M^{2+}Fe_2O_4$ will have saturation magnetizations determined by the M^{2+} ion, since the Fe^{3+} ions appear in equal numbers on a and b sites and cancel. The saturation magnetization of solid solutions (such as $Mn_{1-x}Zn_xFe_2O_4$, a widely used commercial ferrite) and ferrites with fractional degrees of inversion can be readily computed by simple summation as long as the site occupancy is known. One should also keep in mind the relative strength of site-to-site interactions. Several exercises at the end of the chapter address this issue. Figure ST11 shows the change in saturation

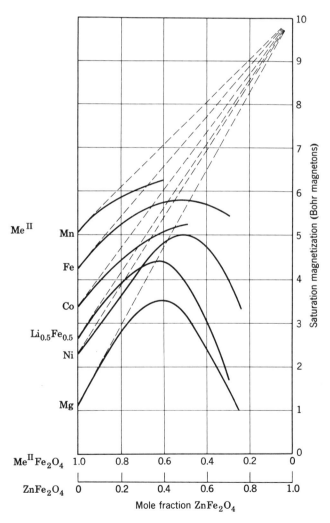

Fig. ST11 Change in saturation magnetization of several ferrites with zinc ferrite added in solid solution. [From E. W. Gorter, *Nature* (London), 165, 798 (1950).]

56 Chapter 1 / Structure of Ceramics

magnetization upon adding the normal spinel $ZnFe_2O_4$ to various inverse spinels. The initial increase is due to the substitution of nonmagnetic Zn^{2+} onto *a* sites, displacing Fe^{3+} ions to the *b* site where they contribute to the net moment. If this process continued one would end up with all of the *b* site Fe^{3+} ions in $ZnFe_2O_4$ in parallel alignment, and a saturation magnetization of 10 μ_B. However, above about 40% subsitution the antiparallel *b–b* interaction takes over and diminishes the moment toward the antiferromagnetic limit of 0 μ_B.

Ferromagnetic and ferrimagnetic materials share another feature with ferroelectric materials, which is the formation of microstructural domains. In magnetic materials these are known as *Weiss domains*, within which magnetic dipoles are cooperatively aligned. Adjacent domains have different orientations, and the whole domain structure is arranged to keep magnetic flux within the solid as shown in Fig. ST12. The preferred orientation of individual domains is not randomly chosen, but falls along certain low energy crystallographic directions. Like the ferroelectric domain wall, there is a magnetic domain wall or *Bloch wall* across which the domain orientation changes gradually (Fig. ST13). The thickness of the Bloch wall is about 100 nm, far greater than that of a ferroelectric domain wall (≤1 nm), and it also has lower energy (~1 erg/cm²).

The presence of domains also results in magnetic hysteresis, as shown in Fig. ST14. Starting with an unpoled ferrimagnet in which the domains are randomly oriented and there is zero net macroscopic moment, the application of a field *H* results in domain wall motion and an increase in the magnetization. The *magnetic permeability*, μ, characterizes the net induced magnetic flux density *B*, and is the sum of that of the vacuum and that due to the material:

$$B = \mu_0 H + \mu_0 M = \mu H$$

where μ_0 is the permeability of a vacuum. (The permeability and the susceptibility are related by $\mu/\mu_0 = 1 + \chi$.) The initial slope of the *B-H* curve is the

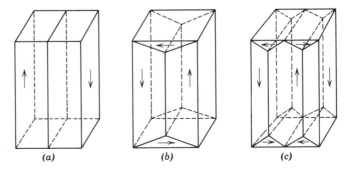

Fig. ST12 Several domain structures of a solid, each having zero net magnetization.

1.3 Ceramic Crystal Structures

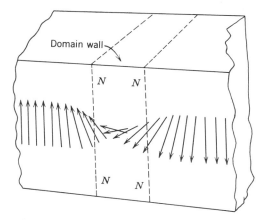

Fig. ST13 Change in magnetic moment orientation through a domain (Bloch) wall.

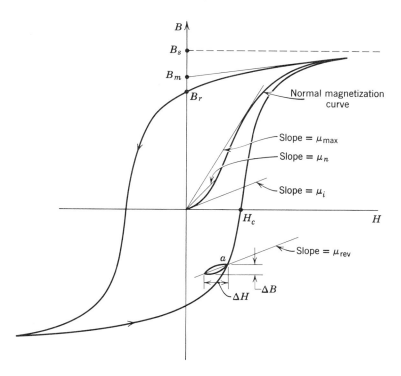

Fig. ST14 Magnetization hysteresis loop caused by domain action.

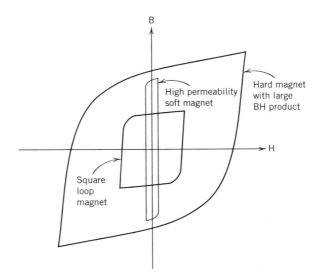

Fig. ST15 B-H loops representative of hard and soft magnetic materials.

initial permeability, μ_i. With increasing field, B increases until saturation is reached. If complete alignment of the material into a single domain is achievable then $B_s = \mu_0 H + \mu_0 M_s$, M_s being the saturation magnetization we discussed above. Normally microstructural factors and *domain wall pinning* by defects such as grain boundaries limit B_s to less than this ideal value. Upon removing the field, B decreases until at zero field a remnant magnetization B_r is left. If the field is reversed ($H<0$), there is a characteristic field necessary to achieve zero B, known as the *coercive field*, H_c.

The shape of the B-H loop characterizes whether a magnet is *hard* or *soft*. Figure ST15 compares several types of B-H loops useful for different magnetic applications. Permanent magnets are hard, and have both a high value of B_r, which is essentially the strength of the poled magnet after removing the field, and a high coercive field H_c, meaning that the magnet is difficult to demagnetize. The maximum of the product B · H, or *energy product* $(BH)_{max}$, is often used to characterize the strength of hard magnets. On the other hand, soft magnets such as the spinel ferrites are used in ac applications where a high saturation magnetization but low coercive field are desirable. These exhibit a narrow B-H loop. *Square loop* ferrites show sharply discontinuous changes in B at the coercive field and are useful in magnetic storage applications such as computer memory, and music and video tapes.

Perovskite / Rocksalt Derivatives: Cuprate Superconductors

In April of 1986, G. Bednorz and A. Mueller at IBM's Zurich laboratory reported signs of superconductivity in Ba-doped La_2CuO_4 at the then-unprecedented high temperature of 40 K. (Prior to this discovery, the highest temperature superconductor was in Nb_3Ge thin films, at 23 K.) Later that year, a group of researchers at the University of Tokyo led by K. Kitazawa reported a confirmation of this result, thereby sparking a world-wide search for new superconducting compounds that within two short years had resulted in the discovery of numerous additional cuprate compounds superconducting at temperatures as high as 125 K. R. J. Cava (pp. 3–8 in *Advances in Superconductivity V*, Y. Bando and H. Yamauchi, editors, Springer-Verlag, Tokyo, 1993) identified 34 structurally distinct cuprate superconductors which had been discovered by the end of 1992. Here, we will show how the family of cuprate superconductors may be considered derivatives of the perovskite and rocksalt structures, following a concise description by D. Smyth (pp. 1–10 in *Ceramic Superconductors II*, Research Update 1988, M. F. Yan, editor, The American Ceramic Society, 1988).

Several compounds of the simple perovskite structure type are superconducting. $SrTiO_3$ has a superconducting critical temperature, T_c, below which it is superconducting and above which it is a "normal" conductor, of less than 1 K. $BaPb_{1-x}Bi_xO_3$ ($T_c \sim 13K$ for $x = 0.25$) and $Ba_{1-x}K_xBiO_3$ ($T_c \sim 30$ K for $x = 0.4$) are two

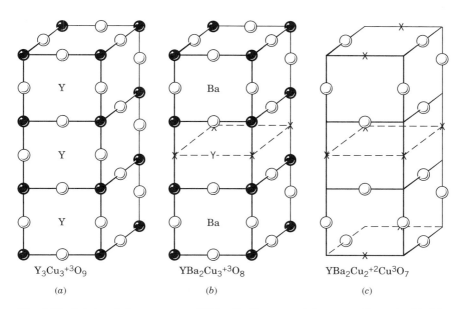

Fig. 1.32 Origin of the structure of $YBa_2Cu_3O_{7-x}$ as a triple-perovskite unit. (D. M. Smyth, pp. 1–10 in *Ceramic Superconductors II*, Research Update 1988, M. F. Yan, Ed. The American Ceramic Society, 1988.)

other examples. The high-temperature cuprate superconductor $YBa_2Cu_3O_{7-x}$, discovered by M.-K. Wu and C.-W. Chu in 1987, has a maximum T_c of about 93K, and is loosely based on a triple unit cell of "$YCuO_3$" perovskite, as shown in Fig. 1.32a. If two of the three Y^{3+} ions in this structure are substituted by Ba^{2+} ions, we have "$YBa_2Cu_3O_8$" and the structure shown in Fig. 1.32b. Charge neutrality has required the removal of one oxygen ion per formula unit, accommodated in this structure by oxygen vacancies around the central Y^{3+} ion. A distinctive characteristic of yttrium barium cuprate and most of the other superconducting cuprates is the existence of some copper in the Cu^{3+} valence state. In yttrium barium cuprate approximately one third of the total copper is trivalent, further reducing the oxygen content from our $Y_3Cu_3O_9$ reference state to $YBa_2Cu_2^{2+}Cu^{3+}O_7$. The additional oxygen vacancies are accommodated in the top and bottom-most copper planes (Fig. 1.32c). The resulting structure is shown in Fig. 1.33. Notice that there are two nonequivalent sites for copper, four-fold coordinated sites resulting in Cu-O "chains" oriented along [100] and five-fold coordinated sites forming layers of square pyramids joined at the corners. The oxygen stoichiometry (which varies proportionally with the copper valence state) is sensitive to temperature and oxygen pressure, and can vary between O_6 and O_7 depending on the annealing conditions. In optimized superconducting compositions the oxygen stoichiometry is about $O_{6.92}$, and is typically achieved by a low temperature (~500°C) annealing treatment in an oxidizing ambient. As the oxygen vacancies order along [100], the crystal symmetry correspondingly changes from tetragonal to orthorhombic. A number of isovalent substitutions can be made for Y and Ba in this structure with relatively modest effects on the superconducting properties. However, substitutions of other metal ions for Cu are generally detrimental.

Most other cuprate superconductors are based on structures that may be viewed as a combination of rocksalt and perovskite units. The most elementary of these are based on La_2CuO_4 and form in the previously known (i.e., prior to 1986) K_2NiF_4 structure type. La_2CuO_4 is not itself superconducting; partial substitutions of alkaline earth ions (Ca, Sr, Ba) are necessary, with a maximum T_c of ~35 K reached for a Sr-substituted composition of about $La_{1.85}Sr_{0.15}CuO_4$. And, La_2NiO_4 and La_2CoO_4 are two isostructural but non-superconducting compounds. The formula unit of La_2CuO_4 reflects the combination of one rocksalt and one perovskite formula unit: $LaO·LaCuO_3$. Beginning with two separate unit cells of $LaCuO_3$, shown in Fig. 1.34a, we can obtain the La_2CuO_4 structure (Fig. 1.34b) by displacing one of the unit cells by $1/2\langle 110\rangle$, and joining it to the other unit cell. This introduces an additional LaO layer between the copper-oxygen layers, with the configuration of half a rocksalt unit cell. That this is so can be seen by examining the sequential (100) layers of the rocksalt unit cell shown in Fig. 1.35. Here the in-plane coordination of cations around anions and vice versa is of either an "A_4O" or "AO_4" square-planar arrangement. The structure follows an A_4O-AO_4-A_4O stacking sequence in the [001] direction. The same LaO_4-La_4O sequence can be seen between Cu-O layers in the La_2CuO_4 structure (Fig. 1.34b). Notice that the copper ion retains a six-fold coordination, whereas the La ions, being coordinated on one

1.3 Ceramic Crystal Structures   61

side by a rocksalt-like structure (CN = 6) and on the other by perovskite (CN = 12), have a nine-fold coordination. La_2CuO_4 is tetragonal in symmetry, but alkaline earth substitutions cause a change to orthorhombic symmetry. The alkaline earth cations substitute for lanthanum in the structure. At lower concentrations (below $x \sim 0.2$ in $La_{2-x}Sr_xCuO_4$), the alkaline earth ions act as acceptors charge-compensated by electron holes (see Chapter 2 for discussion of electronic defects). The presence of an excess of holes is chemically equivalent to converting

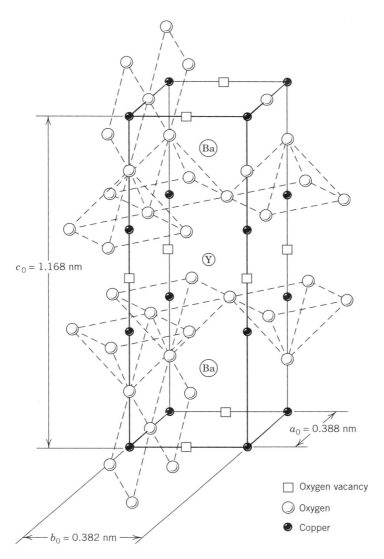

Fig. 1.33 $YBa_2Cu_3O_{7-x}$ structure showing four-coordinated copper in Cu-O chains and five-coordinated copper in Cu-O sheets.

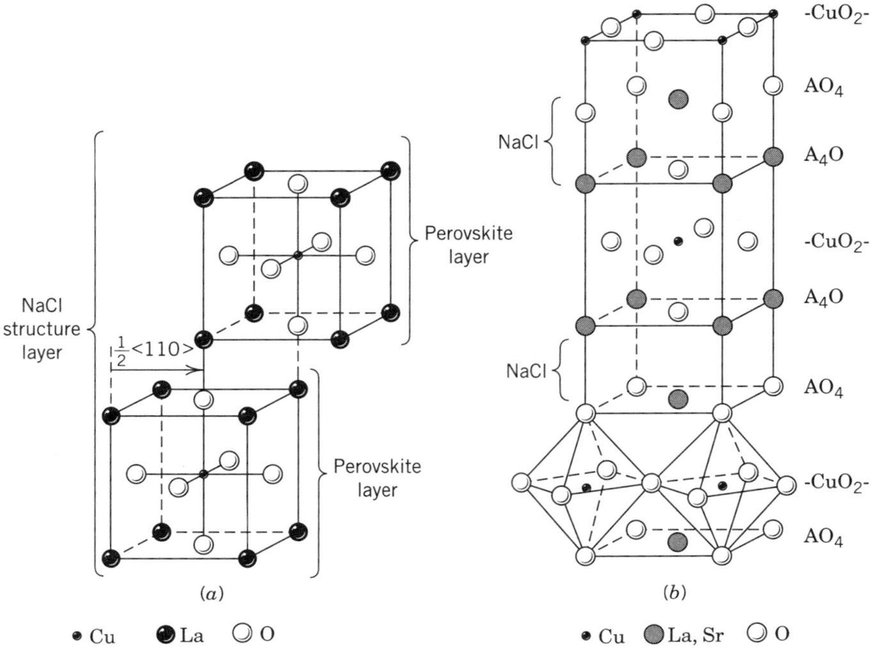

62 Chapter 1 / Structure of Ceramics

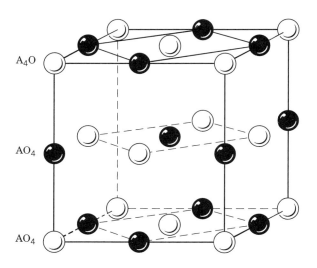

Fig. 1.34 (a) Origin of $La_{2-x}Sr_xCuO_4$ structure, shown in (b), as two perovskite unit cells.

Fig. 1.35 Alternating A_4O and AO_4 units in the (001) plane of the rocksalt structure. (From D. M. Smyth, pp. 1–10 in *Ceramic Superconductors II*, Research Update 1988, M.F. Yan, Ed., The American Ceramic Society, 1988).

1.3 Ceramic Crystal Structures   63

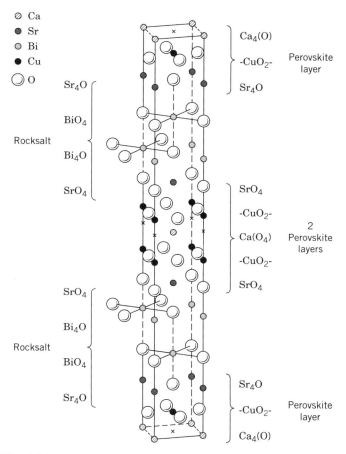

Fig. 1.36 Ideal structure of the $Bi_2Sr_2CaCu_2O_8$ superconductor.

some Cu^{2+} to Cu^{3+}. At higher Sr concentrations, oxygen vacancies are introduced to charge compensate the Sr and simultaneously a degradation of the superconducting transition temperature is observed.

More complex cuprate superconductors involve additional rocksalt and perovskite layers, and many have chemical formulas that can be written generally as $mAO \cdot nABO_3$.[4] Extensively studied examples include the Bi-Sr-Ca cuprates and Tl-Ba-Ca cuprates. For instance, in the $Bi_2Sr_2CaCu_2O_{8-\delta}$ structure ($T_c \sim 85$ K) the coefficients are $m = 3$ and $n = 2$. The unit cell, shown in Fig. 1.36, contains two formula units. A double perovskite unit cell in which copper occupies B sites and Sr and Ca occupy A sites can be seen at the center of the unit cell. There is a single

---

[4] Ruddlesden and Popper identified some time ago [*Acta Met.*, **11**, 54-55 (1958)] a similar class of structures in strontium titanate, based on the formula $SrO \cdot nSrTiO_3$ (i.e., $m = 1$). These include $n = 1$ (Sr_2TiO_4), $n = 2$ ($Sr_3Ti_2O_7$) and $n = 3$ ($Sr_4Ti_3O_{10}$), all the way up to $n = \infty$ (the perovskite $SrTiO_3$).

64 Chapter 1 / Structure of Ceramics

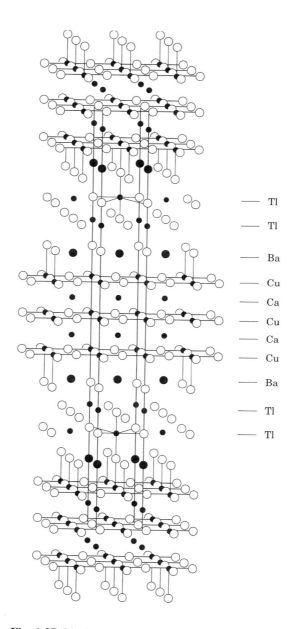

Fig. 1.37 Ideal structure of $Tl_2Ba_2Ca_2Cu_3O_{10}$ (From M. A. Subramanian, *Ceramic Transactions*, Vol. 13, K.M. Nair and E.A. Giess, Eds., The American Ceramic Society, 1990).

perovskite unit at the top and bottom of the unit cell, but counting adjacent unit cells these also form double perovskite layers in the repeat sequence of the lattice. In each double perovskite unit, the central layer of A cations is predominantly Ca and the top and bottom layers predominantly Sr, but considerable substitution of Sr for Ca is known to occur. Each double perovskite unit is displaced in the vertical stacking sequence from the next double perovskite layer by 1/2<110>, similar to the stacking in La_2CuO_4. However, two additional rocksalt layers of Bi-O also appear between the Sr-O layers terminating the double perovskite. In the unit cell depiction in Fig. 1. 36, the rocksalt layers repeat in the sequence $Sr_4O-BiO_4-Bi_4O-SrO_4$.

The next level of complexity involves $m = 3$, $n = 3$ compounds, such as the rather intimidating $Tl_2Ba_2Ca_2Cu_3O_{10}$ structure shown in Fig. 1.37 (T_c ~125 K) or its counterpart $Bi_2Sr_2Ca_2Cu_3O_{10}$ (often stabilized in this structure by partial substitution of Pb for Bi, with T_c ~ 110 K). Both of these compounds are of considerable technical interest for potential applications in thin films (for superconducting electronic devices) and bulk forms (for high current-carrying conductors). The structure contains triple perovskite unit cells layered with four-layer rocksalt sequences like those separating the double perovskite layers in Fig. 1.36. Structures of higher n have also been synthesized with difficulty in both systems; T_c seems to increase with n up to about $n = 5$. Complex crystals such as these, and the layered silicates discussed later in this chapter, can almost always be reduced to variations on a very few structural themes.

SPECIAL TOPIC 1.4

STRUCTURE, CONDUCTIVITY, AND SUPERCONDUCTIVITY

From the structural viewpoint, a common feature of all known cuprate superconductors is the existence of infinitely extending CuO_2 planes, separated by layers of alkaline earth or rare earth oxides. The bonding between copper and oxygen in these layers has a fair amount of covalent character, with electrons being shared between copper $3d$ and oxygen $2p$ orbitals. A second common characteristic is the existence of a significant fraction of copper in the trivalent state. This causes the electron energy band formed between copper and oxygen to be only partly filled; oxidation of copper from Cu^{2+} to Cu^{3+} leaves an electron hole in the conduction band and yields p-type conductivity. Partial band filling combined with weak localization of the valence electrons at atoms (low ionic character) results in a high electronic conductivity within the layer. The layered cuprates all show a strongly anisotropic electrical conductivity in the "normal" or nonsuperconducting state that is much higher in directions within the CuO_2 planes than it is normal to them. The resistivity generally shows a weak temperature dependence, sometimes decreasing with increasing temperature like a semiconductor but more often increasing with temperature like a metal (Fig. ST16 Also see Fig. 3.2). Thus the electronically active layers may be considered to be the CuO_2 planes. The intervening alka-

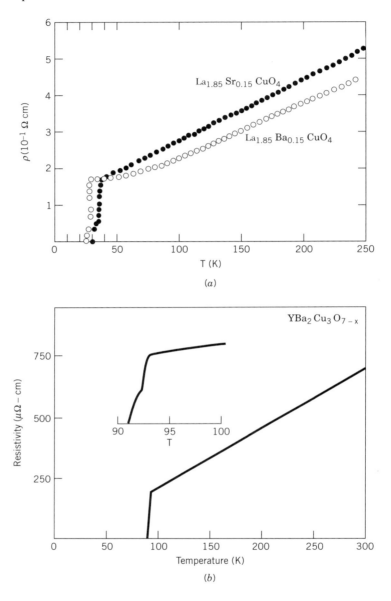

Fig. ST16 Resistivity versus temperature relationship for cuprate superconductors, showing the drop to zero resistance below the superconducting transition temperature, T_c. Early measurements in the field on polycrystalline samples. [(a) $La_{1.85}Sr_{0.15}CuO_4$ and $La_{1.85}Ba_{0.15}CuO_4$, from S. Uchida, H. Takagi, S. Tanaka, K. Nakao, N. Miura, K. Kishio, K. Kitazawa, and K. Fueki, *Jap. J. Appl. Phys.*, 26[3], L196, 1987.] [(b) $YBa_2Cu_3O_{7-x}$, from R. J. Cava, B. Batlogg, R. B. van Dover, D. W. Murphy, S. Sunshine, T. Siegrist, J. P. Remeika, E. A. Rietman, S. Zahurak, and G. P. Espinosa, *Phys. Rev. Lett.*, 58[16] (1676, 1987).]

line earth and rare earth layers are not unimportant; they establish the spacing of the CuO_2 layers and, most importantly, act as "charge reservoirs" that influence the electron hole concentrations within the CuO_2 layers as their compositions and valence states are varied. (see R. J. Cava, *Science*, 247, 656–662, 1990, and references therein for further discussion.)

For particular compositions and oxygen stoichiometries within these systems, superconducting behavior is observed. The metallic resistivity followed by the transition to a state of zero resistance for $YBa_2Cu_3O_{6.9}$ is shown in Fig. ST16. Metallically conducting oxides are not particularly uncommon (in addition to the cuprates, a few examples include RuO_2, CrO_2, TiO, V_2O_3, and $LiTi_2O_4$), but only a limited set of these are superconducting. While the theory of oxide superconductivity is not sufficiently advanced to predict from first principles which oxides will be superconducting or at what temperature, the mechanism of conduction, like that in previously known superconductors, is believed to be based on a pairing of electrons into an ordered state below the superconducting transition temperature T_c. This widely accepted theory was proposed by J. Bardeen, L. H. Cooper, and J. R. Schrieffer in 1957 (*Phys. Rev.*, 106, 162, 1957) and is referred to as the BCS theory; it won the Nobel prize for the authors in 1972. An elementary physical description is as follows (some references for additional reading are given at the end of Chapters 1 and 3). The origin of electrical resistance is the scattering of a moving electron from a latttice undergoing thermal vibration. However, at a low temperature when random thermal vibrations have become sufficiently quiet, the interaction between electrons and lattice vibrations can take on a different character. The motion of an electron under an electrical potential causes a momentary polarization of positive ion cores towards it as it passes, setting up wave or *phonon* motion in the lattice. Immediately after the passage of the first electron, another electron in the vicinity is attracted to the polarized region. Quantum mechanical calculations by Cooper (*Phys. Rev.*, 104, 1189, 1956) showed that such a pair of electrons interacting via the phonons have a lower energy than isolated electrons. The interaction results in an attractive binding of electrons into what are known as *Cooper pairs*. The temperature below which they form is the superconducting transition temperature. Electrical conduction of Cooper pairs is resistanceless since the electrons move in a coordinated manner, coupled (by rather than scattered by) lattice vibrations. Thus in different ways the electron-phonon interaction is both the origin of resistance and of superconductivity. Once a sufficient number of the electrons in the solid have condensed into Cooper pairs, zero resistance and other electromagnetic properties characteristic of superconductors are observable. While the BCS theory was successful in explaining superconductivity in materials known prior to 1987, it did not predict the existence of higher temperature superconductors. What is presently unknown is the microscopic mechanism of coupling which allows formation of Cooper pairs at higher temperatures in oxide superconductors; this remains a hotly debated topic among researchers seeking to explain known cuprates and to predict just how high a superconducting temperature is ultimately possible.

Covalent Ceramics

Many of the hardest, most refractory, and toughest ceramics (for particular engineered polycrystalline microstructures) have structures in which covalent bonding predominates. Boron nitride (BN) is much like carbon, appearing in a graphite-like hexagonal form, which is platy and lubricious, and a cubic zincblende form, which is extremely hard like diamond and is useful as an abrasive. Silicon carbide, the various polytypes of which have been mentioned, is widely used in abrasives, refractories, as a hard and wear-resistant structural ceramic, and potentially as a high-temperature semiconductor. Silicon nitride is currently used in a variety of structural ceramic applications, the most complex and demanding of which are as rotors for automotive turbochargers and high-temperature gas turbines. Oxynitrides are extensive solid solutions between nitrides (Si_3N_4, AlN) and oxides (e.g., SiO_2, Al_2O_3, Y_2O_3, MgO), which also have received much attention as structural ceramics. Titanium nitride (TiN) and zirconium nitride (ZrN) are rocksalt-structure compounds that are electrically conductive as well as extremely hard, and are used in monolithic form as cutting tools or as coatings for wear-resistant applications. High-purity aluminum nitride, diamond, and diamond-like carbons produced by vapor deposition possess a combination of high thermal conductivity and electrical insulation useful in applications as heat sinks and "packaging" for semiconductor devices.

Despite the generality of the term "covalent ceramics," few compounds are entirely covalent (or, for that matter, entirely ionic). Using the Pauling electronegativity difference between the components of various binary oxides, nitrides, and carbides as a rough guide, one finds that SiC, B_4C, and BN are highly covalent (12%, 6%, and 22% ionic character), nitrides such as AlN, Si_3N_4, and TiN are less covalent (43%, 30%, and 43% ionic, respectively), SiO_2 is approximately 51% ionic, and alkaline earth oxides such as MgO, CaO, and BaO, and alkali halides such as NaCl, KCl, LiF, NaF are more than 70% ionic. As discussed earlier, many highly covalent compounds form structures with a tetrahedral bonding coordination promoted by the sp^3 hybridization of valence electron orbitals. Examples include silicon and carbon in the diamond cubic structure and in the various structures of silicon carbide. For compounds with significant covalency and ionicity, however, it is characteristic to find structures that satisfy both the directionality of covalent bonding and local electrical neutrality as defined by Pauling's second rule. SiO_2 is a prime example, discussed separately in the following section. The solid solutions of Si_3N_4 are another, in which compatibility of the structure with ionic bonding is illustrated by the extensive solid solutions that can form with various oxides.

Silicon Nitride In Si_3N_4, the coordination of nitrogen around silicon is tetrahedral (Fig. 1.38a), and the SiN_4 tetrahedra are joined so that each nitrogen is coordinated by three silicon atoms (which satisfies sp^2 hybridization of N, in addition to sp^3 hybridization of Si). Thus it is an open structure with large interstices. The coordination of atoms also satisfies Pauling's second rule for local charge neutrality if partial ionic charges are assumed for Si and N in the proportion 4/3. Silicon nitride occurs in two known polytypes, the α and β phases. Both are hexagonal, but differ in the sequence in which Si-N layers are arranged in the c-axis direction. β-Si_3N_4 has an *abab* stacking sequence of layers with an unit cell of formula Si_6N_8, while α-Si_3N_4 has an *abcd* stacking sequence of layers and twice the unit cell formula, $Si_{12}N_{16}$. Since in α-Si_3N_4 the stacking repeats after twice the number of layers, it has approximately twice the c-axis lattice parameter of the β phase (c_0 = 0.2909nm for β and 0.5618nm for α; the ratio is not exactly a factor of two, due to distortions of the tetrahedral arrangement. The a_0-axes are 0.7753nm and 0.7606nm for β and α, respectively). Fig. 1.38*b* shows the atomic arrangement in a single *ab* layer sequence. The N atoms indicated by the filled triangles and the layer of Si atoms shown as filled circles lie in a plane forming the base of the unit cell. The Si atoms in this layer are indicated by "0" in the schematic in Fig. 1.38*c*. The next layer of N and Si atoms are shown as open circles and triangles, and are located at the "1/2" position in Fig. 1.38*c*. Notice that a ring-like configuration of Si-N-Si-N bonds exists, with either 8 or 12 membered rings (similar configurations exist in the crystalline silicates). In β-Si_3N_4 the next *ab* layer is superimposed directly over the first, so that a continuous interstitial channel parallel to the c-axis forms around the site surrounded by 12 Si-N bonds. In α-Si_3N_4, the *cd* layer shown in Figure 1.38*d* is superimposed on the *ab* layer. This isolates the large interstices formed by the 12-membered rings. The height of Si atoms above the base of the unit cell in the α phase is shown in Fig. 1.38*e*. The availability of large interstitial sites in both structures facilitates the formation of oxynitride solid solutions.

The β phase is the more stable of the two at high temperatures, but the transformation from α to β is reconstructive and occurs slowly unless a liquid phase is present. In sintered and hot-pressed silicon nitrides, a small amount of oxynitride liquid phase is usually present due to additives or oxygen impurity. If the starting material is of the α phase, solution–precipitation growth of the β phase in the form of acicular (needle-like) grains can be promoted, leading to an interlocking grain microstructure with high fracture toughness. This microstructure is an important objective in the processing of polycrystalline silicon nitrides, as discussed further in Chapter 5.

Oxynitrides: Charge-Compensating Solid Solutions. *Oxynitrides* are solid solutions between nitrides and oxides, the most well-known of which are the "sialons," an acronym for Si-Al-O-N solid solutions. Sialons first received attention in the early 1970s when extensive alloying of Al_2O_3 with β-Si_3N_4 was discovered.[5] Other oxide structures (spinels, eucryptite, apatite) were later found to dissolve substan-

---

[5] Y. Oyama and O. Kamigaito, *Japan. J. Appl. Phys.*, **10**, 1937, (1971), and K.H. Jack and W.J. Wilson, *Nature Physical Science*, **283**, 28 (1972).

70 Chapter 1 / Structure of Ceramics

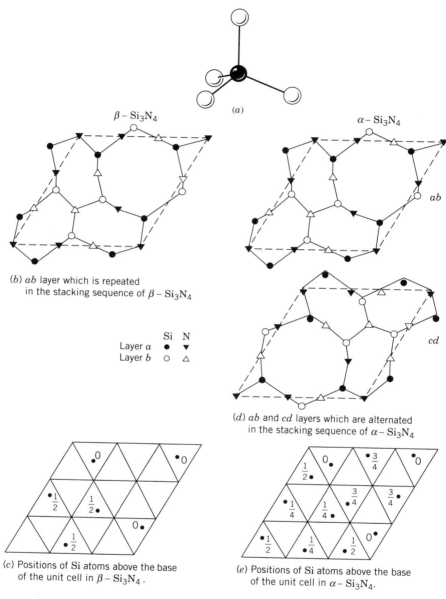

Fig. 1.38 (a) SiN$_4$ tetrahedron, the basis of crystalline Si$_3$N$_4$. (b) *ab* layer, which is repeated in the *abab* stacking sequence of b-Si$_3$N$_4$; (c) positions of Si atoms above the base of the hexagonal unit cell in β-Si$_3$N$_4$;. The hatched lines show edges of SiN$_4$ tetrahedra, which lie parallel to the basal plane and at the same level as the indicated Si atoms. (d) *ab* and *cd* layers which are alternated in the stacking sequence of α-Si$_3$N$_4$; (e) positions of Si atoms above the base of the unit cell in α-Si$_3$N$_4$. [After K. H. Jack, pp. 45-60 in *Progress in Nitrogen Ceramics*, F. L. Riley, Ed., Martins Nijhoff Publishers, Boston, 1983, and S. N. Ruddlesden and P. Popper., *Acta Cryst.*) 11, 465 (1958)]

tial amounts of nitrogen at high temperatures as well, and numerous other oxynitride intermediate compounds with structures distinct from both the parent nitride and oxide have been discovered. A number of the oxynitrides of interest for structural applications are derivatives of phases which occur in the system Si_3N_4-AlN-SiO_2-Al_2O_3. A special type of square-planar phase diagram known as a "reciprocal salt" diagram, discussed in Chapter 4, is often used to show phase equilibria in systems such as these where there is simultaneous substitution of cations and anions.

The solid solution phases in the Si-Al-O-N system (also refer to the phase diagram in Fig. ST49, Chapter 4) are the following. Two are based on oxide substitution into the α and β Si_3N_4 structures and are known, respectively, as α' and β' sialons. These are the most widely studied sialons, and β' sialon in particular has been commercially developed as a structural ceramic. A series of "polytypoids" based on AlN-rich compositions exists, given the designations in Ramsdell notation 8H, 15R, 12H, 21R, 27R, and $2H^\delta$. These phases are analogous to the polytypes discussed for SiC, but only form in AlN when it is alloyed with Si and O. (Since they differ in composition as well as the stacking sequence of a two dimensional layer, strictly speaking they are not polytypes.) Si_2N_2O is an intermediate oxynitride phase, which also has extensive solid solubility for Al_2O_3 and goes by the designation O' phase. A separate triclinic phase known as X phase (designated as such, prior to conclusive identification, due to its mysterious presence amongst other phases) forms as a solid solution between Si_2N_2O and $3Al_2O_3 \cdot 2SiO_2$ (mullite) at a ratio of approximately 1:2. Between AlN and Al_2O_3 there are two intermediate oxynitrides, of compositions Al_3O_3N and $3Al_2O_3 \cdot AlN$, both of which tolerate some variation in O/N ratio.

A common feature of the α', β', and O' sialons is the substitution of aluminum for silicon in tetrahedral coordination, with simultaneous substitution of oxygen for nitrogen (or vice versa) for charge neutrality. We may imagine dissolving Al_2O_3 into β-Si_3N_4 by directly substituting Al^{3+} onto Si^{4+} sites, and O^{2-} onto N^{3-} sites. However, since the metal:anion stoichiometries of the two compounds differ (2:3 vs 3:4), some sites must be left vacant in the solid solution if charge neutrality is to be preserved. The creation of these vacancy defects is energetically costly. On the other hand, if we instead dissolve the oxynitride spinel Al_3O_3N, which has the same 3:4 metal-anion ratio as silicon nitride, electrical neutrality and site-stoichiometry are both simultaneously satisfied. A high solubility therefore exists along the join between Si_3N_4 and Al_3O_3N in the phase diagram (Fig. ST46), giving the solid solution known as β' sialon. The resulting composition may be simplified to the formula unit:

$$(2-\frac{z}{3})Si_3N_4 \cdot \frac{z}{3}Al_3O_3N = Si_{6-z}^{(24-4z)+} Al_z^{3z+} O_z^{2z-} N_{8-z}^{(24-3z)-}$$

The extent of solid solution increases with temperature, and is as high as $z = 4$ (two-thirds substitution of Al for Si) at ~1750°C. Since the crystal structure remains essentially that of β-Si_3N_4, the physical and mechanical properties of the oxynitride are also similar. However, there are advantages conferred by the high-alumina substitution, including improved high-temperature oxidation resistance and a lower vapor pressure of the oxynitride spinel compared to silicon nitride. The latter makes it possible to densify β' sialons without an overpressure of nitrogen to prevent decomposition, unlike pure silicon nitride. Other oxide substitutions besides Al_2O_3 are possible in β' sialons, including Y_2O_3, MgO (as Mg_2SiO_4 substituted for Si_3N_4), and BeO ($BeAl_2O_4$ for Si_3N_4).

α' sialons are based on a similar substitution into the α-Si_3N_4 structure, often with charge compensating cations such as Li^+, Ca^{2+}, and the rare-earths occupying the large interstitial sites. O' sialons are solid solutions between Si_2N_2O and Al_2O_3; here the oxide alone has the correct stoichiometry for substitution. The polytypoid sialons based on AlN differ from these in that some of the Al^{3+} takes on an octahedral coordination. With increasing substitution of SiO_2 into the wurtzite AlN structure, the metal-anion ratio decreases and is accommodated by the substitution of Si^{4+} into tetrahedral positions and displacement of Al^{3+} into octahedral layers where it is coordinated by oxygen. These octahedral layers separate tetrahedral (Al, Si)N layers. With increasing substitution of SiO_2, new polytypoid phases are formed at integer ratios of metals to anions (e.g., 1:1 for AlN, 9:10 for 27R phase, 6:7 for 12H phase). Within each polytypoid, a range of solid solution exchanging Al^{3+} and Si^{4+} is possible as long as it is accommodated by a corresponding substitution between O^{2-} and N^{3-} for charge neutrality.

1.4 CRYSTALLINE SILICATES

A great many naturally occurring ores and minerals are silicates. According to the Si^{4+}/O^{2-} radius ratio, tetrahedral coordination is preferred. Pauling's electronegativities also indicate that the Si-O bond is 51% ionic and 49% covalent. The high degree of covalency favors tetrahedral coordination where in the $3s^23p^2$ outer electron shell of the Si atom is stabilized in a bonding configuration with four $3sp^3$ hybridized orbitals, as in the diamond cubic. Together this leads to a very strong preference for the formation of $(SiO_4)^{4-}$ tetrahedra in both crystalline and glassy silicates. In this configuration the bond strength of Si^{4+} is 4/4 = 1, and from Pauling's second rule an oxygen ion must be coordinated by two Si^{4+} in pure SiO_2. This mandates the corner sharing of $(SiO_4)^{4-}$ tetrahedra, a characteristic common to crystalline SiO_2, the silicate minerals, and silicate glasses alike. Corner sharing of oxide tetrahedra also prevents the close-packing of anion layers as in the FCC- and HCP- based oxides, and so crystalline silicates tend to have open structures.

Another feature of crystalline silicates is the ready substitution of other cations in place of the tetrahedrally coordinated Si^{4+}. The most important of these is Al^{3+}, which can adopt tetrahedral or octahedral coordinations. When substituted for Si^{4+}, a charge imbalance results that can be corrected by the substitution of an OH^- ion for O^{2-} (in the clay minerals) or by adding alkaline or alkaline earth cations in local interstitial sites (in glasses and crystalline silicates). Charge-compensating substitutions permit great flexibility in dissolving high concentrations of addi-

1.4 Crystalline Silicates

tives. Also, the $(SiO_4)^{4-}$ tetrahedron may no longer need to share all corners in order to satisfy Pauling's second rule. Silicate minerals can therefore be classified according to the connectivity between their silica tetrahedra.

Oxygen/Silicon Ratio

The O/Si ratio is a useful parameter for characterizing the degree of connectivity between silica tetrahedra in silicate compounds. This classification scheme is summarized in Table 1.4. It is also useful in characterizing the degree of connectivity in silicate glasses, which we discuss later.

Crystalline SiO_2 has an O/Si ratio of 2, which requires sharing of all four corners. It occurs in three basic crystalline polymorphs, cristobalite, tridymite and quartz, transformations between which are reconstructive (and sluggish). Each of these phases also has a *high* and *low* form related by a displacive transformation (tridymite also has a *middle* form) as shown in Table 1.5. The high form always has the higher symmetry. Cristobalite is the highest-temperature polymorph, shown in Fig. 1.39. It has some similarity to the zincblende structure (Fig. 1.16) but with tetrahedra substituted for atoms; note the sharing of all corners. Tridymite is monoclinic (high form) and is the intermediate-temperature structure. Quartz is hexagonal (low quartz is trigonal) and is the lowest-temperature form (Fig. 1.40).

With increasing O/Si ratio the number of shared corners decreases. The physical habit of the compound often refects its crystalline structure. For instance, layered silicate compounds such as talc, mica, and clay (O/Si = 2.5) contain $(Si_2O_5)_n$ sheets in which three corners of each tetrahedron are shared. These are described in greater detail below. Chain silicates, formed at O/Si ratios of 2.75 to 3, often have a fibrous habit. Pyroxenes, examples of which are enstatite ($MgSiO_3$) and jadeite, have single chains of $(SiO_3)_n^{2n-}$. Close examination of the fracture surface of jadeite (the semiprecious gemstone jade) reveals its fibrous nature. Amphiboles have double chains, $(Si_4O_{11})_n^{6n-}$. The most common examples are the fibrous asbestos minerals. In this O/Si range it is also possible to form structures with isolated rings of tetrahedra, such as beryl, $Be_3Al_2Si_6O_{18}$ (the basis of emerald), in which six-tetrahedron $Si_6O_{18}^{12-}$ rings form, and wollastonite, $CaSiO_3$, in which three-tetrahedron $Si_3O_9^{6-}$ rings form. Progressing then to O/Si = 3.5, one finds the pyrosilicates in which double tetrahedra of $Si_2O_7^{6-}$ can be found. Finally, orthosilicates, in which the O/Si ratio is 4, have completely isolated tetrahedra with intervening cations, as in the structures of olivine (forsterite, Mg_2SiO_4, and olivine, Fe_2SiO_4) and zircon ($ZrSiO_4$). Table 1.4 shows the systematic changes in structure between these classes of silicates.

Clay Minerals

Many of the common layered silicates such as clay, talc, and mica may be known to the reader. They are based on hydrated aluminosilicate structures in which $(Si_2O_5)_n$ sheets are joined to $AlO(OH)_2$ layers containing octahedrally coordinated

Table 1.4 Effect of Oxygen-Silicon Ratio on Structure in Silicates

O/Si Ratio	Silicon-Oxygen groups	Silicate structure	Examples
2		SiO_2 (completely interconnected tetrahedra)	Quartz, tridymite, cristobalite
2.5		Si_4O_{10} (sheets)	Talc, mica, kaolinite, montmorillonite clays, vermiculite
2.75		Si_4O_{11} (chains)	Amphiboles (asbestos minerals)
3		SiO_3 (chains or rings)	Pyroxenes (chains), beryl (rings)
3.5	$Si_2O_7^{6-}$	Si_2O_7 (tetrahedra sharing one oxygen ion)	Pyrosilicates
4	SiO_4^{4-}	SiO_4 (isolated tetrahedra)	Orthosilicates (forsterite, olivine, zircon)

1.4 Crystalline Silicates

Table 1.5 Polymorphic Forms of Silica

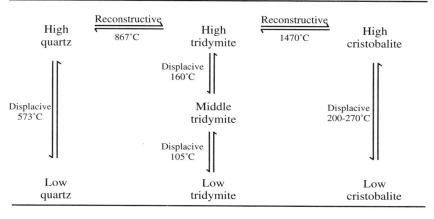

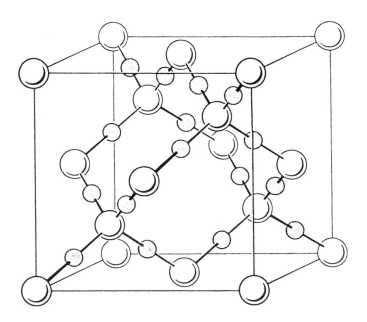

Fig. 1.39 Structure of high cristobalite.

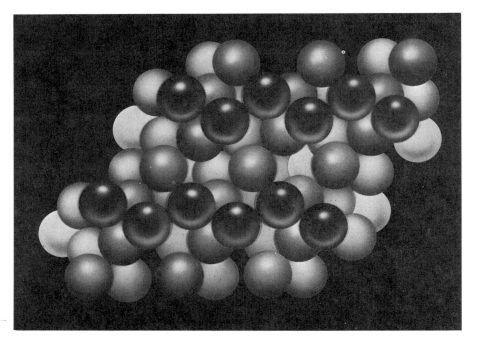

Fig. 1.40 Structure of high quartz, looking down on the basal plane.

Al^{3+}. The hydroxyl ion (OH)⁻ substitutes widely for O^{2-} in such structures. Fig. 1.41a shows the $(Si_2O_5)_n$ sheets in their ideal hexagonal array and the distorted arrangements that actually occur. The simplest clay mineral is kaolinite, in which one $(Si_2O_5)_n$ sheet is joined to one $AlO(OH)_2$ layer, to give the ideal composition $Al_2(Si_2O_5)(OH)_4$, as shown in Figs. 1.41b and 1.42. The $AlO(OH)_2$ layer may be thought of as a gibbsite layer, $Al(OH)_3$, in which one OH ion has been replaced by an O ion from the corner of the silica tetrahedron pointing out of the $(Si_2O_5)_n$ sheet. Viewing normal to the layers in Fig. 1.41b, one finds that the octahedrally coordinated Al^{3+} ions are located over the centers of the rings in the $(Si_2O_5)_n$ sheets of the next kaolinite layer. The misfit between these two layers and consequent bending strain of the crystal promote formation of fine particle sizes in kaolinite, a common clay mineral.

If the $AlO(OH)_2$ layer is joined on the other side to another $(Si_2O_5)_n$ sheet, as shown in Fig. 1.43, we then have the mineral pyrophyllite, $Al_2(Si_2O_5)_2(OH)_2$. Both kaolinite and pyrophyllite readily accept isomorphous substitutions of Al^{3+} and sometimes Fe^{3+} for the tetrahedrally coordinated Si^{4+}, and of Mg^{2+} and Fe^{2+} for the octahedrally coordinated Al^{3+}. Note that these substitutions lead to an excess negative charge on the layers, which is typically compensated by interstitial cations between the layers. The type of substitution and the way in which the charge imbalance is accommodated distinguish the specific clay minerals. For instance, talc is similar to pyrophyllite except that the gibbsite ($Al(OH)_3$) central layer is

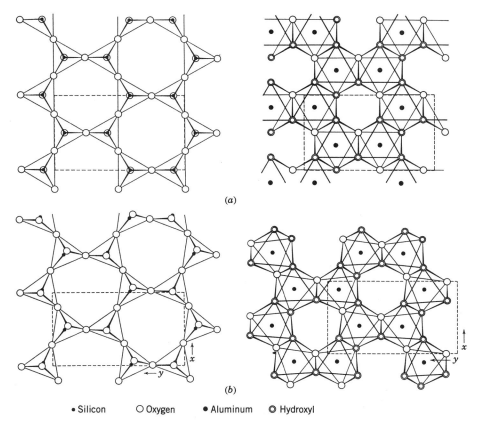

- Silicon   ○ Oxygen   • Aluminum   ◎ Hydroxyl

Fig. 1.41 Atomic arrangements of Si_2O_5 and $AlO(OH)_2$ layers. Pattern (*a*) is idealized, and (*b*) is the distorted arrangement found to occur in kaolinite and dickite by R. E. Newnham and G. W. Brindley, *Acta Cryst.*, 9, 759 (1956); 10, 89 (1957). (From G. W. Brindley, in *Ceramic Fabrication Processes*, W. D. Kingery, Ed., Technology Press, Cambridge, MA, and John Wiley and Sons, New York, 1958.)

instead brucite $(Mg(OH)_2)$; the ideal composition is then $Mg_3Si_4O_{10}(OH)_2$. In mica, Al^{3+} substitutes for Si^{4+} in the pyrophyllite structure where charge compensation is provided by K^+ ions in between the triple layers as shown in Fig. 1.44. Montmorillonite clays have both Al^{3+} substituted for Si^{4+} and Mg^{2+} for octrahedral Al^{3+}, with charge compensation provided by local Na^+ and Ca^{2+}. The interlayer cations in clays are readily exchanged in aqueous suspensions (termed base exchange), whereupon a net charge can form on a clay particle. The resulting electrostatic repulsion between particles greatly improves the plasticity and forming characteristics of a clay-water suspension, as is well-known by potters. (This effect is the historical origin of the electrostatic stabilization of ceramic powder slurries, now understood and practiced in ceramics processing at a highly sophisticated level.)

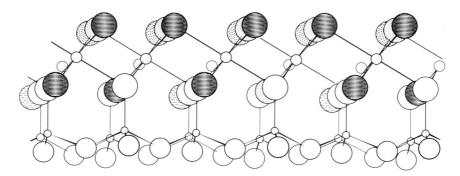

Fig. 1.42 Perspective drawing of kaolinite, $Al_2(Si_2O_5)(OH)_4$, showing Si-O tetrahedra on the bottom half of the layer and Al-O,OH octahedra on the top half. (From G. W. Brindley, in *Ceramic Fabrication Processes*, W. D. Kingery, Ed., Technology Press, Cambridge, MA, and John Wiley and Sons, New York, 1958.)

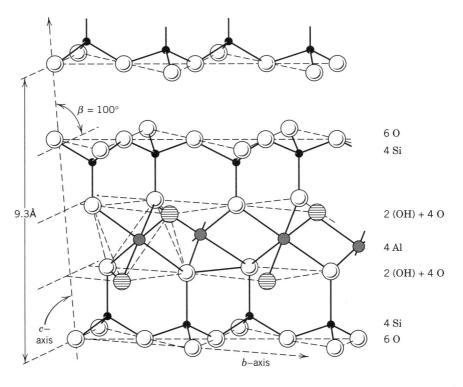

Fig. 1.43 Structure of pyrophyllite, $Al_4(Si_4O_{10})_2(OH)_4$, showing joining of two layers of tetrahedrally coordinated Si-O to a central octahedral Al-O,OH layer.

1.4 Crystalline Silicates 79

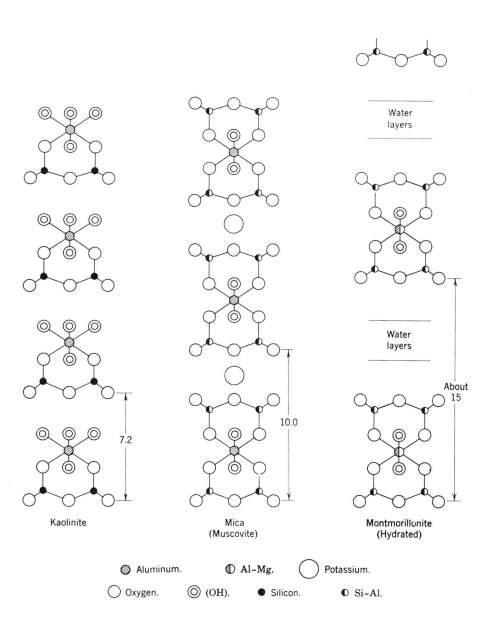

Fig. 1.44 Layer structure of clay and similar materials. (From G. W. Brindley, in *Ceramic Fabrication Processes*, W. D. Kingery, Ed., Technology Press, Cambridge, MA, and John Wiley and Sons, New York, 1958.)

1.5 GLASS STRUCTURE

Glasses, or amorphous solids, lack long-range periodicity in their atomic arrangements. This is the single largest distinction between glasses and crystalline solids. However, not all types of glasses are structurally the same. While all— whether they are ceramics, semiconductors, or metals— are disordered over lengths greater than a few atomic separations, there are important differences between these materials in the level of order at the first few nearest-neighbor distances. We will see that oxide glasses and glassy covalent semiconductors (Si, Ge) have a number of structural similarities and can be described as *continuous random networks* based on the linkage of compact structural units (tetrahedra or triangles) similar to the nearest-neighbor coordinations in their crystalline counterparts. In contrast, glassy metals form in a *random close-packed* atomic array and lack distinct order even at the nearest-neighbor level. The structure of any glass can also vary significantly with its processing history.

To set a context for discussion of glass structure, we will first consider briefly the formation of glassy materials and a few of their properties. The subject of glass formation as related to the avoidance of crystallization is taken up in detail in Chapter 5.

Glass Formation

What causes a material to become glassy? Traditionally, glasses have been processed by cooling a liquid fast enough to prevent detectable crystallization. From this kinetic viewpoint, we can define glass formation as the *avoidance of crystallization*. The kinetics of glass formation are therefore those of crystal nucleation and growth, and in principle any liquid can be rendered glassy given a sufficiently rapid cooling rate. It is the difference in respective rates of crystallization that allows us to form many commercial oxide glasses by cooling at a leisurely rate of a few degrees per minute (°/min), while metallic glasses must be quenched at more than 10^6 degrees per second (°/sec). *Glass-ceramics* are commercially important ceramics with unique thermal shock and mechanical properties, made by controllably nucleating a very high density of crystals in a parent glass body. The processing and microstructure of glasses and glass ceramics are discussed in Chapter 5.

Glasses can also be made by a number of alternative processes, which have in common the aspect of consolidation at low temperatures to defeat crystallization. Condensation of a vapor onto a cold substrate is one method (*physical* or *chemical vapor deposition*), often used for the preparation of electronic thin-films (glassy and crystalline). Another is the gelation or precipitation of a disordered ceramic from liquid chemical solution (referred to as *sol-gel* processing), followed by densification into a glass. Irradiation and ion bombardment can be used to disorder initially crystalline materials and form a glass. Oxidation of silicon typically results in an amorphous film of SiO_2, which has enormous technological importance as an insulating layer in silicon device technology. And, sometimes amor-

1.5 Glass Structure

phous reaction products are formed between two crystalline solids when the equilibrium crystalline product cannot nucleate, in what are known as *solid-state amorphization reactions*. The structure and properties of glasses made by these techniques can differ substantially from those prepared from the melt; for instance, amorphous vapor-deposited silicon films undergo considerable structural rearrangement at temperatures well below those of crystallization or melting. Many of the novel processing techniques are restricted to the preparation of thin films of glass on a supporting substrate; some are not yet commercially important on a widespread basis. For applications that require a monolithic body, the vast majority of glasses continue to be processed from the melt.

When we process glasses by cooling from the melt, the phase transformation from liquid to solid occurs at a *glass transition temperature* (T_g) that lies below the melting temperature T_m at which crystallization would otherwise take place. At T_g there is a transformation in physical properties from those of a liquid to those of a solid; one such property illustrated in Fig. 1.45 is the specific volume. The slope of this curve is directly related to the volume expansion coefficient α ($=\partial V/V\partial T$ at constant composition and pressure). Above T_m one has a liquid; between T_m and T_g there exists a supercooled liquid (which despite references in the earlier literature is not a glass). At T_g, the change in slope in Fig. 1.45 shows a transition to a glassy state where structural rearrangements are no longer able

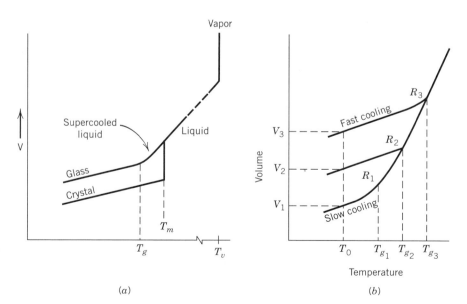

Fig. 1.45 (*a*) Volume-temperature relations for liquid, crystal, and glass phases. The transformation from a supercooled liquid to a glass occurs at the glass transition temperature, T_g. (*b*) Variation in T_g and specific volume of glass as a function of the cooling rate.

to take place on a reasonable time scale, and where the thermal expansivity and other properties become that of a solid. In most systems the specific volume of the glass is greater than the crystal (density is lower). According to the theory of D. Turnbull and M.H. Cohen, upon cooling of a liquid the *free volume*, defined as the volume available for molecules or polyhedra to maneuver without encountering strong interactions with their neighbors, decreases continuously until at the glass transition all free volume has been removed, leaving each molecule trapped in a local energy minimum from which all other positions are inaccessible.

We notice in Fig. 1.45 that the volume changes at crystallization and vaporization are discontinuous. Since volume is a first-derivative function of the free energy G (at constant composition x and temperature T):

$$V = \left(\frac{\partial G}{\partial P}\right)_{x,T}$$

the discontinuities in V at T_m and T_v define melting and vaporization as first-order phase transitions. In contrast, the change in V at T_g is continuous. The thermal expansivity α is a second-derivative function of G (first derivative of V), and since changes in α at T_g as well as other second-derivative functions such as heat capacity are fairly discontinuous, the glass transition has the appearance of a somewhat diffuse second-order phase transition. However, unlike equilibrium second-order phase transitions, the transition to a glass (which is not the ultimate equilibrium form but only metastable) occurs over a range of temperatures depending on the thermal history of the glass, as shown in Fig. 1.45b. A faster rate of cooling freezes structural rearrangement at a higher T_g and specific volume, whereas slower cooling results in a denser glass with a lower T_g. Upon heating, hysteresis in the reverse glass-to-liquid transition can also occur if the heating and cooling rates are not the same. Some difficulty persists in defining the nature of the glass transition. It is perhaps sufficient for this discussion to say that the glass transition is a true phase transition (between a supercooled liquid and a glassy solid), but with characteristics determined primarily by kinetics rather than equilibrium thermodynamics.

Concurrent with the changes in volume shown in Fig. 1.45 are changes in the viscosity of the glass. Anyone who has watched a glassblower at work will recognize that glass viscosity is greatly temperature-dependent; we will see that it is also greatly composition-dependent. Glasses are often characterized by the temperature at which several viscosities of practical interest are achieved. The *working point*, where forming is possible, is defined as the temperature at which $\eta=10^4$ Poise; the *softening point* by $\eta = 10^{7.6}$. The *annealing point* is a viscosity at which internal stresses can be substantially relieved in 15 minutes or so, and is defined as $\eta = 10^{13.4}$ Poise. The viscosity at the glass transition is 10^{13}-$10^{14.5}$ Poise for oxide glasses, and represents a limit above which structural rearrangements cannot take place at normal cooling rates.

Continuous Random Networks

Amorphous SiO_2 (fused silica) is the prototypical oxide glass. As in the crystalline silicates, directional covalent bonding and ionic bonding in accordance with Pauling's rules promote the formation of tetrahedral $(SiO_4)^{4-}$ basic structural units. Therefore, short-range order is identical in crystalline and glassy SiO_2; the coordination of anions around cations is four, and the O-Si-O bond angle within each tetrahedral unit is 109.5°. The O/Si ratio of 2 indicates that all four corners of each tetrahedra are interconnected in glassy SiO_2, as is the case in the various crystalline phases of SiO_2. However, it is possible for complete connectivity to be maintained without crystalline order; and this is the basis of the *continuous random network* structure first proposed by Zachariasen in 1932 (*J. Am. Chem. Soc.*, 54, 3841). In two dimensions, the continuous random network structure of an A_2B_3 glass is shown in Fig. 1.46.

Examining Fig. 1.46, and imagining the three-dimensional counterpart that is SiO_2, we can identify three structural parameters at a medium range of order that allow us to distinguish between the crystalline and continuous random network amorphous structures. These are: (1) the angle of the Si-O-Si bond between tetra-

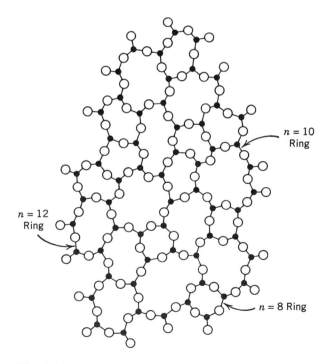

Fig. 1.46 Continuous random network model of an A_2B_3 glass. [From W. H. Zachariasen, *J. Am. Chem. Soc.*, 54, 3841 (1932).]

hedra; (2) the rotational angle between tetrahedra; and (3) the number of Si-O bonds that complete each of the "rings" seen in Fig. 1.46. Each of these three parameters has firmly fixed values in crystalline SiO_2, but varies over a wider range in the corresponding glass (see Table 1.6.) The intertetrahedral Si-O-Si angle θ in amorphous SiO_2 takes on a broad range of values with an rms Δθ of about 15° centered at a mean of 150°. The rotational angle between tetrahedra is random, whereas in crystalline SiO_2 it is either 0° or 60°. And, if we examine the number of Si-O bonds n that form an interconnected ring (ring statistics), we find well-defined distributions in the crystalline silicates with a minimum of $n = 12$ (n must be an even number to avoid the occurrence of energetically unfavorable Si-Si or O-O bonds). In amorphous SiO_2 on the other hand, some $n = 8$ and 10 rings appear to exist (R. J. Bell and P. Dean, *Phil. Mag.*, 25, 1381, 1972).

Similar features are found in other AB_2 compounds with an anion coordination of 4 and cation coordination of 2. In covalent semiconductors such as Si and Ge that take on tetrahedral coordination in their crystalline form (diamond cubic structure), the amorphous form can also be described as a continuous random network of tetrahedra. In contrast to SiO_2, the coordination of every atom is 4, and odd-numbered rings are allowed, although $n < 5$ is not found due to the strain necessary to accommodate it (the minimum in the diamond cubic structure is $n = 6$). There is only one characteristic angle, Si-Si-Si, and it varies by a rms value of 9° about the 109.5° tetrahedral angle. Rotation angles between tetrahedra are again random. Compound covalent semiconductors such as GaAs might be considered as intermediate in structure between Si and SiO_2. Although in any AB_2 compound A-B bonds are favored over A-A or B-B bonds, skewing the distribution of rings toward even n, odd n values are not completely absent in highly covalent compounds such as GaAs. Table 1.6 lists the distinctions between continuous random network SiO_2 and Si and their crystalline counterparts.

Table 1.6 Structural Characteristics of Continuous Random Network Glasses

	SiO_2		Si	
	Glass	Crystal	Glass	Crystal
Number of nearest neighbors	Si : 4 O : 2	Si : 4 O : 2	4	4
Bond angles	109.5° (O-Si-O) 150°+15°rms (Si-O-Si)	109.5° (O-Si-O) 180°(tridymite) 150° (quartz, cristobalite)	109°+9°rms (Si-Si-Si)	109.5° (Si-Si-Si)
Rotation angle between tetrahedra	Random	0° or 60°	Random	0° (diamond cubic)
Ring counts	8, 10, 12...	12	5, 6, 7, 8,...	6, 8, 10....

Random Close-Packing

The nondirectional nature of metallic bonding combined with the absence of charge-neutrality requirements results in a very different structure for metallic glasses. They form in a random close-packed (rcp) structure of somewhat lower density (63.7% for a monatomic solid) than the close-packed FCC and HCP lattices (74.1%). Each atom occupies a local energy minimum position, and while this structure is metastable, reconstruction rather than a continuous change of atom positions is necessary to transform to the FCC and HCP structures. There is less order in the first coordination shell than in continuous random network structures. Common foamed polystyrene (Styrofoam™) is a reasonably good simulation of the rcp topology, being composed of deformable spheres of roughly constant size each of which represents the volume available to an atom. Close examination of compacts of deformable spheres (J. L. Finney, *Proc. Roy. Soc. (London)* A, 319, 479, 1970) shows that the polyhedra have on average 14.3 faces, in which the average number of edges per face is 5.16.[6]

Radial Distribution Function

The radial distribution function $\rho(r)$ characterizes the variation in atom density with distance r from an atom chosen as the origin, and is a useful parameter for illustrating the level of short-, medium-, and long-range order in glasses. The probability of finding an atom between the distances r and $r + dr$ is given by $\rho(r)dr$. For a dilute gas in which the density of atoms is n, $\rho(r)$ is given by the continuous function $4\pi nr^2$, shown in Fig. 1.47a. For a crystal, $\rho(r)$ is given by a series of delta functions:

$$\rho(r) = \sum_i N_i(r)\, \delta(r - r_i)$$

where $N_i(r)$ is the number of atoms at distance r, as shown in Fig. 1.47b. Here the probability is unity (neglecting the effect of thermal vibrations) at the distance of each coordination shell (corresponding to first-, second-, third-nearest neighbor distances and so forth). The variation of order in SiO_2 glass is seen in the radial distribution function in Fig. 1.47c. Peaks appear for the first few coordination shells, and these become increasingly broadened with distance from the origin. The first three peaks correspond to the initial Si-O, O-O, and Si-Si distances. Beyond the first few atomic distances, the radial distribution function

---

[6] These are therefore approximately tetrakaidecahedra (14 faces), which we may compare with the Kelvin tetrakaidecahedra (truncated octahedron) that is the ideal space-filling regular polyhedron representing polycrystalline microstructures obtained by enforcing angles of 120° between all grain boundaries, and 109.5° between the lines that are three-grain junctions. See Chapter 5.

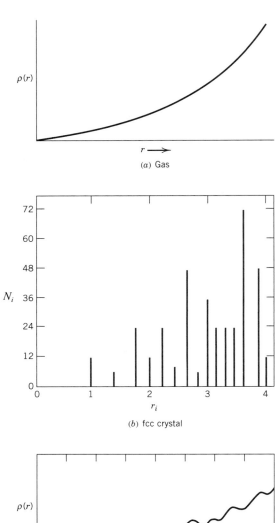

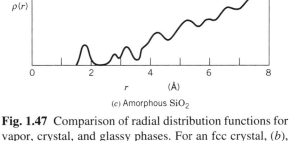

Fig. 1.47 Comparison of radial distribution functions for vapor, crystal, and glassy phases. For an fcc crystal, (b), the probability of finding an atom is unity at each coordination shell, while the number of atoms Z_i varies. (after R. Zallen, *The Physics of Amorphous Solids*, J. Wiley and Sons, 1983.) (c) shows the experimental RDF for amorphous SiO_2 [from R. J. Bell and P. Dean, *Phil. Mag.*, 25, 1381 (1972).]

1.5 Glass Structure

quickly approaches that of a gas. The RDF's for different continuous random network glasses and for random close-packed glasses differ in their quantitative details. It is the careful experimental determination of RDF's, primarily by x-ray, electron, and neutron scattering experiments, that has confirmed the amorphous structures described above.

Oxide Glasses

In his pioneering descriptions of glass structure, Zachariasen recognized that while cation/anion radius ratios in the correct range (<0.4) for tetrahedral coordination were often found in glass-forming oxides, this was a necessary but not sufficient condition for glass formation (i.e., many tetrahedrally coordinated crystalline oxides are not good glass formers). He enumerated the following structural rules, which are analogous to Pauling's rules for crystalline structures, for determining which compounds will form continuous random network oxide glasses. At the time, amorphous semiconductors and metallic glasses had yet to be discovered or characterized. Zachariasen's rules state:

1. Each oxygen atom is linked to no more than two cations.
2. The number of oxygen atoms surrounding the cation must be small, four or less.
3. The oxygen polyhedra share corners rather than edges or faces.
4. At least three corners of each oxygen polyhedra are shared.

These rules correctly predict ease of glass formation for the oxides B_2O_3, SiO_2, GeO_2, P_2O_5, and a few others. These oxides are known as *network formers*. Glasses made solely from network formers often have limited utility. For example, pure B_2O_3 glass ($T_m \sim 450°C$) is not water resistant, and pure SiO_2 glass (fused silica) while valued for its chemical durability, high use temperatures (up to ~1200°C), and thermal shock resistance, must be processed above 1750°C. The great majority of useful glasses contain additives that serve to alter processing and properties. These are commonly termed *network modifiers* and *intermediates*, which we will now discuss. Table 1.7 classifies various cations with respect to their roles as network formers, modifiers, and intermediates in oxide glasses.

Network modifiers provide extra oxygen ions but do not participate in the network, thereby raising the O/Si ratio of the glass. The extra oxygen allows the *bridging* oxygen between two tetrahedra to be disrupted, and two *nonbridging* oxygen to terminate each tetrahedron. For example, in sodium silicates, each Na_2O molecule results in the formation of two *nonbridging* oxygen in SiO_2, with the Na^+ ions providing local charge neutrality, as shown in Fig. 1.48. The fraction of nonbridging oxygen is readily determined from the composition (see Problem 24 at the end of the chapter). The effects of modifiers are directly analogous to the decreasing SiO_4 interconnectivity observed in crystalline silicates with increasing O/Si ratio described earlier. In glasses the loss of connectivity results in greatly decreased viscosities and T_g's for modified silicates and reduces the processing temperatures of silicate glasses

Table 1.7 Coordination Number and Bond Strength of Oxides

	M in MO_x	Valence	Dissociation Energy per MO_x (kcal/mole)	Coordination Number	Single-Bond Strength (kcal/mole)
Glass formers	B	3	356	3	119
	Si	4	424	4	106
	Ge	4	431	4	108
	Al	3	402–317	4	108
	B	3	356	4	89
	P	5	442	4	111–88
	V	5	449	4	112–90
	As	5	349	4	87–70
	Sb	5	339	4	85–68
	Zr	4	485	6	81
Intermediates	Ti	4	435	6	73
	Zn	2	144	2	72
	Pb	2	145	2	73
	Al	3	317–402	6	53–67
	Th	4	516	8	64
	Be	2	250	4	63
	Zr	4	485	8	61
	Cd	2	119	2	60
Modifiers	Sc	3	362	6	60
	La	3	406	7	58
	Y	3	399	8	50
	Sn	4	278	6	46
	Ga	3	267	6	45
	In	3	259	6	43
	Th	4	516	12	43
	Pb	4	232	6	39
	Mg	2	222	6	37
	Li	1	144	4	36
	Pb	2	145	4	36
	Zn	2	144	4	36
	Ba	2	260	8	33
	Ca	2	257	8	32
	Sr	2	256	8	32
	Cd	2	119	4	30
	Na	1	120	6	20
	Cd	2	119	6	20
	K	1	115	9	13
	Rb	1	115	10	12
	Hg	2	68	6	11
	Cs	1	114	12	10

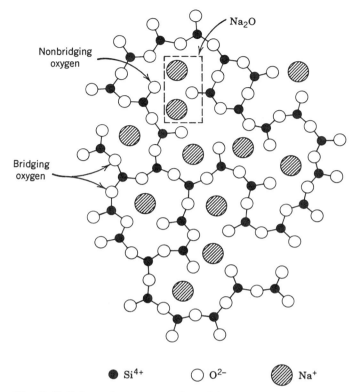

Fig. 1.48 Schematic representation of the structure of a sodium silicate glass. Each Na_2O addition results in the formation of two nonbridging oxygen terminating silica tetrahedra.

into more practical ranges. While alkaline oxides (Na_2O, Li_2O, K_2O) are very effective modifiers, they result in glasses that are not chemically durable. (Sodium silicates are known as *water glasses* as they are water-soluble.). In the commercially important soda-lime-silicates most commonly seen as windows and containers, the chemical durability is greatly improved with the additional modifier CaO. One thus has the reduced processing temperatures of a modified glass without the drawbacks of an alkaline silicate. Container and window glasses have O/Si ratios of 2.3–2.4; in glazes and enamels it varies from 2.25 to 2.75. Other commonly used modifiers in silicate glasses include PbO, MgO, ZnO, and BaO.

Intermediates as the name implies are not clearly modifiers or network formers and may contribute in part to the network structure. They are generally cations with higher valence than the alkalis and alkaline earths, but which do not satisfy Zachariasen's rules for network formers. One of the most interesting intermediates is Al^{3+}, for its structural role depends on the presence and concentration of alkali ions. In discussing clay minerals, we pointed out that Al^{3+} can often substitute for Si^{4+} as long as charge neutrality is achieved by having an adjacent alkali

90 Chapter 1 / Structure of Ceramics

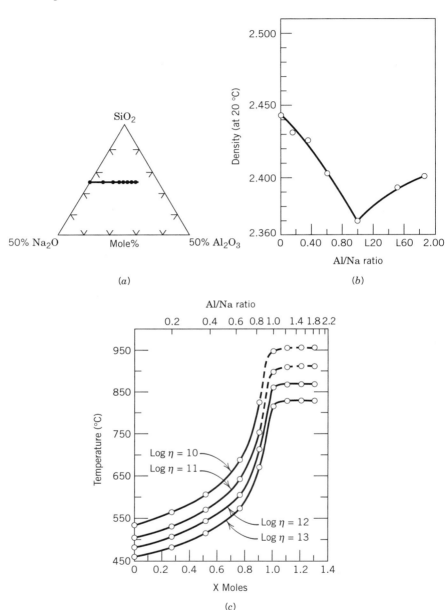

Fig. 1.49 Variation in properties of SiO_2-Al_2O_3-Na_2O glass with Al/Na ratio. (a) Composition series in which Al/Na ratio is varied at constant SiO_2 content; (b) variation in density showing minimum at Al/Na = 1, the equivalence point, where all Al is believed to be tetrahedrally coordinated (D. E. Day and G. E. Rindone, *J. Am. Ceram. Soc.*, 45[10] 489 (1962).; (c) isoviscosity lines showing sharp decrease in viscosity for compositions with excess modifier, Al/Na < 1. [T.D. Taylor and G.E. Rindone, *J. Am. Ceram. Soc.*, 53[1] 692 (1976).]

1.5 Glass Structure 91

ion. In silicate glasses, aluminum behaves in a similar fashion. Figure 1.49 shows the variation in viscosity and density of a sodium aluminosilicate with Al/Na ratio. When the concentration of aluminum is less than that of the alkali, the substitution of $Al^{3+}+Na^{1+}$ units for Si^{4+} can take place, with excess alkali serving in its usual modifier role. For Al/Na ratios above the "equivalence point" (Al/Na =1) however, the excess Al^{3+} acts as a modifier.

Borates and Borosilicates

Few practical glasses are based on B_2O_3 alone as the network former due to its very low melting point and poor chemical durability (possible exceptions are the alkaline borates, useful for their high alkaline ion conductivity). However, when

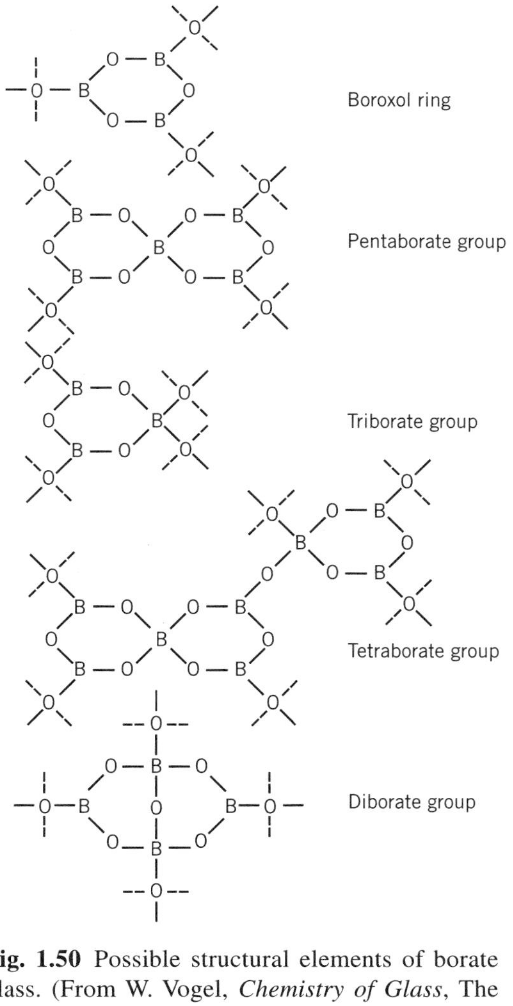

Fig. 1.50 Possible structural elements of borate glass. (From W. Vogel, *Chemistry of Glass*, The American Ceramic Society, Columbus, OH, 1985.)

B_2O_3 is alloyed with SiO_2, and other constituents, the processing temperatures are lowered compared to SiO_2 yet the chemical durability is not lost. Amorphous B_2O_3 is composed of triangular BO_3 units, probably interconnected locally into boroxyl groups as shown in Fig. 1.50. Alkali additions do not act as modifiers, but instead *increase* connectivity by forming tetrahedral BO_4 units with local charge compensation by the alkali, as discussed above for Al^{3+}. The formation of compact tetrahedra causes an increase in the density and, importantly, a decrease in the thermal expansion coefficient. Originally termed the *boron anomaly*, the concentration dependence of α reaches a minimum at about 16 mole% alkaline oxide, as shown in Fig. 1.51, beyond which α again increases due to the loss of simple tetrahedral coordination (fig. 1.52). At R_2O concentrations above 30%, the alkali may begin to act as a modifier, increasing the nonbridging oxygen content between tetrahedra.

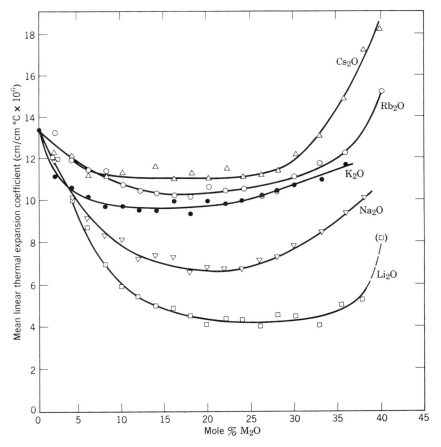

Fig. 1.51 Thermal expansion coefficients of alkali borate glasses as a function of composition. [From R. R. Shaw and D. R. Uhlmann, *J. Non-Cryst. Solids*, 1, 347 (1969).]

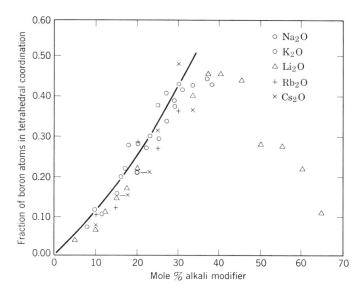

Fig. 1.52 The fraction of boron atoms in BO_4 configurations in alkali borate glasses plotted against the molar percent of alkali oxide. (From P. J. Bray, in *Interaction of Radiation with Solids*, Plenum Press, New York, 1967.)

These structural features are largely retained in borosilicate solid solutions. Commercial low thermal expansion (high thermal shock resistance) glasses of the Pyrex™ or Kimax™ type are sodium borosilicates in which the alkali/boron ratio is close to that of minimum thermal expansion coefficient. Another class of useful glasses from the sodium borosilicate system takes advantage of the low chemical resistance of borates in order to prepare a high-silica glass at much lower temperatures than are normally necessary. These are the Vycor™ type glasses, in which a borosilicate is phase-separated at 500–600°C into high borate and high silica phases. The borate phase is then chemically leached, leaving behind a porous high silica glass that is easily densified at about 1100°C. The processing and microstructure of these glasses discussed in greater detail in Chapter 5.

ADDITIONAL READING

General

W. D. Kingery, H. K. Bowen, and D. R. Uhlmann, *Introduction to Ceramics*, 2nd Edition, John Wiley and Sons, New York, 1976.

A. R. West, *Solid State Chemistry and Its Applications*, John Wiley and Sons, New York, 1984.

94 Chapter 1 / Structure of Ceramics

R. J. Brook, Editor, *Concise Encyclopedia of Advanced Ceramics*, Pergamon Press/MIT Press, Cambridge, Mass., 1991.

Crystallography, Diffraction, and Tensor Properties

B. D. Cullity, *Elements of X-Ray Diffraction*, 2nd Edition, Addison-Wesley, Reading, Mass., 1978.

C. Barrett and T. B. Massalski, *Structure of Metals*, 3rd Edition, Pergamon Press, 1980.

A. Kelly and G. W. Groves, C*rystallography and Crystal Defects*, Addison-Wesley, Reading, Mass., 1970.

M. Buerger, *Elementary Crystallography*, The MIT Press, Cambridge, Mass., 1978.

J. F. Nye, *Physical Properties of Crystals*, Oxford University Press, New York, 1985.

International Tables for Crystallography, Theo Hahn, Ed., D. Reidel Publishing Co., Dordrecht, The Netherlands, 1983.

Electronic Ceramics

A. R. von Hippel, *Dielectrics and Waves*, John Wiley and Sons, New York, 1954.

B. Jaffe, W. R. Cook Jr., and H. Jaffe, *Piezoelectric Ceramics*, Academic Press, London, 1971.

L. L. Hench and J. K. West, *Principles of Electronic Ceramics*, John Wiley and Sons, New York, 1990.

A. J. Moulson and J. M. Herbert, *Electroceramics*, Chapman and Hall, 1990.

R. C. Buchanan, Editor, *Ceramic Materials for Electronics, 2nd Edition*, Marcel Dekker, Inc, New York, 1991.

B. D. Cullity, *Introduction to Magnetic Materials*, Addison-Wesley, Reading, Mass., 1972.

J. Smit and H. P. J. Wijn, *Ferrites*, John Wiley and Sons, New York, 1959.

Cuprate Superconductors

D. M. Smyth, pp. 1–10 in *Ceramic Superconductors II*, Research Update 1988, M. F. Yan, editor, The American Ceramic Society, 1988.

R.J. Cava, *Scientific American*, 42–49, August 1990; *Science*, **247**, 656–662, 1990; pp. 3–8 in *Advances in Superconductivity V*, Y. Bando and H. Yamauchi, editors, Springer-Verlag, Tokyo, 1993.

Glass Structure:

W. Vogel, *Chemistry of Glass*, The American Ceramic Society, Columbus, Ohio, 1985.

R. Zallen, *The Physics of Amorphous Solids*, John Wiley and Sons, 1983.

Glass: Science and Technology, Vols. 1-5, D. R. Uhlmann and N. J. Kreidl, editors, Academic Press, 1980–1990 (Vol. 1, 1983; Vol. 2, 1984; Vol. 3, 1986; Vol. 4A and 4B, 1990; Vol. 5, 1980).

Other Specific Materials

E. Doerre and H. Huebner, *Alumina*, Springer-Verlag, Berlin, 1984.

K. H. Jack, "Sialons and Related Nitrogen Ceramics," *J. Mat. Sci.*, 11, 1135-1158 (1976).

A. H. Heuer, "Transformation Toughening in ZrO_2-Containing Ceramics," *J. Am. Ceram. Soc.*, 70[10], 689, 1987.

A. H. Heuer and L.W. Hobbs, Editors, *Science and Technology of Zirconia, Advances in Ceramics Vol. 3*, The American Ceramic Society, Columbus, Ohio, 1981.

N. Claussen, M. Ruehle, and A. H. Heuer, Editors, *Science and Technology of Zirconia II, Advances in Ceramics, Vol. 12*, The American Ceramic Society, Columbus, Ohio, 1984.

PROBLEMS

1. Verify the limiting radius ratios for 3, 4, 6, and 8-fold coordinations that are given in Figure 1.12.

2. Show that Pauling's first and second rules apply to the compound CsCl. Discuss the connection of polyhedra. Compounds of this structure type are relatively rare. Explain why this might be so, from the viewpoint of what is required of the radius ratio and the relative densities of sites.

3. Calculate the lattice parameter and crystal density for:

    MgO containing 0.1 mole% CaO
    MgO containing 0.05 mole% Al_2O_3

    ZrO_2 containing 10 mole% CaO
    ZrO_2 containing 10 mole% Y_2O_3

    You may either find the lattice parameter of undoped MgO and ZrO_2 in another source or determine it from the ionic radii in Table 1.2.

4. For magnesium oxide:
    (i) Draw the (111) and (110) planes, identifying atoms and empty site locations.
    (ii) Draw several cross sections parallel to the (100) plane of the unit cell at heights where there are atoms and interstitial sites. Label the coordinates of each of the atoms, empty sites, and the height of the cross section drawn.

5. Lithium oxide forms in the antifluorite structure.
    (a) Compute the value of the lattice constant.
    (b) Calculate the density of Li_2O, and the density of an 0.01 mole% SrO solid solution in Li_2O.
    (c) What is the maximum radius of a cation which can be accommodated in the vacant interstice of the anion array?

96 Chapter 1 / Structure of Ceramics

6. Predict the structure of the following oxides, using Pauling's rules for guidance. Show that the structure you choose conforms to Pauling's second rule. Give an example of a real oxide of each structure. If polymorphs are likely to exist, explain which they would be. (Assume throughout an oxygen ion radius of $R(O^{2-})=1.40$ angstroms.)

   (i) M_2O, where $R(M^{1+}) = 0.52$ angstroms.
   (ii) MO, where $R(M^{2+}) = 0.49$ angstroms.
   (iii) ABO_3, where $R(A^{3+}) = 1.25$ angstroms and $R(B^{3+}) = 0.70$ angstroms.
   (iv) ABO_3, where $R(A^{2+}) = 0.73$ angstroms and $R(B^{4+}) = 0.65$ angstroms.

7. Discuss using Pauling's rules why some compounds of ABO_3 formula are of the ilmenite or lithium niobate structures ($FeTiO_3$, $LiNbO_3$) while others form the perovskite structure ($LaGaO_3$, $PbTiO_3$). Note that the cation valences vary.

8. Determine the Madelung constant for a two-dimensional crystal consisting of two interpenetrating square lattices.

9. Calculate the distance between octahedral sites in the *fcc* and *hcp* close-packed structures in units of r_o, the radius of the close-packed atom. Then, explain why there are no common HCP-based ionic structures with all of the octahedral sites *or* all of the tetrahedral sites filled. (That is, the FCC structure is preferred when all sites must be filled.) Use Pauling's rules and your knowledge of the ABA and ABCA stacking sequences. Note: NiAs, which does have a structure based on HCP with all of the octahedral sites filled, is covalently bonded.

10. In the corundum structure, layers of half-filled octahedral sites are stacked in a particular sequence along the c-axis direction (Fig. 1.21). Yet, unlike SiC, polytypes of alumina, in which the stacking sequence of this two-dimensional layer varies, do not form readily. Explain why this is so.

11. Calculate the lattice parameter of the rocksalt structure in terms of the cation and anion radii r_C and r_A. Do the same for the wurtzite and cubic fluorite structures. Using the ionic radii in Table 1.2, calculate the lattice parameters of MgO, ZnO, and CeO_2.

12. Describe the coordination polyhedron of anions around the different cations in the perovskite and the idealized $YBa_2Cu_3O_8$ structures, and apply Pauling's rules.

13. Apply Pauling's second rule to the *inverse* spinel structure. Discuss the linkage between octahedral and tetrahedral sites. Draw a (110) plane for the structure and label the ions and unoccupied sites.

14. Sketch the polarization-electric field (P-E) hysteresis curve for a ferroelectric material and the magnetization-magnetic field (B-H) hysteresis for a "soft" and "hard" ferrimagnetic material. In each case explain the cause of

the nonlinear relationship in terms of the mechanisms involved. How might each loop differ between a single crystal and a polycrystalline material?

15. Describe the polarization mechanisms that exist in a polycrystalline capacitor based on $BaTiO_3$ at a temperature below the Curie point. Draw a curve of dielectric constant and loss factor as a function of frequency from 10 to 10^{20} Hz. In order to obtain a high specific capacitance (small size, large capacitance), what compositional and microstructural features would you aim for? How might you make a capacitor with a flat temperature dependence of the dielectric constant (at low frequency)?

16. $Pb(Sc_{1/2}Ta_{1/2})O_3$ is a *relaxor ferroelectric* compound which is an ordered perovskite below room temperature. The ordering takes place among the B-site cations. Determine the coordination of cations around the oxygen ion in the fully ordered state, and show that Pauling's second rule is satisfied. Relative to the crystal axes, in which direction(s) do you expect ferroelectric polarization?

17. Iron oxide forms in three distinct stoichiometries depending on the oxidation state of iron: FeO, Fe_2O_3, and Fe_3O_4. Answer the following:
    (a) What is the structure of each of these oxides?
    (b) Fe^{2+} has the outer-shell electron configuration $3d^6 4s^2$, while Fe^{3+} has the configuration $3d^5 4s^2$. What is the magnetic moment per ion, in Bohr magnetons (μ_B)?
    (c) Which of these compounds would you expect to be ferrimagnetic, and why? Which would you expect to be ferromagnetic, or, anti-ferromagnetic, and why?

18. When the normal spinel, $CdFe_2O_4$ is added to an inverse spinel such as magnetite, Fe_3O_4, the Cd ions retain their normal configuration. Calculate the magnetic moment for the following $Cd_x Fe_{3-x} O_4$ compositions:
    (a) $x = 0$
    (b) $x = 0.1$
    (c) $x = 0.5$

19. Predict the saturation magnetic moment per unit volume in Bohr magnetons for the following inverse spinel structures:
    (a) $MgFe_2O_4$
    (b) $CoFe_2O_4$
    (c) $Zn_{0.2}Mn_{0.8}Fe_2O_4$
    (d) γ-Fe_2O_3

20. The magnetic moment of a ferrite, $Li_{0.5}Fe_{2.5}O_4$, has been measured and observed to be 2.6 Bohr magnetons per unit of spinel formula. How do you justify this result from the known net spins associated with the ions involved? What position(s) in the crystal lattice does Li^+ occupy? Fe^{3+}?

98 Chapter 1 / Structure of Ceramics

21. In spinel ferrites the exchange interactions between a and b cations are strong and antiparallel, and a-a and b-b interactions are weaker but also antiparallel.

   (i) Magnesium ferrite ($MgFe_2O_4$) shows a higher saturation magnetization when it is quenched from high temperature than when it is slowly cooled. Explain how this can arise, based on the relative cation occupancy of tetrahedral and octahedral sites.

   (ii) Pure $MnFe_2O_4$ has a saturation magnetization of $5\mu_B$ per formula unit. With just this information, can you determine whether it is a normal or inverse spinel? Explain why or why not. Notice also that pure $CoFe_2O_4$ has a larger saturation magnetization than pure $NiFe_2O_4$. Is this consistent with both being normal spinels, or both being inverse spinels (and explain)?

   (iii) Manganese-zinc ferrites are commercially important partly because they have the highest saturation magnetization of the spinel ferrites, as shown in Figure ST11. In this figure we see that pure $ZnFe_2O_4$ has zero magnetization. This is not because it is completely inverse but rather because it is an *antiferromagnetic* normal spinel (the antiparallel b-b interactions happen to dominate). As we add $ZnFe_2O_4$ to the partly inverse $MnFe_2O_4$, Zn^{2+} occupies tetrahedral sites and displaces any Fe^{3+} on those sites to the octahedral sites. Compute the saturation magnetization of a 10% $ZnFe_2O_4$ solid solution for comparison with the experimental data in the figure.

22. Si^{4+} in silicates tends to be tetrahedrally coordinated by oxygen. SiO_2 forms in three different structure types: quartz, crystobalite, and tridymite, none of which are close-packed. Yet, we have discussed other oxides that are based on close-packed anions in which all or half of the tetrahedral sites are filled. Explain why crystalline SiO_2 is not close-packed.

23. The atomic weights of Si and Al are very similar (28.09 and 26.98 g/mole, respectively), yet the densities of SiO_2 and Al_2O_3 are quite different (2.65 and 3.98 g/cm³, respectively). Explain this difference in terms of crystal structure and Pauling's rules.

24. For binary alkaline silicate glasses (SiO_2-R_2O) determine on the basis of the continuous random network/alkaline modifier structural model:

   (a) The nonbridging oxygen fraction (F_{NBO}) for R_2O contents of 5, 10, and 15 mole %.

   (b) The composition at which a continuous interconnected silicate network is no longer possible.

   (c) Propose a plausible structural model for the same glass containing 5% NaF in place of R_2O.

   (d) Do the same as in (a) and (b) for an SiO_2-Al_2O_3-R_2O glass containing 5% Al_2O_3.

25. Sketch the radial distribution function for a diamond cubic crystalline compound such as Si or Ge. Show schematically how it would change for a glassy form of the same material.

26. Water is found to have a marked effect on the viscosity of many silicate glasses including fused silica. How do you expect the viscosity to vary with increasing H_2O content, and why (structurally)?

27. Most asbestos minerals such as tremolite $(OH)_2Ca_2Mg_5(Si_4O_{11})_2$ have a fibrous habit; talc $(OH)_2Mg_3(Si_2O_5)_2$ has a platy habit. Explain this difference in terms of O/Si ratio and bonding between silica tetrahedra.

28. Kaolin has the molecular formula: $Al_2O_3 \cdot 2SiO_2 \cdot 2H_2O$. What is the O/Si ratio of the Si_xO_y sheet? How does this structure differ from that for pyrophyllite? How does the structure of mica differ from that of pyrophyllite?

29. *Silicon oxynitride glasses* are silicate glasses with some dissolved nitrogen, usually a few percent at most. High temperatures ($\geq 1800°C$) and/or elevated nitrogen pressures are necessary to get this amount of nitrogen in solution, since silicate glasses do not usually dissolve detectable amounts of nitrogen. (After all, air is mostly nitrogen, yet oxide glasses are melted and processed in air without becoming nitrides.)

    One reason for interest in the properties of silicon oxynitride glass is that it is a minor but important phase in polycrystalline silicon nitride ceramics. This glass is located at grain junctions in contact with the Si_3N_4 grains, and does dissolve some of the nitride due to the high firing temperatures (1800–2000°C) and the application of nitrogen overpressure to prevent decomposition of Si_3N_4. However, in actual use the glass softens upon heating long before the silicon nitride grains themselves become deformable. Flow of the glass thus limits the highest temperature at which the sintered silicon nitride can be used.

    Does the dissolution of nitrogen raise or lower the viscosity of the silicate glass at a constant temperature? Explain your answer from the viewpoint of the glass structure. It may be useful to refer to the structures of the corresponding crystalline compounds.

Chapter 2

Defects In Ceramics

To this point we have primarily discussed the ideal crystalline state. However, a great many properties of crystals are determined by imperfections. Electrical conductivity and diffusional transport (the topic of the following chapter) in most ceramics is determined by the number and type of point defects. Various optical properties, for instance those giving rise to color and lasing activity, are caused by electronic absorption and emission processes at impurity ions and other point defects. The rates of kinetic processes such as precipitation, densification, grain coarsening, and high-temperature creep deformation are determined by mass transport due to defects.

We will confine most of our discussion to *point defects*,[1] defined as deviations from the perfect atomic arrangement: missing ions, substituted ions, interstitial ions, and their associated valence electrons. Point defects occur (to greater or lesser degrees) in all crystalline materials. A principle difference between point defects in ionic solids and those in metals is that in the former, all such defects can be electrically charged. *Ionic defects* are point defects that occupy lattice atomic positions, including vacancies, interstitials, and substitutional solutes. *Electronic defects* are deviations from the ground state electron orbital configuration of a crystal, formed when valence electrons (i.e., those responsible for bonding) are excited into higher orbital energy levels. Such an excitation may create an electron in the conduction band and/or an electron hole in the valence band of the

---

[1] Point defects may be distinguished from *extended defects*, which also play an important role in determining crystal and polycrystal properties. These include aggregates of point defects in *clusters*, *line defects* such as dislocations, which have their most important role in plastic deformation, as well as various *planar defects*. Planar defects include external surfaces, grain boundaries, stacking faults, and crystallographic shear planes.

crystal. In terms of spatial positioning, these defects may be localized near atom sites, in which case they represent changes in the ionization state of an atom, or may be delocalized and move freely through the crystal. Point defects in ceramic systems can be formed by thermal excitation at high temperature, by the addition of solutes and impurities, or by oxidation or reduction processes which cause a variation in the metal/anion stoichiometry of the compound.

Isolated charged defects in crystals are also able to interact with one another in an analogous way to the interactions which take place between different ions or between ions and electrons in aqueous solutions. In the solid-state analogy, the perfect crystal may be regarded as a neutral medium into which the charged defects are dissolved. This fruitful similarity between solution-chemical interactions and solid-state defect interactions has resulted in the development of the field known as *defect chemistry*, which provides immensely useful tools for understanding the properties of crystals containing point defects.

2.1 POINT DEFECTS

In considering point defect behavior, one is usually concerned with two principal issues: what types of point defects are present, and in what concentration? The answer depends greatly on crystal structure, chemical composition, and bonding, as well as the temperature of interest. Physical properties such as density, melting point, electrical conductivity, diffusion, and optical absorption often provide strong clues to defect behavior. The formation of atomic defects requires the breaking of interatomic bonds, and in the same way that the melting (or decomposition) temperature of a compound scales with the strength of its interatomic bonding, so does the energy to create atomic defects. Within compounds of a given structure type (e.g., rocksalt structure type alkali halides and oxides), the energy to form defects scales with the melting point. Due to the importance of melting temperature as a parameter, defect and diffusional properties are sometimes scaled to the *homologous temperature*, T/T_m, when comparing different materials. At equal temperature, the concentration of defects in MgO (T_m = 2825°C) is many orders of magnitude lower than than those in the alkali halides (NaCl, KCl, LiF, etc.), which have melting points below 1000°C. (Many earlier studies of point defects were conducted in alkali halides due to their simple structures and a relative abundance of defects.) At equal homologous temperature, however, the defect concentrations are rather similar.

Another simple consideration is the availability of defect sites in the crystal structure. A close-packed sublattice of ions does not easily accommodate interstitial defects. For instance, in the rocksalt structure, anion interstitials have a high energy relative to other defects such as anion vacancies, cation vacancies and cation interstitials. Furthermore, since all octahedral sites in the cation sublattice are filled, only the smaller tetrahedral sites are possible interstitial atom locations. Thus in this structure one finds that interstitial defects are unlikely to form unless the atom in question is quite small (such as H or Li). On the other hand, a variety of foreign atoms can substitute in the octahedral sites for the host cations, as long as charge neutrality can be maintained. A corollary is that crystal structures that are not close-packed, such as silicon nitride (Fig. 1.38) or the various crystalline forms of silicon dioxide (Figs. 1.39, 1.40) often readily accommodate interstitial atoms. The extensive solid solutions known as sialons and β-quartz solid solutions (see also the discussion of glass ceramics in Chapter 5) form in these two respective structures due to the availability of interstitial sites for charge-compensating cations.

Some caution is necessary in applying this simple line of reasoning, since unoccupied interstitial sites in close-packed ionic structures are empty for good reason: The atomic arrangement minimizes repulsive electrostatic interaction with second-nearest neighbors of the same kind. For example, corundum (Al_2O_3) has only two-thirds of the octahedral sites filled, yet theoretical calculations and experiments both indicate that the energy to displace an aluminum ion to an interstitial site (Frenkel energy) is higher than the formation energies of other defects. The same is true in rutile, where only one-half of the octahedral cation sites are filled.

The location of solutes and impurities in the lattice often depends on the compatibility in size and valence with host ions. Solutes that differ in valence from the host ion (known as *aliovalent* solutes) must be compensated by additional charged defects in order to maintain overall electrical neutrality in the crystal. An important part of the total energy for solute incorporation is the energy to form the charge-compensating defects, and it can be misleading to consider only the energetics for substituting the solute ion.

Chemical composition will often suggest the presence of certain defects. *Nonstoichiometric* compounds are those in which the metal/anion ratio deviates significantly from the ideal value on which the structure is based, due to the existence of multiple ion valence states. Transition metal oxides (e.g., NiO, FeO, Fe_3O_4, and TiO_2) are often highly defective. For example, wüstite ($Fe_{1-x}O$), which crystallizes in the rocksalt structure type, is cation-deficient due to the presence of a significant fraction (at least 5%) of the iron being in the Fe^{3+} state. One can deduce that there must be either cation vacancies or oxygen interstitials present to accommodate the deviation from a perfect 1:1 stoichiometry. Knowing that the rocksalt structure is based on close-packing of the (larger) anions, it is reasonable to suppose that cation vacancies are more easily formed than are oxygen interstitials. Another type of compositional deviation, which suggests a high defect concentration, is an extensive solid solution with an aliovalent solute. One example is the solid solution of CaO in ZrO_2, where a concentration of oxygen vacancies equal to the calcium concentration is formed.

Electrical conductivity is directly related to the concentration of mobile electronic defects, and a crystal that is observed to be electrically very insulating can be assumed to have negligible concentrations of free electrons and electron holes. Conversely, in semiconducting or metallically conducting ceramics, the concentrations of electrons or holes are comparatively plentiful and are the defects of primary interest. Note that unlike atomic defects, the intrinsic concentration of these charge carriers does *not* scale with the melting point. Instead, the *type* of bonding is much more important, since the extent to which electrons are shared among the orbitals of adjacent atoms determines the ease with which they can be excited. As a rule, highly ionic ceramics have larger bandgaps than do covalent ceramics, and lower intrinsic carrier concentrations. For instance, pure NaCl has a bandgap of 7.3 eV and is much more electrically resistive than pure SiC, which has a bandgap of 2.9 eV, even though the latter is much more refractory and does not decompose until 3000°C. The electrical conductivity of compounds also depends greatly on the concentration of impurities. Furthermore, electrical conductivity in ceramics can also result from ion motion, so that the observation of high conductivity alone is not sufficient to conclude that the concentration of electrons or holes is significant. This is discussed further in Chapter 3. The diffusion rate of atoms is also strongly dependent on defect concentrations, and if the rate of a diffusion-controlled process such as phase separation or sintering or creep deformation is particularly slow or fast, it may suggest that particular ionic defects are negligible or abundant.

From these and other clues, it is often possible to make a reasonable guess as to the dominant defects in a particular material. Since it is difficult to directly observe atomic defects, verification is usually based on correlations between expected changes in physical properties and experimental data. Examples are given later in this chapter and in Chapter 3. Computer simulation methods have also proven useful in evaluating the relative stability of different defect types. While a detailed discussion of these techniques is beyond the scope of our discussion, the basic approach is to calculate the lattice energy of a crystal with and without defects, the difference between which is then the formation energy of the defect.

Intrinsic Ionic Disorder

Starting with a perfect crystal, one can only form atomic defects with an expenditure of energy, which is most commonly thermal, although radiation of various kinds can also displace atoms. The increased energy and amplitude of lattice vibrations at elevated temperatures increase the probability that an atom will be displaced from its lattice position. Thus the formation of atomic defects is a *thermally activated* process, in which the defect formation energy represents the activation barrier. In many ceramic systems, significant concentrations of defects are formed only at temperatures well above half the melting point.

One may ask, if a positive energy expenditure is necessary to form defects in the first place, what allows there to be a finite concentration of defects? We will show that the increase in energy upon forming the defects is counterbalanced by the decrease in free energy which results from an increase in the entropy of the system. This additional entropy is mostly configurational, resulting from the distribution of a small number of defects over a large number of lattice sites. Thus point defects in solids are *entropically stabilized*.

The two most common types of crystalline defects in ionic materials are *Frenkel* and *Schottky* defects. We term these *intrinsic* defects since they can be thermally generated in a perfect crystal, as opposed to *extrinsic* defects, which are formed only by the addition of impurities or solutes. A Frenkel defect (Fig. 2.1) is formed when an atom is displaced from its normal site onto an interstitial site forming a defect pair: a vacancy and an interstitial. In ionic materials, both the cation and the anion can undergo this kind of displacement (the anion Frenkel is sometimes referred to as an anti-Frenkel). In metals and covalent compounds Frenkel defects can also form; they differ from those in ionic compounds only in that the defects need not be electrically charged.

The Schottky defect is unique to ionic compounds and is represented by the simultaneous creation of both cation and anion vacancies, which is illustrated in Fig. 2.2. The vacancies must be formed in the stoichiometric ratio in order to preserve the electrical neutrality of the crystal. Thus in NaCl and MgO, one forms a Schottky *pair*, while in TiO_2 the Schottky defect consists of three defects (one titanium vacancy and two oxygen vacancies), and in Al_2O_3, the Schottky defect is a quintuplet (two aluminum vacancies, three oxygen vacancies). The total number of lattice sites is increased by one formula unit upon formation of a Schottky defect, unlike the Frenkel defect, which conserves the number of lattice sites.

Concentration of Intrinsic Defects. In order to understand the factors that determine the concentration of intrinsic defects, let us consider the change in free energy of a perfect crystal with initial free energy G_0, upon forming n Frenkel defect pairs at an energy expense of Δg_f per pair. The free energy of the crystal becomes:

$$G = G_0 + n\Delta g_f - T\Delta S_c \qquad (2.1)$$

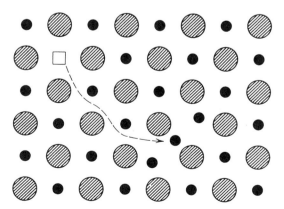

Fig. 2.1 Frenkel disorder. Ion leaving normal site forms an interstitial ion and leaves a vacancy.

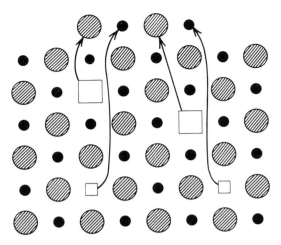

Fig. 2.2 Schottky disorder. Displacement of anion and cation to surface leaves a pair of vacancies.

where ΔS_c is the increase in configurational entropy of the crystal. The change in free energy is therefore:

$$\Delta G = (G - G_0) = n\Delta g_f - T\Delta S_c \qquad (2.2)$$

The magnitude of both of the terms on the right-hand side is dependent on the number of defects n. The *equilibrium* number of defects is found by minimizing ΔG with respect to n. To determine this equilibrium concentration, we must first evaluate the configurational entropy ΔS_c, which is given by:

$$\Delta S_c = k \ln \Omega \qquad (2.3)$$

where Ω is the number of distinct ways in which the defects can be arranged. For n Frenkel pairs we have n_i interstitials and n_v vacancies. If these are arranged on a total of N lattice sites, the vacancies can be arranged in Ω_v ways, where:

$$\Omega_v = \frac{N!}{(N-n_v)! n_v!} \qquad (2.4)$$

assuming a structure in which there are N intersitial sites for every N lattice sites (this is simply for convenience in the current discussion; if for instance there are $2N$ interstitial sites per N lattice sites, we would carry through a factor of 2). The interstitials can be arranged in Ω_i ways:

$$\Omega_i = \frac{N!}{(N-n_i)! n_i!} \qquad (2.5)$$

The total number of configurations is then $\Omega = \Omega_i \Omega_v$. Since we've formed the defects in pairs, $n_i = n_v = n$. The configurational entropy is therefore given by:

$$\Delta S_c = k \ln\left(\frac{N!}{(N-n_v)!n_v!}\right)\left(\frac{N!}{(N-n_i)!n_i!}\right) = 2k \ln\left(\frac{N!}{(N-n)!n!}\right) \quad (2.6)$$

For large numbers N and n (e.g., a mole of atoms) Stirling's approximation can be used: $\ln N! = N \ln N - N$. The configurational entropy is then:

$$\Delta S_c = 2k\left[N \ln N - (N-n)\ln(N-n) - n \ln n\right] \quad (2.7)$$

and the total free energy change is

$$\Delta G = n\Delta g - 2kT\left[N \ln\left(\frac{N}{N-n}\right) + n \ln\left(\frac{N-n}{n}\right)\right] \quad (2.8)$$

To find the equilibrium concentration of defects, we evaluate the derivative $(\partial \Delta G / \partial n)_{T,P} = 0$, keeping in mind that for dilute concentrations of defects, $(N-n) \sim N$. After rearrangement we obtain the following result for the concentration of defects (given as the fraction of the total number of atoms N):

$$\frac{n}{N} = \exp\left(-\frac{\Delta g}{2kT}\right) = \exp\left(\frac{\Delta s}{2k}\right)\exp\left(-\frac{\Delta h}{2kT}\right) \quad (2.9)$$

A most important result given by Eq. 2.9 is that the concentration of defects is exponentially dependent on the formation free energy, Δg, and on temperature. Note that the entropy Δs is not the configuration entropy, which we have already accounted for (ΔS_c), but is the nonconfigurational entropy associated with lattice strains and changes in vibrational frequencies accompanying the defect. It is sometimes assumed that Δs is much less than the configurational entropy, such that $\exp(\Delta s/2k) \sim 1$. However, this assumption can be seriously in error, as values of $\exp(\Delta s/2k)$ ranging from 10^{-4} to 10^4 have been deduced from experimental data. As a result, the absolute concentrations of intrinsic defects can be difficult to determine to a high level of accuracy. Relative differences between different compounds, and the temperature dependence of defect concentrations, are more easily determined. In Table 2.1, defect concentrations are shown as a function of Δh and temperature, assuming that $\exp(\Delta s/2k) \sim 1$. Table 2.2 lists values of Δh for the most prevalent (lowest-energy) intrinsic defects in a number of systems. Comparing these values, we find that concentrations can vary by many orders of magnitude at a constant temperature between systems of the same structure type (e.g., rocksalt). For systems with large Δh, Table 2.1 shows that concentrations change by many orders of magnitude as temperature varies.

Intrinsic versus Extrinsic Behavior. The concentration of intrinsic defects can be exceedingly small in highly refractory ceramics of large defect formation energy. As a result, solutes and especially aliovalent solutes which are accompanied by the formation of extrinsic vacancies or interstitials have a great importance in determining overall defect behavior. Oxidation and reduction pro-

Table 2.1 Defect Concentration at Different Temperatures

$$\frac{n}{N} = \exp\left[-\frac{\Delta g}{2kT}\right] = \exp\left[\frac{\Delta s}{2k}\right]\exp\left[-\frac{\Delta h}{2kT}\right] \approx \exp\left(-\frac{\Delta h}{2kT}\right)$$

Defect Concentration	1eV[a]	2eV	4eV	6eV	8eV
n/N at 100°C	2×10^{-7}	3×10^{-14}	1×10^{-27}	3×10^{-41}	1×10^{-54}
n/N at 500°C	6×10^{-4}	3×10^{-7}	1×10^{-13}	3×10^{-20}	8×10^{-27}
n/N at 800°C	4×10^{-3}	2×10^{-5}	4×10^{-10}	8×10^{-15}	2×10^{-19}
n/N at 1000°C	1×10^{-2}	1×10^{-4}	1×10^{-8}	1×10^{-12}	1×10^{-16}
n/N at 1200°C	2×10^{-2}	4×10^{-4}	1×10^{-7}	5×10^{-11}	2×10^{-19}
n/N at 1500°C	4×10^{-2}	1×10^{-4}	2×10^{-6}	3×10^{-9}	4×10^{-12}
n/N at 1800°C	6×10^{-2}	4×10^{-3}	1×10^{-5}	5×10^{-8}	2×10^{-10}
n/N at 2000°C	8×10^{-2}	6×10^{-3}	4×10^{-5}	2×10^{-7}	1×10^{-9}

[a] 1eV = 23.05 kcal/mole.

cesses also cause point defects to be introduced, the concentrations of which can exceed that of the intrinsic ionic defects.[2] Whether a defect structure is considered "intrinsic" or "extrinsic" is solely based on the relative concentrations. Intrinsic defect concentrations increase with temperature, as discussed above. Extrinsic defects, with the exception of nonstoichiometry, remain largely constant in concentration. Thus at higher temperatures, the likelihood of intrinsic behavior increases. However, in some materials the formation energies of intrinsic defects are so high, and the resulting concentrations so low, that intrinsic behavior is virtually never encountered.

As an example, let us consider the relative importance of intrinsic defects in two compounds of rocksalt structure type, NaCl and MgO. In both, the dominant intrinsic defect is the Schottky defect, as opposed to cation or anion Frenkel defects. For charge neutrality, the concentrations of cation and anion vacancies must be equal. From Eq. 2.9 we can write the concentration of each vacancy as

$$n_v/N \sim \exp[-\Delta h_s/2kT]$$

Table 2.2 shows that the Schottky formation enthalpy for MgO (~7.7 eV) is much higher than that for NaCl (~2.4 eV), consistent with the stronger bonding and higher melting temperature (2825°C vs. 801°C) of the former. At 700°C, MgO clearly has many orders of magnitude fewer intrinsic Schottky defects. Scaled to the homologous temperature, the concentrations are not so different; at its melting point NaCl has a defect concentration of ~2×10^{-6} (2 parts per million, ppm) and MgO has ~4×10^{-7} (0.4 ppm).

---

[2] There is not a consistent formalism that regards defects due to oxidation and reduction, i.e., nonstoichiometry, as either intrinsic or extrinsic. This is primarily a matter of semantics, and for the purposes of this discussion we will consider them to be extrinsic.

Table 2.2 Some Defect Energies of Formation

Compound	Reaction	Energy of Formation Δh (eV)
AgCl	$Ag_{Ag}^x \Leftrightarrow Ag_i^{\cdot} + V'_{Ag}$	1.1
NaCl	null $\Leftrightarrow V'_{Na} + V^{\cdot}_{Cl}$	2.2 – 2.4
KCl	null $\Leftrightarrow V'_K + V^{\cdot}_{Cl}$	2.6
LiF	null $\Leftrightarrow V'_{Li} + V^{\cdot}_F$	2.4 – 2.7
CsCl	null $\Leftrightarrow V'_{Cs} + V^{\cdot}_{Cl}$	1.86
BeO	null $\Leftrightarrow V''_{Be} + V^{\cdot\cdot}_O$	~6
MgO	null $\Leftrightarrow V''_{Mg} + V^{\cdot\cdot}_O$	7.7
CaO	null $\Leftrightarrow V''_{Ca} + V^{\cdot\cdot}_O$	~6
BaO	null $\Leftrightarrow V''_{Ba} + V^{\cdot\cdot}_O$	3.4
MnO	null $\Leftrightarrow V''_{Mn} + V^{\cdot\cdot}_O$	4.6
FeO	null $\Leftrightarrow V''_{Fe} + V^{\cdot\cdot}_O$	6.5
ZnO	$O_O^x \Leftrightarrow O''_i + V^{\cdot\cdot}_O$	2.51
Li$_2$O	$Li_{Li}^x \Leftrightarrow Li_i^{\cdot} + V'_{Li}$	2.28
CaF$_2$	$F_F^x \Leftrightarrow V^{\cdot}_F + F'_i$	2.3 – 2.8
	$Ca_{Ca}^x \Leftrightarrow V''_{Ca} + Ca_i^{\cdot\cdot}$	~7
	null $\Leftrightarrow V''_{Ca} + 2V^{\cdot}_F$	~5.5
UO$_2$	$O_O^x \Leftrightarrow O''_i + V^{\cdot\cdot}_O$	5.1
	$U_U^x \Leftrightarrow V''''_U + U_i^{\cdot\cdot\cdot\cdot}$	~9.5
	null $\Leftrightarrow V''''_U + 2V^{\cdot\cdot}_O$	~6.4
TiO$_2$ (rutile)	null $\Leftrightarrow V''''_{Ti} + 2V^{\cdot\cdot}_O$	5.2
	$O_O^x \Leftrightarrow O''_i + V^{\cdot\cdot}_O$	8.7
	$Ti_{Ti}^x \Leftrightarrow Ti_i^{\cdot\cdot\cdot\cdot} + V''''_{Ti}$	12
α–Al$_2$O$_3$	null $\Leftrightarrow 2V'''_{Al} + 3V^{\cdot\cdot}_O$	20.1 – 25.7 (4.2 – 5.1 eV/defect)
	$Al_{Al}^x \Leftrightarrow Al_i^{\cdot\cdot\cdot} + V'''_{Al}$	10.4 – 14.2 (5.2 – 7.1 eV/defect)
	$O_O^x \Leftrightarrow O''_i + V^{\cdot\cdot}_O$	7.6 – 14.5 (3.8 – 8.3 eV/defect)
MgAl$_2$O$_4$	null $\Leftrightarrow V''_{Mg} + 2V'''_{Al} + 4V^{\cdot\cdot}_O$	29.1 (4.15 eV/defect)

However, there is a great difference in the purity levels attainable in these two materials. With zone-refining procedures, NaCl can be purified to levels below one part-per-million. In contrast, the highest purity MgO currently available has about 50 ppm impurities (because of the much higher processing temperatures necessary, contamination from crucibles and the like is harder to avoid). These impurities are often aliovalent cations. Thus the concentration of "extrinsic" defects, that is, the impurity concentration, is much greater than the intrinsic defect concentration in MgO. As a result, "intrinsic" NaCl can in practice be achieved rather easily, whereas all presently available MgO is likely to be extrinsic at all temperatures up to the melting point.

Units for Defect Concentration. A variety of units are used in the description of defect concentrations, and it is useful to know how to translate between them. The two most common systems of units are the number fraction relative to a particular atom, and the number of defects per unit volume. The concentration n/N discussed above represents the number fraction of defects n relative to the number of possible sites N. This fraction could be expressed in concentration units of atomic fraction (in an elemental solid), cation fraction, or mole fraction. For a compound of 1:1 stoichiometry, the cation fraction, anion fraction, and mole fraction are all equivalent; for other stoichiometries the translation is a simple numerical factor. Frequently the number fractions are expressed as part-per-million (ppm) units, where a fraction of 10^{-6} is 1 ppm. Note that trace chemical analyses are often reported in units of *weight* ppm or weight percent.

Defect concentration units of number per unit volume (e.g., no./cm³ or cm⁻³) are also commonly used, especially in the discussion of electrical properties. The density of atoms in solids is $\sim 10^{23}$ cm⁻³, so a 1 ppm concentration is $\sim 10^{17}$ cm⁻³. An exact conversion from mole fraction to number per unit volume requires knowing the molecular weight (MW) and density (ρ) of the compound in question; the number of formula units per unit volume is

$$\frac{N_a \text{ (no./mole)} \cdot \rho \text{ (g/cm}^3\text{)}}{MW \text{ (g/mole)}}$$

where N_a is Avogadro's number (6.02×10^{23} mole⁻¹). In this and the following chapter, we will use units of both mole fraction and no./cm³, whichever is the more convenient, reflecting common usage in the published literature.

SPECIAL TOPIC 2.1

KRÖGER–VINK NOTATION

A standard notation used for the description of defects in ionic materials is *Kröger-Vink* notation, in which a defect is described by three parts. The main body of the notation identifies whether the defect is a vacancy "V," or an ion such as "Mg." The *subscript* denotes the site that the defect occupies, either the normal atom sites of the host lattice or an interstitial site "i." The *superscript* identifies the *effective charge* (or *relative* charge) of the defect relative to the perfect crystal lattice. For this part of the notation, dots (˙) represent positive effective charges, dashes (´) represent negative charge, and x's are sometimes used to show neutrality. Let's illustrate with some examples:

V_{Mg}'' is a vacant magnesium site; $\underline{V}$ stands for vacancy, the subscript $\underline{Mg}$ shows that it occupies what is normally a magnesium site, and, the superscript " shows that the vacancy has a doubly negative charge relative to the perfect lattice, as there is the *absence* of an Mg^{2+} ion.

$Al_i^{\cdots}$ is an interstitial aluminum ion. The subscript $\underline{i}$ denotes that the aluminum is interstitial, and the superscript ⋯ shows that the normally unoccupied interstitial site now has an excess +3 charge due to the Al^{3+} ion.

Some atoms can accept more than one valence state. This is especially common for the transition metal ions. It is therefore possible to have several distinctly different defects for a particular element on a given site with different "ionization" or "oxidation" states. Iron substituted for Mg^{2+} in MgO can be Fe_{Mg}^x (an Fe^{2+} ion) or $Fe_{Mg}^{\cdot}$ (if it is Fe^{3+}). In FeO, where the iron oxidation state is predominantly 2+, trivalent iron can also be described as a positive defect, $Fe_{Fe}^{\cdot}$.

Clustered defects or *defect associates* are denoted with parentheses that group together the defects that are bound to one another by electrostatic attraction. The net effective charge of the associate is shown with superscripts. For instance, $(V_{Na}' - V_{Cl}^{\cdot})^x$ is a clustered pair consisting of one sodium vacancy and one chlorine vacancy, which together are electrically neutral. $(Al_{Mg}^{\cdot} - V_{Mg}'')'$ is a substitutional aluminum solute (in MgO) bound to a magnesium vacancy. This type of defect is sometimes called a "dimer." The associate $(Al_{Mg}^{\cdot} - V_{Mg}'' - Al_{Mg}^{\cdot})^x$ or "trimer" is a cluster of three defects. This notation generally pertains to defects that occupy neighboring sites, since defects of further separation are not as strongly bound.

The *concentration* of defects is denoted by square brackets, for example $[V_{Mg}'']$, $[Al_{Mg}^{\cdot}]$ and $[(V_{Na}' - V_{Cl}^{\cdot})^x]$. A shorthand often used for the concentration of electrons and holes, $[e']$ and $[h^{\cdot}]$, is n and p, respectively.

Finally, in Kröger-Vink notation we always define defects relative to a "perfect" crystal. For complex ceramics with multiple cations distributed over more than one type of site, such as the spinels (which can have a normal, inverse, or random cation distribution, see Chapter 1), the choice of this reference state can be somewhat arbitrary. Nonetheless, defect chemical notation and the principles discussed herein can be applied as long as the reference system is self-consistent throughout.

Defect Chemical Reactions

We showed previously from a statistical thermodynamic viewpoint how defect concentrations depend on their formation energies and temperature. An equivalent way to view the formation of defects is as a chemical reaction, for which there is an equilibrium constant which is governed by the law of mass action. For example, the Schottky reactions for NaCl and MgO, respectively, can be written using Kröger-Vink notation as:

$$\text{null} \rightarrow V'_{Na} + V^{\cdot}_{Cl} \tag{2.10}$$

and

$$\text{null} \rightarrow V''_{Mg} + V^{\cdot\cdot}_{O} \tag{2.11}$$

where null (sometimes "nil") indicates the creation of defects from a perfect lattice. The respective mass-action equilibrium constants are:

$$K_s = [V'_{Na}][V^{\cdot}_{Cl}] \tag{2.12}$$

and

$$K_s = [V''_{Mg}][V^{\cdot\cdot}_{O}] \tag{2.13}$$

The brackets denote concentration, usually given in mole fraction (equivalent to n_v/N). Writing the equilibrium constant as the product of concentrations implies that the thermodynamic activity of each defect D is equal to its concentration, $a_D = [D]$.[3] The free energy for each of these "quasichemical" reactions is simply the Schottky formation energy, and the equilibrium constant is given by:

$$K_s = \exp\left(-\frac{\Delta g_s}{kT}\right) \tag{2.14}$$

Notice that the equilibrium constant is a function of temperature only. This dictates that the product of the cation and anion vacancy concentrations (Eqs. 2.12 and 2.13) is a constant at fixed temperature. Furthermore, when only the intrinsic defects are present, the concentration of anion and cation vacancies must be equal for charge neutrality, and so we obtain

$$[V'_{Na}] = [V^{\cdot}_{Cl}] = \exp\left(-\frac{\Delta g_s}{2kT}\right) \tag{2.15}$$

which is the same result as in Eq. 2.9.

Defect chemical reactions such as Eqs. 2.10 and 2.11 are written for the formation of defects within a solid and must obey *mass*, *site*, and *charge* balance. In this

---

[3] This assumption, namely that we have an ideal solution, is usually good for dilute solutions (<<1%) and sometimes applies for surprisingly concentrated solutions as well. It fails when interactions between defects become significant. When this is the case, activities should be used in place of concentrations. This has the effect of introducing activity coefficients γ into the equilibrium constants. For instance, Eq. 2.13 becomes $K_s = \gamma_{V_{Mg}}[V''_{Mg}]\gamma_{V_O}[V^{\cdot\cdot}_{O}]$. One method of evaluating these activity coefficients is to use the *Debye-Hückel* theory of electrolytes, discussed briefly in 2.3 and also in references listed at the end of the chapter.

2.1 Point Defects

respect they differ somewhat from ordinary chemical reactions, which must obey only mass and charge balance. (Defect chemical reactions are sometimes referred to as *quasichemical* reactions.) Mass balance simply means that a chemical reaction cannot create or lose mass. In defect chemistry, we may also write balanced reactions involving defects without mass, as in the Schottky reactions of Eqs. 2.10 and 2.11. Site balance means that the *ratio* of cation to anion sites of the crystal must be preserved, although the total number of sites can be increased or decreased. For instance, in Eqs. 2.10 and 2.11, vacancies are formed in the stoichiometric ratios. And, in order to ensure charge balance under the Kröger-Vink notation system we verify that the total *effective charge* is balanced.

As examples of site balance, consider the Schottky reactions for Al_2O_3 and $BaTiO_3$. Cation and anion vacancies must be formed in the stoichiometric ratios, and if so, we find that the effective charges are automatically balanced:

$$\text{null} \rightarrow 2V_{Al}''' + 3V_O^{\cdot\cdot} \quad (Al_2O_3) \quad (2.16)$$

$$\text{null} \rightarrow V_{Ba}'' + V_{Ti}'''' + 3V_O^{\cdot\cdot} \quad (BaTiO_3) \quad (2.17)$$

The Frenkel reaction (Fig. 2.9), written in Kröger-Vink notation for the system AgCl (in which this intrinsic mechanism is dominant) is

$$Ag_{Ag}^x = Ag_i^{\cdot} + V_{Ag}' \quad (2.18)$$

Site balance is maintained here since the formation of interstitials does not create new crystal sites, but rather occupies pre-existing ones. Mass and charge are also automatically balanced.

We may write all processes involving defect formation or reaction in a crystal as defect chemical reactions. And, as long as mass, site, and charge balance are obeyed, the reaction is formally "correct," no matter how improbable or high in formation energy. Like liquid and gas-phase chemical equilibria, many conceivable defect chemical reactions can simultaneously take place in a solid, even though the relative rates of reaction may differ by many orders of magnitude. It is the most probable reactions that are of greatest interest, and of which there are fortunately only a few that usually need to be considered. In addition to the formation of intrinsic defects, reactions of particular interest include the incorporation of solutes, formation of intrinsic *electronic* defects, oxidation and reduction, and defect association and precipitation. We now discuss some of these processes as further examples of defect chemical reactions.

Solute Incorporation. Solutes may enter solid solution in crystals as either substitutional or interstitial species. A simple example of a substitutional solute is NiO in MgO; the two form a complete solid solution (see Fig. 4.2). The defect chemical reaction for the dissolution of NiO in MgO is

$$NiO = Ni_{Mg}^x + O_O^x \quad (2.19)$$

and involves no charged species.

Aliovalent solutes greater or lesser in valence than the host on the other hand, must be charge-compensated in solid solution. This can occur by the formation of additional ionic defects or by liberating electrons and holes. The former is termed ionic compensation; the latter electronic compensation. The two types of compensation are related by oxidation/reduction equilibria, as discussed later with respect to electronic disorder (cf. Eqs. 2.54 and 2.55). Here we will introduce ionic compensation mechanisms.

Consider the dissolution of Al_2O_3 in MgO. Based on the similarity in ionic radii between Al^{3+} and Mg^{2+} in six-fold coordination (see Table 1.2) we may presume that the aluminum will substitute for magnesium. The oxygen ions are likely to occupy additional oxygen lattice sites. We at this point have

$$Al_2O_3 = 2Al_{Mg}^{\cdot} + 3O_O^x$$

which accounts for all of the mass involved. However, we have satisfied neither site nor charge balance. The basis crystal for this solid solution is MgO, which has a 1:1 cation–anion stoichiometry, yet we have just created a 2:3 site ratio. By adding an additional vacant magnesium site, we can correct the situation:

$$Al_2O_3 = 2Al_{Mg}^{\cdot} + 3O_O^x + V_{Mg}'' \tag{2.20}$$

and at the same time satisfy charge balance. (One can also think of Eq. 2.20 as the extension of the crystal by three MgO formula units, in which two of the three new magnesium sites are occupied by aluminum and one is vacant.)

As a third example, consider the incorporation of MgO into Al_2O_3. In this instance, the Mg ions may enter the solid solution substitutionally or interstitially; it is not exactly clear which is the lower energy option. If it is substitutional, a possible reaction is

$$2MgO = 2Mg_{Al}' + 2O_O^x + V_O^{\cdot\cdot} \tag{2.21}$$

and if it is interstitial, we can write:

$$3MgO = 3Mg_i^{\cdot\cdot} + 3O_O^x + 2V_{Al}''' \tag{2.22}$$

A third possibility is that magnesium is *self-compensating* and forms both the interstitial and substitutional defect:

$$3MgO = 2Mg_{Al}' + Mg_i^{\cdot\cdot} + 3O_O^x \tag{2.23}$$

The reader should confirm that mass, site, and charge balance is maintained in each of these instances. Eqs. 2.21-2.23 are *limiting cases*, however, and if the respective energies are similar, no single one will be the dominant mechanism of incorporation. In analogy to chemical equilibria, Eqs. 2.21–2.23 represent chemical reactions that are simultaneously in equilibrium. Experimental data and calculations of the relative energies for these incorporation mechanisms suggest that at high temperatures magnesium is self-compensated (Eq. 2.23) to a large extent, but that some minor compensation by cation vacancies or oxygen interstitials also exists. That is, the net incorporation reaction can be considered to be mostly reaction 2.23, plus some fraction of reactions 2.21 or 2.22.

Using these examples, we also emphasize that the free energy of solution for aliovalent solutes also includes the formation energy of the charge-compensating defects. Although Mg^{2+} and Al^{3+} are relatively close in ionic size, and can exchange for one another with little energy expenditure when both are present in a spinel (see Chapter 1), the energy of forming vacancies and interstitials constitutes a large fraction of the total energy for incorporation according to mechanisms such as 2.20–2.23.

Electrons, Holes, and Defect Ionization. In Kröger-Vink notation, free electrons and electron holes do not themselves occupy lattice sites. The process of forming intrinsic electron-hole pairs is excitation across the bandgap, which can be written as the "intrinsic electronic" reaction:

$$null = e' + h^{\cdot} \tag{2.24}$$

and for which the formation energy is the band gap of the material. An equilibrium constant may also be written for this reaction. The quantitative evaluation of electron and hole concentrations is discussed later.

When electrons and holes are tightly bound to an ion, or otherwise localized ("trapped") at a lattice sight, the whole is considered to be one ionic defect. Thus the valence state of defects such as vacancies and interstitials can vary, in analogy to the valence of a transition metal ion (Fe_{Mg}^x, $Fe_{Mg}^{\cdot}$). For instance, an oxygen vacancy can in principle take on different valence states ($V_O^{\cdot\cdot}$, $V_O^{\cdot}$, V_O^x), as can cation interstitials (for instance, $Zn_i^{\cdot\cdot}$, and $Zn_i^{\cdot}$ in the wurtzite structure compound ZnO). At a given temperature, one valence state is often much more prevalent than the others. Changes in valence take place via ionization reactions such as:

$$V_O^{\cdot\cdot} + e' = V_O^{\cdot} \tag{2.25}$$

or:

$$Zn_i^{\cdot} + h^{\cdot} = Zn_i^{\cdot\cdot} \tag{2.26}$$

for which equilibrium constants can be written:

$$K_1 = \frac{[V_O^{\cdot}]}{[V_O^{\cdot\cdot}]n} = \exp\left(-\frac{\Delta g_1}{kT}\right) \tag{2.27}$$

$$K_2 = \frac{[Zn_i^{\cdot\cdot}]}{[Zn_i^{\cdot}]p} = \exp\left(-\frac{\Delta g_2}{kT}\right) \tag{2.28}$$

The energies for these reactions Δg_1, Δg_2 are effectively the ionization energies of the defect.

116   Chapter 2 / Defects in Ceramics

Oxidation and Reduction Reactions. Equilibration of ionic solids with an ambient gas that is also a constituent of the solid plays an important role in determining defect structure. We may think of the ambient gas (e.g., oxygen, or a halogen, or metal vapor) as another type of solute species. For example, the reduction of an oxide can be written as the removal of oxygen to the gas phase leaving behind oxygen vacancies:

$$O_O^x = \frac{1}{2}O_2(g) + V_O^{\cdot\cdot} + 2e' \tag{2.29}$$

Note that the two electrons that were associated with the O^{2-} ion are liberated within the solid. The equilibrium constant for this reaction is:

$$K_R = n^2 [V_O^{\cdot\cdot}] P_{O_2}^{1/2} = K_R^0 \exp\left(-\frac{\Delta g_R}{kT}\right) \tag{2.30}$$

where the oxygen partial pressure is equivalent to oxygen activity, K_R^0 is a constant, and Δg_R is the free energy of reduction. The concentration of oxygen ions on their proper sites, $[O_O^x]$, has been left out of the demominator in this expression since it is essentially unity (that is, the concentration of vacancies is assumed dilute).

Oxidation can be written as the consumption of oxygen vacancies:

$$\frac{1}{2}O_2(g) + V_O^{\cdot\cdot} = O_O^x + 2h^{\cdot} \tag{2.31}$$

for which the equilibrium constant is:

$$K_O = \frac{p^2}{[V_O^{\cdot\cdot}]P_{O_2}^{1/2}} = K_O^0 \exp\left(-\frac{\Delta g_O}{kT}\right) \tag{2.32}$$

It is important to note that since oxidation and reduction are the same thermodynamic process simply reversed, the reactions that we write to describe them are *not independent*. Examining the reduction reaction (Eq. 2.29) and the intrinsic electronic defect equilibrium (Eq. 2.24), we find that a combination of the two yields the oxidation reaction (Eq. 2.31). The equilibrium constants in Eqs. 2.30 and 2.32 are also not unitless. Typically, the units will reflect whichever pressure and concentration units are in use. Oxygen partial pressure is frequently given in atmospheres (atm) or megapascals (MPa), while the defect concentrations are either in no./cm³ or in mole fraction.

There are a number of ways of writing the oxidation and reduction reactions, which we may choose for convenience in order to show the formation or removal of particular defects. This is illustrated for MgO by the following reactions, each of which shows the formation of cation vacancies upon oxidation. One is

$$\frac{1}{2}O_2(g) = O_O^x + V_{Mg}'' + 2h^{\cdot} \tag{2.33}$$

obtained by adding the Schottky reaction to Eq. 2.31. Another is

$$\frac{1}{2}O_2(g) + 2e' = O_O^x + V_{Mg}'' \tag{2.34}$$

obtained from the previous reaction by adding twice the intrinsic electronic reaction. These schemes (Eqs. 2.29–2.34) are simply alternative representations of the chemical process of oxidation, written to include a particular defect type. It is usually convenient (and less confusing) to choose just one representation that includes the prevailing defects in the system. In some systems where the metal has a significant vapor pressure it may be useful to write the oxidation/reduction reaction as the equilibration of the solid with the metal vapor rather than the anion vapor. This is also strictly a matter of convenience, for the two can be directly coupled by a third reaction representing formation of the compound from its vapors.

Extent of Nonstoichiometry

Compounds vary greatly in the extent to which they can be oxidized or reduced. Highly stoichiometric oxides such as MgO, Al_2O_3, and ZrO_2 have cations of a fixed valence, and correspondingly, a large free energy for reduction or oxidation. Changes in P_{O_2} alone cause little change in defect concentration in such solids. Oxides containing multivalent cations, such as the transition metals, are much more nonstoichiometric. TiO_{2-x}, $BaTiO_{3-x}$, and $SrTiO_{3-x}$ are compounds in which Ti^{4+} can easily be reduced to Ti^{3+}, causing oxygen deficiencies of order 1% within the limits of stability of the oxide (i.e., before decomposition or a transition to a different phase occurs). The transition metal monoxide series, $Ni_{1-x}O$, $Co_{1-x}O$, $Mn_{1-x}O$, and $Fe_{1-x}O$, are oxides in which a fraction of the divalent cations is easily oxidized to the trivalent state, resulting in a cation deficiency. The extent of cation deficiency, x, increases from 5×10^{-4} for $Ni_{1-x}O$ to ~1% for $Co_{1-x}O$ to ~0.1 for $Mn_{1-x}O$ to ~0.15 for $Fe_{1-x}O$. $Fe_{1-x}O$ can never be made stoichiometric, having a minimum cation deficiency of about 0.05. The defect chemistry of these oxides is discussed in greater detail in Chapter 3.

To illustrate the relationship between the previously discussed defect reactions and the extent of nonstoichiometry, let us explicitly calculate the nonstoichiometry for two examples. In MgO, electrical conductivity measurements of doped compositions conducted by D. Sempolinski, et al.[4] allowed them to determine the following equilibrium constant for oxidation, corresponding to the reaction in Eq. 2.33:

$$K_{ox} = \frac{[V_{Mg}'']p^2}{P_{O_2}^{1/2}} = 7.2 \times 10^{63} \exp\left(\frac{-610 \pm 80 \text{ kJ/mole}}{RT}\right) \quad (\text{cm}^{-9}\,\text{MPa}^{-1/2})$$

---

[4] D. Sempolinski, W.D. Kingery, and H.L. Tuller, *J. Am. Ceram. Soc.*, **63**[11-12], 669 (1980).

where oxygen partial pressure has units of MPa and the concentrations are in no./cm^3. This equilibrium constant tells us that MgO can be oxidized at high temperature, producing extrinsic magnesium vacancies and electron holes. If no other defects are present in significant concentration, the concentrations of these two are related, for charge neutrality, by $2[V''_{Mg}] = p$. The actual concentrations are very small, but do vary with temperature and oxygen partial pressure. For example, at 80% of the absolute melting point, or 2478 K, the magnesium vacancy concentration in air atmosphere is only 3×10^{16} cm^{-3} (x in Mg$_{1-x}$O equal to 0.6 ppm), and 2×10^{15} cm^{-3} ($x = 0.04$ ppm) in a reducing atmosphere of $P_{O_2} = 10^{-9}$ MPa.

By comparison, TiO$_{2-x}$, is much more highly nonstoichiometric. The equilibrium constant for reduction has the following value based on conductivity measurements by J. F. Baumard and E. Tani[5]:

$$K_R = \left[Ti_i^{\cdots} \right] n^4 P_{O_2} = 6.55 \times 10^{122} \exp\left(\frac{-960 \text{ kJ/mole}}{RT} \right) \quad (\text{MPa} \cdot \text{cm}^{-15})$$

(It is left to the reader to write the corresponding reduction reaction.) If only two defects are present, electroneutrality requires that $n = 4[Ti_i^{\cdots}]$. At 80% of its melting point (1690 K), TiO$_{2-x}$ has a nonstoichiometry of $x = [Ti_i^{\cdots}] = 93$ ppm in air. The nonstoichiometry becomes more appreciable in reducing ambients, reaching $x = 0.27\%$ at $P_{O_2} = 10^{-9}$ MPa. Reduced TiO$_2$ becomes a good n-type semiconductor due to the high concentration of free electrons.

Electronic Disorder

Disorder in the *electronic structure* of solids influences many electrical, optical, and chemical properties. Unlike intrinsic atomic defects, which are for the most part thermally generated, electronic defects can be thermally or optically excited (by absorption processes). We will focus primarily on thermally-generated electronic defects. Another distinction between atomic and electronic defects is that while atomic defects are generally discussed with reference to physical position in a lattice, and their concentrations can be directly related to the number of atoms, the reference for electronic defects is the density of electron states of a given energy per volume of crystal, which is not a constant quantity but varies with temperature and composition.

When a large number of individual atoms (each of which has its own discrete electron orbital energy levels) are brought together into a crystal, interatomic bonding results when the discrete energy levels of the valence electrons (outer shell electrons) are broadened into continuous energy bands. Perfect electronic order, which is achieved only at a temperature of 0 K, corresponds to the filling by electrons of the lowest possible energy levels under the constraint of the Pauli exclusion principle, which allows at most two electrons per energy level. In this

---

[5] J.F. Baumard and E. Tani, *J. Chem. Phys.*, **67**[3] 857 (1977).

respect a perfect crystal may be regarded as one in which all electrons occupy their "ground state" or lowest energy configuration, and "disorder" is any excitation of electrons to higher energy levels.

Bandgaps. We distinguish metals from semiconductors and insulators by the occupancy of the energy bands (Fig. 2.3). In a metal the highest energy band is only partially filled, even at 0 K, such that there is no barrier to the excitation of electrons to higher energy levels within the band with an increase in temperature. On the other hand, in intrinsic semiconductors and insulators a *bandgap* of energy E_g separates a completely filled *valence band* from a completely empty *conduction band* at 0 K, and thermal or optical excitation of electrons across this energy gap is necessary to occupy the conduction band.[6] The distinction between an intrinsic semiconductor and an insulator is based on the value of E_g and is somewhat arbitrary; well-known semiconductors such as silicon and gallium arsenide have small bandgaps of 1–1.5 eV, and compounds with bandgaps above 3 eV are usually considered insulating. Table 2.3 illustrates the bandgap values for a number of well-known semiconductors and insulators. Quite a few ceramic compounds have bandgaps between 2.5 and 3 eV and are considered large-bandgap semiconductors. Oxides can also have metallic bonding character; examples include various vanadates, ruthenates, and cuprates.

Concentration of Intrinsic Electrons and Holes. What determines the concentration of free electronic defects in a crystal? We may draw a direct analogy to Eqs. 2.9 or 2.15, which express the concentration of intrinsic vacancies as the product of a site density (N) and an exponential term which represents the probability that the site contains a vacancy (Boltzmann probability factor):

$$n_v = N \exp\left(-\frac{\Delta g_v}{2kT}\right) \qquad (2.35)$$

The concentration of electrons with a particular energy level E is given by a function of similar form:

---

[6] Since the broadening of electron energy levels that leads to formation of bands originates from interactions between atomic orbitals, in detail the band energies can vary with crystal direction. An *indirect gap* (which occurs in Si, as one example) occurs when the lowest energy level of the conduction band lies in a different crystal direction than the highest level of the valence band, and a *direct gap* (as occurs in GaAs) when they lie in the same crystal direction. Due to the nondirectional nature of ionic bonding, the band energy tends to vary little with direction in ionic compounds, and direct gap behavior results. The *Fermi surface* of a crystal maps a constant energy level, the Fermi energy, E_f, with direction and also shows the variation in band energy with direction. It is more spherical for ionic than for covalent compounds, in keeping with the non-directional nature of the ionic bond. (For a detailed discussion of band theory, the reader is referred to an introductory solid state physics text, such as C. Kittel, *Introduction to Solid State Physics*, 6th Edition, J. Wiley and Sons, New York, 1986, or G. Burns, *Solid State Physics*, Academic Press, Orlando, Florida, 1985.)

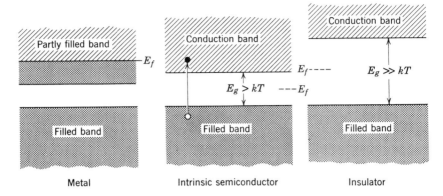

Fig. 2.3 Electron energy band levels for metals with partly filled conduction band, intrinsic semiconductors with a narrow band gap, and insulators with a high value for Eg.

Table 2.3 Bandgap Values for Some Semiconductors and Insulators

Compound	Band Gap (eV)	Compound	Band Gap (eV)
Si	1.11	NaF	6.7
Ge	0.66	KCl	7
Diamond	5.4	NaCl	7.3
InSb	0.17	LiF	12.0
InAs	0.36	BaF_2	8.9
InP	1.27	SrF_2	9.5
GaSb	0.68	CaF_2	10.0
GaAs	1.43	MgF_2	11.8
GaP	2.25	SrO	5.7
CdTe	1.44	MgO	7.8
CdSe	1.74	NiO	4.2
CdS	2.42	CoO	4.0
ZnSe	2.6	MnO	3.7
ZnO	3.2	FeO	2
ZnS	3.6	VO	0.3
PbSe	0.27	Fe_2O_3	3.1
PbTe	0.29	Ga_2O_3	4.6
PbS	0.34-0.37	Al_2O_3	8.8
AgI	2.8	$BaTiO_3$	2.8
AgCl	3.2	TiO_2	3.0
SiC (α)	2.9	UO_2	5.2
BN	4.8	SiO_2	8.5
		$MgAl_2O_4$	7.8

$$n(E) = N(E) \cdot F(E) \tag{2.36}$$

in which $N(E)$ is the volume density of electron levels of energy E, termed the *density of states*, and $F(E)$ is the probability that they are occupied, called the Fermi-Dirac function. In order to determine the volume concentration of electrons with energies in the conduction band (and which are therefore available for electrical conduction), it is necessary to evaluate these two functions.

Notice that, unlike the site density N in Eq. 2.35, we have written the electron density of states $N(E)$ as a function which varies with energy level. It represents the maximum density of electrons of energy E allowed per unit volume of crystal by the Pauli exclusion principle. At the edges of the conduction and valence bands, $N(E)$ is found from band theory to vary approximately parabolically with energy away from either band edge, as depicted schematically in Fig. 2.4b. Integrating over energy near the conduction band edge gives the following approximate result for N_c, the *effective conduction band density of states*:

$$N_c = 2 \left[\frac{2\pi m_e^* kT}{h^2} \right]^{3/2} \approx 10^{19} \text{ cm}^{-3} \ (at\ T = 300\,\text{K}) \tag{2.37}$$

where m_e^* is the *effective mass* of electrons in the conduction band and h is Planck's constant. Similarly, electron holes have available an effective valence band density of states N_v given by:

$$N_v = 2 \left[\frac{2\pi m_h^* kT}{h^2} \right]^{3/2} \approx 10^{19} \text{ cm}^{-3} \ (at\ T = 300\,\text{K}) \tag{2.38}$$

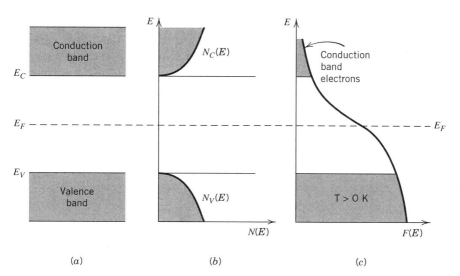

Fig. 2.4 (*a*) Conduction band and valence band in intrinsic semiconductor or insulator, (*b*) Density of states function N(E). (*c*) Fermi-Dirac distribution function.

where m_h^* is the effective mass of holes. m_e^* and m_h^* are generally greater than the mass of a free electron, by factors of approximately 2 to 10 in oxides. Notice that on a per volume basis, the density of states is about 10^4 less than the typical atom density of solids (~10^{23} cm^{-3}). This becomes important later when we compare absolute numbers of ionic and electronic defects.

The Fermi–Dirac function $F(E)$ now has to be evaluated. It is given by

$$F(E) = \frac{1}{1 + \exp\left[\dfrac{E - E_f}{kT}\right]} \qquad (2.39)$$

which when plotted against energy has the form shown in Fig. 2.5. At 0 K, all electron levels are occupied up to a maximum energy E_f, called the Fermi energy, so that the probability function has a step-like shape, with a value of unity below E_f and zero above. At temperatures above 0 K, some electrons are now excited to energy levels above E_f, and the probability function becomes more diffuse, as shown in Fig. 2.5. Notice that at the Fermi energy, the probability function remains $F(E) = 1/2$. Applying this function to the energy band diagram for a semiconductor or insulator, Fig. 2.4b, one finds that the product of $N(E)$ and $F(E)$ in the conduction band gives a finite concentration of electrons occupying the conduction band and an equal concentration of electron holes in the valence band. When the energy difference ($E-E_f$) is much greater than kT, the Fermi function is given to reasonable approximation by the simpler Boltzmann distribution function:

$$F(E) \approx \exp\left[-(E - E_f)/kT\right] \qquad (2.40)$$

This approximation is reasonable in the tail portion of the Fermi–Dirac function, where the conduction band lies, Fig. 2.4c. With these results, the concentration of conduction band electrons and valence band holes for an intrinsic semiconductor or insulator are given by:

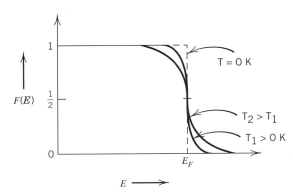

Fig. 2.5 Variation in Fermi probability function, $F(E)$, as a function of electron energy E.

$$\frac{n_e}{N_c} = \exp\left[-\frac{(E_c - E_f)}{kT}\right] \tag{2.41}$$

$$\frac{n_h}{N_v} = \exp\left[-\frac{(E_f - E_v)}{kT}\right] \tag{2.42}$$

A final step necessary to obtain actual concentrations is to evaluate the Fermi level E_f. Since for charge neutrality in this intrinsic semiconductor we must have $n_e = n_h$ in the bulk of the solid, equating Eqs. 2.41 and 2.42 with the aid of Eqs. 2.37 and 2.38 gives:

$$E_f = \frac{E_c + E_v}{2} + \frac{3}{4}kT \ln\left(\frac{m_h^*}{m_e^*}\right) \tag{2.43}$$

where $E_g/2 = (E_c+E_v)/2$. The second term in Eq. 2.43 is usually small compared to the first, and therefore the Fermi level of an intrinsic semiconductor or insulator is generally close to the middle of the bandgap.

Note the similarity between Eqs. 2.41 and 2.42 and Eq. 2.9 for the concentration of lattice defects. The density of states may be thought of as the electronic analogy to the density of lattice sites. Writing the excitation of electrons across the bandgap as a defect chemical reaction,

$$\text{null} \rightarrow e' + h^\cdot \tag{2.24b}$$

The equilibrium constant is given by

$$K_i = n_e n_h = N_c N_v \exp\left(-\frac{E_g}{kT}\right)$$

$$\approx 10^{38} \exp\left(-\frac{E_g}{kT}\right) \text{cm}^{-6} \quad (\text{at } 300\,\text{K}) \tag{2.44}$$

Notice that as expressed here, this equilibrium constant is not unitless, in contrast to the Schottky constant (Eqs. 2.14 and 2.15). This is simply because we have elected to write the electron and hole concentrations in units of number per cubic centimeter. If the semiconductor is intrinsic, then $n_e = n_h$, and

$$n_e = n_h = (N_c N_v)^{1/2} \exp\left(-\frac{E_g}{2kT}\right) \tag{2.45}$$

In defect chemistry calculations where ionic defect concentrations are commonly given in units of mole fraction, it becomes important to compare concentrations of electronic and ionic defects on an equivalent basis. For example, electron and hole concentrations can also be expressed in mole fraction relative to the molar density of the compound; alternatively, equilibrium constants may be written with all defect concentrations in no./cm^3, as was done earlier for the oxidation and reduction equilibrium constants.

The Fermi level has a special significance. Formally it is defined as the energy at which the probability of electron occupation is one-half (see Fig. 2.5). In a metal, E_f is the maximum occupied energy at 0 K (Fig. 2.3), but in intrinsic semiconductors and insulators, the energy representing $F(E) = 1/2$ lies very near the middle of the energy gap. Thermodynamically, the Fermi level represents the chemical potential of electrons in the system. When electron donors are added to the system and increase the conduction band electron density, the chemical potential of electrons is raised toward the conduction band (Fig. 2.6). Conversely, acceptor dopants lower the Fermi level toward the valence band edge. When dissimilar materials are joined, electrons flow from one system to the other until the Fermi levels are equilibrated. Thus the joining of an n-type and a p-type semiconductor results in *band bending* in the vicinity of the interface in order to accomodate the equalization of the Fermi level, Fig. 2.7. The resulting *p-n junction* has rectifying properties (electron conduction is energetically uphill from the n to p but downhill from p to n) and is the basis for many semiconductor devices. Another instance of band-bending at an interface occurs when acceptor levels or "traps" are present at the grain boundary of an n-type semiconductor. Equilibration of the Fermi levels of the acceptor-rich grain boundary and donor-doped grains causes band bending at the interface, which results in barriers to electron conduction across the boundaries. These conduction barriers, which can be overcome with sufficient electric field, are the basis for nonlinear conducting devices based on semiconducting oxides, such as zinc oxide varistors and barium titanate positive-temperature-coefficient thermistors (discussed in Chapter 3).

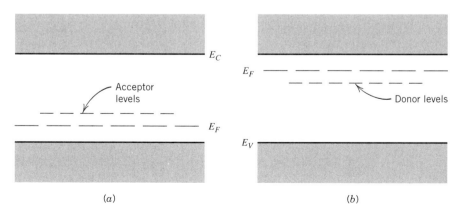

Fig. 2.6 (*a*) Fermi level in extrinsic *p*-type semiconductor, (*b*) Extrinsic *n*-type semiconductor.

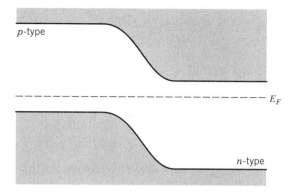

Fig. 2.7 Band bending across a p-n junction upon equilibration of the Fermi level.

Example: *Intrinsic Ionic and Electronic Defect Concentrations in MgO and NaCl*

Let us now compare the relative concentrations of intrinsic ionic and electronic defects in two systems where the formation energies are rather accurately known. The lowest energy intrinsic defect in MgO is the Schottky pair, with an enthalpy estimated to be 7.7 eV. The bandgap for MgO is rather close in value, being 7.65 eV at room temperature. However, its value decreases with increasing temperature at a rate of about 1 meV/K.[7] We may ask, in the ideal case of an absolutely pure and stoichiometric MgO, which defects are present in higher concentration? (We neglect for the moment the fact that MgO is unlikely to be intrinsic because of its high Schottky formation energy and the presence of impurities.) Choosing somewhat arbitrarily a high temperature of 1400°C (1673K), the Schottky defect concentration is given by:

$$[V''_{Mg}] = [V_O^{\cdot\cdot}] = K_s^{1/2} \approx \exp\left(-\frac{\Delta h_s}{2kT}\right)$$
$$= 2.5 \times 10^{-12} \text{ mole fraction} \tag{2.46}$$

The electron and hole concentrations are given by

---

[7] The temperature dependence of the bandgap in oxides (e.g., -1 meV/K for magnesia and alumina) is larger than in semiconductors such as Si and Ge (-0.22 and -0.44 meV/K respectively). Combined with the much higher temperatures at which oxides may be used, the reduction in bandgap is significant for high temperature electrical properties. See R.H. French, *J. Am. Ceram. Soc.*, **73**[3], 477 (1990) for additional discussion.

$$n = p = K_i^{1/2} = (N_c N_v)^{1/2} \exp\left(-\frac{E_g}{2kT}\right) = 2\left(\frac{kT}{2\pi\hbar^2}\right)^{3/2} (m_e^* m_h^*)^{3/4} \exp\left(-\frac{E_g}{2kT}\right) \quad (2.47)$$

The effective masses m_e^* and m_h^* have been determined[8] to be 0.38 m_o and 0.77 m_o, respectively, where m_o is the free electron mass (9.11×10^{-31} kg). Taking into account temperature dependence, at 1673K the bandgap energy is $E_g \sim 6.28$ eV. This gives

$$n = p = (1.3 \times 10^{20} \text{ cm}^{-3})\exp(-6.28 \text{ eV}/2kT) = 4.6 \times 10^{10} \text{ cm}^{-3}$$

To complete the comparison we must convert to equivalent units. Multiplying the vacancy concentration by $\rho_{MgO} N_a / MW_{MgO}$, where ρ_{MgO} is the density of MgO (3.58 g/cm³), N_a is Avogadro's number, and MW_{MgO} is the molecular weight (40.31 g/mole), the vacancy concentrations are:

$$[V_{Mg}''] = [V_O^{\cdot\cdot}] = 1.4 \times 10^{11} \text{ cm}^{-3}$$

We see that at this temperature, despite a considerably higher value of the Schottky energy, the Schottky defect concentration is still slightly greater than that of electrons and holes, due to the difference between the conduction band and valence band density of states ($N_c \sim N_v \sim 10^{20}$ cm⁻³) and the density of lattice sites (5.3×10^{22} cm⁻³). However, to emphasize an earlier point, notice that the concentrations of *either* intrinsic electronic or ionic defects is so low, $\sim 10^{-12}$ mole fraction, that it is exceedingly unlikely that the purity in real materials would be high enough for intrinsic behavior to be observed.

If we compare these results to a similar calculation in NaCl, which has a Schottky energy of 2.2-2.4 eV and a bandgap of 7.3 eV, it becomes apparent that anywhere below its melting point (801°C) the intrinsic electronic carrier concentration in NaCl is truly infinitesimal. At T_m, the intrinsic electron and hole concentrations are only of order 10^4 cm⁻³ ($\sim 10^{-19}$ mole fraction). The Schottky defect concentration is comparatively much higher, being ~1 ppm near the melting point, due to the much smaller value of the Schottky energy. Thus, in contrast to MgO, intrinsic NaCl can be achieved and is overwhelmingly dominated by ionic defects—for this reason, it has long been regarded as a model ionic solid.

Donors and Acceptors

Given the large bandgaps and high impurity levels of many ceramics, it is no surprise that electronic carrier concentrations are virtually always extrinsically determined. In elemental and compound semiconductors such as Si, Ge, GaAs, and InSb, impurities are commonly used to increase the concentrations of electrons or holes and thereby the *n*- or *p*-type conductivity. The same is true of semiconducting ceramics, but in addition, electrically active defects such as vacancies and interstitials

---

[8] D. Sempolinski, W.D. Kingery, and H.L. Tuller, *J. Am. Ceram. Soc.*, 63, 669 (1980).

take on greater importance due to their comparatively high concentrations. Semiconducting oxides with appreciable conductivity at room temperature are usually *n*-type, and contain cation dopants or defects due to reduction.

Solutes, vacancies, and interstitials all perturb the band structure to some extent, and can introduce localized energy levels within the bandgap, as shown in Figs 2.6 and 2.8. A defect with energy level located near the conduction band may be able to donate an electron to the conduction band and is considered a *donor*. Examples of donors in silicon are pentavalent cations such as As^{5+}, P^{5+}, or Sb^{5+}, which upon ionization contribute an extra electron to the conduction band (Fig. 2.8). Defects with energy levels near the valence band which can accept an electron are likewise known as *acceptors*. Examples of acceptor impurities in silicon are trivalent cations such as B^{3+} and Al^{3+}. In terms of defect chemical reactions, the ionization of donors and acceptors in silicon can be written

$$As \text{ (metal)} = As_{Si}^{\cdot} + e' \tag{2.48}$$

and

$$B \text{ (metal)} = B_{Si}' + h^{\cdot} \tag{2.49}$$

The probability that a donor impurity at an energy level E_D or an acceptor at energy E_A will be ionized is also given by Fermi statistics. If the ionization energy is of the same order as kT, then the probability of ionization is high and the defect is known as a *shallow* donor or acceptor. At room temperature (where $kT = 0.025$ eV) impurities with ionization energies of ≤ 0.05 eV are considered shallow; these dopants are effective at increasing the concentration of free electronic carriers. A *deep* donor or acceptor refers to a solute with a high ionization energy. At the temperature of interest, if the solute is not ionized it is essentially electrically inactive. When both donor and acceptor solutes are present, they are also able to *compensate* for one another. By this we mean that the donor solute contributes its

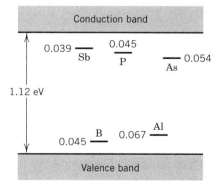

Fig. 2.8 Donor and acceptor levels for selected dopants in silicon.

electron to the acceptor solute rather than to the conduction band. In this instance, the net doping is determined by the amount of dopant that is in excess.

In an ionic solid, *all* ionic defects with nonzero effective charge can be viewed as either a donor or acceptor species. From Eq. 2.29 we see that the process of reduction introduces oxygen vacancies, which if ionized (Eq. 2.25) produce charge-compensating electrons. Therefore, the oxygen vacancy may be regarded as a donor specie, Fig. 2.9(b). In the same vein, oxidation of MgO (Eq. 2.33) introduces magnesium vacancies and compensating holes; the magnesium vacancy is therefore an acceptor. Defects with a positive effective charge are donors; these have given up an electron in order to become ionized positively relative to the perfect lattice site. Correspondingly, defects with negative effective charge are acceptors, having accepted electrons relative to the perfect lattice. Thus, in MgO, $Al_{Mg}^{\cdot}$ and $Cl_O^{\cdot}$ are donor solutes while Na_{Mg}' and N_O' are acceptor solutes; $V_O^{\cdot\cdot}$ and V_{Mg}'' are donor and acceptor defects, respectively. In the semiconducting oxide TiO_2, pentavalent solutes such as $Nb_{Ti}^{\cdot}$ and $Ta_{Ti}^{\cdot}$ as well as the defects $V_O^{\cdot\cdot}$ and $Ti_i^{\cdot\cdot\cdot\cdot}$ are donors, and trivalent solutes such as Al_{Ti}' and Ga_{Ti}' and the defects V_{Ti}'''' and O_i'' are acceptors. In the semiconducting ternary oxide $BaTiO_3$, cation solutes can substitute for either the barium or titanium site. Thus larger trivalent dopants such as $La_{Ba}^{\cdot}$ that substitute for Ba are donors, while smaller ones which substitute for titanium, such as Al_{Ti}' and Fe_{Ti}', are acceptors. Yttrium is a special case which can be either a donor ($Y_{Ba}^{\cdot}$) or an acceptor (Y_{Ti}') depending on which cation it substitutes for; the ionic radius of Y^{3+} is intermediate between that of Ba^{2+} and Ti^{4+}. The site occupancy of Y^{3+} is also sensitive to the A/B cation ratio of the overall composition. Multiple ionization states are possible for some solutes, in which case the energy level of each state will differ.

In the bandgap scheme, the energy levels of donor and acceptor defects are by convention labeled for the state of the defect after gaining an electron. For donor defects, the energy level is labeled with the charge state *before* ionization, while for acceptor defects, it is the charge state *after* ionization. Referring to Fig. 2.9(b), which shows various defect levels in MgO, we see that the donor level labeled V_O^x corresponds to the ionization process

$$V_O^x = V_O^{\cdot} + e' \qquad (\Delta h = 0.5 \text{ eV}) \qquad (2.50)$$

and the level labeled $V_O^{\cdot}$ is for the process

$$V_O^{\cdot} = V_O^{\cdot\cdot} + e' \qquad (\Delta h \sim 2 \text{ eV}) \qquad (2.51)$$

The acceptor energy levels labeled V_{Mg}' and V_{Mg}'' are, respectively, for the ionization process

$$V_{Mg}^x + e' = V_{Mg}' \qquad (\Delta h = 0.5 \text{ eV}) \qquad (2.52)$$

and

$$V_{Mg}' + e' = V_{Mg}'' \qquad (\Delta h \sim 1.5 \text{ eV}) \qquad (2.53)$$

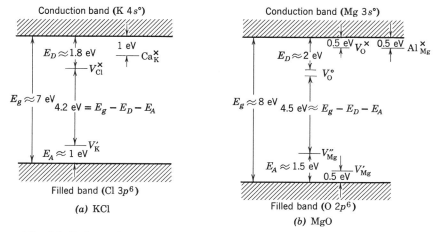

Fig. 2.9 Estimated defect energy levels in MgO.

In comparison to the dopants shown for silicon in Fig. 2.8, all of these defect levels should be regarded as deep.

Electronic versus Ionic Compensation of Solutes

Another important complexity in the electrical behavior of solutes in ionic ceramics arises from the existence of simultaneous equilibria. In a covalent semiconductor such as silicon, the effectiveness of donor and acceptor solutes at a given temperature is solely determined by their ionization energies. In oxide semiconductors, the effectiveness of even shallow dopants also depends on the extent of oxidation and reduction. This is due to the fact that an aliovalent solute in an ionic compound can be charge-compensated by ionic defects (ionically compensated) or by electrons and holes (electronically compensated), or by a combination of the two (mixed compensation). This effect is of particular interest in electronic ceramic systems, where large variations in electrical conductivity can result with changes in P_{O_2} and temperature at a constant doping level, due to changes in the compensation mechanism.

The results are best visualized by using the defect chemical formalism to write alternative mechanisms of solute compensation. In Eqs. 2.20–2.23, we wrote solute incorporation reactions in which the charge-compensating defects were fully ionized vacancies or interstitials, representing the ionic compensation limit of behavior. Let us now consider the incorporation of Nb_2O_5 as a solute in TiO_2, for which both ionic and electronic compensation can occur. The ionic compensation of niobium is given by:

$$2\,Nb_2O_5 = 4\,Nb^{\cdot}_{Ti} + 10\,O^x_O + V''''_{Ti} \tag{2.54}$$

The alternative electronic compensation mechanism is:

$$2\,Nb_2O_5 = 4\,Nb_{Ti}^{\cdot} + 8\,O_O^x + O_2(g) + 4\,e' \tag{2.55}$$

where one molecule of oxygen gas has been liberated upon incorporation. Which of these reactions dominates, and under what conditions? Notice that Eqs. 2.54 and 2.55 are interrelated by a third defect chemical reaction representing the oxidation/reduction equilibrium, which upon subtracting Eq. 2.55 from Eq. 2.54 is:

$$O_2(g) + 4\,e' = 2\,O_O^x + V_{Ti}'''' \tag{2.56}$$

Examining Eqs. 2.54–2.56, one finds that the prevailing compensation mechanism will depend on solute concentration, oxygen pressure, and temperature. The oxidation reaction in Eq. 2.56 tends toward the right at higher oxygen pressure and lower temperature. Thus niobium tends to be compensated by titanium vacancies when the niobium concentration is high, temperature low, and oxygen pressure high. In the bandgap scheme, ionic compensation is the condition where the niobium donor is completely compensated by the vacancy acceptor. At lower concentrations, higher temperature, and lower oxygen pressure, the compensating defect is the electron. This corresponds to the introduction of niobium donor levels in the energy gap without compensating ionic defects.

The reactions in Eqs. 2.54 and 2.55 are limiting cases, and there is also a transition region between the two where niobium is partially compensated by both vacancies and electrons $\left[Nb_{Ti}^{\cdot} \right] = n + 4\left[V_{Ti}'''' \right]$. Fig. ST19 on page 144 shows that the n-type electrical conductivity of Nb-doped TiO_2 varies with oxygen pressure at a constant temperature of 1350°C, due to changes in the compensation mechanism of Nb (and native reduction of the oxide, at much lower oxygen pressures). At room temperature, the n-type electrical conductivity can differ by several orders of magnitude depending on whether compensation is electronic or ionic. The same effect occurs in more complex titanates such as $BaTiO_3$ and $SrTiO_3$, and is an important consideration in selecting compositions and firing conditions for specific applications.

While most semiconducting oxides that are highly conductive at room temperature are n-type, acceptor dopants can also undergo changes between ionic and electronic compensation. It is left as an exercise to the reader to show that the electronic compensation of an acceptor in TiO_2 is promoted at high oxygen pressures (in fact, much higher than 1 atmosphere). Due to the fact that highly stoichiometric oxides such as Al_2O_3 and MgO are difficult to reduce, aliovalent solutes are generally ionically compensated. That is, oxidation reactions such as Eq. 2.56 tend very strongly to the right throughout all accessible ranges of temperature and oxygen activity. Thus even highly doped compositions are not electrically conductive at room temperature.

SPECIAL TOPIC 2.2

POINT DEFECTS AND CRYSTAL DENSITY IN ZrO_2

As one example of how the point defect chemistry of a compound can be determined from physical properties, let's consider CaO-doped ZrO_2. Calcia is one of a number of additives which allow stabilization of the higher temperature tetragonal or cubic polymorphs of zirconia to room temperature (as a metastable phase). The oxygen vacancies formed to compensate these stabilizers also result in a high ionic conductivity and allow many applications of cubic zirconia as a solid electrolyte. When CaO is dissolved substitutionally into ZrO_2, it must be charge-compensated by a defect of positive effective charge. Two possible incorporation reactions are:

$$CaO(s) \xrightarrow{ZrO_2} Ca''_{Zr} + O^x_O + V^{\cdot\cdot}_O \tag{D1}$$

and

$$2CaO(s) + Zr^x_{Zr} \xrightarrow{ZrO_2} 2Ca''_{Zr} + 2O^x_O + Zr^{\cdot\cdot\cdot\cdot}_i \tag{D2}$$

A change in the crystal density results for each incorporation mechanism. This change can be calculated from the mass change due to substituting Ca for Zr, removing oxygen (for reaction D1), and adding cation interstitials (reaction D2). At the same time, the change in lattice parameter due to the substitution of Ca^{2+} for Zr^{4+} can be calculated from the ionic radii in Table 1.2. Actual crystal densities can be determined by pycnometric measurements, and the lattice parameter measured using x-ray diffraction. A comparison of the change in density expected for compensation by oxygen vacancies and by zirconium interstitials with experimental data is shown in Fig. ST17. It so happens that both models give rise to a decrease in crystal density with increasing CaO content. In samples quenched from 1600°C, the measured densities are in agreement with those calculated for the oxygen vacancy model. At 1800°C, there is an apparent change in the compensating defect from cation interstitials to oxygen vacancies as the concentration of CaO increases. The temperature dependence of the compensating mechanism indicated by these data suggests that in samples slowly cooled to room temperature, it will be the oxygen vacancy which predominates.

One uncertainty in calculations of this type is how to account for the volume change due to an interstitial or vacancy. It is perhaps reasonable to expect that an interstitial will expand the lattice, but the effects of a vacancy are not so clear. It is tempting to believe that the lattice will shrink around a missing ion, but we must keep in mind that the removal of a cation causes the nearest neighbor anions to experience a charge repulsion between one another, unshielded by the cation. Thus it is also pos-

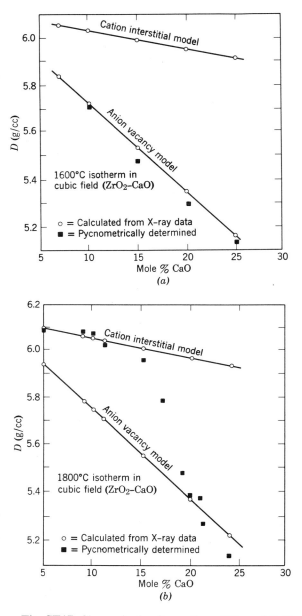

Fig. ST17 Change in density on the addition of CaO to ZrO$_2$ in solid solution quenched from (a) 1600°C and (b) 1800°C. At 1600°C each Ca^{2+} addition is accompanied by the formation of a vacant lattice site. At 1800°C, there is an apparent change in defect type with composition. [From A. Diness and R. Roy, *Solid State Communications*, 3, 123 (1965).]

2.1 Point Defects 133

sible that the vacancy will expand. There seems to be no clear trend; vacancies may have higher or lower volume than the ion which is removed. The good correspondence in Fig. ST17 between the measured and calculated densities (which do not account for defect volume changes) suggests that in zirconia, at least, the effects are small.

SPECIAL TOPIC 2.3

COLOR AND COLOR CENTERS

Point defects and solutes are important in determining the color of ceramic materials, and those that absorb light in the visible spectrum resulting in a perceivable change in color are often called "color centers." One interesting example is the various colors that can be achieved for impurities in Al_2O_3. Rubies, blue sapphires, and star sapphires are all based on single crystals of corundum. (These come in a variety of colors including ruby and blue, and derive their special asterism from the precipitation of crystallographically oriented rutile TiO_2.) We shall examine the effect of Cr or (Fe+Ti) additions which are responsible for the colors of ruby and blue sapphire respectively.

Color in Ruby: Cr^{3+} in Al_2O_3.

The brilliant red gemstone known as ruby consists of corundum with chromium impurity level of about 1%. The bright red color is a result of electronic transitions occurring entirely at the site of the chromium ions in the lattice. Chromium substitutes into alumina as individual Cr^{3+} ions on Al^{3+} sites. Using Kröger-Vink notation, the chromium point defect is written: Cr_{Al}^{x}. Since there is no charge imbalance, no additional charge balancing defects are created by the substitution.

The electronic transitions that cause the red color are interlevel transitions between different chromium d-orbitals. Chromium, like other transition elements, has incompletely filled d-orbitals; in the 3+ valence state the outer-shell electron configuration is $3d^3$. For a free ion the five d orbitals (d_{xy}, d_{yz}, d_{xz}, $d_{x^2-y^2}$, d_{z^2}) are *degenerate*, meaning they all have the same energy level. However, as discussed in Chapter 1, the surrounding oxygen ions form an octahedron around the aluminum site where the chromium is substituted. This octahedral environment interacts with the d-orbitals, raising some of their energies and lowering others, creating an energy-level splitting that then allows transitions between various d levels separated by relatively small energies corresponding to frequencies in the visible spectrum. The splitting depends on the strength of the surrounding bonding environment (i.e. the "ligand field"). Figure ST18a depicts the effect of ligand field strength on the possible d-orbital transitions, with a vertical dashed-line showing the level found for Cr^{3+} in ruby. Figure ST18b shows schematically the absorptions, emissions, and fluorescences possible between d-orbitals in ruby. In otherwise clear crystals, color is primarily a transmission phenomenon resulting from the portions of the visible spectrum which are not absorbed. Notice that the two key absorptions are at energies of about 2.2eV and 3.0eV, so that a major fraction of the blue and green end of the visible spectrum (shown in Fig. ST18c) is simply removed and our eye perceives the color as being bright red. The two emissions lines ending at the ^2E level are at low enough energies that they have no influence on the visible spectrum, but the last emission yields visible photons at 1.79eV which is at the red end of the visible spectrum. Under normal illumination this fluorescence makes little contribution to our perceived color, but under ultraviolet lamp illumination, red fluorescence may be observed.

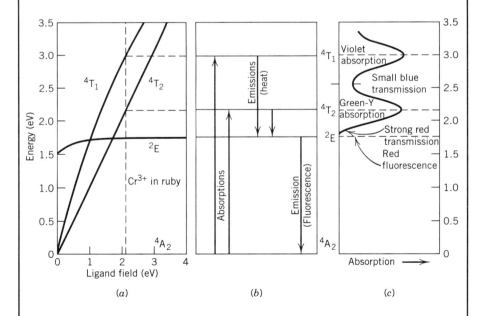

Fig. ST18 (a) Effect of distorted octahedral ligand field on possible d-orbital transitions for Cr^{3+} in ruby. (b) Resulting energy levels and transitions in ruby, and (c) The resulting absorption spectrum and fluorescence of ruby. (From K. Nassau, *The Physics and Chemistry of Color: The Fifteen Causes of Color*, John Wiley and Sons, New York, 1983.)

The Ruby Laser.

The energy levels shown in Fig. ST18b are also important because they enable the use of ruby as a laser. A broad-band white-light lamp is used to "pump" the electrons from their ground state into excited energy levels. The ruby crystal can absorb a great deal of the blue and green end of the spectrum, promoting many electrons into the $^4T_1$ and $^4T_2$ levels and reaching a state of *inversion* where there are more electrons in the excited levels than in the ground states. These can then decay to the ^2E level and emit red light — the characteristic color of the ruby laser.

Other characteristics that are important for making a laser are that the material must not absorb the emitted red light and that the red emission can be easily "stimulated" by other photons of the same wavelength passing through the crystal. To enhance the stimulation effect, the ruby is fabricated into a cylinder with partially silvered ends causing reflection back and forth along the cylinder axis, thus strengthening the light intensity along the axis. Then, photons emitted from one color center stimulates fluorescence from other centers which is coherent to (in phase with) the first, and ultimately amplification occurs and laser light has been created.

Blue Sapphire.

A similar effect is responsible for the blue coloration found in gemstones of blue sapphire. In sapphire the dopants iron and titanium are added together to Al_2O_3. These impurities can take on various valence states (Fe^{2+}, Fe^{3+}, Ti^{3+}, Ti^{4+}) depending on their concentrations, and the temperature and oxygen pressure at which the sapphire was formed. When the Fe^{2+} and Ti^{4+} ions substitute for two nearest-neighbor Al^{3+} ions (maintaining charge neutrality as a pair), a broad absorption band is formed at the red end of the spectrum causing the crystal to appear blue to the naked eye. This absorption results from electronic transitions between neighboring ions rather than among the d levels of a single ions, and is known as *charge transfer* color. The process responsible is the excitation of an electron from the Fe^{2+} into an energy-level of the Ti^{4+}. This causes a simple oxidation/reduction process:

$$Fe^{2+} + Ti^{4+} \xrightarrow{h\upsilon} Fe^{3+} + Ti^{3+}$$

For this charge transfer process to occur, the iron and titanium must be close neighbors in the lattice, and must have the right relative valence states (e.g., if both are fully oxidized to Fe^{3+} and Ti^{4+}, the transition cannot occur). Part of the trick of getting a good blue color when making synthetic blue sapphire, or in the common practice of heat-treating natural stones to enhance their color, is to control the relative valence states via the firing atmosphere and temperature. (Many other interesting examples can be found in *The Physics and Chemistry of Color: The Fifteen Causes of Color* by K. Nassau, J. Wiley, New York, 1983.)

2.2 SIMULTANEOUS DEFECT EQUILIBRIA: THE BROUWER DIAGRAM

We have now discussed a wide variety of individual defect reactions that can occur in a solid. Let us consider the net effect when all of these reactions take place simultaneously! Since any one defect can participate in a number of defect reactions, in a real system multiple defect equilibria will occur. In order to determine the defect concentrations, we must simultaneously solve a number of defect reactions. In principle, the number of defect reactions that can be written for any system is nearly limitless; to account for all of them would appear to be a daunting task. Fortunately, the number of defects that must be included for an adequate description of defect-related properties in a system is usually small. Which defects are important? The majority defects are always important; but in addition, certain minority defects may have direct relevance to a property of interest, such as the electrical conductivity. For example, in large bandgap materials where ionic defects are in the majority, electrons and holes may nonetheless determine electrical conductivity. Conversely, in a small bandgap compound where electronic defects dominate, minority ionic defects remain important for diffusional processes.

Most often we are interested in the variation of defect concentrations with temperature, solute concentration, and ambient gas activity (usually P_{O_2}). The defect formation reactions of interest will generally include those for:

1. Predominant intrinsic ionic defects (Schottky or Frenkel).
2. Intrinsic electronic defects.
3. Oxidation and reduction.
4. Incorporation of any significant solutes or impurities.

These reactions will involve a total of N defects, the concentrations of which are N variables to be solved for. If we are careful to write only independent reactions, one finds that the equilibrium constants for the above reactions always result in $(N-1)$ independent equations.

Using intrinsic MgO as an example, there are four defects of interest V_{Mg}'', $V_O^{\cdot\cdot}$, e' and $h^{\cdot}$, for which the relevant defect-forming reactions are:

$$null \rightarrow V_{Mg}'' + V_O^{\cdot\cdot} \qquad (2.11)$$

$$null \rightarrow e' + h^{\cdot} \qquad (2.24)$$

$$O_O^x = \frac{1}{2}O_2(g) + V_O^{\cdot\cdot} + 2e' \qquad (2.29)$$

with the respective equilibrium constants:

$$K_s = [V_{Mg}''][V_O^{\cdot\cdot}] \qquad (2.13)$$

$$K_i = np \qquad (2.44)$$

$$K_R = n^2 [V_O^{\cdot\cdot}] P_{O_2}^{1/2} \qquad (2.30)$$

2.2 Simultaneous Defect Equilibria: The Brouwer Diagram

In order to solve for all defect concentrations, one additional equation is always necessary. This is provided by the requirement of bulk electrical neutrality. In the *electroneutrality condition* each defect contributes a charge equal to its concentration times the effective charge per defect. In the current example this is:

$$2[V_{Mg}''] + n = 2[V_O^{\cdot\cdot}] + p \qquad (2.57)$$

The system of equations can now be solved to give the concentration of each defect at any temperature and P_{O_2} of interest, provided that the equilibrium constants, which are the parameters that characterize the thermodynamics of defect formation, are known. Note that neutral defects are not included in the electroneutrality condition. Their concentrations can nonetheless vary with temperature, solute concentration, and oxygen pressure through ionization and association reactions involving charged defects.

Solving this set of simultaneous equations using the full electroneutrality condition (Eq. 2.57) is usually a fairly straightforward procedure of solving a polynomial. The task is further simplified when the complete electroneutrality condition is reduced to a *Brouwer approximation* in which there is just one positive defect and one negative defect. (One then ends up with a simple second-or third-order monomial.) This assumption is valid when a single defect of each sign has a concentration much higher than others of the same sign. This condition is often satisfied due to the strong (exponential) dependence of defect concentrations on temperature and formation energy (i.e., each concentration term in Eq. 2.57 may differ by orders of magnitude from the others). For pure MgO, four Brouwer approximations are in principal possible:

(1) $n = 2[V_O^{\cdot\cdot}]$
(2) $2[V_{Mg}''] = p$
(3) $[V_{Mg}''] = [V_O^{\cdot\cdot}]$
(4) $n = p$

Only one of the last two of these will be important, depending on the relative values of the Schottky energy and the bandgap. In MgO, we found earlier that the intrinsic Schottky concentration (3) is greater than the electronic carrier concentration (4) at high temperature. For simplicity, let us consider (3) to be dominant, in which case there are only three *Brouwer regimes* to be considered, those defined by (1), (2), and (3). Each of these regimes will be represented by a different range of oxygen activity. Since reduction raises the concentration of oxygen vacancies and electrons (Eq. 2.29), Brouwer approximation (1) is likely to be important at the lowest range of oxygen pressure. Conversely, at high oxygen pressures the electron concentration is suppressed and the hole concentration p increases ($K_i = np$). Brouwer approximation (2) may then apply.

The *Brouwer diagram* (also referred to as a Kröger-Vink diagram) is a convenient way to represent the variation in defect concentrations with changes in the

138 Chapter 2 / Defects in Ceramics

activity of a component of the compound, usually oxygen pressure. It is a plot of log [concentration] against log P_{O_2} showing variations in defect concentrations *at constant temperature* in different Brouwer regimes. The Brouwer diagram for a binary metal oxide such as MgO is shown in Fig. 2.10. Consider the regime defined by the Brouwer approximation $[V''_{Mg}] = [V_O^{\cdot\cdot}]$, which lies in the center of Fig. 2.10. Over this regime the concentration of both vacancies is constant; $[V''_{Mg}] = [V_O^{\cdot\cdot}] = K_s^{1/2}$. Note that $K_i^{1/2}$ lies below $K_s^{1/2}$ since electronic defects are in the minority. Rearranging Equation 2.30, we obtain the P_{O_2} dependence of the electron concentration:

$$n = K_R^{1/2} K_s^{-1/4} P_{O_2}^{-1/4} \qquad (2.58)$$

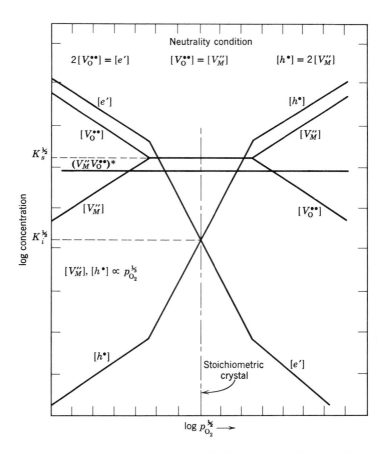

Fig. 2.10 Schematic representation of defect concentrations as a function of oxygen pressure for a pure oxide that forms predominantly Schottky defects at the stoichiometric composition.

2.2 Simultaneous Defect Equilibria: The Brouwer Diagram

and upon taking the logarithm of both sides, we have

$$\log n \propto -\frac{1}{4}\log P_{O_2}$$

In the central region of Fig. 2.10, the line representing n therefore has a slope of $-1/4$. Since $np = K_i$ and is constant at fixed temperature, the hole concentration follows a P_{O_2} dependence opposite to that of the electron concentration:

$$p = \frac{K_i}{n} = K_i K_R^{-1/2} K_s^{1/2} P_{O_2}^{1/4} \qquad (2.59)$$

and shows a $+1/4$ slope on the Kröger-Vink diagram.

With sufficient reduction, the Brouwer approximation $n = 2[V_O^{\cdot\cdot}]$ may come into play; this is the leftmost region of Fig. 2.10. By substituting the Brouwer approximation into Eq. 2.30 and rearranging, one obtains

$$n = 2[V_O^{\cdot\cdot}] = (2K_R)^{1/3} P_{O_2}^{-1/6} \qquad (2.60a)$$

and both n and $[V_O^{\cdot\cdot}]$ show a $-1/6$ slope, displaced from one another by a factor of $\log 2$. Lines for p and $[V_{Mg}'']$ show the opposite slope, being coupled to n and $[V_O^{\cdot\cdot}]$ through the intrinsic electronic (Eq. 2.44) and Schottky equilibria (Eq. 2.13). At the intersection between Brouwer regimes there is continuity since n increases with decreasing P_{O_2} until it becomes a majority defect. With increasing P_{O_2} the hole concentration p increases, until the Brouwer approximation $2[V_{Mg}''] = p$ comes into play. In this regime, substitution of the Brouwer approximation into the set of Eqs. 2.13, 2.44, and 2.30 results in the $+1/6$ slopes in Fig. 2.10. Notice that at all times $[V_O^{\cdot\cdot}]$ and $[V_{Mg}'']$ are symmetric about the horizontal line given by $K_s^{1/2}$, and n and p about $K_i^{1/2}$. This is required by the Schottky and intrinsic electronic equilibria.

The temperature dependence of defect concentrations are not apparent in Brouwer diagrams since they are isothermal, but can be obtained from the equilibrium constants, which have the form

$$K = \exp\left(-\frac{\Delta G}{kT}\right) = \exp\left(\frac{\Delta S}{k}\right)\exp\left(-\frac{\Delta H}{kT}\right) \qquad (2.60b)$$

Thus, in the Brouwer regime where $[V_{Mg}''] = [V_O^{\cdot\cdot}]$, from Eq. 2.58 the electron concentration n varies exponentially with temperature with an activation energy $(\Delta h_R/2k - \Delta h_s/2k)$. In the $n = 2[V_O^{\cdot\cdot}]$ Brouwer regime, from Eq. 2.60 the activation energy is $\Delta h_R/3k$.

Several other Brouwer diagrams are shown in Figs. 2.11-2.13. Fig. 2.11 is a diagram for an oxide dominated by oxygen Frenkel defects, and Fig. 2.12 represents a similar oxide in which the bandgap is small enough, or the oxygen Frenkel energy large enough, for electronic defects to dominate ($K_i > K_F$). Fig. 2.13 shows a more complex diagram which represents the binary metal oxide in Fig. 2.10 when it is dominated by trivalent impurities ($[F_M^{\cdot}] > K_s^{1/2}$). Defect associates (discussed in Eq. 2.3) are included in this diagram, but remain in the minority. The solution of these diagrams is left as an exercise for the reader.

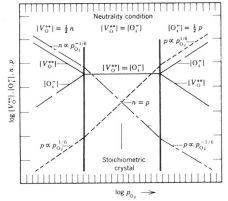

Fig. 2.11 Kröger–Vink diagram for an oxide in which oxygen Frenkel defects dominate the intrinsic defect structure.

Finally, it should be emphasized that while a number of regimes can be constructed for any Brouwer diagram, not all of these may be relevant to real defect behavior. The width of each Brouwer regime is temperature-dependent; and the experimentally accessible range of P_{O_2} may only cover one or two Brouwer regimes. With sufficient knowledge of the defect thermodynamics (that is, the equilibrium constants) the locations of the P_{O_2} boundaries between Brouwer regimes can be quantitatively determined. However, this is only the case for a few well-

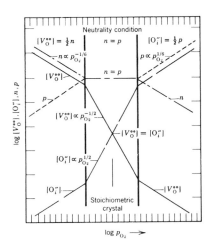

Fig. 2.12 Kröger-Vink diagram for oxide similar to Fig. 2.11 except that electronic defects dominate ($K_i > K_F$).

2.2 Simultaneous Defect Equilibria: The Brouwer Diagram

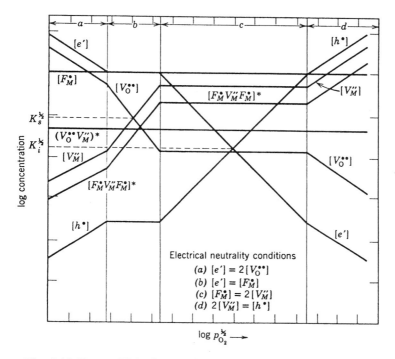

Fig. 2.13 Kröger-Vink diagram for An oxide that forms Schottky defects but contains cation impurities $[F_M^{\bullet}] > K_s^{1/2}$.

studied systems. Also, the Brouwer diagram does not account for the range of P_{O_2} over which the phase in question actually exists. As an example, at high temperature the oxides of iron undergo phase changes from $Fe_{1-x}O$ to Fe_2O_3 to Fe_3O_4 with increasing P_{O_2}. While it is possible to construct separate Brouwer diagrams to describe the defect behavior of each phase, in reality the P_{O_2} range over which any one diagram applies will be determined by the boundaries between the oxide phases.

SPECIAL TOPIC 2.4

SIMPLE PROCEDURES FOR CONSTRUCTING A BROUWER DIAGRAM

1. How many defect species are relevant? A great many defects can in principle exist, but one can decide which ones will be the dominant ones based on knowledge of the crystal structure, solute concentration, and auxiliary information such as electrical conductivity or diffusion rates. For instance, in a close-packed structure Frenkel defects are often unlikely ($K_F \ll K_S$) and one can neglect interstitials.

142 Chapter 2 / Defects in Ceramics

2. Write independent defect chemical reactions to account for the formation of those defects, and their equilibrium constants according to the law of mass action. (For instance, the reduction and oxidation reactions are not independent, being related through the intrinsic electronic reaction.) Usually the reactions of importance will include the intrinsic defect formation mechanism of lowest energy, intrinsic electronic defect formation, oxidation or reduction (but not both), and the incorporation of any solutes present in significant concentrations.
3. You should end up with $(N-1)$ equations, where the N defect concentrations are your unknowns. In order to solve the system of equations, one more equation is necessary. This last one is the electroneutrality approximation, or Brouwer approximation, which contains only one negative and one positive defect. From the complete electroneutrality condition relating all N defects, individual Brouwer approximations can be isolated.
4. Each Brouwer approximation defines a region of P_{O_2} (for oxides) on the Kröger-Vink diagram. Begin at a convenient point such as stoichiometry. Label the log concentration axis with K's for the intrinsic electronic and ionic defects (evaluate which is the greater). The Brouwer approximation of this region is determined by the higher of these two equilibrium constants. Use the Brouwer approximation in your mass-action relationships to solve for the P_{O_2} dependence of each defect concentration.
5. Observe which defect concentrations are increasing with decreasing or increasing P_{O_2}. This will help identify which of the remaining Brouwer approximations apply in adjacent regimes. For these adjacent regimes, again substitute the Brouwer approximation into the mass-action equilibrium constants to solve for the P_{O_2} dependence of defect concentrations.

SPECIAL TOPIC 2.5

OXYGEN SENSORS BASED ON NONSTOICHIOMETRIC TiO$_2$

In automobiles, increasing emphasis on fuel efficiency and pollution control has led to the development of combustion control systems based on oxygen pressure sensors. In order to minimize exhaust pollutants such as nitrous oxides, carbon monoxide, and hydrocarbons, catalysts which decompose these compounds are used in the exhaust system. However, catalytic converters are only able to operate efficiently over a narrow range of combustion conditions determined by the air/fuel ratio of the intake system. Above the "stoichiometric" air/fuel ratio of ~14.5, hydrocarbons and CO are

2.2 Simultaneous Defect Equilibria: The Brouwer Diagram

efficiently decomposed, but NO_x is not; whereas below the stoichiometric ratio, the reverse becomes true. Since the oxygen activity of the exhaust gas also varies dramatically as one traverses the stoichiometric air/fuel ratio, from about 10^{-9} to 10^{-21} atm (10^{-4} to 10^{-16} Pa), this provides an opportunity to use oxygen activity as a measurement of combustion efficiency. Virtually all modern automobiles have a feedback system in which an oxygen sensor is used to measure the P_{O_2} of the exhaust stream, and provide an electrical input to the fuel injection system to rapidly optimize the intake air/fuel ratio as driving conditions change. These oxygen sensors are typically conductive ceramics, and operate on one of two principles. One type is based on the Nernst effect, where an electrical potential is created across an oxygen conducting ceramic when there is a gradient in oxygen pressure. This is discussed in Chapter 3. The other type, which we consider here, is based on the oxygen pressure dependence of electronic conductivity. TiO_2 is one material that has been developed for this purpose.

The discussions of defect chemistry in this chapter showed us that the concentrations of defects can vary by many orders of magnitude with changes in temperature and oxygen pressure. As discussed in greater detail in Chapter 3, one of the properties of ceramics which is profoundly affected by these changes is the electrical conductivity. Because the electrical conductivity is generally proportional to the concentration of current carriers, the variation in electron concentration of an oxide with P_{O_2}, illustrated in the Brouwer diagrams in Figs. 2.10–2.13, results in a corresponding variation in the electrical conductivity. One material which has an electron conductivity which is both sufficiently high and P_{O_2} dependent for gas sensor applications is TiO_2. Figure ST19 shows that the electrical conductivity of TiO_2 increases with decreasing P_{O_2} by about a factor of 10 for every four decades change in P_{O_2}. This is a suitable sensitivity for automotive oxygen sensor applications, given that the exhaust P_{O_2} changes by about 12 decades across the operating air/fuel ratio.

The results in Fig. ST19 are for TiO_2 samples containing various donor (Nb) doping levels, at a temperature of 1350°C. This is a higher temperature than the range at which practical oxygen sensors operate, 350–850°C, but the dependence on oxygen pressure is similar. Notice that with decreasing oxygen pressure, the P_{O_2} dependence of conductivity for all samples reaches a limiting situation where the slope of the data is $-1/5$. How can we understand this in terms of the defect chemistry? First, we note that since the electrical conductivity is proportional to the electron concentration, $\sigma \propto n$, this is a Brouwer regime where $n \propto P_{O_2}^{-1/5}$.* Reduction increases the oxygen deficiency, x, in TiO_{2-x}. This deficiency could be accommodated by oxygen

---

*For the higher donor concentrations, regions also appear in which σ is independent of P_{O_2} or proportional to $P_{O_2}^{-1/4}$; it is left as an exercise to the reader to determine the Brouwer approximations of these regimes, and to prove that TiO_2 is an n-type semiconductor over the entire composition and range shown in Fig. ST19.

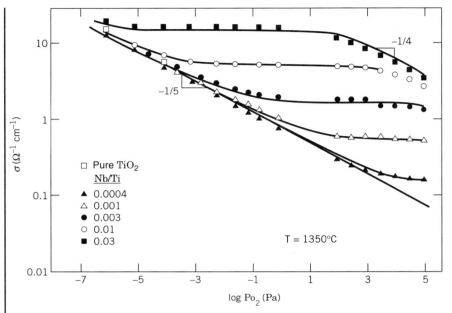

Fig. ST19 Oxygen partial pressure dependence of electrical conductivity in Nb-doped TiO_2 at 1350°C. [From J. F. Baumard and E. Tani, *J. Chem. Phys.* 67[3] 857(1077)].

vacancies or titanium interstitials. In order to determine which is dominant, we consider whether or not these reduction mechanisms result in $n \propto P_{O_2}^{-1/5}$. From Eqs. 2.29, 2.30, and 2.60, we saw that when reduction of an oxide (any oxide) is accommodated by formation of $V_O^{\cdot\cdot}$, the electron concentration follows the relation $n \propto P_{O_2}^{-1/6}$; this therefore does not appear to be the right mechanism. However, if reduction is accomodated by the formation of $Ti_i^{\cdot\cdot\cdot\cdot}$,

$$Ti_{Ti}^x + 2O_O^x \Leftrightarrow Ti_i^{\cdot\cdot\cdot\cdot} + 4e' + O_2(g), \quad \text{(Eq. OS1)}$$

and reduction dominates the defect chemistry, the Brouwer approximation is

$$4[Ti_i^{\cdot\cdot\cdot\cdot}] = n. \quad \text{(Eq. OS2)}$$

Since the equilibrium constant for reaction OS1 is

$$K_{Ti_i} = [Ti_i^{\cdot\cdot\cdot\cdot}]n^4 P_{O_2}, \quad \text{(Eq. OS3)}$$

upon combining Eqs. OS2 and OS3, the oxygen pressure dependence of electron concentration is given by

$$n = \left(\frac{4K_1}{P_{O_2}}\right)^{-1/5} \quad \text{(Eq. OS4)}$$

showing the $-1/5$ dependence we are seeking. While this is not a "proof," the data in Fig. ST19 *are consistent with* a reduction mechanism in which the primary ionic defects are $Ti_i^{\cdot\cdot\cdot\cdot}$. (It is left as an exercise for the reader to show that if the interstitial is trivalently charged, $Ti_i^{\cdot\cdot\cdot}$, the P_{O_2} dependence is again $-1/4$, in which case one could not discriminate between oxygen vacancy and titanium interstitial mechanisms.)

Regardless of the actual compensation mechanism, the important point for practical applications is that the conductivity of reduced TiO_2 exhibits a useful P_{O_2} dependence. In an actual oxygen sensor, Fig. ST20, the conductivity of a small piece of polycrystalline TiO_2 exposed to the exhaust gas is continuously monitored. In order to shorten the response time of conductivity with changes in oxygen activity, a porous TiO_2 element is desirable. This increases the surface area for gas exchange, and decreases the effective cross section across which the nonstoichiometry must change.

One potential difficulty presented by a resistive oxygen sensor is that the conductivity is temperature-dependent as well as P_{O_2}-dependent, due to the fact that defect chemical equilibrium constants are exponentially dependent on temperature (e.g., in Eq. OS4 n is also proportional to $K_1^{-1/5}$, and K_1 is thermally activated.) Since a range of operating temperatures are encountered in use, some form of temperature compensation is necessary if the

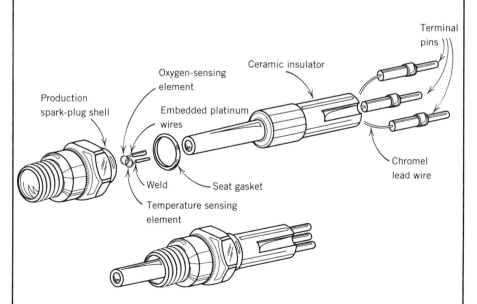

Fig. ST20 Temperature-compensated titania exhaust sensor. (J. G. Rivard, pp. 37-42 in *Automotive Sensors, Ceramic Proceedings*, Vol. 1, L. H. Van Vlack, ed., The American Ceramic Society, Westerville, Ohio, 1980.)

sensor output is to be accurate. One engineering solution is to incorporate a heater to keep the sensor at a constant temperature which is above the range of exhaust temperatures. Another, developed by Ford Motor Company, is to use a piece of dense TiO_2, which operates at the same temperature as the porous sensor, but does not equilibrate quickly with the gas stream, to provide a reference point for the conductivity (other types of thermistors can be used as well). A comparison of the resistivities of the two ceramic elements then allows the oxygen pressure dependence of conductivity in the porous TiO_2 to be isolated.

2.3 DEFECT ASSOCIATION AND PRECIPITATION

It is useful at this juncture to remind ourselves of a critical assumption in the preceding treatment of point defects: that the concentration of defects is sufficiently dilute that the thermodynamic activity of defects is equal to their concentration. We have assumed that quasichemical equilibria can be established within the time scale of an experiment (e.g., after heating to high temperature), but that defects otherwise form an ideal solid solution. Under what conditions is this assumption valid? While a precise answer is not presently possible, it is clear that important factors include the concentration of defects, the charge of the defects, and the dielectric properties of the medium in which they are embedded.

Some materials exhibit behavior consistent with idealized point defect models at surprisingly high defect concentrations, of up to several percent. As concentrations increase, the first kind of a modification that can be introduced is the Debye-Hückel correction for the activity of defects in concentrated solution, which has its origins in the theory of liquid electrolytes. A brief description is given below. However, in many heavily doped or highly nonstoichiometric ceramics, defect concentrations can be so high that defects become by definition associated, being literally unable to avoid one another. Evidence for strong defect interactions in highly defective oxides such as $Fe_{1-x}O$, cubic stabilized ZrO_2, and pyrochlore structure ionic conductors are discussed in Chapter 3. Defect interactions under highly concentrated conditions are not well understood. Here we will discuss defect association interactions which take place in dilute solutions (e.g., less than 1%), where a point defect formalism is still applicable.

Point Defect Association

Point defects in solid solution often experience a binding energy with other defects that leads to the formation of defect associates. For defects that are electrically charged, this attraction (or repulsion) is primarily coulombic in character. Figure 2.14 shows a {100} plane of MgO in which several types of defect associates are shown. Each of these defects has a number of possible spatial orientations, Z. Vacancy pairs ($V_O^{\cdot\cdot} - V_{Mg}''$) occupy nearest-neighbor sites and can take

2.3 Defect Association and Precipitation

on six distinct $\langle 100 \rangle$ orientations, so $Z = 6$. When trivalent solutes are present, cation vacancies are introduced for charge compensation and the solutes and vacancies can associate. Two different orientations of $(Al_{Mg}^{\cdot} - V_{Mg}^{\prime\prime})^{\prime}$ "dimer" associates are possible, ones in the $\langle 110 \rangle$ directions for which $Z = 12$ (there are 12 cation nearest neighbors in the rocksalt structure), and ones of $\langle 100 \rangle$ orientation bridging an oxygen ion for which $Z = 6$ (second nearest cation neighbors). Also shown is the "trimer" associate $(Al_{Mg}^{\cdot} - V_{Mg}^{\prime\prime} - Al_{Mg}^{\cdot})^{x}$, which can have a number of different orientations both linear and "kinked."

Association can be written as a defect reaction, which for the dimer is:

$$Al_{Mg}^{\cdot} + V_{Mg}^{\prime\prime} = (Al_{Mg}^{\cdot} - V_{Mg}^{\prime\prime})^{\prime} \tag{2.61}$$

with an equilibrium constant (neglecting Debye-Hückel corrections) given by:

$$K_a = \frac{\left[(Al_{Mg}^{\cdot} - V_{Mg}^{\prime\prime})^{\prime}\right]}{\left[Al_{Mg}^{\cdot}\right]\left[V_{Mg}^{\prime\prime}\right]} = Z \exp\left(-\frac{\Delta g_a}{kT}\right) = Z \exp\left(\frac{\Delta s_a}{k}\right) \exp\left(-\frac{\Delta h_a}{kT}\right) \tag{2.62}$$

The nonconfigurational entropy Δs_a is difficult to evaluate accurately and generally assumed small, so that $\exp(\Delta s_a/k) \sim 1$. The enthalpy for association Δh_a is

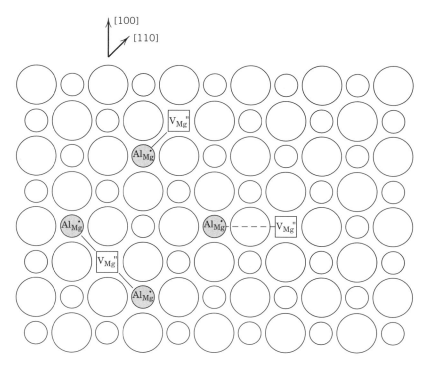

Fig. 2.14 (100) plane of MgO showing several configurations of defect associates: $(Al_{Mg}^{\cdot} - V_{Mg}^{\prime\prime})^{\prime}$ along $\langle 110 \rangle$, $(Al_{Mg}^{\cdot} - V_{Mg}^{\prime\prime})^{\prime}$ along $\langle 100 \rangle$, and $(Al_{Mg}^{\cdot} - V_{Mg}^{\prime\prime} - Al_{Mg}^{\cdot})^{x}$ along $\langle 110 \rangle$.

more easily determined and has been studied widely. The simplest estimate equates the association enthalpy to the coulombic attraction between defects

$$\Delta h_a = \frac{q_i q_j}{4\pi\varepsilon\varepsilon_0 R} \tag{2.63}$$

where q_i and q_j are the charges of the respective defects i and j, $\varepsilon\varepsilon_0$ is the static dielectric constant and R the separation between the defects. Experimental values of association enthalpies can sometimes be determined from transport measurements. Values can also be calculated using a variety of computer simulation techniques (similar to those used to calculate defect formation enthalpies, such as those in Table 2.2). Table 2.4 lists defect association enthalpies estimated using Eq. 2.63, compared with some values calculated using more refined computer simulation routines. We see that Eq. 2.63 is a reasonable estimate, particularly for simple associates involving only two defects. For complex associates, such as the trimers, the values listed are for the binding of an additional defect such as $Al^{\cdot}_{Mg}$ to $(Al^{\cdot}_{Mg} - V''_{Mg})'$, in which case q and R in Eq. 2.63 are rather ill-defined and a high level of accuracy should not be expected.

With decreasing temperature, the concentration of associates rises. If the dimer is the only associate present, we can calculate the concentration of dimers from the expression in Eq. 2.63 keeping in mind that $[Al]_{total} = [Al^{\cdot}_{Mg}] + [(Al^{\cdot}_{Mg} - V''_{Mg})']$ For the formation of vacancy pairs the equilibrium constant is

$$K_a = \frac{[(V''_{Mg} - V^{\cdot\cdot}_O)^x]}{[V^{\cdot\cdot}_O][V''_{Mg}]} = 6\exp\left(-\frac{\Delta g_a}{kT}\right) = 6\exp\left(\frac{\Delta s_a}{k}\right)\exp\left(-\frac{\Delta h_a}{kT}\right) \tag{2.64}$$

Since the product $[V^{\cdot\cdot}_O][V''_{Mg}]$ is given by the Schottky equilibrium constant K_s, the associate concentration has a temperature dependence that includes the Schottky formation enthalpy:

$$[(V''_{Mg} - V^{\cdot\cdot}_O)^x] = 6\exp\left(\frac{\Delta s_s}{k}\right)\exp\left(\frac{\Delta s_a}{k}\right)\exp\left(-\frac{\Delta h_s}{kT}\right)\exp\left(-\frac{\Delta h_a}{kT}\right) \tag{2.65}$$

Precipitation

Formation of larger defect clusters also frequently occurs, which may grow and eventually lead to the formation of a precipitate. For example, when the solid solution limit is exceeded upon cooling of MgO containing aluminum, precipitation of the equilibrium phase $MgAl_2O_4$ spinel occurs. The precipitation reaction is

$$Mg^x_{Mg} + 2Al^{\cdot}_{Mg} + V''_{Mg} + 4O^x_O = MgAl_2O_4 \,(ppt) \tag{2.66}$$

with an equilibrium constant given by

$$K_{ppt} = \frac{a_{MgAl_2O_4}}{[Al^{\cdot}_{Mg}]^2[V''_{Mg}]} \propto \exp\left(-\frac{\Delta h_{ppt}}{kT}\right). \tag{2.67}$$

2.3 Defect Association and Precipitation

Table 2.4 Defect Association Energies

	κ	R (nm)	Simple Coulombic Estimate (Eq. 2.63) $-\Delta h_a = q_i q_j / 4\pi\varepsilon_o R$ (eV)	Atomistic Simulations $-\Delta h_a$ (eV)
NaCl	5.62			—
$V'_{Na} - V^{\cdot}_{Cl}$		0.282	0.9	—
$Ca^{\cdot}_{Na} - V'_{Na}$		0.399	0.6	—
CaF$_2$	8.43			—
$F'_i - V^{\cdot}_F$		0.274	0.6	—
$Y^{\cdot}_{Ca} - V''_{Ca}$		0.386	0.9	—
NiO	12.0			—
$V''_{Ni} - V^{\cdot\cdot}_O$		0.209	2.3	—
$V''_{Ni} - Ni^{\cdot}_{Ni}$		0.295	0.8	—
$Li'_{Ni} - Ni^{\cdot}_{Ni}$		0.295	0.4	—
UO$_2$	~15			—
$O''_i - V^{\cdot\cdot}_O$		0.209	0.5	—
MgO	9.8			—
$V''_{Mg} - V^{\cdot\cdot}_O$		0.211	2.8	—
$Fe^{\cdot}_{Mg} - V''_{Mg}$ (110)		0.298	1.0	0.85
$Fe^{\cdot}_{Mg} - V''_{Mg} - Fe^{\cdot}_{Mg}$ (110)		0.298	1.75	1.42
$Fe^{\cdot}_{Mg} - V''_{Mg}$ (110)		0.424	0.71	1.13
$Fe^{\cdot}_{Mg} - V''_{Mg} - Fe^{\cdot}_{Mg}$ (110)		0.424	1.24	2.20
$Al^{\cdot}_{Mg} - V''_{Mg}$ (110)		0.298	1.0	0.68
$Al^{\cdot}_{Mg} - V''_{Mg}$ (110)		0.424	0.71	0.86
$Al^{\cdot}_{Mg} - V''_{Mg} - Al^{\cdot}_{Mg}$ (110)		0.298	1.75	1.32
$Al^{\cdot}_{Mg} - V''_{Mg} - Al^{\cdot}_{Mg}$ (110)		0.424	1.24	1.68
$(4 V''_{Mg} - Mg^{\cdot\cdot}_i - 6 Al^{\cdot}_{Mg})^x$ (the 4:1 cluster)		—	—	4.77
Al$_2$O$_3$	8.6 ($\perp c_o$) 10.6 ($//c_o$)			
$Ti^{\cdot}_{Al} - V'''_{Al}$		0.275	1.87	1.9
$Ti^{\cdot}_{Al} - V'''_{Al} - Ti^{\cdot}_{Al}$		0.275	3.43	3.5

The activity of the spinel phase, $a_{MgAl_2O_4}$, is constant if it is at equilibrium (unity if it is the standard state). If defect association is not extensive, then $[Al^{\cdot}_{Mg}] \sim 2[V''_{Mg}]$ and the free vacancy concentration will vary with temperature as:

$$[V''_{Mg}] \propto \exp\left(+\frac{\Delta h_{ppt}}{3kT}\right). \tag{2.68}$$

150   Chapter 2 / Defects in Ceramics

The formation of $MgAl_2O_4$ precipitates in MgO is facilitated by the fact that both compounds are based on cubic close-packing of oxygen. Equation 2.66 shows four formula units of MgO combining to form one formula unit of $MgAl_2O_4$. A rearrangement of the cations within the same oxygen sublattice can lead to a local spinel structural unit within the MgO lattice. Since $MgAl_2O_4$ is of the normal spinel structure (Figs. 1.30, 1.31), Mg^{2+} ions must be displaced from their octahedral sites in the MgO solid solution to tetrahedral sites upon forming the spinel. Close examination of the spinel structure in Fig. 1.30 shows that the closest cation sites to a tetrahedral Mg^{2+} ion are four octahedral vacancies at the corners of a tetrahedron. These are the four octahedral cation sites around a tetrahedral interstitial site in the FCC lattice (Fig. 1.8). Beyond this are six aluminum ions, which relative to the MgO lattice are $Al^{\cdot}_{Mg}$ defects. It is likely that this "4:1" cluster, which relative to the MgO lattice is the defect associate $(4V''_{Mg} - Mg^{\cdot\cdot}_i - 6Al^{\cdot}_{Mg})^x$, is a precursor to the coherent spinel precipitate (more complex clusters can also be identified that are subunits of the spinel unit cell). Table 2.4 shows that on a per-vacancy basis, the energy of this cluster is not very different from that of the simple dimer.

Small precipitates often form with a shape that minimizes their strain energy against the anisotropic elastic constants of the crystal matrix. For instance, $MgAl_2O_4$ precipitates in MgO tend to form as thin plates with their faces along {100} planes (Fig. 2.15). On the other hand, $MgFe_2O_4$ precipitates have virtually no difference in lattice parameter with MgO and form as octahedra with the faces along {111} planes. In either case the spinel nucleates from a defect cluster, and then grows into a larger, but still coherent, precipitate in which the continuity of lattice planes across the interface with the matrix phase is retained. As the precipitate grows, however, it eventually may be unable to maintain a coherent interface against the increasing lattice strain, whereupon the interface becomes incoherent. The pres-

Fig. 2.15 Morphology of $MgAl_2O_4$ precipitates in MgO. (Courtesy of A. F. Henriksen.)

2.3 Defect Association and Precipitation   151

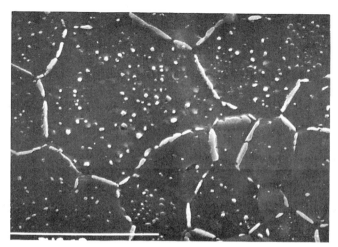

Fig. 2.16 Precipitation of Sc_2O_3 at grain boundaries and within grains in MgO. (Courtesy of A. F. Henriksen.)

ence of a lattice discontinuity such as grain boundaries and dislocations frequently aids precipitation by serving as a heterogeneous nucleation site. Fig. 2.16 shows that precipitates form preferentially at grain boundaries in MgO. Fig. 2.17 illustrates precipitation at the nodes of a dislocation network in MgO.

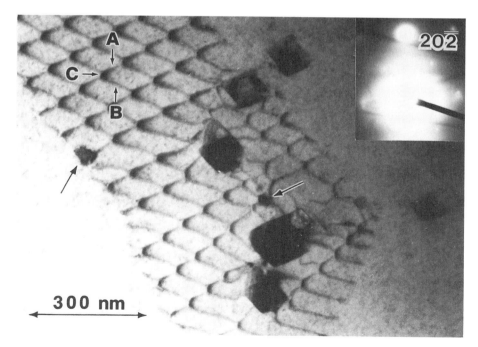

Fig. 2.17 $Mg_{1-x}Ca_xAl_2O_4$ spinel precipitated at the nodes of a dislocation network in MgO. (Courtesy of M. Fujimoto.)

Debye–Hückel Corrections

As defect concentrations increase, there is an increasing tendency for each defect to be surrounded by a diffuse cloud of opposite charged defects, which "screen" the defects from one another and lessens their interaction. The net effect is that the activity of individual defects is less than their concentration (activity coefficient less than one). The Debye–Hückel theory provides a means of correcting the activity of the defects for this interaction. The screening distance over which the excess charge of a defect is effectively neutralized by opposite charges is given by $1/\ell$, where ℓ is given by

$$\ell^2 = \frac{8\pi n_d (z_i e)^2}{\varepsilon kT} \tag{2.69}$$

n_d is the bulk concentration of charged defects in units of cm^{-3}, $z_i e$ is the defect charge, and ε is the dielectric constant. As the dielectric constant increases, or defect concentration decreases, the screening length increases. The change in energy per pair of defects formed is

$$H_{DH} = -\frac{(z_i e)^2 \ell}{\varepsilon(1+\ell R)} \tag{2.70}$$

where R is the distance of closest approach of the charged defects. The activity a of the defects is reduced, with the activity coefficient given by

$$\gamma = \frac{a}{[c]} = \exp\left[-\frac{z^2 e^2 \ell}{kT(1+\ell R)}\right] = \exp\left[\frac{H_{DH}}{2kT}\right] \tag{2.71}$$

Debye–Hückel screening lowers the formation enthalpy of defects, so that more defects form at a given temperature than would be expected in the absence of interactions. Also, fewer defect associates are formed than expected from simple association theory due to the shielding. The concentration at which Debye–Hückel corrections become appreciable clearly depends on the system in question; for alkali halides, effects are noticeable at concentrations greater than 1000 ppm.

SPECIAL TOPIC 2.6

CATION NONSTOICHIOMETRY, DISORDER, AND DEFECT ENERGETICS IN LITHIUM NIOBATE

Point defects, at a high enough concentration, have a noticeable influence on the overall composition or nonstoichiometry of a compound. Conversely, the size and composition of solid solution regions can give hints about the point defects which are present. Nonstoichiometric transition metal oxides are one example; TiO_{2-x} is anion-deficient, which indicates the presence of either oxygen vacancies or titanium interstitials. $Ni_{1-x}O$, $Co_{1-x}O$, $Mn_{1-x}O$, and

2.3 Defect Association and Precipitation

$Fe_{1-x}O$ are all to varying degrees cation-deficient, indicating the presence of cation vacancies. Compounds with extensive cation nonstoichiometry are also "defective". $MgAl_2O_4$ exists over a wide range of aluminum-rich solid solution (as high as $MgO \cdot 6Al_2O_3$), but a much narrower range of magnesium-rich nonstoichiometry (Fig. 4.4b). This is an indication that the defects that can possibly compensate for magnesium excess, such as the creation of oxygen vacancies or cation interstitials, have higher energy than those compensating for aluminum excess, namely the formation of cation vacancies. One might a priori guess that this is possible due to the existence of a γ-Al_2O_3 polymorph, a highly defective spinel with composition $Al_{8/3}O_4$. Of course, these hints must be combined with knowledge derived from spectroscopic and other techniques to confirm the true solid state defect state in a material.

Another interesting and technologically important example is the compound lithium niobate (with an ideal formula of "$LiNbO_3$"), which exhibits a wide range of stable nonstoichiometric compositions (see Fig. ST21). Our

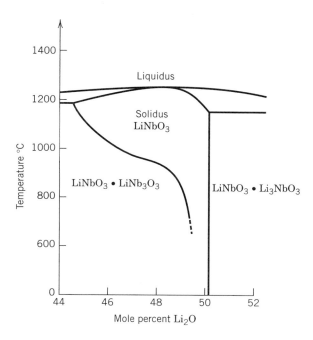

Fig. ST21 Phase diagram for Li_2O-Nb_2O_5 near the composition of $LiNbO_3$, showing extensive Nb-rich solid solution field at elevated temperatures.

goal in this discussion is to determine the Brouwer approximation which can explain the nonstoichiometry observed. Notice first of all that the nonstoichiometry extends mainly to niobium-rich compositions. We therefore seek defect combinations consistent with a nobium-rich solid solution, and suspect that possible defects, which might yield lithium-rich solid solution, will have relatively high formation energies.

One key defect has been identified as the niobium *antisite* defect ($Nb_{Li}^{....}$). Antisite defects are possible whenever more than one host cation (or anion) is present. In the temperature range where a large Nb-rich solid solution exists, it is believed that the defects providing charge balance for $Nb_{Li}^{....}$ are lithium vacancies (V_{Li}'), yielding the Brouwer approximation

$$4[Nb_{Li}^{....}] = [V_{Li}']$$

This defect picture, suggested by the asymmetry of the solid solution phase field, is supported by computer simulation studies [Donnerberg et. al., *Phys. Rev. B.*, **40** 11909 (1989)], which show that the niobium antisite defect is of very low energy. This alone does not indicate a low energy for the overall incorporation reaction; the energy to form the charge-compensating defect V_{Li}' must also be included. Of possible intrinsic defect reactions in this system, only the lithium Frenkel defect is found to be low in energy. Since the lithium Frenkel energy is composed of formation energies for a lithium vacancy and lithium interstitial, the fact that the sum of the two is low also supports the notion that the lithium vacancy is the lowest energy ionic defect of negative effective charge. We might expect this to be the case from simple charge considerations; a missing monovalent cation is likely to disrupt the local ionic bonding configuration less than, say, a missing pentavalent cation.

However, a puzzle is that room-temperature single-crystal x-ray diffraction has revealed niobium vacancies along with the niobium antisite defects. It may be that the relative energies for compensating niobium antisite defects by lithium vacancies and niobium vacancies are not very different, and upon cooling from high temperatures, the defect compensation changes to include V_{Nb}'''''. This can take place by a simple exchange of lattice Nb ions (Nb_{Nb}^x) with Li vacancies, creating in the process a new defect associate, $(Nb_{Li}^{....} - V_{Nb}''''')'$ in place of V_{Li}'. Coulombic attraction may preserve the new defect in an associated state. (It is left to the reader to write proper defect chemical equations for these various reactions.) Notice that since this process involves only an exchange between cation nearest-neighbor sites, it does not result in a change in charge balance (which would require the formation of additional charged defects). It is consistent with the very low formation energy of the niobium antisite defect. The existing x-ray diffraction results cannot distinguish whether or not the niobium vacancies are in fact adjacent to the niobium antisite defects, as suggested by this model.

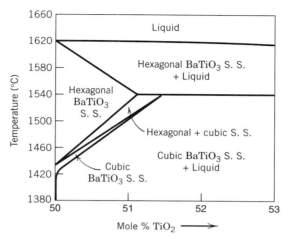

Fig. ST22 High temperature phase diagram for barium titanate, showing a Ti-rich solid solution field for the hexagonal polymorph.

It is possible that in other ternary compounds, the high temperature nonstoichiometry tends toward an excess of the higher-valence cation for similar reasons. For example, in barium titanate there is a titanium-rich nonstoichiometry in the high temperature hexagonal polymorph, Fig. ST22. However, the high-temperature defect structure of this phase has not been as extensively studied.

2.4 INTERACTIONS BETWEEN POINT DEFECTS AND INTERFACES

We now consider some details of where and how defects are formed. While a Frenkel defect can be formed internally within the lattice (Fig. 2.1), the formation of Schottky defects (Fig. 2.2) requires a lattice discontinuity to which atoms can be displaced, such as a surface, grain boundary, or dislocation, Fig. 2.18. This discontinuity is referred to as a *source* or *sink* for defects, being a site from which defects are emitted or absorbed. It is sometimes assumed to be "perfect" in the sense that there is no limit to the rate at which defects can be formed or removed. This assumption is sometimes violated in kinetic processes such as diffusional creep and solid-state reactions, where defect formation or annihilation at the interface can be rate-limiting. At thermal equilibrium, however, the lattice concentration of defects reaches a constant value as the rates of defect formation and annihilation reach a steady state.

Ionic Space Charge

An important effect that results when ionic defects equilibrate with a surface is the formation of a surface electrical potential, associated with an excess of surface ionic charge. The existence of this potential is a major distinguishing characteristic between interfaces (surfaces, grain boundaries, and dislocations) in ionic solids and those in metals. We can see how this potential arises in Fig. 2.18. The Schottky energy may be separated into individual cation and anion vacancy formation energies; for MgO this is $g_s = g_{V_{Mg}} + g_{V_O}$. Because the creation of a Schottky defect is not site-conserving, it is necessary to have a lattice discontinuity at which ions can be absorbed in order to form the defects. Thus the formation energy of an individual defect is defined as that for bringing the ion to or from the lattice discontinuity, as shown in Fig. 2.18. Although the formation of a Frenkel defect is site-conserving (Fig. 2.1), it can nonetheless be separated into the virtual processes of: (1) bringing an ion to the surface, and (2) bringing the surface ion to an interstitial site. Thus the Frenkel energy is also the sum of two energies: $g_f = g_v + g_i$.

The individual defect energies are likely to differ from one another; for instance, in an oxide based on a close-packed anion sublattice the cation vacancy formation energy may be less than the oxygen vacancy formation energy. In the fluorite structure where the cations are close-packed, the reverse is likely to be true. Furthermore, the energy of a surface ion depends greatly on its local atomic environment, and it is expected that the formation energies of individual defects will differ between surfaces of different orientations and grain boundaries and dislocations of different structures. (Little is currently known about the details of this relationship.) Differences in the formation energies for individual defects cause the concentrations of ions on the surface to differ from those in the bulk. The surface (or grain boundary or dislocation) becomes nonstoichiometric and bears a net charge. For overall electrical neutrality, the surface charge is compensated by an adjacent *space-charge* layer, which decays with distance into the crystal over

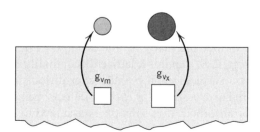

Fig. 2.18 Formation of anion and cation vacancies at a free surface. A difference in the free energies of formation g_{v_m} and g_{v_x} can lead to a nonstoichiometric cation/anion ratio at the surface and the formation of excess surface charge.

2.4 Interactions Between Point Defects and Interfaces

dimensions that are typically a few to several hundred angstroms. Within the bulk, electroneutrality is preserved despite the redistribution of defects at the surface. At equilibrium, there then exists an electrostatic potential difference between the surface and the interior.

Intrinsic Potential

What is the magnitude of this electrical potential difference? In an ideal, pure compound in which Schottky defects are dominant, such as MgO, the equilibration of the cation and anion with the surface can be written as the following defect reactions (see Fig. 2.18):

$$Mg_{Mg}^x = Mg_{surf}^{\cdot\cdot} + V_{Mg}'' \qquad [Mg_{surf}^{\cdot\cdot}][V_{Mg}''] = K_{S1} \qquad (2.72)$$

$$O_O^x = O_{surf}'' + V_O^{\cdot\cdot} \qquad [O_{surf}''][V_O^{\cdot\cdot}] = K_{S2} \qquad (2.73)$$

Notice that the energy for the sum of these two reactions is the Schottky energy; that is,

$$Mg_{Mg}^x + O_O^x = Mg_{surf}^{\cdot\cdot} + O_{surf}'' + V_{Mg}'' + V_O^{\cdot\cdot}$$

is the same as

$$null = V_{Mg}'' + V_O^{\cdot\cdot}$$

if the surface ions simply extend the perfect lattice. The energies for Eqs. 2.72 and 2.73 are the vacancy formation energies, $g_{V_{Mg}}$ and g_{V_O}. If the energy for the Mg vacancy formation reaction (Eq. 2.72) is the lower, the surface is enriched in the cation [$Mg_{surf}^{\cdot\cdot}$], and, through the Schottky equilibrium $K_s = [V_{Mg}''][V_O^{\cdot\cdot}]$, the surface oxygen concentration [O_{surf}''] will at the same time be depressed. The concentration of each type of vacancy then varies with distance in the space-charge layer according to a spatially-varying potential $\phi(x)$, as shown in Fig. 2.19

$$[V_{Mg}''](x) = \exp\left[\frac{-g_{V_{Mg}} + 2e\phi(x)}{kT}\right] \qquad (2.74)$$

$$[V_O^{\cdot\cdot}](x) = \exp\left[\frac{-g_{V_O} - 2e\phi(x)}{kT}\right] \qquad (2.75)$$

For convenience, we use a reference for the potential $\phi(x)$ that is zero at the surface and ϕ_∞ far into the bulk, whereupon the sign of the space-charge potential is the same as that of the accumulated defects. Since far from the surface the electroneutrality condition $[V_{Mg}'']_\infty = [V_O^{\cdot\cdot}]_\infty$ still holds, from Eqs. 2.74 and 2.75 we have:

$$e\phi_\infty = \frac{1}{4}\{g_{V_{Mg}} - g_{V_O}\}. \qquad (2.76)$$

158   Chapter 2 / Defects in Ceramics

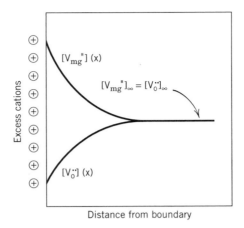

Fig. 2.19 Spatial distribution of defects in the near-surface region of intrinsic MgO. It is assumed that $g_{V_{Mg}} < g_{V_O}$, such that the surface (or grain boundary or dislocation) has an excess of cations and the space charge an excess negative charge.

Therefore, in the intrinsic case the potential difference between the surface and bulk is determined by the *difference* in vacancy formation energies, and can be approximately a fraction of a volt. In MgO $e\phi_\infty$ is probably negative in sign, although accurate values for individual defect energies are not readily available. The spatial distribution of defects in this potential is shown in Fig. 2.19.

Extrinsic Potential

Because the origin of the surface charge is the equilibration of surface ions with lattice defects, changes in the bulk defect structure cause changes in the sign and magnitude of this potential. For a material in which Schottky defects dominate, the space charge potential changes when bulk concentrations of vacancies are altered by extrinsic solutes or nonstoichiometry. For instance, a trivalent substitutional cation in MgO raises the lattice concentration of cation vacancies $\left(eg., \left[Al_{Mg}^{\cdot}\right] = 2\left[V_{Mg}''\right]\right)$. Referring to Figure 2.18, we see that an increase in the bulk vacancy concentration will, upon equilibration, depopulate the surface of cations, resulting in an excess negative surface charge (composed of an excess of oxygen ions) and a reversal in the sign of the electrostatic potential. Since the spatial distribution of cation vacancies is still given

2.4 Interactions Between Point Defects and Interfaces

by Eq. 2.74, far from the surface we have

$$[V''_{Mg}]_\infty = \frac{1}{2}[Al^{\cdot}_{Mg}]_\infty = \exp\left[\frac{-g_{V_{Mg}} + 2e\phi_\infty}{kT}\right] \quad (2.77)$$

and the potential difference between surface and bulk is given by:

$$e\phi_\infty = \frac{g_{V_{Mg}}}{2} + kT \ln[Al^{\cdot}_{Mg}] - kT \ln 2 \quad (2.78)$$

which for extrinsic concentrations of $[Al^{\cdot}_{Mg}]$ and reasonable values of $g_{V_{Mg}}$ will be positive. The defect distribution across the space charge in this extrinsic case is shown in Fig. 2.20; of special importance is the segregation of the solute $Al^{\cdot}_{Mg}$ in the near-surface space-charge region. This accumulation is distinct from the adsorption of solutes and impurities that can also occur at a surface or grain boundary due to the elastic strain energy resulting from size misfit, or other chemical driving forces. Figure 2.21 shows that $Sc^{\cdot}_{Mg}$, a solute with a different valence but virtually no size misfit, is segregated over a distinctly wider distance near the

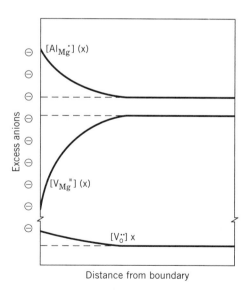

Fig. 2.20 Spatial distribution of solute and defects near the surface of MgO with extrinsic doping levels of Al_2O_3. The sign of the surface and space charge respectively are reversed from the intrinsic case in Fig. 2.19, and segregation of takes place in the space charge.

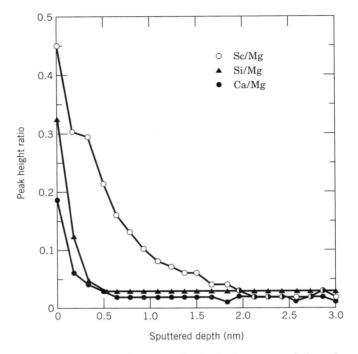

Fig. 2.21 Concentration versus depth of solutes at a grain boundary in MgO. Results from Auger spectroscopy during ion-sputtering of a grain boundary exposed by fracture. A wider distribution is observed for the trivalent solute $Sc_{Mg}^{\cdot}$ segregated in a space charge layer, compared to Ca and Si, which adsorb at the core of the boundary. [From Y. -M. Chiang, A. F. Henriksen, D. Finello, and W. D. Kingery, *J. Am. Cer. Soc.*, 64[7] 383 (1981).]

grain boundary in MgO than the solutes Ca and Si, which are presumed to adsorb at the structurally disordered "core" of the boundary.

Based on this discussion, it may be apparent that the addition of extrinsic oxygen vacancies will result in the depopulation of anions from the surface, and the formation of a larger negative potential $e\phi_\infty$ than in the intrinsic case. At a single temperature, the potential can vary from positive to negative with defect concentration. Somewhere between the negative intrinsic potential and the positive extrinsic potential lies a point of zero potential known as the *isoelectric point*. (This is directly analogous to the isoelectric point that exists for surfaces of powder particles in a ceramic slurry. The surface potential controls the dispersion of electrically charged particles in a suspension.) The isoelectric point is dependent on temperature as well as solute concentration through relationships analogous to Eq. 2.78. In Fig. 2.22 are shown results for the migration of a grain boundary of simple tilt misorientation in an NaCl bicrystal. This is a "grain boundary electrophoresis" experiment, in which the direction of motion of the boundary directly

2.4 Interactions Between Point Defects and Interfaces 161

reveals the excess charge on it. From the observation that the direction of migration is reversed between an experiment at 600°C and 640°C, it can be deduced that the isoelectric point for this boundary (bulk impurity content ~2 ppm) lies between these two temperatures.

Another system in which the effects of a space-charge are particularly clear is rutile, TiO_{2-x}. Segregation of solutes of positive and negative effective charge (i.e., donors and acceptors) depends on the overall doping level, temperature, and oxygen pressure. Figure 2.23 shows a variety of segregation results, which have been observed in samples in which a donor and acceptor are simultaneously present, but the relative amounts of the two are varied to change the net doping level. Notice that even a very pure TiO_2 cannot be "intrinsic" in the same sense as a stoichiometric oxide such as MgO; it is still defective due to nonstoichiometry, with an anion deficiency largely accommodated by titanium

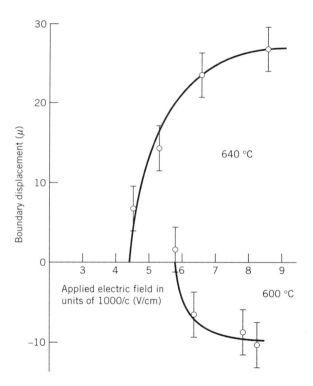

Fig. 2.22 Migration of a tilt grain boundary (edge dislocation array) in NaCl under applied electric field. The change in direction of motion between 600°C and 640°C reflects a change in the sign of the boundary electrical potential. [From R. J. Schwensfir and C. Elbaum, J. Phys. Chem. Solids, 28, p. 297 (1967).]

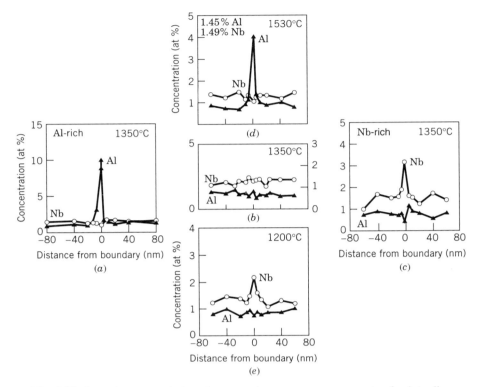

Fig. 2.23 Scanning transmission electron microscope measurements of solute distributions across grain boundaries in TiO_2 containing varying amounts of the co-dopants Al and Nb. The effective charge of the segregated/depleted solute varies with the net doping and temperature. (b) represents a sample very near the isoelectric point. [From J. A. S. Ikeda, Y.-M. Chiang, A.J. Garratt-Reed and J.B. Vander Sande, *J. Am. Ceram. Soc.*, 76[10], 2447 (1993).]

interstitials (with possibly some oxygen vacancies). At high temperatures, the excess of cation interstitials in undoped TiO_2 results in a cation-rich grain boundary, and a negative space-charge potential. The same is true of an acceptor-doped TiO_2, in which the acceptors are compensated by cation interstitials. In the negative space charge, defects of negative effective charge such as acceptors and electrons are accumulated, while defects of positive effective charge such as donors and cation interstitials are depleted, Fig. 2.24a and 2.24b. This results in the acceptor segregation observed in Fig. 2.23a and 2.23d (slight donor depletion can also be detected). On the other hand, when there is an excess of the donor Nb_2O_5, it is compensated in the lattice by titanium vacancies, causing a reversal in the sign of the potential, Fig. 2.24c. In this kind of composition the segregation of $Nb^{\cdot}_{Ti}$ is seen (see Figs. 2.23c and 2.23e). Finally, when the solute concentration and temperature are varied so as to "titrate" the grain boundary potential to the isoelectric point, segregation of neither solute occurs, as shown

2.4 Interactions Between Point Defects and Interfaces

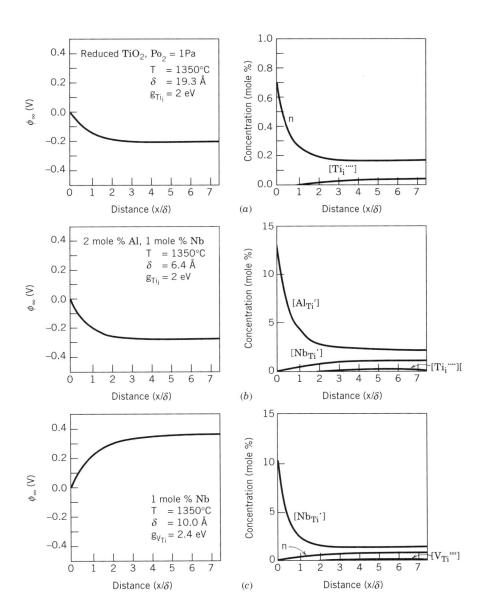

Fig. 2.24 Calculated space charge potential and defect spatial distributions for TiO_2. [From J.A.S. Ikeda and Y.-M. Chiang, *J. Am. Ceram. Soc.*, **76**[10], 2437 (1993).]

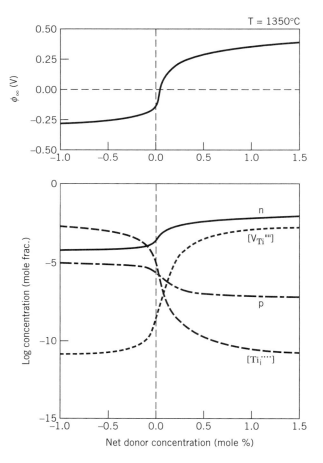

Fig. 2.25 Lattice defect concentration versus acceptor and donor doping in TiO_2 at 1350°C in air. The space charge potential, shown at top, is predicted to vary from negative to positive sign with increasing donor doping, showing an isoelectric point at slight donor doping. [From J.A.S. Ikeda and Y.-M. Chiang, *J. Am. Ceram. Soc.*, 76[10], 2437-2446 (1993).]

in Fig. 2.23*b*. At a constant oxygen activity and temperature, the concentrations of defects and the space charge potential varies with the nature of the doping as shown in Fig. 2.25.[9] Experimental observations, Fig. 2.26, show that the potential varies with doping as indicated by the defect model. Notice that the isoelectric point appears slightly to the donor-doped side. This is because at high temperatures there is sufficient intrinsic reduction, even in quite oxidizing ambient conditions, that some donor doping is necessary to compensate for the defects introduced by reduction. Thus the position of the isoelectric point, and the mag-

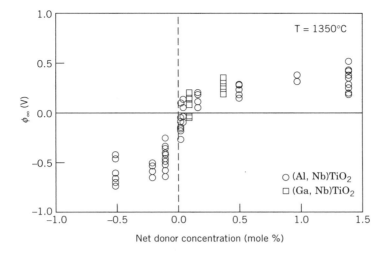

Fig. 2.26 Experimental measurements of the space charge potential in TiO_2 corresponding to theoretical model in Fig. 2.25. [From J. A. S. Ikeda, Y. -M. Chiang, A. J. Garratt-Reed and J. B. Vander Sande, *J. Am. Ceram. Soc.*, **76**[10], 2447 (1993).]

nitude of the electrical potential, varies systematically with oxygen activity and temperature due to changes in the lattice defect structure.

The existence of a space charge and the accumulation of defects and solutes at interfaces impact a number of interfacial properties, some of which are only now being explored. These include surface and grain boundary diffusion, interfacial energies, grain growth rates, hardness in plastically deformed ionic systems (especially alkali halides), and the electrical properties of grain boundaries.

2.5 LINE AND PLANAR DEFECTS

Several types of extended, larger scale imperfections are of importance in crystalline ceramics. Dislocations are line defects, which are generally less common in ceramics than in metals due to a higher energy of formation, but are nonetheless important to a number of physical properties, most notably high-temperature plastic deformation. Many varieties of planar defects exist in single crystalline and poly-

---

[9] To continue the analogy between the ionic space charge potential and the potential at the surfaces of colloidal particles, in the solid state, the electrostatic potential varies with doping due to equilibration of the surface with lattice defects in solid solution, while in colloid chemistry, the "zeta potential" at the surfaces of particles in aqueous solution varies with pH, and with salt additions, due to chemical equilibration of the surface with ions in liquid solution. Titration curves for the zeta potential as a function of pH are very much like the titration curve as a function of acceptor/donor doping shown in Fig. 2.26.

crystalline materials. The free surface is a planar defect at which the lattice terminates, with an atomic structure and composition that generally deviates from that of the lattice. Internal planar defects include twin boundaries, stacking faults, grain boundaries (the interface between two like crystals), interphase boundaries (between dissimilar crystals), and crystal-glass interfaces. Many grain boundaries in ceramics include a thin layer of amorphous material separating the crystals. The structure, composition, and properties of surfaces and grain boundaries are particularly important to the development of microstructure and properties in polycrystalline and polyphase ceramics, as discussed in Chapter 5.

As with point defects, some expenditure of energy is necessary to form a line or planar defect, resulting in an excess of energy per unit length or per unit area (the latter is the interfacial energy). However, unlike point defects there is no equilibrium concentration of line or planar defects since the additional entropy created is not large enough compared to the formation energy for an overall lowering of free energy. The field of study encompassing interfacial structure and properties is virtually as large and as varied as that concerning crystalline materials, and is impossible to discuss in much detail in a text of this size. In this section, we give a basic physical description of the main types of line and planar defects encountered in ceramics.

Dislocations

A dislocation is a displacement in a perfect crystal lattice, which result from a virtual shearing process, as illustrated in Fig. 2.27. An edge dislocation appears as an extra half-plane of atoms, Fig. 2.27*a*, in which the dislocation *line* is the edge of the half-plane. Upon completing a "stepwise circuit" around the edge dislocation, one falls short by a vector **b** (Fig. 2.27*a*); this vector is termed the *Burgers vector*, and for an edge dislocation is normal to the dislocation line. A screw dislocation is arrived at by the virtual shearing process shown in Fig. 2.27*b*. Completing the Burgers circuit, one finds the Burgers vector to be *parallel* to the dislocation line. The relative directions of **b** and the dislocation line therefore define a pure edge or pure screw dislocation. However, dislocations of mixed edge and screw character are also possible, as shown in Fig. 2.27*c*.

Dislocations are generally formed by stresses or crystal growth accidents. Plastic deformation under an externally applied stress results in the formation and/or propagation of dislocations. Internal stresses due to, for instance, thermal expansion mismatch between a precipitate and a crystal can result in the creation of dislocations. Dislocations are often grown into crystals as a continuation of dislocations in the starting or "seed" crystal, and the presence of screw dislocations can facilitate the growth of certain crystals and thin films by a spiral growth process requiring no nucleation of a new phase (see Section 5.6). The type and density of dislocations that result in any material are directly related to the energy required to form the dislocation. In the immediate vicinity of the dislocation line, bonds are highly strained, and atomic positions are displaced in comparison to the prefect

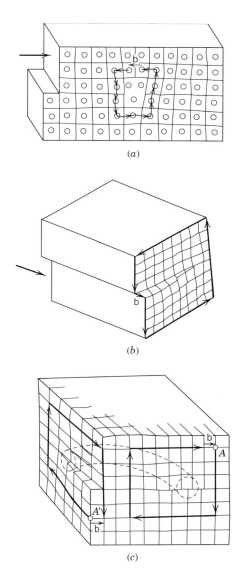

Fig. 2.27 (*a*) Pure edge and (*b*) pure screw dislocations occurring during plastic deformation. (*c*) Combination edge and screw dislocation. Burgers vector **b** shown for pure screw and for pure edge. Dislocation line connecting these is shown.

crystal, resulting in an excess line energy per unit length, or line tension. If we treat the region surrounding the dislocation as a strained elastic medium (there is also an excess energy due to the disordered "core"), the strain energy is given approximately by

$$E = \alpha G b^2 \qquad (2.79)$$

where G is the shear modulus of the material, **b** is the magnitude of the Burgers vector, and α is a factor that depends on the type of dislocation, having a value in the range of 0.5–1.0. Since the dislocation energy is proportional to b^2, of the various dislocations which may form in a material, the preferred ones will be those of shortest Burgers vector. This relationship in Eq. 2.79 allows us to understand why dislocations are less often seen in ceramic materials than in metals. In ionic solids, the Burgers circuit must be completed in a way that does not leave a net charge on the dislocation, which requires that the extra half-plane or planes contain both anions and cations. The shortest Burgers vector in the rocksalt structure is thus 1/2 of the 110 vector, as illustrated in Fig. 2.28, and is larger than it would be for a comparable fcc metal. Dislocations of large **b** sometimes separate into *partial dislocations* with Burgers vectors, the vector sum of which is **b**, in order to lower the excess energy. Fig. 2.29 shows the Burgers vector in the basal plane of corundum, given by **b**. Separation of **b** into the partial dislocations **b'** and **b''** lowers the net elastic energy given by Eq. 2.79. However, this separation also

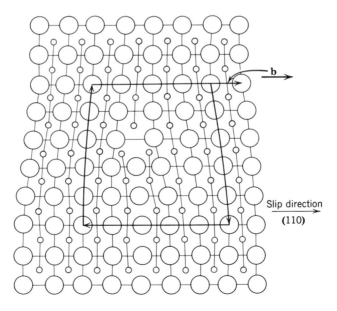

Fig. 2.28 Edge dislocation structure in MgO showing slip direction, Burger's circuit, and Burger's vector **b**.

2.5 Line and Planar Defects   169

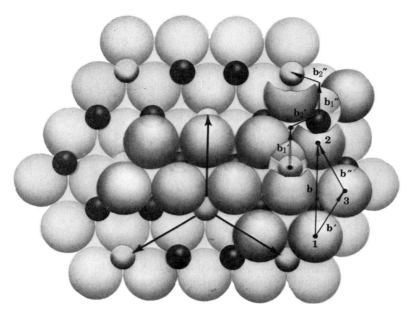

Fig. 2.29 Structure of Al_2O_3 showing two layers of large oxygen ions with hexagonal array of Al^{3+} and vacant octahedral interstitices. Slip directions and Burgers vector **b** for basal plane slip are indicated.

results in a defective region between the partials, known as a stacking fault, which contains an error in the stacking sequence of atomic layers. The net energy balance must include the increase in energy due to this planar defect; and the appearance of partial dislocations is an indication that the net energy is lowered.

The fact that a greater stress is necessary to introduce dislocations into a ceramic crystal than into metals, combined with crystallographic restrictions on the directions in which dislocations can move in a complex crystal, cause crystalline ceramics to be less ductile than most metals at equivalent fractions of the melting point. Nonetheless, the deformation processes are identical. The process known as slip, illustrated in Fig. 2.30, results from the *glide* of dislocations across the crystal when a critical stress is reached. This critical stress, called the Peierls stress, is related to the strength of atomic bonding but is much less than a theoretical shear stress. For an edge dislocation, glide can be visualized as the sequential displacement of the extra half-plane of atoms in the direction of the Burgers vector, one lattice distance at a time, across the crystal. The net result is a shear displacement of the half-crystal by one Burgers vector relative to the other half-crystal, without requiring the simultaneous separation of many bonds, a process of prohibitively high energy. After some amount of slip the surface of a single crystal will often show surface features known as slip bands oriented normal to the direction of slip, Fig. 2.31. Another process of dislocation motion

170   Chapter 2 / Defects in Ceramics

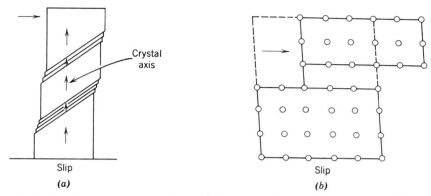

Fig. 2.30 Schematic representation of (a) macroscopic and (b) microscopic slip.

is *climb*, Fig. 2.32, which for an edge dislocation is the shortening or extension of the half-plane of atoms by the diffusion of atoms to or from the dislocation line. Climb is a slower process than glide since it is controlled by atom diffusion, and is generally important at higher temperatures.

Slip in an ionic crystal is easier in some crystal directions than others, due to electrostatic considerations. In the rocksalt structure, the $1/2\langle 110 \rangle$ Burgers vector is the shortest, but an edge dislocation with this Burgers vector can terminate on a variety of crystal *glide planes*, two of which are illustrated in Fig. 2.33. In Fig. 2.33a, glide occurs on the $\{110\}$ plane and in the $\langle 110 \rangle$ direction. This combination of a glide plane and direction is known as a slip system: In this case it is $\{110\} \langle 110 \rangle$. Fig. 2.33b shows glide on the $\{100\}$ plane: The slip system is $\{100\} \langle 110 \rangle$. A close comparison of the two shows that in glide on the $\{110\}$ plane, cations slip over anions, and vice versa, throughout the translation process. In glide on the $\{100\}$ plane, however, cations must pass over cations, and anions over anions, halfway through the Burgers vector translation. This results in an unfavorable electrostatic repulsion. Thus $\{110\} \langle 110 \rangle$ slip is preferred, although at higher temperatures the $\{100\} \langle 110 \rangle$ and other slip systems also become active in the rocksalt structure.

Slip occurs when the Peierls stress for dislocation glide is exceeded. When a stress is applied to a crystal with a randomly chosen orientation, it is only that component of the stress resolved onto the slip plane and in the slip direction that is useful in causing slip. In a polycrystalline material where the orientation of all crystalline grains is randomly determined, the ease of deformation thus depends on the number of glide planes and glide directions that are available. Rocksalt structure ceramics (e.g., NaCl, MgO, TiC) tend to become ductile at lower homologous temperature than ceramics of more complex structure and lower symmetry (e.g., corundum), due to the availability of multiple, independent slip systems at lower temperatures.

2.5 Line and Planar Defects 171

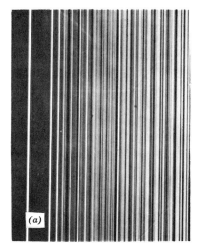

Fig. 2.31 (*a*) Deformation markings on surface of bent MgO crystal showing group of (110) deformation offsets in region of increasing deformation, going from left to right (175X). (Courtesy of M. L. Kronberg, J. E. May, and J. H. Westbrook.) (*b*) Etched cross section of (110) deformation bands in bent MgO crystal (130X). (Courtesy T. L. Johnston, R. J. Stokes, and C. H. Li.)

Grain Boundaries

The simplest type of internal planar defect is the *twin boundary*, which can be introduced into a perfect crystal by the homogeneous shear process illustrated in Fig. 2.34. The orientation of crystal planes is symmetric about a twin boundary and there is no loss of continuity of the lattice planes. A *stacking fault* is another simple type of planar defect across which no atomic bonds are broken, resulting when the normal stacking sequence of atomic layers is disturbed. For example, if the *abcabc* stacking sequence of close-packed layers in the FCC struc-

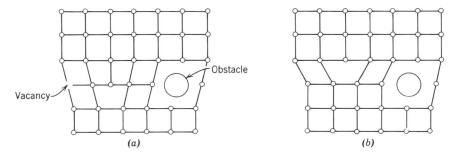

Fig. 2.32 By absorbing vacancies, a dislocation can climb out of its slip plane to where its glide is not hindered by an obstacle.

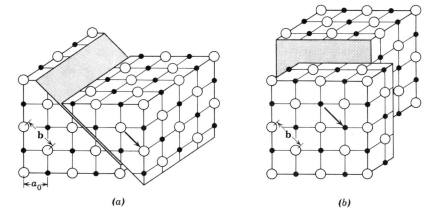

Fig. 2.33 Translation gliding in the <110> direction and on (*a*) the {110} plane and (*b*) the {100} plane for crystals with the rocksalt structure. {110} <110> glide is preferred.

ture is momentarily disturbed by a layer with the *abab* stacking sequence characteristic of the HCP structure, giving *abcabc**abab**cabc*, a stacking fault has resulted.

Grain boundaries are the interfaces between like crystals, at which atomic planes are always disrupted to some extent. Grain boundaries are often classified as either *special boundaries* or *general boundaries*, being defined by the crystallographic uniqueness of the relative orientations of the two grains. The interface between the crystals is defined by the relative rotational misorientation of the two grains, the orientation of the grain boundary plane between them, and any translations of the two crystals relative to one another. In synthesizing a grain boundary from two crystals, there is little control over the last of these, but one can often select both the misorientation between grains and the initial orientation of the boundary plane (e.g., by cutting each crystal on a particular crystal plane and joining the two).

Special Boundaries Special grain boundaries are those in which there is a particular misorientation relationship between the two crystals and boundary plane orientation allowing a simplified structural description of the boundary. In a microstructure where all misorientation relationships are chosen at random, these occur with lesser statistical frequency than do general, high-angle grain boundaries (i.e., those of no special misorientation), but since they are simpler and more easily defined, they have been widely studied. One type of special grain boundary is a low-angle boundary, which results when the two crystals have only a slight misorientation relative to one another. A low-angle *tilt boundary* is composed of an array of parallel edge dislocations, illustrated in Fig. 2.35, and has an excess

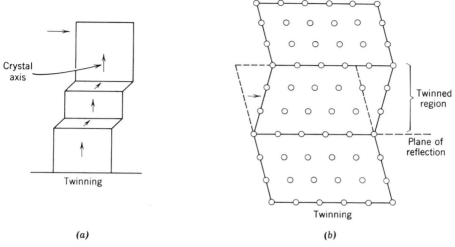

Fig. 2.34 Schematic representation of (a) macroscopic and (b) microscopic twinning.

energy given by the sum of the dislocation energies. Geometrically, the spacing between edge dislocations is given by

$$D = \frac{\mathbf{b}}{\sin\theta} \approx \frac{\mathbf{b}}{\theta} \qquad (2.80)$$

where θ is the misorientation angle. Up to a tilt angle of about 20°, the model of a low-angle tilt boundary as an array of dislocations is reasonable. At higher angles where the dislocations are no longer well separated, this model becomes physically unrealistic. A low-angle *twist boundary* occurs when two crystals are rotated slightly about a common axis which is normal to the plane of the boundary (Fig. 2.36), and for a cubic crystal results in the formation of a square net of screw dislocations separating regions of good lattice matching (notice that for an equal misorientation angle, twist boundaries have twice the dislocation density of tilt boundaries). Low-angle grain boundaries with purely tilt and purely twist misorientations are limiting cases; it is also possible for a low-angle boundary to have both tilt and twist components. A dislocation network like that shown in Fig. 2.17 can be considered to be a low-angle boundary. Often these appear in a deformed and annealed single crystal or within the grains of a polycrystal, and are also known as subgrain boundaries.

Another type of special grain boundary that is not restricted to low-misorientation angles is that containing a high density of coincident atomic sites. With increasing tilt misorientation, an angle is periodically reached at which a relatively high number of the atoms across the boundary plane fall

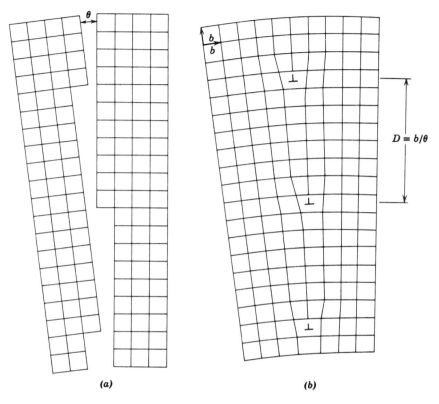

Fig. 2.35 Low-angle tilt boundary. (From W. T. Read, *Dislocations in Crystals*, McGraw-Hill: New York, 1953.)

into coincidence. The same is true for twist boundaries. A commonly used nomenclature for boundaries of this type is the CSL description, standing for Coincident Site Lattice, in which the notation "Σ" gives *the reciprocal of the fraction of sites which are coincident*. The CSL is actually a three-dimensional lattice, obtained when two interpenetrating crystals are rotated with respect to one another. Fig. 2.37a shows two simple cubic lattices (closed and open points, respectively) that have been rotated about an $\langle 001 \rangle$ axis by 36.87°, giving a CSL lattice in which one out of every five lattice sites of each grain is coincident with the other; i.e., $\Sigma = 5$. From this three-dimensional lattice, the limiting case of a pure twist boundary is obtained when the boundary plane is normal to the twist axis, as shown in Fig. 2.37b (a $\Sigma = 5$ twist boundary), and a symmetric tilt boundary is obtained when the plane of the grain boundary is parallel to the rotation axis as shown in Fig. 2.37c (a $\Sigma = 5$ symmetric tilt boundary). In an ionic structure it is also necessary to avoid too close an approach of ions of like charge. Fig. 2.38 shows a schematic of a possible ar-

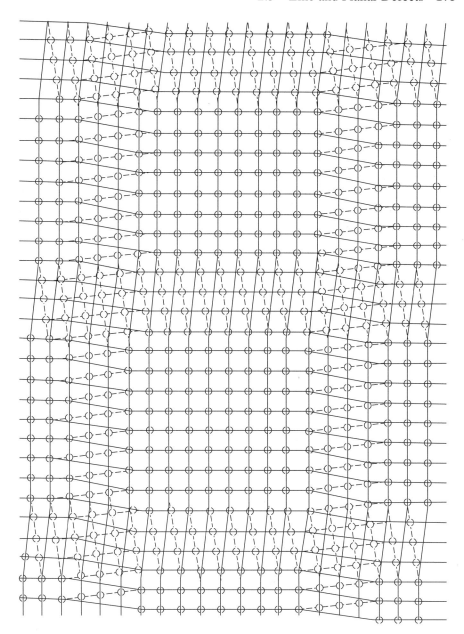

Fig. 2.36 A pure twist boundary, in which the boundary plane is parallel to the plane of the figure. The two grains have a small relative rotation about their cube axis, which is normal to the boundary. Regions of coincidence between atoms in the two grains are separated by two sets of screw dislocations running in horizontal and vertical directions. (From W. T. Read, *Dislocations in Crystals*, McGraw-Hill: New York, 1953.)

rangement for the $\langle 001 \rangle$, $\phi = 36.8°$, $\Sigma = 5$ tilt boundary in the rocksalt structure, the atom positions indicated by a computer simulation for NiO, and a high-resolution transmission electron microscope image of an actual boundary. In detail, the observed boundary has lower expansion normal to the boundary plane and greater relative translation of the two crystals parallel to the boundary than the theoretical construct. Special tilt boundaries have a characteristic structural periodicity that is apparent in Fig. 2.38, and also a lower boundary energy than non-coincident boundaries. A classic experiment demonstrating that CSL boundaries posses lower energy was conducted by Chaudhari and Matthews [*J. Appl. Phys.*, **42**, 3063, (1971)], (see Fig. 2.39). When magnesium is burned to form small cubic MgO smoke particles, the particles aggregate with considerable freedom to form CSL twist misorientations, illustrating that they are low-energy configurations.

General Boundaries

General boundaries are defined by what they are not: They are boundaries with no special misorientation relationship or boundary plane orientation. The adjoining atomic planes are *incommensurate* in their spacing, meaning each is not a multiple of the other, and as a result the structure of general grain boundaries is aperiodic. While the misorientation angle does not need to be high in order to satisfy these requirements, the most frequently encountered boundaries in polycrystalline materials are high-angle general boundaries. Little is currently known about the atomic level structure of general grain boundaries, but it seems certain that not all general grain boundaries are alike; additional classifications in terms of the structure, energy, and properties of this class of grain boundary are likely to emerge with further study.

Boundary Films

It was once thought that grain boundaries contained a highly disordered glasslike layer, and that the region of disorder extended as far as a micron from the boundary plane. As the resolutions of observations have improved, primarily through the development of high-resolution transmission electron microscopy, this view has been shown to be largely incorrect. At both special and general boundaries, the atomic order of the lattice has been found to be preserved up to within approximately a unit cell of the dividing plane. Thus the "core," or disordered region of a grain boundary, is typically only 0.5–1 nm wide, although it does vary somewhat with the type of boundary and the crystal lattice periodicity. However, high-resolution imaging has also shown that there is another type of grain boundary that frequently occurs in ceramic materials. These contain a thin film of amorphous material of a different composition, often a silicate glass. Grain boundaries containing a thin (0.5–2 nm) thick amorphous silicate film have been observed in

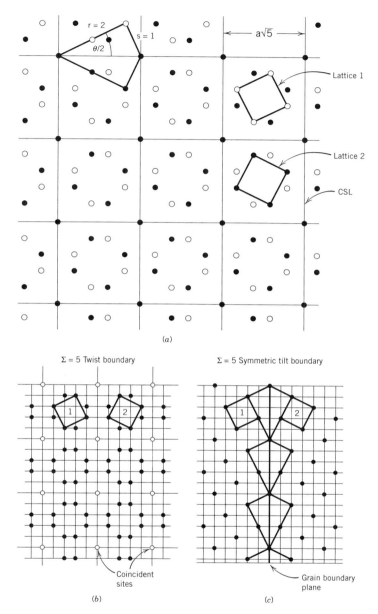

Fig. 2.37 (a) Σ=5 coincident site lattice, viewed along [001], obtained by a 36.87° rotation about the [001] direction of two interpenetrating crystals initially with exact coincidence. Atoms in the two crystals are indicted by filled and solid circles, respectively. (b) A Σ=5 twist boundary is represented by a plane parallel to the figure above which only one crystal and below which only the other crystal is present. (c) A Σ=5 symmetric tilt boundary is represented by a plane normal to the figure bisecting the angle of misorientation between the two crystals.

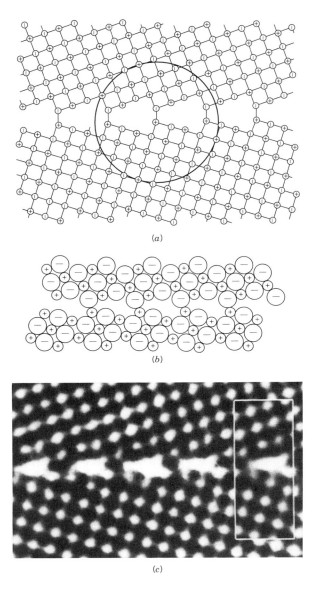

Fig. 2.38 (a) An idealized model of a $\Sigma = 5$ symmetric tilt boundary in the rocksalt structure, in which the rotation axis is [001] and the boundary plane is the (310) plane of the respective crystals, and (b) The relaxed structure of the $\Sigma = 5$, (310)/[001] tilt boundary in NiO as calculated by D. M. Duffy and P. W. Tasker [Phil. Mag. A, 48, 155 (1983)] showing a small relative translation of the two grains. (c) High-resolution transmission electron microscope image of a real $\Sigma = 5$ grain boundary in NiO, which shows less expansion normal to the boundary plane and greater lateral translation than the model in (b). The inset shows the image averaged over many of the periodic structural units. (Courtesy K. L. Merkle, Argonne National Laboratory.)

2.5 Line and Planar Defects

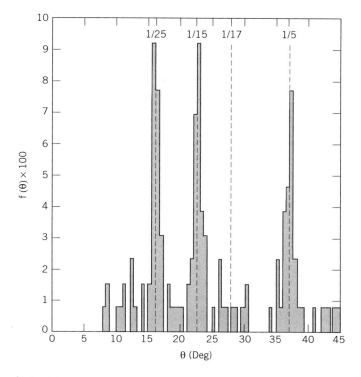

Fig. 2.39 Histogram of f(θ), the fraction of twist boundaries at angle θ, against θ, for MgO smoke particles collected above hot and rapidly burning magnesium rods. Twist boundaries corresponding to θ=0 are not included. The histogram is drawn for 1/2° intervals. [From P. Chaudhari and I. W. Matthews, *J. Appl. Phys.*, **42**, 3063 (1971).]

a number of materials, including silicon nitride (Fig. 2.40), lead ruthenate thick film resistors (Fig. 2.41), aluminum oxide, and zirconia.

There is evidence that the thickness of these films sometimes reaches a constant, equilibrium value. That is, regardless of the volume fraction of silicate glass in the system, the thickness of the amorphous layer remains the same (the excess accumulates elsewhere in the microstructure, such as at three- and four-grain junctions). An explanation has been proposed [D. R. Clarke, *J. Am. Ceram. Soc.*, **70**[1], 15–22 (1987)], which is based on an energy minimum being reached at a finite thickness due to a balance between attractive van der Waals force between the grains and a short-range (1-nm scale) repulsion presented by the resistance to deformation of the silicate glass structure. In silicon nitride and ruthenate-glass resistors, respectively, the high-temperature mechanical properties and the electrical properties are profoundly influenced by the presence and the properties of

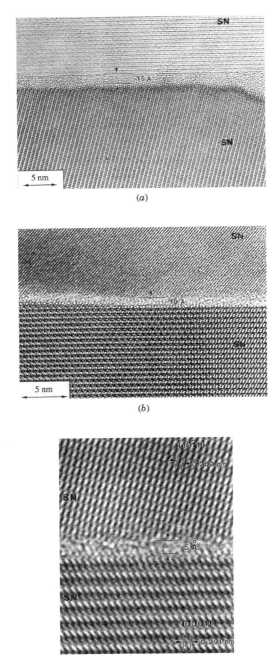

Fig. 2.40 High-resolution electron microscope images showing thin amorphous films at grain boundaries in sintered silicon nitride. The oxide additives and corresponding thickness of the intergranular films are (a) Y_2O_3, 1.5 nm; (b) Yb_2O_3, 1.0 nm; and (c) ZrO_2, 0.5 nm, respectively. [From H. J. Kleebe, Hoffman, and Rühle, Z. Metallkd., **83**, 8, (1992).]

2.5 Line and Planar Defects   181

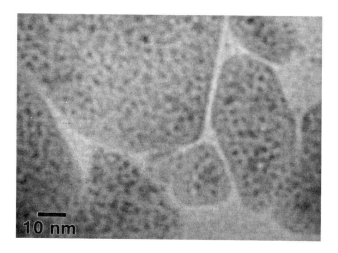

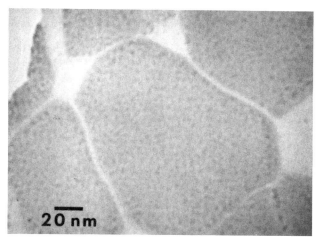

Fig. 2.41 Thin, glassy films (1–2-nm thickness) separating conductive lead-ruthenate grains in a thick-film resistor containing 20 vol % particles in highly modified silicate glass. [From Y. -M. Chiang, L. A. Silverman, R. H. French and R. M. Cannon, *J. Am. Ceram. Soc.*, **77**[5], 1143 (1994).]

this thin glass film. In a variety of other materials, thicker glass films (10 nm to several microns) are also observed, which are also important for microstructure development and properties. However, these appear to represent a different regime of behavior where the film thickness is not a constant, but fluctuates with the total amount of glass present.

ADDITIONAL READING

General

F. A. Kröger, *The Chemistry of Imperfect Crystals*, North-Holland Publishing Company, Amsterdam, 1964.

W. D. Kingery, H. K. Bowen, D. R. Uhlmann, *Introduction to Ceramics,* 2nd Edition, John Wiley and Sons, New York, 1976.

W. D. Kingery, Editor, *Structure and Properties of MgO and Al_2O_3 Ceramics, Advances in Ceramics Vol. 10*, The American Ceramic Society, Columbus, Ohio, 1984.

A. Kelly and G. W. Groves, *Crystallography and Crystal Defects*, Addison-Wesley, Reading, Massachusetts, 1970.

K. Nassau, *The Physics and Chemistry of Color*, John Wiley and Sons, New York, 1983.

Band Theory and Electronic Structure

C. Kittel, *Introduction to Solid State Physics*, 6th Edition, John Wiley and Sons, New York, 1986.

G. Burns, *Solid State Physics*, Academic Press, New York, 1985.

L. L. Hench and J.K. West, *Principles of Electronic Ceramics*, John Wiley and Sons, New York, 1990.

S. M. Sze, *Physics of Semiconducting Devices,* 2nd Edition, John Wiley and Sons, New York, 1981.

Computational Methods for Defects

C.R.A. Catlow, Editor, *Defects and Disorder in Crystalline and Amorphous Solids*, Kluwer Academic Publishing, Dordrecht, The Netherlands, 1994.

C. R. A. Catlow and W. C. Mackrodt, Editors, *Computer Simulation of Solids, Lecture Notes in Physics,* vol. 166, Springer, Berlin, 1982.

Nonstoichiometry

Per Kofstad, *Nonstoichiometry, Diffusion, and Electrical Conductivity in Binary Metal Oxides*, Krieger Publications, Malabar, Florida, 1983.

O. Toft Sorensen, *Nonstoichiometric Oxides*, Academic Press, New York, 1981.

Dislocations and Grain Boundaries

D. Hull, *Introduction to Dislocations, 2nd Edition*, Pergamon Press, New York, 1975.

J. P. Hirthe and J. Lothe, *Theory of Dislocations*, McGraw Hill, New York, 1968.

A.P. Sutton and R.W. Balluffi, *Interfaces in Crystalline Materials*, Oxford University Press, Inc., New York, 1995.

D. Wolf and S. Yip, Editors, *Materials Interfaces: Atomic Level Structure and Properties*, Chapman and Hall, London, 1992.

PROBLEMS

1. Write formally correct defect incorporation reactions and the corresponding mass action equilibrium constants for the following. Refer to the table of ionic radii (Table 1.2) for guidance. If you think more than one incorporation mechanism is possible, briefly discuss why and show the different possibilities.
   (i) Solid solution of $CaCl_2$ in $NaCl$.
   (ii) Solid solution of CaO in ZrO_2.
   (iii) Reduction of ZrO_2.
   (iv) Solid solution of $SrTiO_3$ in $BaTiO_3$.
   (v) Solid solution of Y_2O_3 in $BaTiO_3$.
   (vi) Solid solution of Nb_2O_5 in $BaTiO_3$.
   (vii) Solid solution of Al_2O_3 in $BaTiO_3$.

2. In the text, we discuss the transition between vacancy and electron compensation for Nb_2O_5 in TiO_2. Write defect chemical reactions for the limiting cases of ionic and electronic compensation of a trivalent solute, such as Al_2O_3, in TiO_2. Discuss the factors that will determine whether ionic or electronic compensation occurs.

3. LiF has a Schottky formation energy of 2.6 eV and a bandgap of 12 eV. At 500°C, estimate the relative concentrations of ionic and electronic defects, and determine which are dominant on an absolute concentration basis.

4. Construct the Brouwer diagram for a pure binary oxide MO of the rocksalt structure for which, at the temperature of interest, $K_i \gg K_s$. (That is, the bandgap is small relative to the Schottky energy.)

5. Uranium dioxide, UO_2, is of the fluorite structure. The oxygen ions occupy tetrahedral sites in an FCC array of uranium ions, leaving octahedral interstices empty. Some relevant defect formation energies and other useful information are:

$$O_O^x = V_O^{\cdot\cdot} + O_i'' \quad\quad 3.0 \text{ eV}$$
$$U_u^x = V_u'''' + U_i^{\cdot\cdot\cdot\cdot} \quad\quad 9.5 \text{ eV}$$
$$\text{null} = V_u'''' + 2V_O^{\cdot\cdot} \quad\quad 6.4 \text{ eV}$$
$$E_g = 5.2 \text{ eV}$$

   (i) What are the predominant intrinsic ionic defects in *stoichiometric* UO_2 at 1600°C? Calculate their concentrations.
   (ii) Calculate the intrinsic electron and hole concentrations at the same temperature. (Assume that the bandgap decreases with temperature at a rate of ~1 meV/K.)

(iii) UO_2 is easily rendered extrinsic by reduction (to UO_{2-x}) or oxidation (to UO_{2+x}) at high temperatures. Write defect chemical reactions to illustrate these processes. Are your results independent of one another? Explain.

(iv) For undoped UO_2, write the full electroneutrality condition at 1600°C.

(v) What is the Brouwer approximation for *stoichiometric* UO_2 at 1600°C? For strongly reduced UO_{2-x}? Strongly oxidized UO_{2+x}?

(vi) Construct a Brouwer diagram for undoped UO_2 at 1600°C.

6. Determine the Brouwer diagram corresponding to the conductivity data shown for Nb-doped TiO_2 in Fig. ST19.

7. Barium titanate is frequently doped with a small amount of a donor additive such as Nb or La in order to raise the conductivity. It does not matter whether the donor substitutes for Ba or Ti, as long as it is a *shallow* donor. Experiments on a large single crystal of $BaTiO_3$ doped with 0.2% donor show that a good *n*-type semiconductor can be obtained at room temperature after first annealing the crystal at 1400°C in air. However, when the experiment is repeated with a 100-nm-thick film of $BaTiO_3$ with the same dopant and concentration, the thin film is found to be insulating at room temperature. Explain, using the appropriate defect equilibria, why this might have happened.

8. LiF crystallizes in the rocksalt structure type and is highly stoichiometric (see Problem 3). Predict the sign of the excess surface charge and the nonstoichiometry of the surface in highly pure material heated to high temperature. Will it be the same for LiF containing 0.1 mole% CaF_2? Explain.

Chapter 3

Mass and Electrical Transport

Diffusion and electrical conductivity are among the most important properties of ceramic materials; indeed, of many inorganic solids. The diffusion of atoms is generally necessary for changes in microstructure to take place in processes such as the densification of powder compacts, creep deformation at high temperatures, grain growth, and the formation of solid-state reaction products. The electrical conductivity of ceramics varies over a tremendous range, from the most insulating compounds known to man to the most conductive (superconductors). Electrical conduction in ceramics can furthermore occur by the motion of ions as well as electrons. *Ionic conductors* are the basis for many electronic ceramics such as chemical and gas sensors, solid electrolytes, and fuel cells. A widely used example of such a device is the oxygen sensor used to monitor combustion in virtually all modern automobiles, where an ion-conducting ceramic (usually zirconia) placed in the exhaust manifold separates the reducing exhaust gas from the external air, the latter acting as a "reference" gas of fixed oxygen pressure. A voltage is developed across the ceramic due to the difference in oxygen pressure, and this voltage is used in a feedback loop to adjust the fuel/air ratio for optimum combustion.

Atomic diffusion rates and electrical conductivity are largely determined by the type of defects present and their concentrations. As discussed in Chapter 2, a great variety of defects are possible in ceramics, the concentrations of which can vary by many orders of magnitude with changes in composition, temperature, and atmosphere. Consequently, the transport properties of ceramics are considerably more complex than those in simpler solids such as metals and covalent semiconductors. The range of diffusion behavior seen in ceramics is illustrated in Fig. 3.1, where the *diffusion coefficient*, D, of a number of ceramic materials is plotted against temperature. At a given temperature, the diffusion coefficient in ceramics can vary by over 10 orders of magnitude. The temperature dependence, or activation energy, can also vary by over an order of magnitude, from about 0.5 eV to more than 8 eV. The electrical conductivity of ceramic materials varies over an even greater range, about 25 orders of magnitude at room temperature, as shown in Fig. 3.2.

In this chapter, we aim to understand this kind of behavior by considering the relationship between crystal structure, defect structure, and mechanisms of charge and mass transport at the atomic level. We first discuss basic principles governing macroscopic mass and charge transport, and then the diffusion of individual defects and ions. Some of the unique consequences of transport due to charged ionic defects are discussed later in the chapter.

3.1 CONTINUUM DIFFUSION KINETICS

Diffusive mass transport takes place when there is a gradient in the chemical potential (or electrochemical potential) and when the species in question has sufficient mobility. Viewed from a distance, the diffusion process can be characterized in terms of gradients in potential and transport coefficients, without taking into account the atomic nature of defects and ions. This "macroscopic" treatment is known as *continuum diffusion* and is mathematically quite similar to heat transfer. Mass transport occurs in the solid, liquid, and vapor phase, and in addition to being diffusive, can be convective in nature or limited by chemical reaction rates. In the solid state, mass transport processes are generally controlled by atom diffusion, or, less frequently, by the rate of reactions at interfaces.

In continuum diffusion, transport is described by solutions to Fick's first and second laws, under geometric and concentration boundary conditions, which are determined by the experimental configuration. Fick's first law (in one dimension):

$$J = -D\left(\frac{dC}{dx}\right) \tag{3.1}$$

states that the particle flux J (number per unit area per unit time) at a steady state is proportional to the concentration gradient dC/dx. The proportionality constant D is termed the *diffusion coefficient* or *diffusivity*, and is usually written in units of cm^2/sec. The diffusion coefficient is a materials property and is perhaps the most useful parameter for characterizing the rate of diffusive mass transport. It is usually a strong function of temperature and, generally speaking, is a function of composition as well, although in certain limiting cases D can be taken to be independent of concentration.

3.1 Continuum Diffusion Kinetics 187

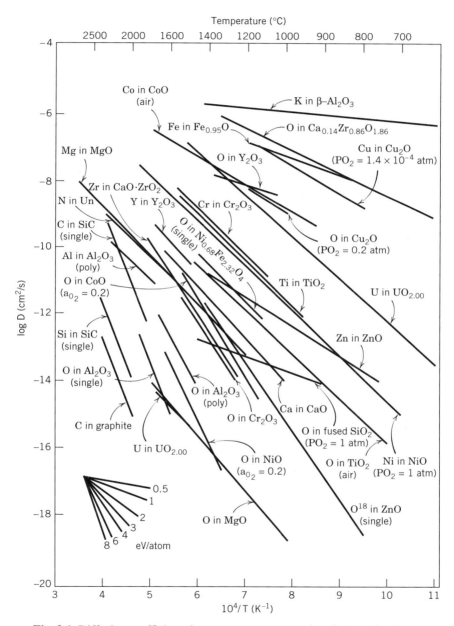

Fig. 3.1 Diffusion coefficients in some common ceramics. Reported values vary over some 10 orders of magnitude. The activation energy Q can be estimated from the slope and the insert.

188 Chapter 3 / Mass and Electrical Transport

Table 3.1. Spatial and Time Distribution of Concentration, $C(x,t)$, for a Few Configurations, Assuming a Concentration-independent Diffusivity, D.

(1) Finite, thin film source containing initial amount of solute M (moles/cm^2), diffusing into an infinite volume of zero initial concentration:

$$C(x,t) = \frac{M}{2\sqrt{\pi Dt}} \exp\left[-\frac{x^2}{4Dt}\right]$$

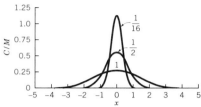

Concentration–distance curves for an instantaneous plane source. Numbers on curves are values of Dt.

(2) Initial, thin film source of width $2h$ and concentration C_o:

$$C = \frac{C_o}{2}\left[\operatorname{erf}\frac{h-x}{2\sqrt{Dt}} + \operatorname{erf}\frac{h+x}{2\sqrt{Dt}}\right]$$

(where erf is the error function)

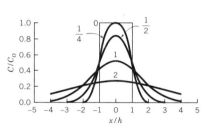

Concentration–distance curves for an extended source of limited extent. Numbers on curves are values of $(Dt/h^2)^{\frac{1}{2}}$.

(3) Finite line source of amount M (moles/cm) diffusing into infinite volume of zero initial concentration:

$$C(r,t) = \frac{M}{4\pi Dt}\exp\left(-\frac{r^2}{4Dt}\right)$$

(4) Finite point source of amount M (moles), diffusing into infinite volume of zero initial concentration:

$$C(r,t) = \frac{M}{8(\pi Dt)^{3/2}}\exp\left(-\frac{r^2}{4Dt}\right)$$

(5) Spherical source of initial concentration C_o and radius a diffusing into infinite medium of zero initial concentration:

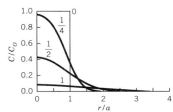

Concentration distributions for a spherical source. Numbers on curves are values of (Dt/a^2).

$$C = \frac{C_o}{2}\left[erf\frac{a-r}{2\sqrt{Dt}} + erf\frac{a+r}{2\sqrt{Dt}}\right] - \frac{C_o}{r}\left[\left(\frac{Dt}{\pi}\right)\left\{\exp\left(\frac{-(a-r)^2}{4Dt}\right) - \exp\left(\frac{-(a+r)^2}{4Dt}\right)\right\}\right]$$

(6) Semi-infinite source of initial concentration C_o, in contact with semi-infinite volume of zero initial concentration:

$$C(x,t) = \frac{C_o}{2}\,erfc\left[\frac{x}{2\sqrt{Dt}}\right]$$

(where *erfc* is the complementary error function, $erfc\,Z = 1 - erf\,Z$)

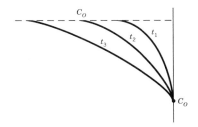

(7) Semi-infinite source of initial concentration C_o; surface maintained at constant concentration C_s:

$$\frac{C(x,t) - C_s}{C_o - C_s} = erfc\left[\frac{x}{2\sqrt{Dt}}\right]$$

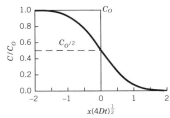

Concentration–distance curve for an extended source of infinite extent.

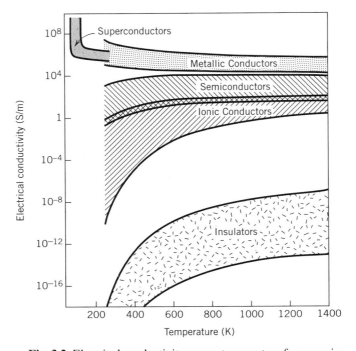

Fig. 3.2 Electrical conductivity versus temperature for ceramic materials with various types of conduction behavior. [From W.J. Weber, H.L. Tuller, T.O. Mason, and A.N. Cormack, *Mat. Sci. Eng. B,* 18, 52 (1993).]

Fick's second law describes the accumulation or depletion of concentration, C, when steady-state conditions are not achieved, and is obtained from the spatial derivative of the flux:

$$\frac{\partial C}{\partial t} = -\frac{\partial J}{\partial x} = D\left(\frac{\partial^2 C}{\partial x^2}\right) \tag{3.2}$$

Fick's first and second laws are the mass-transport analogs to heat transfer (Fourier's law), and the mathematical solutions are similar. A frequently encountered objective in continuum diffusion is the determination of the spatial concentration distribution for a particular species as a function of time. This we can do by solving Fick's second law subject to the boundary conditions of the experiment. There are many configurations that arise in practice, discussed in detail in references listed at the end of this chapter. Table 3.1 gives solutions for the spatial and time variation of concentration, $C(x,t)$, for a few simple geometries and initial conditions. The diffusion coefficient is assumed to be independent of concentration for the solutions given. A close examination of these solutions shows the strong dependence of $C(x,t)$ on the value of D. The diffusion coefficient is a materials parameter, which characterizes the rate of the diffusion process for particular atomic species in a particular phase. In order to understand why D has certain values and can vary over such a wide range (Fig. 3.1), it is necessary to consider the atomistic processes of diffusion.

3.2 ATOMISTIC DIFFUSION PROCESSES

Diffusion on a microscopic scale is illustrated in Fig. 3.3. If two miscible components are brought together, there is a gradual intermingling until a uniform solid solution is achieved. The rate of reaching the final state depends on the mobility of the individual atoms. If a compound is formed between A and B, continuation of the reaction requires that material diffuse through the intermediate layer. The speed of this diffusion process can limit the rate of the reaction. In each instance the mobility of atoms is determined by the crystal structure, composition, and defect structure of the medium in which diffusion is taking place. We may use the defect structure formalism presented in Chapter 2 to discuss the processes that determine atom mobility.

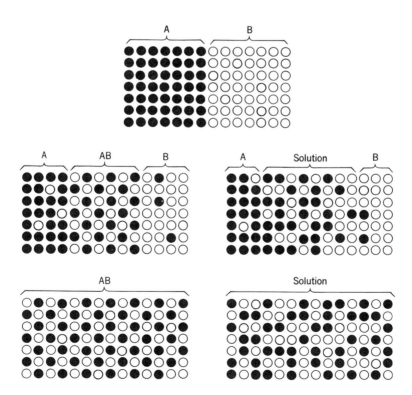

Fig. 3.3 Diffusion processes to form a new ordered compound AB or a random solid solution from pure starting materials A and B.

Random-Walk Diffusion

Random-walk diffusion is the process whereby an atom is able to jump from its present site to any neighboring site with equal probability. Using the one-dimensional random walk in Fig. 3.4, we can show that random-walk diffusion in the presence of a concentration gradient will obey Fick's first law. Several atom planes are depicted in Fig. 3.4, each of which contains a different concentration (n_i) of the atom in black. A linear concentration gradient dC/dx has been imposed, with the concentration decreasing from left to right. The atoms are then allowed to jump at random between planes. At a steady state, Fick's first law (Eq. 3.1) indicates that there will be a constant flux of atoms down this concentration gradient.

To show that this is the case, let us first assign to the atoms a jump frequency Γ (in jumps per second) with which they exchange with the adjacent atom planes; Γ is the same in the $+x$ and $-x$ directions when the concentration of the diffusing atoms is dilute. Considering specifically the two planes labeled 1 and 2, the frequency with which the black atoms in plane 1 jump to plane 2 is $1/2\, n_1\Gamma$, and the frequency for the reverse exchange from plane 2 to

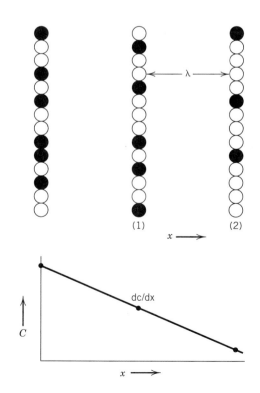

Fig. 3.4 Planes of atoms with a gradient in concentration (black atoms).

plane 1 is $1/2\, n_2\Gamma$. The net flux from plane 1 to plane 2 is the difference in jump frequency:

$$J = \frac{1}{2}(n_1 - n_2)\Gamma \qquad (3.3)$$

Since the concentrations on the two planes labeled 1 and 2 are, respectively, $C_1 = n_1/\lambda$ and $C_2 = n_2/\lambda$, where λ is the separation between planes, the concentration gradient in the x-direction is given by $-dC/dx = (C_1 - C_2)/\lambda = (n_1 - n_2)/\lambda^2$. Equation 3.3 can then be rewritten in the form of Fick's first law:

$$J = -\left(\frac{1}{2}\lambda^2\Gamma\right)\frac{dC}{dx}$$

from which we see that the diffusion coefficient is given by

$$D = \frac{1}{2}\lambda^2\Gamma \qquad (3.4)$$

The result in Eq. 3.4 gives the general form of solid-state diffusion coefficients: They are the product of a geometric factor (here, 1/2), the jump distance squared (λ^2), and the jump frequency (Γ). In a three-dimensional solid, the geometric factor depends on the neighboring coordination of atoms. For example, in a simple cubic lattice where atom jumps can take place in three orthogonal directions, the jump frequency in any one direction becomes $1/6\,\Gamma$, and the diffusion coefficient is $D = 1/6\,\lambda^2\Gamma$. More generally, we can write the numerical factor that accounts for the lattice geometry as $\gamma = 1/N$ where N is the number of neighboring sites into which the atom can jump. λ is simply the distance between these sites. For instance, in the rocksalt structure cations can only jump to another cation site, and anions to another anion site. Since the coordination number of like ions in this structure is 12 (second nearest neighbors), and the jumps are of a distance equal to the length of the $1/2<110>$ vector, we have

$$D = \gamma\lambda^2\Gamma = \frac{1}{12}\left(\frac{a_o\sqrt{2}}{2}\right)^2\Gamma \qquad (3.5)$$

where a_o is the lattice constant. The geometric factor and the jump distance in Eq. 3.5 do not change greatly between different compounds and crystal systems, yet the measured diffusion coefficients in Fig. 3.1 vary over many orders of magnitude. This large variability is directly related to the dependence of the jump frequency Γ on such factors as the energy for migration, temperature, and the availability of jump sites. The latter is directly related to lattice defect concentration.

Diffusion as a Thermally Activated Process

In Fig. 3.1 we see that diffusion coefficients exhibit a straight line when plotted as log D against $1/T$, showing that they can be expressed, at least over some limited temperature range, as

$$D = D_o \exp\left(\frac{-Q}{RT}\right) \tag{3.6}$$

where D_o is a constant and Q is an experimentally determined *activation energy*. This relationship indicates that atom diffusion is a thermally activated process, which from reaction rate theory[1] is consistent with: (1) a single activated state for the atom, as shown in Fig. 3.5; and (2) a driving force for ion motion that is small compared to the thermal energy kT (this driving force may be, for example, chemical potential $\Delta\mu$ or electrical potential $ze\phi$). The jump frequency under these circumstances is given by

$$\Gamma = \upsilon \exp\left(\frac{-\Delta G^*}{kT}\right) = \upsilon \exp\left(\frac{\Delta S^*}{k}\right)\exp\left(\frac{-\Delta H^*}{kT}\right)$$

where υ is the vibrational or attempt frequency, and ΔG^* is the activation energy for migration, equal to $\Delta H^* - T\Delta S^*$, where ΔH^* is the enthalpy of migration and ΔS^* is the entropy associated with the migration process. The diffusion coefficient in Eq. 3.4 becomes

$$D = \gamma\lambda^2 \Gamma = \gamma\lambda^2 \upsilon \exp\left(\frac{-\Delta G^*}{kT}\right) = \gamma\lambda^2 \upsilon \exp\left(\frac{\Delta S^*}{k}\right)\exp\left(\frac{-\Delta H^*}{kT}\right) \tag{3.7a}$$

which, by comparison with Eq. 3.6, indicates that the pre-exponential term is

$$D_o = \gamma\lambda^2 \upsilon \exp\left(\frac{\Delta S^*}{k}\right) \tag{3.7b}$$

Approximate numerical values for the terms in Eq. 3.7b are $\gamma \approx 0.1$, $\lambda \approx 0.2$ nm = 2×10^{-8} cm, and $\upsilon \approx 10^{13}$ sec^{-1}. Assuming $\Delta S^*/k$ to be a small positive number, D_o for vacancy and interstitial diffusion mechanisms has values in the range 10^{-3} to 10^1 cm^2/sec.

Notice that the influence of defect concentration has not yet been taken into account; it has been assumed that all adjacent sites are available to the diffusion atom. As shown below, D is often proportional to the defect concentration, which in many solids is itself exponentially dependent on temperature. As a result, the experimentally measured Q for a diffusive process often contains other energy terms such as the defect formation or association energy in addition to the migration enthalpy ΔH^*.

---

[1] S. Glasstone, K.J. Laidler, and H. Eyring, *The Theory of Rate Processes*, McGraw-Hill Book Company, Inc., New York, 1941.

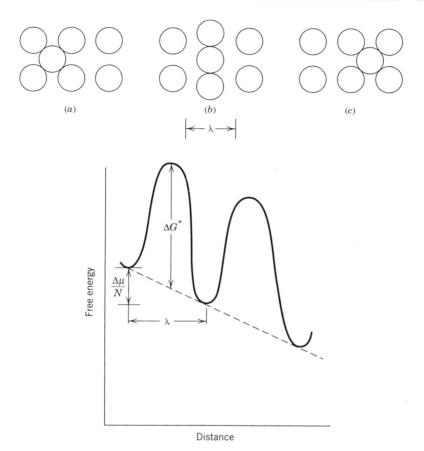

Fig. 3.5 Thermally activated diffusion in a potential gradient $\Delta\mu$ where ΔG^* is the activation energy and λ the jump distance.

Types of Diffusion Coefficients

There are several terminologies used to specify diffusion coefficients of different species, or under different experimental configurations. They may describe the diffusion of a specific defect (e.g., vacancies or interstitials), of a lattice ion or solute (self- and impurity diffusion), a particular atomistic path for diffusion (e.g., lattice, surface, or grain boundary diffusion), or the diffusion coefficient relevant to a specific process (e.g., chemical diffusion, ambipolar diffusion, or interdiffusion). For the present, we limit our discussion to lattice diffusion and the role of defects.

Defect diffusivities refer to the motion of a particular point defect. We can, for instance, specify the diffusion coefficient of a vacancy or interstitial (D_v, D_i), of a bound impurity-vacancy pair (D_{assoc}), or some other type of defect complex. In dilute solutions where defects do not interact with one another, the diffusivity of a

196 Chapter 3 / Mass and Electrical Transport

defect is generally *independent of its concentration*. The diffusion coefficient is given by the expression in Eq. 3.7. The absence of a concentration dependence for diffusion of a vacancy can be understood as follows. A migrating vacancy can only exchange with an atom, and not with another vacancy. Each vacancy-atom exchange results in the motion of the atom in the opposite direction to that of the vacancy. When the concentration of vacancies is dilute, the adjacent lattice sites are virtually always occupied by atoms. Thus the *probability of the adjacent site being available* for exchange is very close to unity. As a result, each vacancy can migrate without interacting with another vacancy, causing no dependence of vacancy diffusion on its own concentration. The same is true for a dilute solution of interstitial atoms; in this instance adjacent unoccupied interstitial sites are virtually always available to the diffusing interstitial atom. Similar analogies follow for other defect diffusivities.

In contrast, the diffusivities of lattice ions (including host ions, solutes, and radio- and mass-isotope tracers) are strongly dependent on the concentrations of the defects by which they migrate. The *self-diffusion* coefficient characterizes the motion of the host lattice ions, and is the important coefficient in many microstructure development processes including sintering and diffusional creep. For instance, when a cation vacancy gradient is imposed across the solid,[2] there is an opposing gradient in the ion concentration (Fig. 3.6). The flux of vacancies down its concentration gradient results in an equal counter-flux of lattice cations. This flux is characterized by the self-diffusion coefficient, D_{self}. If this diffusion process occurs by a vacancy mechanism the self-diffusion coefficient is given by the product of the vacancy diffusivity D_v and the vacancy concentration [V]:

$$D_{self} = [V]D_v. \tag{3.8}$$

The dependence on concentration [V] may be thought of as *the probability that an adjacent site is vacant* and available for exchange. If, instead, the interstitial diffusion coefficient of the ion is much faster, it will provide the majority of the transport, in which case

$$D_{self} = [M_i]D_i \tag{3.9}$$

where $[M_i]$ is the interstitial concentration. The concentration factor gives the probability that at any instant a host ion is in an interstitial position and is therefore free to migrate. Due to the proportionality between atom diffusivity and defect concentration expressed in Eqs. 3.8 and 3.9, lattice diffusion coefficients are sensitive to the same extrinsic parameters that determine the lattice defect structure; namely, solute concentration, atmosphere, and temperature.

---

[2] A gradient in defect concentration may be introduced in a number of ways. One frequently encountered in microstructure development is a stress gradient. The local defect concentration changes with applied stress since work is done against the volume of the defect; the change follows LeChatelier's principle. For instance, a tensile hydrostatic stress at the left end of Fig. 3.6 would increase the local vacancy concentration, and a compressive hydrostatic stress at the other end would decrease it. The net result would be a gradient in vacancy concentration across the material.

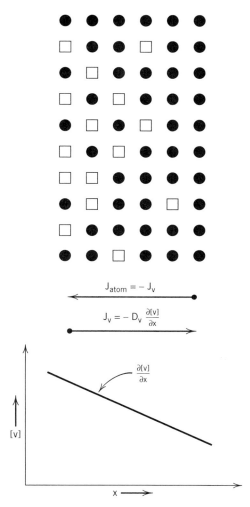

Fig. 3.6 Self-diffusion of cations A (flux J_A) counter to a flux of vacancies J_v down an imposed vacancy concentration gradient.

Direct measurements of self-diffusion coefficients are difficult.[3] It is more common to measure the transport of *tracers*, such as radio- or mass-isotopes, that are chemically identical to the host and can be detected by analytical methods. The *tracer diffusion coefficient* measured in this type of experiment is close to, but not exactly the same as, the self-diffusion coefficient. The distinction lies in the way

---

[3] Techniques are generally based on measuring the rate of a well-understood process that is self-diffusion limited, such as shrinkage of a dislocation loop, and sometimes the rate of sintering or diffusion creep when the rate-limiting mechanism is known.

that one observes the transported specie. In the case of cation self-diffusion under a vacancy flux as shown in Fig. 3.6, the ions being transported are not individually "marked"; all host ions have an equal probability of encountering a vacancy. The ions which actually move are randomly selected from the host lattice. If, for example, at the end of the self-diffusion process we have observed a net mass transport, the specific atoms that moved are unknown. In a tracer experiment, however, we observe the motion of a small fraction of marked tracer ions which are distributed among many host ions. Atomistically, the difference in these two processes is that the motion of a tracer ion is biased by the probability that once a lattice jump has been made, the next jump simply returns it to its former position (Fig. 3.7). Successive jumps are then considered *correlated*. Due to this "wasted" motion, tracer diffusion coefficients are less than self-diffusion coefficients, by a *correlation factor, f*:

$$D_{tracer} = f D_{self}. \tag{3.10}$$

The value of the correlation factor depends on the local atomic coordination as well as the relative frequencies with which the vacancy exchanges with the tracer and the lattice ion. If these exchange frequencies are identical, as is often assumed to be the case of radiotracers and mass isotopes (but not generally for solutes), the correlation factor can be determined from purely geometric considerations. For vacancy diffusion a rigorous calculation shows that f has the values 0.781,

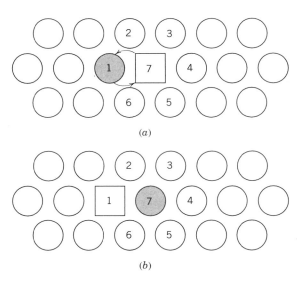

Fig. 3.7 Schematic depicting correlation effect for diffusion in a 2-D hexagonal lattice. After the atom at 1 exchanges with the vacancy at 7, there is a finite probability that the next jump will return the atom to its original position. If the atom is chemically indistinguishable from the host lattice atoms, as for a tracer atom, this probability is 1/6 in this lattice.

0.721, and 0.655, for the FCC, HCP, BCC, and primitive cubic structures, respectively. This is a relatively small correction and is often negligible in comparison to the absolute accuracy with which a diffusion coefficient can actually be measured. Notice that in the case of self-diffusion by an interstitial mechanism, the correlation factor is unity since there is no preference for the interstitial to move to any one adjacent site, all being empty (in the dilute solution limit).

Solute diffusivities differ from self- and tracer diffusivities, first of all because the activation energy for migration will be different even if the mechanism is the same, due to differences in ion size and valence. A difference in activation energy leads to a difference in jump frequency: $\Gamma = \upsilon \exp(-\Delta G^*/kT)$. Second, there may be an association (or even repulsion!) between the solute and a vacancy, which drastically alters the probability that a vacancy will appear on an adjacent site. As a result, the solute diffusion coefficient may be greatly enhanced relative to what it is in the absence of association. If the bound solute-vacancy associate is the fastest diffusing defect species that contains the solute atom, the solute diffusion coefficient is essentially given by that of the defect associate.

When association is prominent, the diffusion coefficient of a solute depends on the probability of association and the diffusion coefficient of the associated species itself. The correlation factor must be taken into account in order to determine the associate diffusivity. Consider the instance where a solute and vacancy appear on adjacent sites, as depicted in Fig. 3.8. If association causes the two to remain on adjacent sites, highly correlated exchanges are clearly possible. Let us consider possible scenarios by first identifying three elementary exchange frequencies: Γ_1, Γ_2, and Γ_3. Γ_1 is the exchange frequency of a host atom and the vacancy; Γ_2 is the exchange frequency between the solute and the vacancy; and Γ_3 is the frequency of a dissociative exchange between vacancy and solute. Firstly, if the impurity and vacancy are strongly bound, the rate of dissociative exchanges is much less than those that keep the vacancy as a nearest neighbor of the impurity: $\Gamma_3 \ll \Gamma_1, \Gamma_2$. Second, notice that a solute-vacancy exchange alone (Γ_2) does not by itself allow the associate to diffuse; also required is the exchange of the vacancy with a neighboring host atom (Γ_1), allowing the vacancy to work its way around the solute into a position where a forward exchange can take place. The rate at which the solute-vacancy associate diffuses is therefore dependent on the relative values of Γ_1 and Γ_2. This is a matter of evaluating a correlation factor.

For the two-dimensional hexagonal lattice in Fig. 3.8, the diffusion coefficient of the impurity atom is, ignoring geometric constants, given by[*]:

$$D_{solute} = \lambda^2 \Gamma_2 \left(\frac{\Gamma_1}{\Gamma_1 + \Gamma_2} \right) p = D_{assoc} p \qquad (3.11)$$

where p is the probability that a vacancy appears on a site adjacent to the solute. If there are equal numbers of solutes and vacancies (as, for example in an extrinsic

---

[*]P. Shewmon, *Diffusion in Solids*, 2nd edition, TMS, Warrendale, PA 1989.

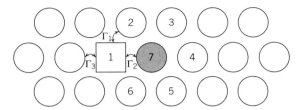

Fig. 3.8 Vacancy-atom exchange frequencies in the vicinity of a solute-vacancy associate. The vacancy is located at 1 and the solute at 7. Γ_1 is the frequency with which the vacancy exchanges position with a host ion, while remaining bound to the solute. Γ_2 is the rate of exchange between solute and vacancy. Γ_3 is the frequency of dissociation, after which the vacancy is no longer a nearest neighbor of the solute.

CaCl$_2$-doped NaCl), and they are strongly associated, then $p \sim 1$ and the solute diffusion coefficient is that of the associate species. If there is only partial association, then p is equivalent to the fraction of the total impurity (Im) which is bound in associates, $p = [(\text{Im}-\text{V})]/[\text{Im}]_{\text{total}}$.

The term in brackets in Eq. 3.11 has the form of a correlation factor (cf. Eq. 3.10):

$$f = \left(\frac{\Gamma_1}{\Gamma_1 + \Gamma_2} \right) \qquad (3.12)$$

and can vary between an upper limit of unity and a very small number, depending on the relative values of Γ_1 and Γ_2. Let us consider two limiting cases.

$\Gamma_1 \gg \Gamma_2$ In this limit many host-vacancy exchanges are able take place before a solute-vacancy exchange occurs. In essence the vacancy circles around the solute many times while waiting for a forward exchange. The correlation factor is unity, $f \sim 1$, and the solute diffusion coefficient is given by

$$D_{solute} \approx \lambda^2 \Gamma_2 p \qquad (3.13)$$

showing that the solute-vacancy exchange is the rate-limiting step. Notice that if p is of order unity, the solute diffusion coefficient is much higher than it would be in the absence of association.

$\Gamma_1 \ll \Gamma_2$ In this limit the vacancy-impurity exchange is highly correlated, with most of the exchanges being wasted motion that does not lead to diffusion. The correlation factor is small, $f \sim \Gamma_1/\Gamma_2$, and the solute diffusion coefficient in Eq. 3.11 reduces to

$$D_{solute} \approx \lambda^2 \Gamma_1 p \qquad (3.14)$$

Now solute diffusion is rate-limited by the frequency of the host ion-vacancy exchange, rather than the impurity-vacancy exchange. Even so, the diffusion coefficient of the solute is much higher than that of the host ion if the probability of association is high.

In three dimensions, similar results are obtained. Lidiard (*Phil. Mag.* 46, 1218, 1955) has analysed diffusion in an fcc lattice (which is applicable to fcc sublattices in ionic crystals) and obtains the result

$$D_{solute} = \frac{2\lambda^2 \Gamma_2 (2\Gamma_1 + 7\Gamma_3)}{3(2\Gamma_1 + 2\Gamma_2 + 7\Gamma_3)} p \qquad (3.15)$$

where the assumption that the dissociation frequency Γ_3 is negligible has been relaxed. The reader can easily verify that Eq. 3.15 reduces to limiting cases like those given in Eqs. 3.13 and 3.14 in the case of strong association.

Diffusion in Lightly Doped NaCl

Let us now consider some specific examples. Sodium chloride containing a small quantity of a divalent cation such as $CdCl_2$ has an extrinsic concentration of sodium vacancies introduced according to the defect reaction:

$$CdCl_2 \xrightarrow{NaCl} Cd^{\cdot}_{Na} + 2Cl^x_{Cl} + V'_{Na}$$

In addition, depending on the temperature, there may be intrinsic sodium vacancies formed as a result of the Schottky reaction. The sodium diffusion coefficient is given by

$$D_{Na} = [V'_{Na}] \gamma \lambda^2 \upsilon \exp\left[\frac{-\Delta G^*_{Na}}{kT}\right] \qquad (3.16)$$

in which ΔG^*_{Na} is the migration free energy for sodium vacancies. Experimental data, Fig. 3.9, shows two regimes of Arrhenius behavior, with distinctly different temperature dependences. At low temperature there is an extrinsic regime in which the concentration of vacancies is a constant, being determined by the solute concentration, $[Cd^{\cdot}_{Na}] = [V'_{Na}]$. The sodium diffusion coefficient is given by

$$D_{Na} = [V'_{Na}] \gamma \lambda^2 \upsilon \exp\left[\frac{-\Delta G^*_{Na}}{kT}\right] = [CdCl_2] \gamma \lambda^2 \upsilon \exp\left[\frac{\Delta S^*_{Na}}{k}\right] \exp\left[\frac{-\Delta H^*_{Na}}{kT}\right] \qquad (3.17a)$$

and exhibits a temperature dependence of $-\Delta H^*_{Na}/k$. At higher temperature, an intrinsic regime appears, in which the temperature dependence is steeper since the activation energy includes both migration and Schottky formation enthalpies:

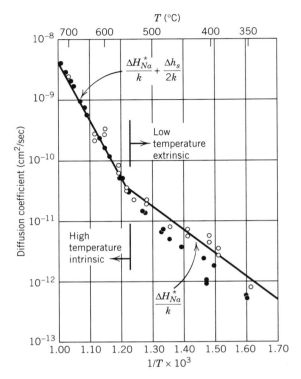

Fig. 3.9 Diffusion coefficients measured directly (open circles) and calculated from the electrical conductivity data (closed circles) for Na⁺ in sodium chloride. [From D. Mapother, H.N. Crooks, and R. Maurer, *J. Chem. Phys.*, 18, 1231 (1950).]

$$D_{Na} = [V'_{Na}]\gamma\lambda^2 v \exp\left[\frac{-\Delta G^*_{Na}}{kT}\right] = \gamma\lambda^2 v \exp\left[\frac{\Delta S_s}{2k}\right]\exp\left[\frac{\Delta S^*_{Na}}{k}\right]\exp\left[\frac{-\Delta h_s}{2kT}\right]\exp\left[\frac{-\Delta H^*_{Na}}{kT}\right] \quad (3.17b)$$

and has a slope of $-(\Delta h_s/2k) - (\Delta H^*_{Na}/k)$. At the break in the curves between the two regimes, the concentration of intrinsic vacancies equals that due to the extrinsic dopant. (It is left as an exercise for the reader to estimate from the results in Fig. 3.9: first, the Schottky energy, and second, the concentration of $CdCl_2$ in the sample.)

Diffusion in a Highly Stoichiometric Oxide: MgO

As discussed in Chapter 2, the Schottky formation energy is high enough for MgO that in practice the defect structure is extrinsic, being dominated by aliovalent impurities. The Brouwer diagram in Fig. 2.13 shows the defect structure expected for MgO containing an extrinsic amount of a trivalent impurity such as Al^{3+}, Fe^{3+}, or Cr^{3+}. Over a broad range of oxygen partial pressure, the magnesium vacancy

3.2 Atomistic Diffusion Processes

concentration should be constant and proportional to the solute concentration, e.g., $[Al'_{Mg}] = 2[V''_{Mg}]$. D. Sempolinski et al. (*J. Am. Ceram. Soc.*, 63[11–12], 664 (1980)) have determined the magnesium vacancy diffusion coefficient from ionic conductivity measurements in MgO doped with known amounts of trivalent impurities (Fe^{3+}, Sc^{3+}), and obtain

$$D_{V''_{Mg}} = (0.38 \pm 0.15) \exp\left[\frac{-2.29 \pm 0.21\, eV}{kT}\right] \quad (cm^2/sec) \quad (3.18)$$

From this determination the magnesium self-diffusion coefficient as a function of solute concentration can be determined using the relation

$$D_{Mg} = [V''_{Mg}] D_{V''_{Mg}} \quad (3.19)$$

The magnitude and temperature dependence of the magnesium self-diffusion coefficient measured in other samples, Fig. 3.10, can be explained using this result and assuming reasonable levels of background impurities. These results therefore represent regimes of extrinsic cation diffusion where the activation energy is that for vacancy migration.

It is useful to consider what can happen in an extrinsic material with further decreases in temperature, such that the impurity precipitates. In Al-doped MgO, the equilibrium phase that precipitates is spinel, $MgAl_2O_4$. When the temperature is low enough that we enter the two-phase field, the amount of solute in solution becomes temperate dependent. The precipitation reaction was given by Eqs. 2.66 and 2.67:

$$Mg^x_{Mg} + 2Al'_{Mg} + V''_{Mg} + 4O^x_O \longrightarrow MgAl_2O_4 \,(ppt)$$

$$[V''_{Mg}][Al'_{Mg}]^2 \propto \exp\left[\frac{+\Delta H_{ppt}}{kT}\right]$$

If the Brouwer approximation in the impurity dominated regime remains $[Al'_{Mg}] = 2[V''_{Mg}]$ (i.e., the concentration of associates remains in the minority), the concentration of unassociated cation vacancies in solution varies as

$$[V''_{Mg}] \propto \exp\left[\frac{+\Delta H_{ppt}}{3kT}\right] \quad (3.20)$$

D_{Mg} will then take on a new temperature dependence of $-(\Delta H^* + \Delta H_{ppt}/3)/k$ in this regime.

In contrast to the relatively ideal picture regarding cation diffusion, anion self-diffusion in MgO has not been well understood. The close-packed oxygen sublattice suggests that oxygen sublattice defects will have a high energy of formation and migration compared to cation sublattice defects. The very low value of the oxygen diffusion coefficient, at least three orders of magnitude lower than that of magnesium at a given temperature, is consistent with a low anion defect concentration and high activation energy for motion. However, the mechanism of diffusion has

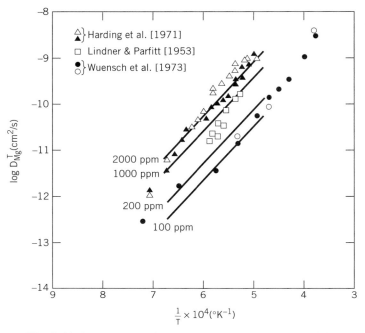

Fig. 3.10 Comparison of measured magnesium tracer diffusion coefficient in MgO (data points) with that calculated from the magnesium vacancy diffusion coefficient (solid curves).

remained a mystery. The most likely mechanisms are oxygen interstitials or oxygen vacancies. Since most compositions are extrinsic and dominated by cation vacancies, it is expected that the anion vacancy concentration will be suppressed to very low levels by the Schottky equilibrium:

$$K_s = [V''_{Mg}][V_O^{\cdot\cdot}] = \exp\left(\frac{-\Delta g_s}{kT}\right)$$

However, if the anion vacancy concentration becomes too low, the vacancy diffusion mechanism may no longer provide the dominant means of transport. (In the alkali halides, an anion vacancy diffusion mechanism is operative despite the presence of extrinsic cation vacancies since the Schottky energies are comparatively much lower.) The oxygen interstitial concentration is expected to simultaneously increase due to the anion Frenkel equilibrium

$$K_{aF} = [V_O^{\cdot\cdot}][O''_i] = \exp\left(\frac{-\Delta g_{aF}}{kT}\right)$$

and may instead become dominant. (Note that the oxygen Frenkel energy is also very high, having been calculated as (~15 eV.) This is the interpretation reached by

B. J. Wuensch et al. (*Ceram. Trans.*, Vol. 24, pp. 79–89, 1991), based on the following experimental evidence. First, ^{18}O mass tracer experiments showed that oxygen self-diffusion lacks a significant P_{O_2} dependence, suggesting an extrinsic mechanism. Second, systematic increases in the aliovalent cation concentration caused a corresponding increase in oxygen self-diffusion coefficient, which is as expected for an interstitial diffusion mechanism. The details of the mechanism remain unexplained; the observed activation energies do not agree well with calculated migration energies for either the interstitial or interstitialcy[4] mechanism.

Diffusion in Cation-Deficient Oxides: The Transition Metal Monoxides

A comparison of cation diffusion in the transition metal monoxides provides a good illustration of the effects of increasing defect concentration and of defect interactions. These are rocksalt-structure oxides, in which the tendency toward partial oxidation of the transition metal cation to a 3+ valence results in a cation deficiency, which is in each case accommodated by cation vacancies. The nonstoichiometry, x, increases for this series of compounds in the order $Ni_{1-x}O$, $Co_{1-x}O$, $Mn_{1-x}O$, and $Fe_{1-x}O$. $Ni_{1-x}O$ is the most ideal of the series in having the lowest cation vacancy concentration, $x \sim 3 \times 10^{-4}$ at 1300°C in air. Under the same conditions CoO has $x \sim 10^{-2}$, still quite dilute and treatable from the viewpoint of dilute point defect chemistry. MnO and FeO are stable only under reducing ambients at high temperature; they oxidize to Mn_3O_4 and Fe_3O_4 at higher oxygen activity. Toward the oxidizing limit of the stability fields for the monoxide, x reaches 0.1 for $Mn_{1-x}O$ and 0.15 for $Fe_{1-x}O$. These defect concentrations are too high for an assumption of non-interacting defects to be valid.

These defect structures are sufficiently complex that confidence in a particular defect model is gained only when the model can self-consistently explain a number of physical properties, such as the deviation from stoichiometry (often measured thermogravimetrically), electrical conductivity, and cation diffusion. In $Ni_{1-x}O$ and $Co_{1-x}O$, the more ideal behavior has permitted a point defect interpretation of these properties. We will use $Co_{1-x}O$ as an example. The nonstoichiometry varies with oxygen activity according to

$$\frac{1}{2}O_2(g) = O_O^x + V_{Co}^x ; \qquad K_1 = \frac{[V_{Co}^x]}{a_{O_2}^{1/2}}. \qquad (3.21)$$

Further ionization of the vacancies are also possible, according to

---

[4] The *interstitialcy mechanism* refers to the process whereby an interstitial migrates by displacing a lattice atom from its site to an interstitial site. In the process the interstitial atom replaces the lattice atom. It is sometimes proposed for atoms which cannot migrate by a straightforward interstitial mechanism, perhaps due to size or other considerations which give a high interstitial migration energy.

$$V_{Co}^x = V'_{Co} + h^\cdot \ ; \qquad K_2 = \frac{[V'_{Co}]p}{[V_{Co}^x]} \qquad (3.22)$$

and

$$V'_{Co} = V''_{Co} + h^\cdot \ ; \qquad K_3 = \frac{[V''_{Co}]p}{[V'_{Co}]} \qquad (3.23)$$

for each of which the corresponding mass-action equilibrium constant is shown. The total nonstoichiometry is given by the sum of vacancy concentrations, $x = [V_{Co}^x] + [V'_{Co}] + [V''_{Co}]$. The electrical conductivity is p-type due to the extrinsic electron holes introduced to compensate the cation vacancies (Eqs. 3.22 and 3.23). Analyzing experimental measurements of nonstoichiometry, diffusion, and electrical conductivity as a function of temperature and oxygen pressure, it has been found that in $Ni_{1-x}O$ only doubly and singly charged vacancies need be included to explain the data, whereas in $Co_{1-x}O$ (and $Mn_{1-x}O$), inclusion of neutral vacancies produces a better fit to data. As an illustration of this kind of modeling, consider the P_{O_2} dependence of Co tracer diffusion in $Co_{1-x}O$, shown in Fig. 3.11a. Notice that the results show a slight upward curvature, which suggests that a single Brouwer approximation is inadequate. Fig. 3.11b shows the point defect model which has been fit to the available data in this system, plotted as a Brouwer diagram. The three cation vacancy ionization states, V_{Co}^x, V'_{Co}, and V''_{Co}, have been included. Notice that at higher oxygen activity, the nonstoichiometry is primarily due to V'_{Co} in this model, while at lower oxygen activity V''_{Co} defects play a more important role. Neutral vacancies V_{Co}^x are in the minority, being only 10% of the total concentration at 1200°C in 1 atm oxygen. Upon taking the measured Co tracer diffusion coefficient and dividing by the total vacancy concentration, it is found that the diffusional mobility exhibits a singly activated temperature dependence, Fig. 3.11c, with an activation energy of 1.41 eV. Thus it appears that the cation diffusion coefficient varies with temperature and oxygen activity in a relatively simple way; it is proportional to the cation vacancy concentration, as discussed previously for a vacancy mechanism.

As the nonstoichiometry increases along the transition metal monoxide series, deviations from ideal point defect behavior become apparent. In $Mn_{1-x}O$, the deviation from stoichiometry (which ranges from 10^{-5} to 10^{-1}) with temperature and P_{O_2} in detail fits neither a noninteracting point defect model nor simple models invoking larger defect clusters (for a detailed review, see A. Atkinson, *Advances in Ceramics*, Vol. 23, p. 3, The American Ceramic Society, 1987). Nonetheless, the Mn tracer diffusion coefficient varies in direct proportion to the deviation from stoichiometry, with a migration energy of 1.73 eV (M. Keller and R. Dieckmann, *Ber. Bunsenges.*, 89, 883, 1985), suggesting that whichever vacancy or vacancies facilitate Mn diffusion, their concentration and mobility remains constant in the face of vacancy aggregation. A smaller-than-expected correlation factor also suggests defect clustering (N. L. Peterson and W. K. Chen, *J. Phys. Chem. Solids*, 43[1] 29, 1985).

3.2 Atomistic Diffusion Processes 207

In $Fe_{1-x}O$, the deviation from stoichiometry seems finally so large (having a *minimum* value of x~0.05) that defect clustering has a clear influence on cation diffusion. Experimental measurements show that the Fe tracer diffusion coefficient is, remarkably, *independent* of x over virtually the entire stability field of $Fe_{1-x}O$ (Fig. 3.12). Nonetheless, the activation energy of 1.3–1.5 eV seems rea-

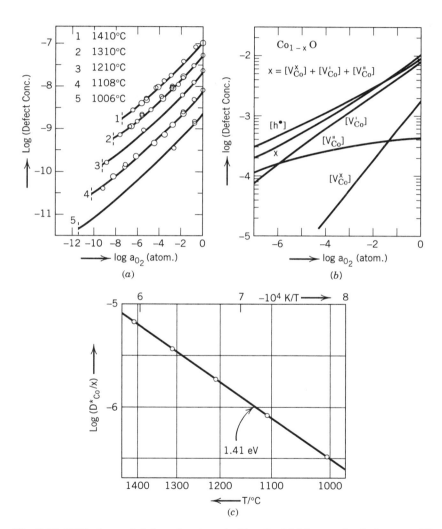

Fig. 3.11 Diffusion and defect chemistry in $Co_{1-x}O$. (a) Measured cobalt tracer diffusion coefficient as a function of temperature and oxygen activity. (b) Concentrations of various defects as a function of oxygen activity, at 1200°C. (c) Effective diffusion coefficient of cobalt vacancies as a function of temperature. [From R. Dieckmann, *Z. Physik. Chemie N.F.*, 107, 189 (1977).]

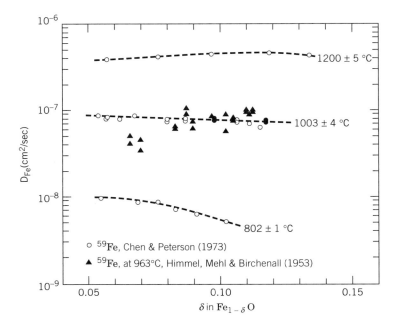

Fig. 3.12 Diffusion coefficient of ^{59}Fe tracer in $Fe_{1-x}O$ as a function of the deviation from stoichiometry. Notice the weak dependence on defect concentration. [From N.L. Peterson, *Mat. Sci. Forum*, 1, 85, (1984).]

sonable for a simple vacancy exchange in comparison with values for the other oxides in the series. The diffusion coefficient of Mn impurity in $Fe_{1-x}O$ has also been found to be approximately independent of x and has a similar activation energy. To reconcile these observations with clear evidence for defect aggregation in $Fe_{1-x}O$, it has been suggested that simple vacancy exchange with a minority population of free vacancies takes place in a defect structure otherwise dominated by clusters. However, it is not clear why the concentration of mobile vacancies should remain constant. A completely satisfactory model for this obviously complex situation has yet to be developed.

Diffusion in a Highly Doped Oxide: Cubic Stabilized ZrO_2

We discussed in Chapter 1 the cubic, tetragonal, and monoclinic symmetry phases of ZrO_2, which are interrelated through displacive transformations of the basic fluorite structure. For electrical applications as gas sensors and solid electrolytes it is primarily the cubic phase that is of interest. This is mostly a coincidence; the high dopant concentrations necessary to obtain a high concentration of oxygen vacancies also happen to stabilize the cubic structure, which is retained at room

3.2 Atomistic Diffusion Processes

temperature as a metastable phase. Commonly used stabilizers such as CaO and Y_2O_3, are lower than 4+ in valence and upon substituting for Zr^{4+} are charge-compensated by oxygen vacancies, for example,

$$CaO \xrightarrow{ZrO_2} Ca''_{Zr} + O^x_O + V^{\cdot\cdot}_O$$

Since stabilizer concentrations are typically in the range of 8–15%, the solute overwhelms the electroneutrality condition (e.g., the Brouwer approximation is $[Ca''_{Zr}] = [V^{\cdot\cdot}_O]$) and yields a concentration of vacancies that is unusually high compared to defect densities in most oxides. The high defect concentrations are also a consequence of unusually high solid solubilities on the order of several tens of percents. Aliovalent solute concentrations of this magnitude are simply not soluble in many refractory oxides such as MgO or Al_2O_3. A high vacancy concentration combined with a small energy of migration (~1 eV) results in an unusually high oxygen diffusion coefficient, shown in Fig. 3.1 for the composition $Ca_{0.14}Zr_{0.86}O_{1.86}$. Cubic zirconia is thus useful as a *fast-ion conductor*. At such high vacancy concentrations, the effects of interacting point defects also become evident.

The variation in oxygen self-diffusion coefficient with temperature for $Ca_{0.15}Zr_{0.85}O_{1.85}$ is shown in greater detail in Fig. 3.13. In the highest temperature regime, the diffusion coefficient is given by:

$$D(O) = 1.0 \times 10^3 \exp[-0.84 \, eV/kT] \, cm^2/sec.$$

Despite the high vacancy concentration, $[V^{\cdot\cdot}_O] = 0.15$, the behavior in this regime is not very different from that expected for diffusion via unassociated vacancies. The measured activation energy is of the same order as theoretical estimates for isolated vacancy migration.

However, there is an increase in the activation energy (to ~1.5 eV) with decreasing temperature (Fig. 3.13). This does not appear to be directly related to the cubic→tetragonal phase transformation (although in detail the migration energy for oxygen vacancies should differ in each phase). The change in activation energy occurs even above the transformation temperature, and is more likely related to defect clustering. The difference in activation energy of 0.66 eV between low and high temperatures may reflect an enthalpy of association between defects or defect clusters containing Ca''_{Zr} and $V^{\cdot\cdot}_O$. A similar effect is shown in Fig. 3.14 for Y_2O_3 stabilized ZrO_2 of two different concentrations, measured over a somewhat higher temperature range than the data in Fig. 3.13. Here the Brouwer approximation is given by $[V^{\cdot\cdot}_O] = 1/2[Y'_{Zr}]$. In a 12% Y_2O_3 composition (6% vacancy concentration) an activation energy of ~0.50 eV is measured, whereas in a 30% Y_2O_3 material (15% vacancies), there is an increase in activation energy to 1.32 eV. Notice that in contrast to solute diffusivities, which we typically expect to *increase* upon pairwise association, here defect interactions *decrease* D(O) by decreasing the concentration of free oxygen vacancies. For ionic conductor applications, the composition must be selected with these varia-

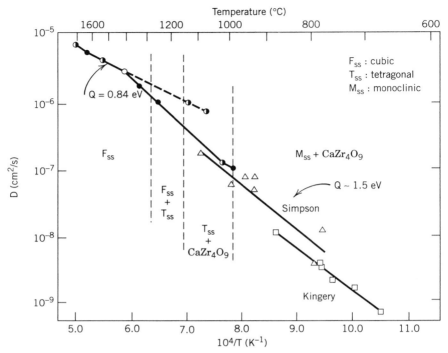

Fig. 3.13 Oxygen self-diffusion coefficients for $Zr_{0.85}Ca_{0.15}O_{1.85}$. Regimes of varying activation energy are observed in this highly concentrated system. (From Y. Oishi and K. Ando, pp. 189–202 in *Transport in Nonstoichiometric Compounds*, G. Simkovich and V. S. Stubican, eds., Plenum Press, 1985.)

tions in mind. In zirconia as well as other fluorite structure oxides such as CeO_2, the activation energy for oxygen diffusion typically reaches a minimum, and the ionic conductivity a maximum, some intermediate concentration of dopant.

Thus, in highly doped or highly nonstoichiometric systems the defect chemistry and corresponding transport properties can deviate far from that of an ideal point defect solution. Beyond simple association or formation of compact clusters, the ordering of defects into "extended defects" is common. For example, in highly reduced TiO_{2-x} ordering of oxygen vacancies or titanium interstitials into extended planar defect arrays eventually leads to collapse and shear of the crystal lattice along those planes. This gives rise to planar faults that themselves order into a series of crystallographic shear phases known as Magneli phases. Sudden decreases in diffusion coefficients result upon formation of these phases, for the point defects are no longer as free to move. In the pyrochlore structure oxides, which are essentially defective fluorites with a built-in oxygen deficiency whereby 1/8 of the oxygen sites are vacant (the ideal composition is $A_2B_2O_7$; e.g., $Gd_2Zr_2O_7$ and $Gd_2Ti_2O_7$), long-range ordering of oxygen vacancies also causes the oxygen ion conductivity to decrease markedly. The behavior is more complex than even highly doped zirconia, since ordering on the oxygen sublattice is coupled to a simultaneous ordering on the cation sublattice.

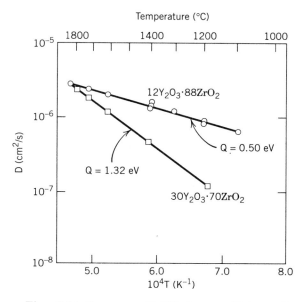

Fig. 3.14 Oxygen self-diffusion coefficients for $Zr_{0.88}Y_{0.12}O_{1.94}$ and $Zr_{0.70}Y_{0.30}O_{1.85}$. (From Y. Oishi and K. Ando, pp. 189–202 in *Transport in Nonstoichiometric Compounds*, G. Simkovich and V. S. Stubican, eds., Plenum Press, 1985.)

3.3 ELECTRICAL CONDUCTIVITY

While one often thinks of ceramics as highly insulating materials, in reality the range of measured electrical conductivities in ceramics covers 25 orders of magnitude, from the most insulating to the most conductive known solids. We have already discussed the approximately 10 orders of magnitude range over which the diffusion coefficient of ionic species varies. Electrical conductivity due to the motion of ionic species, that is, *ionic conduction*, is proportional to diffusivity and covers a similar range of values. Many of the electrical applications of ceramics as sensors, electrochemical pumps, and solid electrolytes in fuel cell and battery systems are only possible due to the existance of systems with high ionic conductivity. When the (comparatively) much higher mobility of electronic defects is also taken into account, the range of conductivities seen in ceramic materials broadens greatly. Figure 3.2 and Table 3.2 show the range of conductivities for common conductive and insulating ceramics. In some oxides metallic bonding (overlap of the energy levels of bonding and nonbonding orbitals) occurs, leading to room temperature electrical conductivities as high as that in some metals. Some complex layered oxides such as the cuprate superconductors ($La_{2-x}Sr_xCuO_4$, $YBa_2Cu_3O_{7-\delta}$, $Bi_2(Sr,Ca)_3Cu_2O_{8+\delta}$, etc.) fall into this category. The metallic bonding occurs along certain directions of the crystal structure (especially within Cu-O planes), while

Table 3.2 Electrical Resistivity of Some Materials at Room Temperature

Materials	Resistivity (ohm-cm)
Metals:	
Copper	1.7×10^{-6}
Iron	10×10^{-6}
Molybdenum	5.2×10^{-6}
Tungsten	5.5×10^{-6}
ReO_3	2×10^{-6}
CrO_2	3×10^{-5}
Semiconductors:	
Dense silicon carbide	10
Boron carbide	0.5
Germanium (pure)	40
Fe_3O_4	10^{-2}
Insulators:	
SiO_2 glass	$> 10^{14}$
Steatite porcelain	$> 10^{14}$
Fire-clay brick	10^8
Low-voltage porcelain	10^{12}–10^{14}

the bonding elsewhere has greater ionic or covalent character, resulting in highly anisotropic electrical conduction. Many ceramic systems (e.g., $BaTiO_3$, TiO_2, ZnO, SiC, $LiNbO_3$) are large bandgap semiconductors, exhibiting E_g values in the range of 2.5–3.5 eV. And, some of the most insulating compounds known are ceramics in which the bandgap energy is in excess of 7 eV.

Relationship between Mobility and Diffusivity

As the first step toward understanding the electrical conductivity provided by mobile defects, we examine the relationship between diffusivity and mobility. Mobility is defined as the velocity of an entity per unit driving force (M = v /F). Depending on the process under consideration, this driving force may be a chemical potential gradient, an electrical potential gradient (electric field), an interfacial energy gradient, elastic strain energy gradient, and so forth. It is possible to define mobilities for atoms, electrons, dislocations, grain boundaries or other entities. The units for mobility change accordingly, as shown in Table 3.3.

For atomic transport, Einstein first pointed out that the most general driving force is the virtual force which acts on a diffusing atom or ion due to the negative gradient of the chemical potential or partial molal free energy:

$$F_i = \frac{1}{N_a}\left(\frac{d\mu_i}{dx}\right) \quad \left(\text{units}: \quad \text{mole} \cdot \frac{\text{ergs/mole}}{\text{cm}} = \frac{\text{ergs}}{\text{cm}}\right) \quad (3.24)$$

3.3 Electrical Conductivity

Table 3.3 Dimensional Units for Mobility

Absolute mobility

$$B_i = \frac{V_i}{(1/N)(\partial \mu_i / \partial x)} = \frac{cm^2}{erg \cdot sec}$$

V_i = cm/sec
μ_i = ergs/mole = 10^{-7} J/mole
N = Avogadro's number = atoms/mole
x = cm

Chemical mobility

$$B'_i = \frac{V_i}{\partial \mu_i / \partial x} = \frac{mole \cdot cm^2}{J \cdot sec}$$

V_i = cm/sec
μ_i = J/mole
x = cm

Electrical mobility

$$\mu_i = \frac{V_i}{\partial \phi / \partial x} = \frac{cm^2}{V \cdot sec}$$

V_i = cm/sec
ϕ = V
x = cm

$B_i = NB'_i$
$\mu_i = z_i F B'_i$

z_i = valence = equiv/mole
F = Faraday const
  = 96,500 C/mole

$$\frac{\partial \mu_i}{\partial x} = z_i F \frac{\partial \phi}{\partial x}$$

1J = C · V = 10^7 ergs = 0.2389 cal = 6.243 × 10^{18} eV.

where μ_i is the chemical potential of i and N_a is Avogadro's number. The *absolute mobility* B_i due to the gradient in this driving force is given by

$$-B_i = \frac{velocity(cm/sec)}{force(ergs/cm)} = \frac{v_i}{[(1/N_a)d\mu_i/dx]} \quad (cm^2/erg \cdot sec) \quad (3.25)$$

To obtain the relationship between mobility and diffusivity, we first write the flux in a general form as the product of concentration and velocity:

$$J_i = c_i v_i = c_i B_i F_i \quad \text{(units: moles/cm}^2 \cdot \text{sec)} \quad (3.26)$$

Chapter 3 / Mass and Electrical Transport

rather than in terms of concentration gradients as in Fick's first law. Substituting for F_i we have:

$$J_i = -\frac{1}{N_a}\left(\frac{d\mu_i}{dx}\right)B_i c_i \qquad (3.27)$$

For an ideal solution with unit activity for species i, $\mu_i = \mu_i^o + RT \ln c_i$, and the change in chemical potential is

$$d\mu_i = RT\, d\ln c_i = \frac{RT}{c_i} dc_i \qquad (3.28)$$

and the gradient is

$$\frac{d\mu_i}{dx} = \frac{RT}{c_i}\left(\frac{dc_i}{dx}\right) \qquad (3.29)$$

Upon substituting this expression into Eq. 3.27 we have

$$J_i = -\frac{RT}{N_a} B_i \frac{dc_i}{dx} \qquad \text{(units: moles/cm}^2\text{ sec)} \qquad (3.30)$$

which when compared with Fick's first law (Eq. 3.1) shows that the diffusion coefficient is directly proportional to the atomic mobility:

$$D_i = kT\, B_i \qquad (3.31)$$

where $k = R/N_a$ is Boltzmann's constant. In the nonideal case it is necessary either to define these equations in terms of activity coefficients or to include an activity coefficient term in Eq. 3.28. The expression in Eq. 3.31 is an important result called the *Nernst-Einstein* relation.

In discussing electrical properties it is often convenient to write the driving force not as the gradient in chemical potential, but instead in terms of the more easily measured electrical potential ϕ. In terms of the electrical potential the absolute driving force is

$$F_i \text{ (electrical)} = z_i e \frac{d\phi}{dx} = z_i e E \qquad (3.32)$$

where $z_i e$ is the particle charge. The flux is, upon substituting $B_i = D_i/kT$ for the absolute mobility:

$$J_i = c_i B_i F_i = c_i z_i e\left(\frac{D_i}{kT}\right) E$$

Since $J_i = c_i v_i$, the velocity is

$$v_i = \left(\frac{z_i e D_i}{kT}\right) E.$$

Defining the *electrical mobility* μ_i as the velocity per unit electric field

$$\mu_i = \frac{v_i}{E}$$

we have

$$\mu_i = \frac{z_i e D_i}{kT} \quad (\text{cm}^2/\text{V} \cdot \text{sec}) \tag{3.33}$$

(Unfortunately, μ_i, which we used earlier to denote chemical potential, is also the most commonly used symbol for electrical mobility.) Eq. 3.33 is a form of the Nernst–Einstein relation, which is particularly useful in determining the mobility of charged defects and in determining diffusion coefficients from electrical conductivity data, or vice versa.

While the Nernst–Einstein relationship can be applied to electronic as well as ionic species, allowing us to state an equivalence between diffusivity and mobility, the motion of electrons is fundamentally quite different. In metals, semiconductors, and high-mobility ceramics the electrons and holes are considered to be quasi-free particles with a *drift* velocity under electric field that is limited by scattering from the lattice. The drift velocity, which is much less than the instantaneous velocity of the random moving particles, is given by the following equation of motion under an applied electrical force $F = zeE$:

$$m\left(\frac{dv}{dt} + \frac{v}{\tau}\right) = zeE \tag{3.34}$$

where m is the particle mass (or more properly the effective mass m_e^* and m_h^* when there are interactions between the charge carrier and the lattice). The first term $m(dv/dt)$ in Eq. 3.34 is equivalent to Newton's law ($F = ma$) and applies when v is changing with time. The second term mv/τ may be thought of as a damping force where τ is the characteristic relaxation time for the velocity to decay to its steady state value. Once steady state is reached under the applied field E, $dv/dt = 0$ and the mobility is given by

$$\mu = \frac{v}{zE} = \frac{e\tau}{m^*} \tag{3.35}$$

The temperature dependence of μ is determined principally by the variation of τ with temperature. Thermal vibrations of the lattice (phonons) decrease the mobility with a $T^{-3/2}$ dependence; impurity scattering on an absolute level decreases the mobility but its effect diminishes with increasing temperature as $T^{+3/5}$. The net effect is that the mobility decreases with increasing temperature, but only weakly so.

The effective mass in Eq. 3.35 depends on the level of interaction between the electron or hole and the periodic lattice potential. In ionic systems the proximity of the carrier to the ions results in a polarized region (termed *polaron*), which may be large or small in comparison to the unit cell size. If the system contains large polarons reflecting a weak interaction between the carrier and ion, the electronic mobility is given by Eq. 3.35 and the effective mass may be close to the free electron mass. However, in a number of ionic systems (including $LiNbO_3$, Fe_3O_4, $CoFe_2O_4$, FeO, CeO_2) the interaction is strong, the effective mass is large, and for these *small-polaron* systems the motion of electrons and holes is thermally activated (in what is also termed a *hopping mechanism*). The electronic carrier mobility in these systems can then be very small (0.1 cm²/V·sec and less), orders of magnitude lower than in good semiconductors, as shown in Table 3.4.

Table 3.4 Approximate Carrier Mobilities at Room Temperature

Crystal	Mobility (cm²/V sec) Electrons	Mobility (cm²/V sec) Holes	Crystal	Mobility (cm²/V sec) Electrons	Mobility (cm²/V sec) Holes
Diamond	1800	1200	Pbs	600	200
Si	1600	400	PbSe	900	700
Ge	3800	1800	PbTe	1700	930
InSb	10^5	1700	AgCl	50	
InAs	23,000	200	KBr (100°K)	100	
InP	3400	650	CdTe	600	
GaP	150	120	GaAs	8000	3000
AlN	...	10	SnO_2	160	
FeO			$SrTiO_3$	6	
MnO			Fe_2O_3	0.1	
CoO	...	~0.1	TiO_2	0.2	
NiO			Fe_3O_4	...	0.1
GaSb	2500–4000	650	$CoFe_2O_4$	10^{-4}	10^{-8}

Ionic and Electronic Conductivity

Electrical conductivity is defined as the charge flux per unit electric field, and usually is written in units of $(\Omega\text{-cm})^{-1}$ or Siemens per meter, S/m, where $S = \Omega^{-1}$. It is given by

$$\sigma_i = \frac{J_i z_i e}{E} = \frac{c_i z_i e v_i}{E} = c_i z_i e \mu_i \qquad (3.36)$$

where J_i is the flux of particle i. For ionic species we can substitute the Nernst–Einstein relationship (Eq. 3.33) and obtain

$$\sigma_i = c_i z_i e \mu_i = \frac{c_i z_i^2 e^2 D_i}{kT} \qquad (3.37)$$

Since more than one mobile charged particle can contribute to the electrical conductivity of a material, we refer to the *partial conductivity* σ_i as that which is attributable to a particular defect. The total electrical conductivity is the sum of all such contributions:

$$\sigma_{total} = \sigma_1 + \sigma_2 + \sigma_3 + \cdots \sigma_i = \sum \sigma_i \qquad (3.38)$$

The fraction of the total conductivity carried by each charged species is termed the *transference number*, t_i:

$$t_i = \frac{\sigma_i}{\sigma_{total}} \qquad (3.39)$$

and the sum of all transference numbers in a system is unity:

$$t_1 + t_2 + t_3 + \cdots t_i = 1 \qquad (3.40)$$

From Eq. 3.37, it is apparent that the partial conductivity of any species depends on the product of mobility and concentration. Since the mobilities of electronic carriers are often orders of magnitude greater than those of ionic defects, a correspondingly greater concentration of the latter is necessary for significant ionic conduction. When the majority of electrical current is carried by electrons or holes, the material is known as an *electronic conductor* ($t_{elect.} \sim 1$), and when ionic defect populations and mobilities are sufficiently high that the ionic current overwhelms the electronic, *ionic conduction* ($t_{ionic.} \sim 1$) results. *Mixed conduction* refers to those circumstances under which significant fractions of the conductivity are carried by both electronic and ionic defects ($t_{elect.} \sim t_{ionic}$).

The temperature dependence of conductivity (or its reciprocal, the resistivity, $\rho = 1/\sigma$) depends on the temperature dependence of both concentration and mobility. In metals, where the free electron concentration does not vary significantly with temperature, the resistivity depends only on mobility and increases with temperature. In comparison, the temperature dependence of electronic conductivity in semiconductors and insulators is often dominated by the temperature dependence of the carrier concentration, which may increase exponentially with temperature. Compounds are frequently referred to as metallic or semiconducting based on the temperature dependence of the electronic conductivity. Ionic conduction, on the other hand, always includes a thermally activated mobility, while the temperature dependence of carrier concentration may be strong or weak, depending on the defect structure. The differences in the temperature dependence of conductivity are shown in Fig. 3.2.

We next illustrate the relationship between defect chemistry and electrical conduction in a few specific systems. As discussed in Chapter 2, pure oxides that are easily reduced, such as TiO_2, SnO_2, ZnO, $BaTiO_3$, and $SrTiO_3$, are *n*-type semiconductors, while oxides that are easily oxidized, such as the transition metal monoxides, are *p*-type semiconductors. We earlier discussed the *n*-type conductivity in reduced TiO_2 with respect to applications as oxygen sensors. Here we give examples of *p*-type electronic conductivity, entirely ionic conductivity, and mixed electronic-ionic conductivity.

Cobalt Oxide and Nickel Oxide: *p*-Type Electronic Conductors

One of the principal experimental methods for studying defect structures in ceramics is the measurement of electrical conductivity. In the transition metal monoxides, the electronic conductivity is *p*-type due to native oxidation (Eqs. 3.21–3.23). Furthermore, at all temperatures the electron hole mobility is much greater than the ionic mobility, such that the conductivity is entirely electronic. If the electron hole mobility is known (e.g., from a Hall measurement) the electrical conductivity can be used to determine the hole concentration, and therefore the ionized vacancy concentration. Figure 3.11 showed the Co diffusion coefficient and Brouwer diagram for CoO. The electrical conductivity corresponding to these results is shown in Fig. 3.15. Notice the slight upward curvature in the log conductivity–log oxygen pressure curves, which parallel the Co diffusion results shown in Fig. 3.11a. This reflects the fact that the electroneutrality condition is governed by $p = [V'_{Co}] + 2[V''_{Co}]$. In NiO, similar results are obtained since its defect structure is also dominated by cation vacancies. The electrical conductivity of NiO as a function of oxygen pressure is shown in Fig. 3.16a. This p-type electronic conductivity has been interpreted using a point defect model which corresponds to the Brouwer diagram shown in Fig. 3.16b.

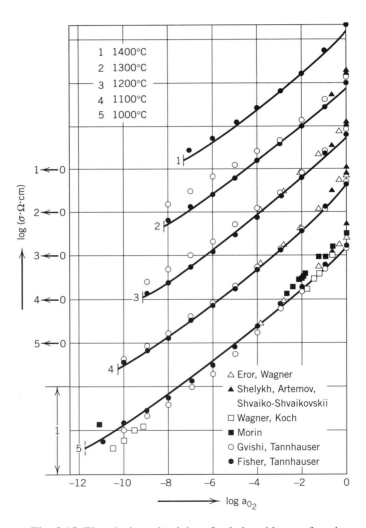

Fig. 3.15 Electrical conductivity of cobalt oxide as a function of oxygen activity at various temperatures. [From R. Dieckmann, Z. Physik. Chemie N.F., 107, 189 (1977).]

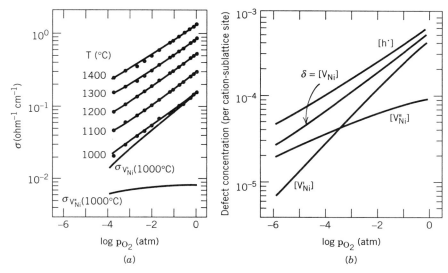

Fig. 3.16 (*a*) p-type electrical conductivity of NiO as a function of oxygen partial pressure, at several temperatures. The relative contributions of electron holes charge-compensating singly and doubly ionized nickel vacancies are shown. (*b*) Corresponding defect concentrations as a function of oxygen partial pressure, at 1400°C. [From N.L. Peterson, *Mat. Sci. Forum*, 1, 85 (1984).]

Mixed Electronic-Ionic Conduction in MgO

In section 2.1 we evaluated the relative concentrations of intrinsic ionic and electronic defects in MgO. In the pure intrinsic case, would this system be an ionic or electronic conductor at elevated temperatures? It is necessary to consider the relative mobilities of ionic and electronic defects as well as the concentrations. We showed earlier that the concentration of intrinsic Schottky defects ($\sim 1.4 \times 10^{11}$/cm^3) is about 40 times greater than those of electronic defects ($\sim 3.6 \times 10^9$ cm^{-3}) at a temperature of 1600°C. The ionic current will be carried primarily by the cation vacancies as they are more mobile than oxygen vacancies; the diffusivity of the magnesium vacancy has been found by D. Sempolinski et al. (*J. Am. Ceram. Soc.*, 63[11–12] 664 (1980)) to be

$$D_{V_{Mg}''} = 0.38 \exp\left(\frac{-2.29 \text{ eV}}{kT}\right) \quad (\text{cm}^2/\text{sec}) \quad (3.41)$$

From the Nernst-Einstein equation, the electrical mobility of the vacancy at 1600°C is

$$\mu_{V_{Mg}''} = \frac{2eD_{V_{Mg}''}}{kT} = 3.3 \times 10^{-6} \quad (\text{cm}^2/\text{V} \cdot \text{sec}) \quad (3.42)$$

However, over this temperature range the electron and hole mobilities (24 and 7 cm^2/V·sec respectively) are $\sim 10^6$ greater than the cation vacancy mobility. Thus, we expect intrinsic MgO to be an electronic conductor despite the deficit in concentrations of electrons and holes relative to those of cation and anion vacancies.

Let's now consider a more realistic, impurity-dominated case. Data are available (Fig. 3.17) for MgO with an aluminum content of 400 mole ppm (2×10^{19} cm^{-3}). The Brouwer approximation is $[Al_{Mg}^{\cdot}] = 2[V_{Mg}'']$. Barring any association, the ionic conductivity will again be mostly carried by the cation vacancies. The ionic conductivity is

$$\sigma_{ionic} = \sigma_{V_{Mg}''} = (z_{V_{Mg}''} e)(C_{V_{Mg}''})(\mu_{V_{Mg}''})$$
$$= 2\,e\,1\times10^{19}\text{ cm}^{-3})\,(7.6\times10^{-6}\text{ cm}^2/\text{V}\cdot\text{sec})$$
$$= 2.4\times10^{-5}\,\Omega^{-1}\text{cm}^{-1} \quad (3.43)$$

The electronic conductivity can be either *p*-type or *n*-type depending on the ambient oxygen pressure. These are minority species, but we see from the Brouwer diagram in Fig. 2.13 that electrons are higher in concentration at low oxygen pressure, while holes are higher in concentration at high oxygen pressure. In air ($P_{O_2} = 0.21$ atm), the electronic conductivity is *p*-type. We can determine the hole concentration using the oxidation reaction:

$$\frac{1}{2}O_2 \rightarrow V_{Mg}'' + O_O^x + 2h^{\cdot} \quad (3.44)$$

for which the equilibrium constant for this reaction has been determined to have a value of

$$K_O = \frac{[V_{Mg}'']p^2}{P_{O_2}^{1/2}} \approx 10^{64}\exp\left[\frac{-6.24\text{ eV}}{kT}\right] \quad (3.45)$$

in defect concentration units of number/cm^3 and oxygen pressure units of megapascals (MPa). The concentration of magnesium vacancies is pinned by the concentration of extrinsic dopant. Using Eq. 3.45, we can calculate the hole concentration *p* at 1600°C in air and use it to obtain the electronic conductivity:

$$\sigma_{elect} \approx \sigma_h = pe\mu_h = (5\times10^{13}\text{ cm}^{-3})e(7\text{ cm}^2/\text{V}\cdot\text{sec})$$
$$= 5.6\times10^{-5}\,\Omega^{-1}\text{cm}^{-1} \quad (3.46)$$

From the results in Eqs. 3.43 and 3.46 the ionic and electronic transference numbers t_{ionic} and t_{elect} are 0.3 and 0.7, respectively. We see that MgO is a *mixed conductor* under these conditions, even though the ionic defect concentration is more than 10^5 greater that the electronic (2×10^{19} cm^{-3} vs 5×10^{13} cm^{-3})!

As the oxygen pressure is reduced, Eqs. 3.45–3.46 and the Kröger–Vink diagram in Fig. 2.13 indicate that the *p*-type electronic conductivity will decrease, and eventually some measure of *n*-type conductivity will be achieved. Since the ionic mobility remains unchanged throughout (the Brouwer approximation remains dominated by the dopant), t_{ionic} should first increase and then decrease as P_{O_2} decreases. Experimental measurements of the electronic conductivity in MgO containing 400 ppm Al show the expected v-shaped *n*-to-*p* transition, Fig. 3.17. The ionic transference number is also shown as a function of P_{O_2}, and exhibits a peak, which corresponds to the minimum in the electronic conductivity.

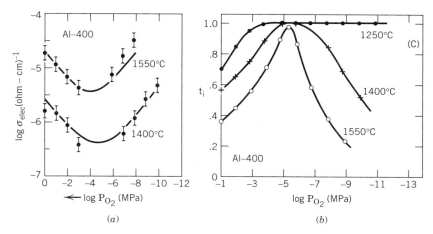

Fig. 3.17 Measurements of (a) electronic conductivity and (b) ionic transference number in MgO containing 400 ppm Al. [From D. Sempolinski, W.D. Kingery, and H.L. Tuller, *J. Am. Ceram. Soc.*, 63[11-12], 669 (1980), and D. Sempolinski and W.D. Kingery, *J. Am. Ceram. Soc.*, 63[11-12] 664 (1980).]

Ionic Conduction in Cubic ZrO_2

We discussed earlier the high diffusion coefficient of oxygen in cubic ZrO_2 that results from oxygen vacancies introduced to charge-compensate stabilizer dopants (Ca, Y). The bandgap of cubic ZrO_2 is ~5.2 eV, and a calculation as done above for MgO shows that electrical conductivity should be overwhelmingly ionic. Because conduction is carried by the most mobile ionic defect, the diffusion coefficient of concern is that for the vacancy $V_O^{\cdot\cdot}$, not oxygen ion *per se*, even though each atomic jump of a vacancy does result in counter-diffusion of an oxygen ion. It is left as a homework assignment to calculate the ionic conductivity due to oxygen vacancies; Fig. 3.18 shows some results for the oxygen diffusion coefficient determined from measurements of the ionic conductivity.

Ionic conductivity in cubic zirconia initially increases with stabilizer concentration, reaching a maximum at 12–13% of CaO stabilizer and 8–9% of Y_2O_3, but then decreases again at higher concentrations. As discussed previously, the reason that it does not increase without limit, even well before the solubility limit is reached, is that defect interactions become inevitable. The concentrations can be so high that the defects literally cannot avoid one another. A useful way to view defect association in these systems is as a competition between the stabilizer cation and Zr^{4+} for coordinating oxygen vacancies (I. W. Chen, P. Li, and J. Penner-Hahn, *Mat. Res. Soc. Symp. Proc.*, 307, 27, 1993). EXAFS (x-ray absorption fine structure) studies probing the local environment of the cations indicate that the oxygen vacancy tends to remain as a nearest-neighbor of Zr when the stabilizer dopant is larger than Zr, and to be preferentially coordinated with the stabilizer cation when it is smaller. The decrease in mobility of the oxygen vacancy with

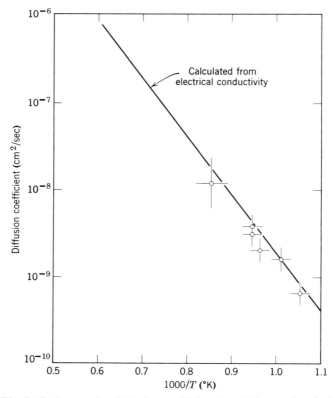

Fig. 3.18 Oxygen ion diffusion coefficient in relation to electrical conductivity in $Zr_{0.85}Ca_{0.15}O_{1.85}$. Solid line is calculated from conductivity data. Experimental points are direct measurements.

increasing concentration is probably related to this kind of local trapping, with corresponding variations in activation energy (Figs. 3.13, 3.14).

As a practical matter, results in polycrystalline zirconia can also be complicated by microstructural effects. Zirconia grain boundaries frequently contain a thin siliceous film that is much more resistive than the lattice. The total resistance of a polycrystal can be dominated by the grain boundaries, in which case separation of lattice and interface effects by methods such as impedance spectroscopy are necessary for a clear understanding of electrical properties.

Conductivity in $SrTiO_3$

Perovskites such as barium titanate and strontium titanate have two cation sublattices onto which aliovalent solutes can substitute, in addition to being easily reduced. Both are, however, line compounds from the cation stoichiometry viewpoint; the Ba/Ti and Sr/Ti ratios are very near unity. The maximum deviation in the Ba/Ti and Sr/Ti ratios at temperatures less than 1200°C is ~100 parts-per-million. In order to obtain a good n-type semiconductor, solutes such as La^{3+}, Nb^{5+}, or Ta^{5+} are commonly added to $BaTiO_3$, and form shallow donor levels in the bandgap. On the other hand, undoped materials typically contain trivalent cation impurities such as Fe^{3+} and Al^{3+}, which when substituted for Ti^{4+} act as acceptor dopants.

In this example, we consider a typical lightly acceptor-doped $SrTiO_3$. The incorporation of aluminum impurity can be written:

$$Al_2O_3 + 2SrO = 2Al'_{Ti} + 2Sr^x_{Sr} + 5O^x_O + V^{\cdot\cdot}_O \quad (3.47)$$

Notice that either the absorption of SrO or rejection of TiO_2 is necessary to maintain the stoichometric cation ratio. Equation 3.47 shows the ionically compensated extreme of acceptor doping; if this is the dominant mechanism, the Brouwer approximation is given by $[Al'_{Ti}] = 2[V^{\cdot\cdot}_O]$. It is also possible for the aluminum acceptor to be entirely electronically compensated at another extreme:

$$Al_2O_3 + 2SrO + \frac{1}{2}O_2(g) = 2Al'_{Ti} + 2Sr^x_{Sr} + 6O^x_O + 2h^{\cdot} \quad (3.48)$$

in which case the Brouwer approximation is: $[Al'_{Ti}] = p$. (Eqs. 3.47 and 3.48 are related by the oxidation reaction $1/2\,O_2(g) + V^{\cdot\cdot}_O = O^x_O + 2h^{\cdot}$.) Experimental observations such as those discussed below show that in $SrTiO_3$ and $BaTiO_3$, acceptors are predominantly ionically compensated as in Eq. 3.47.

G.-M. Choi and H.L. Tuller (*J. Am. Ceram. Soc.*, 71[4] 201, 1988) have measured the electrical conductivity of a single crystal close to pure $SrTiO_3$ in composition ($Ba_{0.03}Sr_{0.97}TiO_3$), and obtain the results shown in Fig. 3.19. No intentional dopants were added (except for the 3% $BaTiO_3$, which has no significant effect on the defect chemistry). Three distinct regimes are seen in the log conductivity ver-

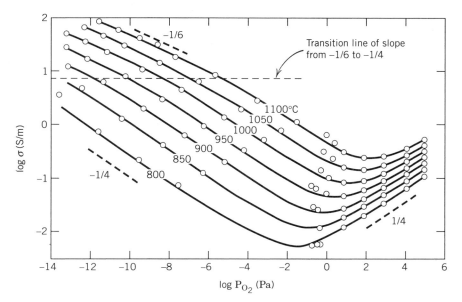

Fig. 3.19 Electrical conductivity in $Ba_{0.03}Sr_{0.97}TiO_3$ as a function of oxygen partial pressure, at various temperatures. [From G.-M Choi and H.L. Tuller, *J. Am. Ceram. Soc.*, 71[4], 201 (1988).]

sus log P_{O_2} curve, with slopes of -1/6, -1/4, and +1/4, respectively, from which we can understand the defect structure of this system.

At low oxygen partial pressures and high temperatures, we might expect native reduction of the compound to be the dominant form of disorder. Since the perovskite structure does not easily accommodate cation interstitials, reduction is likely to be accommodated by the formation of oxygen vacancies:

$$O_O^x = \frac{1}{2} O_2(g) + V_O^{\cdot\cdot} + 2e'$$

for which the equilibrium constant is

$$K_R = [V_O^{\cdot\cdot}] n^2 P_{O_2}^{1/2} = K_R^0 \exp\left[\frac{-\Delta H_R}{kT}\right] \quad (3.49)$$

Choi and Tuller report numerical values of $K_R^0 = 4.81 \times 10^{71}$ atm$^{-1/2}$cm^9 and $\Delta H_R = 5.19$ eV. Since the Brouwer approximation is $2[V_O^{\cdot\cdot}] = n$, the oxygen pressure dependence of n is, from Eq. 3.49, $n \propto P_{O_2}^{-1/6}$. The slope of -1/6 at the lowest oxygen pressure and highest temperatures in Fig. 3.18 is consistent with this result.

With increasing oxygen partial pressure, the conductivity decreases, as does the oxygen vacancy concentration (since $2[V_O^{\cdot\cdot}] = n$), until a transition to the region of −1/4 slope occurs. This slope is consistent with the result expected for a oxygen vacancy concentration no longer determined by reduction, but pinned at a constant level by background acceptors, e.g., $[Al'_{Ti}] = 2[V_O^{\cdot\cdot}]$. If $[V_O^{\cdot\cdot}]$ is indeed fixed, the equilibrium constant in Eq. 3.49 yields $n \propto P_{O_2}^{-1/4}$. These results confirm that vacancy compensation of the acceptor is preferred. If electronic compensation (Eq. 3.48) occurred instead, we expect to see a regime of P_{O_2} over which the conductivity is p-type and constant, being determined by the Brouwer approximation $[Al'_{Ti}] = p$. In contrast to this example, complete electronic compensation does occur in donor-doped BaTiO$_3$ and is important for its applications as a conductive ceramic.

The final regime in Fig. 3.19 has a slope of +1/4. Notice that if n *decreases* as $P_{O_2}^{-1/4}$, p must *increase* as $P_{O_2}^{+1/4}$ (since $np = K_i$). Within the same Brouwer regime, a transition from n to p type conductivity can occur. The conductivity in the high P_{O_2} regime of Fig. 3.19 appears to be p-type, and the oxygen pressure dependence is consistent with the Brouwer regime $[Al'_{Ti}] = 2[V_O^{\cdot\cdot}]$.

One must be cautious in interpreting defect structures from the P_{O_2} dependence of transport properties, for two different defect models may have identical dependences. For instance, when TiO$_2$ is reduced, the defects may be oxygen vacancies or titanium interstitials. A $P_{O_2}^{-1/4}$ dependence of the n-type conductivity can arise when fully ionized oxygen vacancies ($V_O^{\cdot\cdot}$) are formed to compensate accidental acceptors. Yet, the same $P_{O_2}^{-1/4}$ dependence can result when reduction forms interstitial $Ti_i^{\cdot\cdot\cdot}$ defects compensated by electrons, or if accidental acceptors are compensated by fourfold ionized interstitials, $Ti_i^{\cdot\cdot\cdot\cdot}$. (It is left to the reader to verify that these dependences are correct.) The P_{O_2} dependence alone cannot distinguish between these alternative defect models. For the results shown in Fig. 3.19, consistency of the interpretation over a wide range of oxygen partial pressure and two different Brouwer regimes presents a compelling story. In other instances the difference in slope between alternative defect models may be too slight compared to the precision of experimental data (e.g., 1/6 vs. 1/5, or 1/5 vs. 1/4 slopes) for a definitive interpretation.

SPECIAL TOPIC 3.1

NONLINEAR ELECTRICAL CERAMICS: VARISTORS AND THERMISTORS

The useful properties of a number of electronic ceramics are based on resistive barriers between grains in a polycrystal. Some "passive" examples include barrier-layer capacitors based on $SrTiO_3$, and $BaTiO_3$, and X7R dielectrics based on compositionally graded $BaTiO_3$ grains, discussed in Chapter 1. Silicon carbide to which some BeO has been added is another example, used as a support for semiconductor devices (termed a "packaging" material). Pure SiC has a high thermal conductivity that is useful for heat dissipation but is too electrically conductive for use as a packaging dielectric. The BeO additions segregate to grain boundaries and form insulating layers which render the polycrystal as a whole a good electrical insulator.

Here we will discuss two types of widely used electronic ceramics in which grain boundaries play a more active role: varistors and thermistors. Both are polycrystalline materials in which the electrical properties are highly nonlinear as a result of grain boundary electrical barriers. Varistors (i.e., *variable resistors*) exhibit highly nonlinear current-voltage characteristics useful for switching functions; they act like solid-state circuit breakers (and do not need to be reset). Thermistors (*thermal resistors*) show strong variations in resistance with temperature useful for sensors or as self-regulating heater elements. While all materials show some variation in resistivity with temperature (see Fig. 3.2), not all are practical thermistors. *Positive temperature coefficient* (PTC) thermistors based on $BaTiO_3$ solid solutions show a particularly large increase in resistivity, of up to a factor of 10^7, over a few degrees near the Curie temperature of the material. This behavior is related to the formation and electrical compensation of conduction barriers at grain boundaries. There is also a class of *negative temperature coefficient* (NTC) thermistors, which are not based on grain boundary barriers. These are transition metal oxides [e.g., Li-doped monoxides such as MnO, CoO, NiO or their solid solutions; $(Ni,Mn)_3O_4$ or $(Ni, Mn, Co)_3O_4$ spinels, and $(Fe,Ti)_2O_3$ hematite solid solutions] containing multivalent cations in which the semiconductivity is quite temperature dependent due to the thermally activated or "hopping" mechanism of conduction between cations of differing valence on equivalent sites (e.g. between Ni^{2+}/Ni^{3+} or Mn^{4+}/Mn^{3+}). Compared to the PTC thermistors, these materials exhibit a more gradual change in resistance over a broader temperature regime (about 10^4 change in resistivity over ~100°C).

Varistors

Figure ST23 shows the current-voltage characteristics of a varistor. At low voltages a varistor is ohmic (linear I-V relation), but above a certain *threshold* or *breakdown* voltage, large amounts of current are passed and the apparent resistivity becomes extremely low. In this breakdown regime the current varies with voltage approximately as the power law:

$$I \propto V^\alpha$$

The exponent α is a measure of how rapidly current increases with applied voltage and is often used as a figure of merit; good varistors may have values as high as 50. At still higher voltages a second ohmic regime occurs. Unlike dielectric breakdown, varistor breakdown is reversible, and upon decreasing voltage below the breakdown threshold the varistor becomes ohmic again (although degradation can result if a varistor is held in the breakdown regime and large amounts of resistive self-heating are allowed to occur). A simple application of a varistor is as a surge protector connected in parallel with the device to be protected. In this mode, overvoltages are shunted through the varistor to ground before they can damage the apparatus. Varistors are made on widely varying scales depending on the application, ranging from low-voltage devices with a few boundaries exhibiting breakdown at a couple of volts to million-volt varistors a few feet in length used as lightening arrestors or for

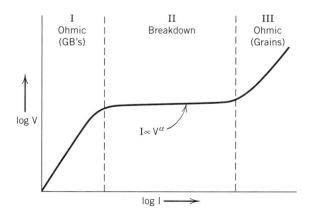

Fig. ST23 Current-voltage characteristics of a varistor.

3.3 Electrical Conductivity 227

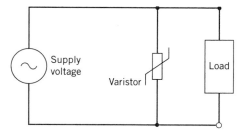

Fig. ST24 Use of a varistor in a parallel circuit with load, for overvoltage protection.

power distribution circuits. Most common are discrete elements with breakdown voltages of tens to hundreds of volts. A typical circuit application of a varistor is shown in Fig. ST24, where the varistor in parallel with the electrical device acts as a shunt to protect against overvoltages.

Varistors can be made from a variety of semiconducting ceramics including SiC, ZnO, TiO_2, and $SrTiO_3$, but the most widely used commercial compositions are based on ZnO. Fig. ST25 shows the microstructure of a typical commercial varistor. Most contain several metal oxide additives and have at least a couple of impurity phases in addition to the primary phase ZnO. One

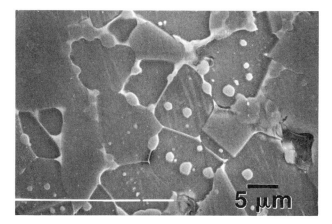

Fig. ST25 Polyphase microstructure of a zinc oxide varistor with multiple oxide additives.

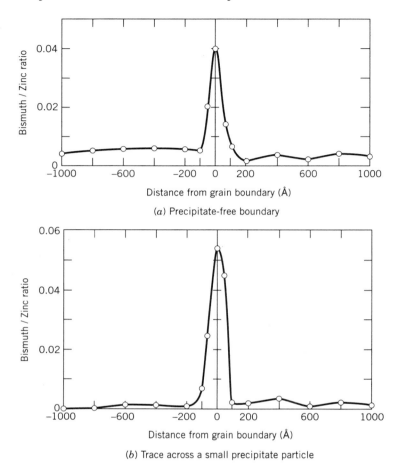

Fig. ST26 Bismuth to zinc ratio measured across (*a*) a grain boundary with no precipitate, showing the segregation of Bi within the solid phase, and (*b*) a small precipitate particle. Measured using scanning transmission electron microscopy. [From W.D. Kingery, J.B. Vander Sande, and T. Mitamura, *J. Am. Ceram. Soc.*, 62 [3–4] 221 (1979).]

type of important additive is a donor dopant such as Co, Sb, or Fe, since zinc oxide is a ~3.2-eV bandgap semiconductor that is not highly conductive when pure and stoichiometric. In either a slightly reduced or donor-doped state, zinc oxide becomes an extrinsic *n*-type semiconductor. The most critical dopant in a varistor, however, is one which segregates to the grain boundaries and results in the formation of the conduction barrier. Bismuth is the prototypical dopant (praseodymium is another example), which when added to ZnO concentrates in a partial monolayer at the grain boundary (Fig. ST26). A simple varistor can be made by painting some Bi_2O_3 powder on a sintered ZnO disc and firing above 820°C to allow some transport of molten Bi_2O_3 along the grain boundaries. Electrically speaking, the segregation of bis-

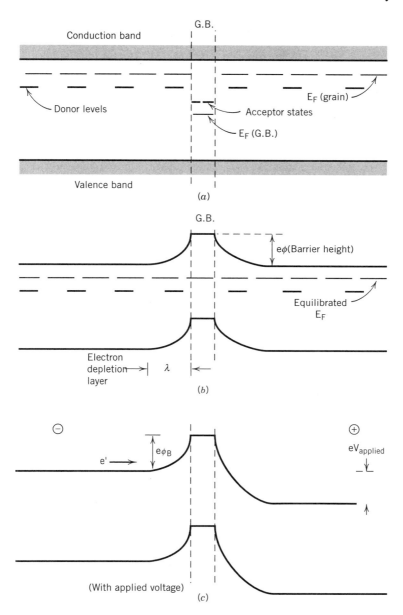

Fig. ST27 (*a*) Acceptor states located at the grain boundary of a zinc oxide varistor lower the Fermi level relative to the donor-doped or nonstoichiometric grains. (*b*) Equilibration of the Fermi level results in band-bending and the formation of a barrier to electron conduction across the boundary. (*c*) Under applied electrical bias the barrier is systematically reduced until the breakdown voltage is reached, permitting unimpeded electron conduction across the boundary.

muth results in the formation of spatially localized acceptor states or traps at energy levels within the bandgap (Fig. ST27a). It is energetically favorable for these to become filled with conduction band electrons. The Fermi level of the grain boundary region is then depressed from the Fermi level in the grain. However, at equilibrium the Fermi level (chemical potential of electrons) is equalized throughout the solid. When this occurs, the conduction and valence band bend, as shown in Fig. ST27b. Conduction band electrons must surmount the potential barrier at grain boundaries for electrical conduction to occur. These barriers are joined in series across the varistor, and when a macroscopic voltage is applied, the potential gradient is primarily supported by the boundaries rather than across the conductive grains (Fig. ST27c). When the voltage across each boundary is low compared to the barrier height, conduction is provided by thermal activation of electrons across the barrier, and the resistivity is high but ohmic (region I in Fig. ST23). As the voltage across the grain boundary increases, the effective barrier height approaches zero and a massive increase in electron conduction occurs. This is the breakdown region (II) of the curve in Fig. ST23. Region III is again ohmic, due to the intrinsic resistance of the grains.

The total breakdown voltage of a varistor is the series sum of all the individual barrier heights, each of which is approximately 3 V in zinc oxide varistors. Some variation in barrier height from boundary to boundary does occur, and the current is expected to follow the lowest potential path. In a uniform varistor, the breakdown voltage can be estimated from the grain size and thickness (e.g., a 1-mm-thick varistor with 10-µm grains will have a breakdown voltage of about 300 V).

The physics of varistor operation have been widely studied and have elements in common with the Schottky barriers formed at metal-semiconductor junctions; varistor boundaries are often referred to as back-to-back Schottky barriers. The relationship of the electrical barriers to the chemistry of varistor boundaries is less well understood. It is also known that oxidation of the grain boundary during the post-sintering cooling cycle plays an important role in activating the grain boundary states. In fact, a quenched varistor exhibits weak or zero nonlinearity. However, the way in which segregation of solutes such as Bi cause acceptor state formation remains unclear despite many years of study (note that Bi^{3+} is, after all, a donor solute). One hypothesis is that large, misfitting solutes stabilize acceptor defects of another kind, such as zinc vacancies, at the boundary.

Positive Temperature Coefficient (PTC) Thermistors

The very high positive temperature coefficient of resistance near the Curie point of polycrystalline barium titanate was first discovered at Philips Laboratories in 1955[5]. Although metals have a positive temperature coefficient of resistance, it is exceedingly weak in comparison to that seen in PTC barium titanate, (Fig. ST28). Over a small range of temperature near the Curie point,

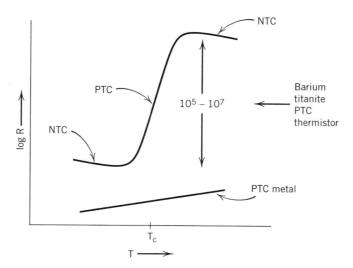

Fig. ST28 Resistance-temperature characteristics of a BaTiO$_3$ positive-temperature-coefficient (PTC) thermistor.

the resistance of a commercial PTC thermistor increases by more than a factor of 10^5 (as much as 10^7 in some laboratory results). This property has proven useful in many temperature sensing applications, including temperature-regulated cooking appliances, automatic chokes in automobiles (where the thermistor provides the electrical feedback to change the fuel/air mixture once the engine is warm), and self-regulating household heaters where sintered barium titanate is the resistively heated element. Heaters based on PTC barium titanate cannot overheat and create a fire hazard since the maximum temperature is determined by the Curie point, above which the high resistivity prevents further heating.

An explanation for this unusual property was first proposed by W. Heywang in 1961 and remains widely accepted in its essence today.[5] PTC compositions are donor-doped in order to achieve n-type semiconducting grains, like the varistors discussed above. These donors may occupy either the Ba sublattice (such as La^{3+}, Gd^{3+}, or any number of large, trivalent cations) or the Ti sublattice (including Nb^{5+} and Ta^{5+}). Then, a firing cycle that includes oxidative cooling is necessary, in order to oxidize the grain boundary regions within the sintered compact (the grain boundary diffusion of oxygen in this system, like many oxides, is much faster than lattice diffusion of oxygen). This oxidation process results in the formation of grain boundary electrical barriers analogous to those in varistors. As in the case of zinc

---

[5] For a review, see J. Daniels, K.H. Haerdtl, R. Wernicke, "The PTC Effect of Barium Titanate," *Philips Tech. Rev.*, 38[3], 73-82 (1978/79).

oxide varistors, a quenched thermistor will show poor or no grain boundary activity. The barrier arises from trapped electrons at the boundary, adjacent to which form electron depletion layers a few tenths of a micron wide. The high resistance in the region above the Curie point (T_c) results from these grain boundary barriers. Notice that there is a negative temperature coefficient in this region (Fig. ST28), characteristic of a semiconductor.

What then causes the massive decrease in resistance upon cooling through the Curie point? Heywang explained this as a compensation of the grain boundary electrical barrier by the preferential polarization of ferroelectric domains in the near-boundary electric field. We discussed in Chapter 1 the formation of ferroelectric domains in $BaTiO_3$, which in the absence of applied electric field are oriented so as to yield zero net polarization. However, the electric field adjacent to the grain boundary can be as high as 10^5 V/cm (the barrier height is 2–3 V, and the space-charge depletion layer width is typically 0.1–0.2 microns). Upon cooling through the Curie point, this field is sufficient to pole the ferroelectric. The domain structure in the near-boundary region resulting from this process is shown schematically in Fig. ST29. Notice that poling causes a compensation of the excess negative charge on the boundary, for the surfaces of a poled ferroelectric must bear an excess charge. Charge compensation lowers the grain boundary conduction barrier and is responsible for the large decrease in resistivity. Below the Curie point, the PTC thermistor once again shows a slightly negative temperature coefficient due to the semiconducting grains that now dominate the material's resistivity.

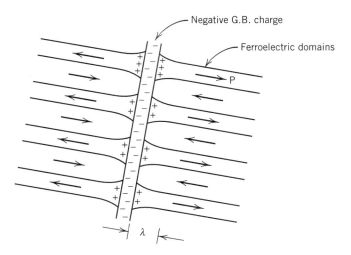

Fig. ST29 Preferential polarization of ferroelectric domains adjacent to charged boundary compensates the electrical potential barrier to conduction at temperatures below the Curie temperature.

The switching temperature of a PTC thermistor therefore corresponds closely to the Curie point of the base composition. $BaTiO_3$ is useful in this regard because the Curie point (140°C for pure $BaTiO_3$) can be varied over quite a wide range by forming solid solutions with other perovskites. Alloying with $PbTiO_3$ raises the Curie point, while alloying with $SrTiO_3$ lowers the Curie point.

3.4 THE ELECTROCHEMICAL POTENTIAL

To this point, we have considered separately the effects of gradients in chemical potential and electrical potential on the mobility of defects. In ionic systems both driving forces are often simultaneously present. Even in the absence of externally applied electric fields, internal electric fields can arise due to a nonuniform distribution of charged species. Consequently the total driving force for mass transport is the *electrochemical potential* rather than the chemical potential alone.[6] The electrochemical potential of a species i, denoted η_i, is the sum of the chemical potential μ_i and the electrical potential ϕ acting upon it:

$$\eta_i = \mu_i + z_i F\phi \tag{3.50}$$

where z_i is the effective charge and F is Faraday's constant (F = eN_a = 96,500 C/mole). Taking the virtual force on the particle to be the negative gradient of η_i rather than of μ_i alone,

$$F_i = -\frac{1}{N_a}\left(\frac{d\eta_i}{dx}\right) \tag{3.51}$$

it becomes apparent that the flux is a function of the electric field $d\phi/dx$:

$$J_i = -\frac{c_i B_i}{N_a}\left(\frac{d\eta_i}{dx}\right) = -\frac{c_i B_i}{N_a}\left(\frac{d\mu_i}{dx} + z_i F\frac{d\phi}{dx}\right) \tag{3.52}$$

Consideration of Eq. 3.52 shows that even a relatively modest electric field acting on an ion can offset the effects of a large concentration gradient in the opposite direction. The electric field may be externally applied, or it may be an internal field that arises when there is a motivating force tending to separate charge carriers. This internal field leads to the preservation of charge neutrality in a solid during *ambipolar diffusion*, which is the coupled transport of species of different charge. These may be ionic or electronic in nature, and charge neutrality may be preserved by transport of like-charged species in a parallel direction, or by the counter-diffusion of oppositely charged species. Examples are given later.

---

[6] Additional driving forces such as stress or temperature gradients are also able to act simultaneously on charged species, but we will neglect these in the current discussion.

The Nernst Equation and Applications of Ionic Conductors

Since the electrochemical potential (Eq. 3.50) is the sum of a chemical potential and electrical potential it is possible to vary one independently to induce a change in the other: An applied chemical potential gradient can be used to generate a voltage, and vice versa. This principle is the basis for fuel cells and batteries, electrochemical sensors, ion pumps, and ion activity probes.

Most automobiles utilize a galvanic oxygen sensor in the exhaust manifold to measure the oxygen pressure of the exhaust gas (Fig. 3.20). This type of sensor separates two gases of differing oxygen activity with an oxygen ion conductor. When the electrochemical potential is equilibrated, the difference in oxygen chemical potential induces a voltage across the sensor. The voltage output of the sensor is sent in a feedback loop to control the air/fuel mixture for optimal combustion and minimal exhaust pollution (see also the discussion of resistive TiO_{2-x} sensors in Chapter 2).

Ion transport is necessary for the imposed oxygen pressure gradient to establish a gradient in oxygen concentration. Simultaneously, electronic conduction must be avoided to prevent short-circuiting of the electrolyte. Accordingly, the system must be an ionic conductor ($t_i=1$). If we bring the system to equilibrium, the electrochemical potential is the same on both sides of the electrolyte:

$$\eta_i \text{ (side I)} = \eta_i \text{ (side II)}.$$

The electrical potential across the electrolyte is related to the chemical potential difference by

$$\mu_i(side\ I) - \mu_i(side\ II) = \int_I^{II} -z_i F \frac{d\phi}{dx} = -z_i F \Delta\phi \tag{3.53}$$

The chemical potential of oxygen is given by $\mu_O = \mu_O^o + 1/2\, RT \ln P_{O_2}$. If we impose an oxygen pressure $P_{O_2}^I$ on one side and $P_{O_2}^{II}$ on the other, upon substituting into Eq. 3.53 and rearranging we obtain the voltage across the sample:

$$\Delta\phi = \frac{RT}{4F} \ln \frac{P_{O_2}^I}{P_{O_2}^{II}} \tag{3.54}$$

The measured voltage directly yields the ratio of oxygen partial pressure $P_{O_2}^I / P_{O_2}^{II}$ across the solid electrolyte. In automotive applications one side is usually air, at a reference $P_{O_2}^I$ of ~0.21 atmospheres, and the other side is a much more reducing exhaust gas at a $P_{O_2}^{II}$ of 10^{-9} to 10^{-21} atmospheres. At a typical operating temperature of 350°C, the voltage across the sensor is 0.25–0.6 V.

Equations 3.53 and 3.54 are forms of the *Nernst equation* (not to be confused with the Nernst-Einstein relation, Eq. 3.33). At equilibrium, Eq. 3.53 relates the standard free energy change of the virtual process (no current flow) to the voltage, $\Delta G^\circ = -z_i F \Delta\phi$. Eq. 3.54 gives the voltage under equilibrated, open circuit condi-

3.4 The Electrochemical Potential

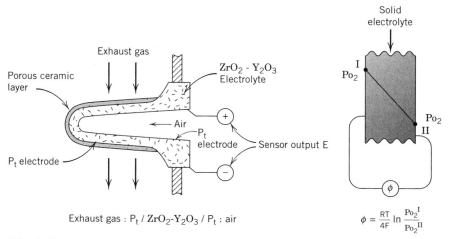

Fig. 3.20 Schematic view of a zirconia automotive exhaust sensor and principle of operation.

tions where no charge flows across the electrolyte. Away from equilibrium, the flux (Eq. 3.52) is no longer zero, and the voltage is reduced.

High-temperature fuel cells, considered a promising technology for clean and efficient energy production, utilize the same principle. By flowing a reducing gas such as hydrogen or hydrogen/carbon monoxide mixtures derived from fossil fuels over one side of a zirconia cell, and supplying air or oxygen to the other, an electrical potential is generated. However, fuel cells are intended to operate far from open-circuit conditions. As oxygen diffuses through the electrolyte and oxidizes the fuel (to water or carbon dioxide), the free energy of the chemical reaction is converted to an equivalent electron flow, that is used to do useful work in an external circuit. The operating voltage is less than the ideal Nernst value since significant current is drawn. A high ionic conductivity is necessary to minimize resistive losses in the fuel cell. A related application is the use of solid electrolytes as electrochemical pumps where an applied voltage is used to drive ions across a ceramic membrane. Another type of electrochemical device that utilizes cationic ceramic conductors is the storage battery, in which a solid electrolyte separates two reactants (e.g., sodium and sulfur) and utilizes the chemical reaction between the two to generate electrical power.

The Nernst effect takes place in any ionic conductor; from Eq. 3.53 we see that the potential generated depends on the chemical species and not the specific electrolyte. However, for most applications, a low resistance as well as completely ionic conduction is desirable. The search for materials of this type has uncovered many systems including fast oxygen conductors such as zirconia and ceria as well as rapid alkali-ion conductors such as β-alumina and borate glasses in which Na^+

Chapter 3 / Mass and Electrical Transport

and Li^+ are the conducting species, and metal-iodides such as AgI in which Ag^+ is the fast diffuser. Various fast ionic conductors and their current carrying species are listed in Table 3.5.

When the conductivity is not completely ionic, the open-circuit potential is reduced from the value given in Eq. 3.54. Detailed analysis shows that it is proportional to the ionic transference number:

$$\Delta \phi = t_i \cdot \frac{RT}{4F} \ln \frac{P'_{O_2}}{P''_{O_2}} \qquad (3.55)$$

In fact, the deviation of the induced voltage from the ideal Nernst relationship can be used as a measure of the ionic transference number. Figure 3.21 shows results for an experiment on extrinsic MgO, in which the potential difference across a thin sample separating a reference gas (1 atm oxygen) and a reducing gas was measured. In Fig. 3.21a the conductivity has been measured at various temperatures as a function of P_{O_2} and shows a flattened region at intermediate P_{O_2} which is characteristic of extrinsic ionic conductivity. In Fig. 3.21b the quantity $4\Delta\phi F/RT$ is plotted against log P_{O_2}, and it is seen that the actual voltage falls below that given by Eq. 3.54. From the slope of the experimental curve is obtained the ionic transference number, plotted in Fig. 3.21c. The values of t_i show that this material is a mixed conductor under these conditions.

Ambipolar Diffusion

The need to avoid long-range charge separation during mass transport in ionic solids leads to coupling of charged species through a common internal electric field that has the effect of slowing down the faster species and accelerating the slower species. Both then diffuse with a common diffusivity in order to main-

Table 3.5 Fast Ion Conducting Ceramics

Composition	Current Carrying Ion
Cubic ZrO_2	O^{2-}
δ-Bi_2O_3	O^{2-}
α-Ta_2O_3	O^{2-}
$Gd_2(Zr_xTi_{1-x})_2O_7$ (pyrochlore structure)	O^{2-}
Li_2O-$LiCl$-B_2O_3 glasses	Li^+
$Li_4B_7O_{12}Cl$ (boracite)	Li^+
Ag_2O-AgI-B_2O_3 glasses	Ag^+
AgI	Ag^+
β-$NaAl_{11}O_{17}$	Na^+
β-Al_2O_3	alkali ions

3.4 The Electrochemical Potential 237

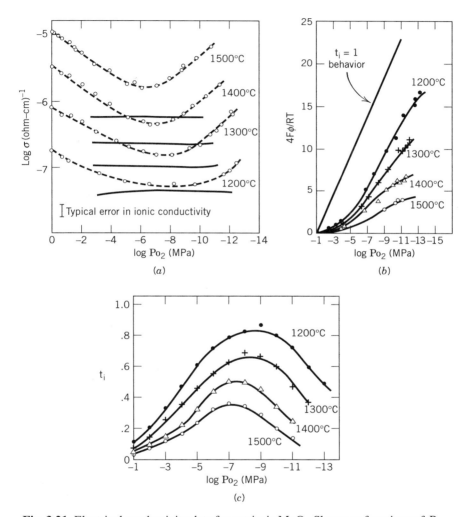

Fig. 3.21 Electrical conductivity data for extrinsic MgO. Shown as functions of P_{O_2} are (a) total and ionic (solid curves) conductivity; (b) Nernst potential. Notice the deviation from the ideal Nernst slope, indicating partial electronic conductivity. From the slope of the curves in (b) the ionic transference number is obtained, as plotted in (c). [From D. R. Sempolinski and W.D. Kingery, *J. Am. Ceram. Soc.*, 63[11-12], 664 (1980).]

238 Chapter 3 / Mass and Electrical Transport

tain overall charge neutrality. The chemical or ambipolar diffusion coefficient,[7] $\tilde{D}$, is an apparent diffusion coefficient that characterizes the overall rate of transport. Its value is bounded at high and low extremes by the diffusivities of the individual species involved. Examples of processes where coupled transport is required include the oxidation or reduction of an oxide, formation of an oxide film on a metal (corrosion), reactions between solids, and mass transport under stress gradients during sintering, diffusional creep, and hot-pressing. We can determine $\tilde{D}$ for these processes through a formalism that is illustrated in the two following examples.

Equilibration of Defect Structures   If the ambient oxygen pressure is changed from one value to another, the oxygen/metal ratio and defect structure of an oxide will change accordingly, the magnitude of this change being determined by defect equilibrium considerations discussed earlier. One manifestation of the transient stage in the approach to a new defect structure is a time-dependent relaxation of electrical conductivity upon changing atmosphere or temperature. Another graphic demonstration of a changing defect structure is shown in Fig. 3.22. The sequence of photographs shows the progressive migration of a clear boundary layer into a crystal of initially reduced TiO_2 as it is oxidized. The central, reduced region of this crystal remains dark due to optical absorption from the free electron carriers formed during high-temperature reduction. As discussed in Chapter 2, the ionic defects due to reduction of TiO_2 are shallow donors, which are ionized at room temperature. The clear boundary layer in this instance represents a different Brouwer approximation; it is transparent because under oxidizing conditions, background acceptor impurities present in the TiO_2 dominate the defect chemistry and are ionically compensated. Thus the boundary layer represents a transition between two limiting defect equilibria, similar to crossing the boundary between Brouwer regimes. The migration of the color boundary proceeds at a rate determined by the ambipolar diffusion coefficient.

Let's consider the simplified case where an oxide forms oxygen vacancies and free electrons upon reduction $\left(O_O^x \rightarrow \frac{1}{2} O_2(g) + V_O^{\cdot\cdot} + 2e' \right)$, resulting in the bulk electroneutrality condition $2\left[V_O^{\cdot\cdot} \right] = n$. Upon subsequent exposure to an oxidizing ambient, there is an outward flux of oxygen vacancies as shown schematically

---

[7] The two terms are sometimes used interchangeably in ceramics. Ambipolar diffusion is perhaps the more specific expression, as it implies transport of coupled charges, whereas chemical diffusion refers to the more general process of diffusion under a chemical potential gradient.

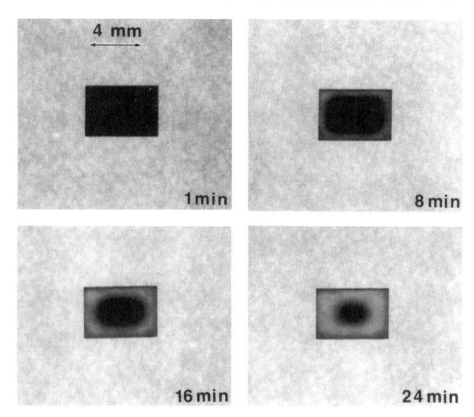

Fig. 3.22 Migration of a color front into an initially reduced TiO_2 single crystal as it is oxidized at 750°C in air. [From J.A.S. Ikeda, Y.-M. Chiang, and B.D. Fabes, *J. Am. Ceram. Soc.*, 73[6], 1633 (1990).]

in Fig. 3.23. Upon arrival at the surface, oxygen vacancies can be removed through the oxidation reaction:

$$\frac{1}{2}O_2(g) + V_O^{\cdot\cdot} + 2e' \rightarrow O_O^x. \quad (3.56)$$

For electroneutrality, this flux of oxygen vacancies must be matched by an equivalent charge flux of electrons outward, holes inward, or some combination of the two. Since the electron mobility is usually the higher of the two, let us for simplicity assume that they are the sole compensating species, whereupon the flux condition for maintaining electroneutrality is

$$2J_{V_O^{\cdot\cdot}} = J_{e'} \quad (3.57)$$

If there is a common ambipolar diffusion coefficient $\tilde{D}$ because the two spe-

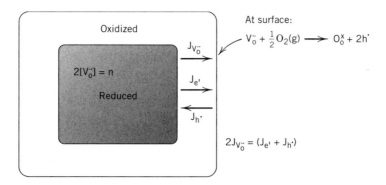

Fig. 3.23 Schematic showing outward flux of oxygen vacancies along with compensating flux of either electrons (outward) or electron holes (inward) upon oxidation of an initially reduced oxide in which the Brouwer approximation is $2[V_O^{\cdot\cdot}] = n$. (Compare with Fig. 3.22).

cies are coupled, then we may write the individual fluxes in terms of Fick's first law as:

$$J_{V_O^{\cdot\cdot}} = -\tilde{D}\left(\frac{dC_{V_O^{\cdot\cdot}}}{dx}\right) \tag{3.58}$$

and

$$J_{e'} = -\tilde{D}\left(\frac{dn}{dx}\right) \tag{3.59}$$

Equating the fluxes to satisfy Eq. 3.57, one finds that not only is charge neutrality maintained, but the concentration *gradients* of the two coupled species must obey:

$$2\left(\frac{dC_{V_O^{\cdot\cdot}}}{dx}\right) = \left(\frac{dn}{dx}\right) \tag{3.60}$$

where the concentration terms are in no./cm³.

We can now determine how $\tilde{D}$ depends on the two individual defect diffusivities, $D_{V_O^{\cdot\cdot}}$ and $D_{e'}$. We carry out a short derivation by rewriting Fick's first law in terms of the electrochemical potential acting on the two defects separately, and equating the fluxes to solve for the internal field, that develops as a result of the defect coupling. This electric field is used in Fick's first law to obtain an expression for $\tilde{D}$.

3.4 The Electrochemical Potential

The flux of each defect can be expanded as

$$2J_{V_{\ddot{O}}} = -\frac{2C_{V_{\ddot{O}}} D_{V_{\ddot{O}}}}{RT}\left(\frac{\partial \mu_{V_{\ddot{O}}}}{\partial x} + 2F\frac{\partial \phi}{\partial x}\right) \qquad (3.61)$$

and

$$J_{e'} = -\frac{nD_e}{RT}\left(\frac{\partial \mu_e}{\partial x} - F\frac{\partial \phi}{\partial x}\right) \qquad (3.62)$$

where Eq. 3.52 has been used to write the flux in terms of the electrochemical potential gradient and the Nernst–Einstein relation (Eq. 3.33) expresses the defect mobility. Notice that the electrical potential gradient (the internal field) acts in the opposite sense for the two defects; they bear opposite charges. Thus, the flux of one is raised and the other slowed by the presence of the internal field.

The internal field, $\partial \phi / \partial x$, is obtained by rewriting Eqs. 3.61 and 3.62 in terms of concentration gradients, equating the two according to Eq. 3.60, and rearranging terms. The result is:

$$\frac{\partial \phi}{\partial x} = \frac{RT}{F}\left[\frac{D_e \frac{\partial n}{\partial x} - 2D_{V_{\ddot{O}}} \frac{\partial C_{V_{\ddot{O}}}}{\partial x}}{nD_e + 4C_{V_{\ddot{O}}} D_{V_{\ddot{O}}}}\right] \qquad (3.63)$$

Upon substituting this result back into Eqs. 3.61 (or equivalently, Eq. 3.62), and making use of Eq. 3.60 and the Brouwer approximation $2C_{V_{\ddot{O}}} = n$, Eq. 3.61 can be rewritten as

$$J_{V_{\ddot{O}}} = -\left(\frac{3D_e D_{V_{\ddot{O}}}}{D_e + 2D_{V_{\ddot{O}}}}\right)\frac{\partial C_{V_{\ddot{O}}}}{\partial x} \qquad (3.64)$$

By comparing this result with Eq. 3.58, it can be seen that the term in brackets is the ambipolar diffusion coefficient $\tilde{D}$, written in terms of the two defect diffusivities.

In the limiting case where $D_e >> D_{V_{\ddot{O}}}$, we have the simple expression $\tilde{D} = 3D_{V_{\ddot{O}}}$. The ambipolar diffusion rate is controlled by the slower species, but the ambipolar coupling causes the rate to be enhanced by a factor of 3. At the other extreme, where $D_e << D_{V_{\ddot{O}}}$, we find that $\tilde{D} = 3D_e/2$; again the slower specie is rate-limiting, but the coupling increases the effective diffusion coefficient. In either case, the ambipolar diffusion coefficient is greater than that of the slower defect, due to charge-coupling to the faster one.

242 Chapter 3 / Mass and Electrical Transport

It is sometimes convenient to express $\tilde{D}$ in terms of transference numbers. The partial conductivities of the two defects are given by

$$\sigma_e = \frac{ne^2 D_e}{kT}$$

and

$$\sigma_{V_O^{\cdot\cdot}} = \frac{4 C_{V_O^{\cdot\cdot}} e^2 D_{V_O^{\cdot\cdot}}}{kT}.$$

The electronic transference number is

$$t_e = \frac{\sigma_e}{\sigma_e + \sigma_{V_O^{\cdot\cdot}}} = \frac{D_e}{D_e + 2 D_{V_O^{\cdot\cdot}}} \tag{3.65}$$

so that

$$\tilde{D} = 3 t_e D_{V_O^{\cdot\cdot}}. \tag{3.66}$$

In a purely electronic conductor ($t_e=1$) we again have $\tilde{D} = 3 D_{V_O^{\cdot\cdot}}$. In a partially ionic conductor where $t_i>0$, the chemical diffusivity is accordingly reduced in value from this limit.

In transition metal oxides of the rocksalt structure type such as $Fe_{1-\delta}O$, $Ni_{1-\delta}O$, and $Co_{1-\delta}O$, the metal deficiency is accommodated by cation vacancies $V_M^{\alpha'}$, where α' is the effective charge of the cation vacancy. In this case, cations are much more mobile than oxygen ions and oxidation is accomplished through an outward flux of cations vacancies along with a charge-compensating electron hole flux:

$$\alpha J_{V_M^\alpha} = J_{h^{\cdot}}. \tag{3.67}$$

according to the reduction equilibrium:

$$\frac{1}{2} O_2 (g) \rightarrow O_O^x + V_M^{\alpha'} + \alpha h^{\cdot} \tag{3.68}$$

A derivation similar to the one above yields for the ambipolar diffusion coefficient:

$$\tilde{D} = (\alpha + 1) D_{V_M^{\alpha'}} \tag{3.69}$$

This result is again greater than that of the vacancy alone; for singly charged cation vacancies $\tilde{D} = 2 D_{V_M^{\alpha'}}$ whereas for doubly charged vacancies $\tilde{D} = 3 D_{V_M^{\alpha'}}$.

Ambipolar Diffusion in Sintering The formalism used above to derive the ambipolar diffusion coefficient for defect equilibration also allows us to determine the effective diffusion coefficient when cations and anions are flowing in the same direction. Consider the ionic solid under uniaxial stress depicted in Fig. 3.24. The applied stress leads to transport of both cations and anions from a common source to a common sink, resulting in a shape change. Transport under such an applied stress takes place during hot-pressing (Fig. 3.25) and high-temperature diffusional creep, and is also the the primary driving force for mass transport in sintering, where the stress arises from the capillary action of a curved surface. Under stress the local chemical potentials of ions and defects change, and defect concentration gradients facilitate transport of ions down their chemical potential gradients. In hot-deformation, transport occurs from regions of higher to lower compressive stress. However, the stability of ionic compounds is usually sufficient so that the overall stoichiometry and starting phase are maintained during transport, rather than decomposing the solid into anion and cation rich compositions at the source and sink (for exceptions, see Special Topic 3.3). The simultaneous preservation of electrical neutrality and stoichiometry must come from the coupled, parallel diffusion of anions and cations from the same source to the same

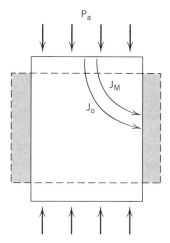

Fig. 3.24 Schematic view of ambipolar diffusion of anions and cations from a common source to a common sink during hot-deformation. Unlike Fig. 3.23, here the stoichiometry of the compound as well as charge neutrality must be preserved.

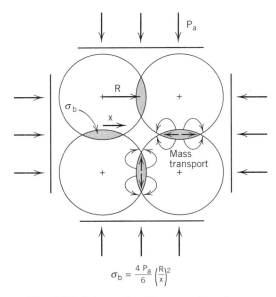

Fig. 3.25 Schematic of hot-pressing of powder particles, showing the enhanced stress at particle contacts and the direction of mass transport. The diffusion of cations and anions out of the grain boundary between contacting particles to the nearby surface is ambipolar in nature.

sink. This is accomplished when the cation and anion charge fluxes are equal; for a binary metal oxide such as MgO this condition is

$$|Z_O|J_O = |Z_{Mg}|J_{Mg} \tag{3.70}$$

or in terms of the electrochemical potentials

$$|Z_O|C_O B_O \left(\frac{\partial \mu_O}{\partial x} + Z_O F \frac{\partial \phi}{\partial x} \right) = |Z_{Mg}|C_{Mg} B_{Mg} \left(\frac{\partial \mu_{Mg}}{\partial x} + Z_{Mg} F \frac{\partial \phi}{\partial x} \right) \tag{3.71}$$

We can again solve for the internal field $d\phi/dx$, and a derivation similar to that used above shows that the flux of MgO is given by

$$J_{MgO} = -\left(\frac{D_{Mg} D_O}{D_{Mg} + D_O} \right) \frac{dC_{MgO}}{dx} \tag{3.72}$$

giving the ambipolar diffusion coefficient

$$\tilde{D} = \frac{D_{Mg} D_O}{D_{Mg} + D_O} \tag{3.73}$$

3.4 The Electrochemical Potential

From this expression we see that the ambipolar diffusion coefficient is bounded by the diffusivities of the two ions and that the *slower specie is rate-limiting*. If $D_{Mg} \gg D_O$, then $\tilde{D} = D_O$, and if $D_O \gg D_{Mg}$, $\tilde{D} = D_{Mg}$. When the ions are not too disimilar in diffusivity, an intermediate value of the ambipolar diffusion coefficient results.

A further complication appears when each ion may have more than one diffusion path. For instance, grain boundary diffusion or surface diffusion along the wall of a pore may be as much as a factor of 10^6 faster than lattice diffusion. Then, the s*lowest species along its fastest path* becomes rate-limiting. Differences in the effective area available for transport along boundaries and in the lattice must be taken into account, and for a given ion, transport by grain boundary diffusion will tend to dominate over lattice diffusion as the grain size decreases. These are important considerations for the rates of solid state sintering and diffusional creep. The ambipolar diffusion coefficient in the presence of multiple diffusion paths is discussed in Special Topic 3.2.

SPECIAL TOPIC 3.2

DIFFUSIONAL CREEP AS AN EXAMPLE OF AMBIPOLAR DIFFUSION

Ceramics that are subjected to stress at high temperatures can deform plastically in a variety of ways. As discussed in Chapter 2, dislocation plasticity is not as common in ceramics as it is in metals due to the high energy required to form and move dislocations of large Burger's vector. Nonetheless, in some ceramics and especially single crystals, it is an important mechanism at high temperatures. In polycrystalline ceramics, *diffusional creep* is the main mode of deformation at elevated temperatures. This is a process in which diffusional transport of atoms occurs between grain boundaries that are under varying states of stress. Grain boundaries are lattice discontinuities at which defects can be formed or eliminated; they act as defect *sources* and *sinks*. In a polycrystal these sources and sinks are separated by a grain diameter or so (Fig. ST30), and the relatively short-ranged diffusion between them allows shape changes in individual grains, which when accommodated over the many grains of a polycrystal results in a macroscopic deformation. This diffusional process is ambipolar in nature, since in order to avoid decomposition of the compound, all constituents must be transported from the source to the sink in the stoichiometric ratio. A similar process occurs during the sintering of powder particles, in which the capillary stress at particle contacts drives mass transport (see Chapter 5).

For a shape change to occur in each grain, it is necessary to have atoms diffuse from regions under higher to lower compressive stress (Fig. ST31). Using the vacancy diffusion mechanism as an example, the creation of a

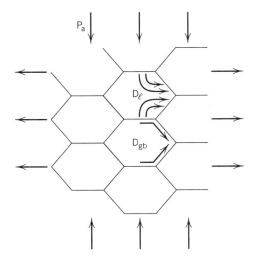

Fig. ST30 Under stress at high temperature, a polycrystalline material may undergo diffusion creep by lattice diffusion or grain boundary diffusion.

vacancy under compressive stress σ requires an additional energy expenditure of σΩ, where Ω is the vacancy volume. At equilibrium the local concentration of vacancies is reduced from the stress-free value, C_o, determined by intrinsic thermal excitation or extrinsic solutes. The new value under compressive stress is given by $C_v = C_o \exp(-\sigma\Omega/kT)$. Forming the vacancy under a tensile stress correspondingly reduces the net formation

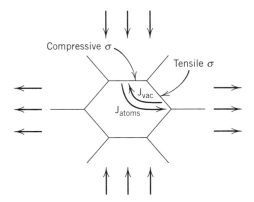

Fig. ST31 A flux of vacancies from a grain boundary under tensile stress (defect source) to a grain boundary under compressive stress (defect sink) causes a counter-flux of atoms, leading to a change in grain shape.

energy by $-\sigma\Omega$ and raises the equilibrium concentration to a new value $C_v = C_o \exp(\sigma\Omega/kT)$. When a gradient in stress is imposed, the diffusion of vacancies down the resulting concentration gradient from a source under tensile stress to a sink under compressive stress causes an equal flux of atoms in the opposite direction (this result is not exclusive to vacancy diffusion; identical conclusions are reached for other types of diffusion mechanisms). As a result, grains elongate in the direction of tensile stress and shrink in the direction of compressive loading. By observing the relative positions of grain centers as grains deform (Fig. ST32), it is furthermore apparent that in order to accommodate the shape change without opening up voids at the grain junctions (i.e., causing cavitation) the grains must slide with respect to one another. Thus the diffusional creep process is always accommodated by grain boundary sliding. (Alternatively, the creep of polycrystals can be thought of as a grain boundary sliding process, which is accommodated at the grain junctions by either volume-conserving diffusional transport or cavitation.)

Atoms can be transported between the source and sink along either the lattice or the grain boundary (Fig. ST30). Relations for the deformation rate have been determined by considering the diffusive flux between the source and sink in grains of simple shape. Nabarro-Herring creep refers to lattice diffusion limited creep (F. R. N. Nabarro, *Report of a Conference on the Strength of Solids* (The Physical Society, London, 1948), p. 75, and C. Her-

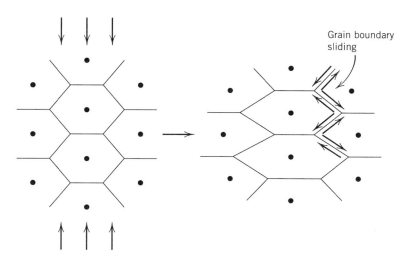

Fig. ST32 Grain shape changes during diffusional creep are accommodated by sliding of grains relative to one another.

ring, *J. Appl. Phys.*, 21, 437, 1950), in which the creep rate at a steady state is

$$\dot{\varepsilon} = \frac{14\Omega D_l \sigma}{kTd^2} \qquad \text{(Eq. 1)}$$

where d is the grain diameter and D_l is the lattice difffusion coefficient. If grain boundary diffusion is rate-limiting, we have *Coble creep* (R. L. Coble, *J. Appl. Phys.*, **34**[6], 1679, 1963) where the creep rate is given by

$$\dot{\varepsilon} = \frac{47\Omega\delta D_{gb}\sigma}{kTd^3} \qquad \text{(Eq. 2)}$$

and δ is the grain boundary "width" (a few angstroms) and D_{gb} the grain boundary diffusivity. Notice that both Eqs. 1 and 2 show a proportionality between the strain rate and stress ($\dot{\varepsilon} \propto \sigma^n$, where $n=1$) characteristic of viscous flow.[1] The deformation rates are also proportional to the rate-limiting diffusion coefficient, and inversely proportional to the grain size squared or cubed. Therefore diffusional creep rates always increase with increasing temperature and decreasing grain size. We can show from the inverse dependence on grain size that diffusional creep is not viable as a deformation mechanism for large single crystals, where the grain size is essentially the sample size. On the other hand, diffusional creep in ultrafine grained or "nanocrystalline" ceramics of a few hundred angstroms grain size may be 10^2 to 10^4 higher in rate than it is in conventional polycrystals (grain sizes of a few tenths to tens of microns) based on this grain size scaling. *Superplastic ceramics* are fine-grained ceramics which are unusually ductile at high temperature. They can undergo large tensile elongations of several hundred percent without fracture due to a high rate of diffusional creep. First observed in zirconia,[2] this phenomenon has now been reported in fine-grained alumina, silicon nitride, and other ceramics, and is of general interest for the deformation processing of ceramics.[3] Superplastic polycrystals frequently contain a small amount of grain boundary liquid or glass to facilitate boundary diffusion and/or sliding.

Since diffusional creep requires an ambipolar coupling of all the constituents in the ceramic, while each atom can diffuse either through the lattice or along the grain boundary, there are twice as many paths as the

---

[1] This proportionality is not expected when the creation or annihilation of defects at the interfaces becomes limiting. This is termed *interface-controlled* creep; if in this process defects are accomodated by the climb of grain boundary dislocations, it is predicted that $n=2$.

[2] F. Wakai, S. Sakaguchi, and Y. Matsuno, *Adv. Ceram. Mater.*, **1**[3] 259-63 (1986).

[3] For a review of superplastic ceramics see I.-W. Chen and L.A. Xue, "Development of Superplastic Structural Ceramics, *J. Am. Ceram. Soc.*, **73**[9] 2585-2609 (1990).

3.4 The Electrochemical Potential

number of species. Any one of these mechanisms can be rate-limiting. It is a paradigm in ceramics that where parallel diffusional processes occur, the rate-limiting step will be *the slowest species along its fastest path*. Figure ST33 is a plot of log $\dot{\varepsilon}$ against log grain size at constant temperature and stress for alumina, a system in which the relative values of grain boundary and lattice diffusion coefficients are known. The two lines of –2 slope show the Nabarro-Herring creep rate that result if Al and O are the respective rate-limiting ions, and the two lines of –3 slope show the same calculated for Coble creep. The diagram has been separated by vertical

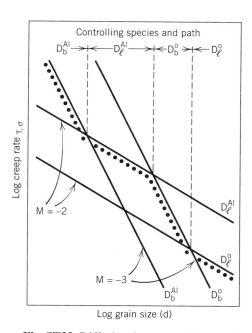

Fig. ST33 Diffusional creep mechanisms in Al_2O_3 (schematic). The logarithm of creep rate is plotted against the logarithm of grain size. Lines with slopes of -2 correspond to lattice diffusion limited creep; slopes of -3, grain boundary diffusion limited creep. The dotted line traces the rate-limiting mechanism, given by the slower ion along its fastest path, across the grain size range. [From R.M. Cannon and R.L. Coble, "Paradigms for Ceramic Powder Processing," pp. 151-170 in *Processing of Crystalline Ceramics*, Mat. Sci. Res. Vol. 11, H. Palmour III, R.F. Davis, and T.M. Hare, Eds., Plenum Press, New York, (1978).]

lines into four regimes, in each of which the creep rate of the rate-limiting species is given by the dotted line. Notice that in each of these regimes the topmost line, giving the absolute highest creep rate, is not rate-limiting; it is the second line, representing the other ion transported along its fastest path, which is.

The ambipolar diffusion coefficient for the process can be evaluated as follows: If the lattice and grain boundary diffusion-accommodated creep rates are simply additive (R. Raj and M. F. Ashby, *Metall. Trans.*, 2[4], 1113, 1971), the creep rate for an elemental solid is

$$\dot{\varepsilon} = \left(\frac{14\Omega}{kTd^2}\right)\left[D_l + \frac{\pi\delta D_{gb}}{d}\right]\sigma \qquad \text{(Eq. 3)}$$

The term in square brackets is an effective diffusion coefficient taking into account both diffusion paths. For a binary compound of the formula M_xO_y, R. S. Gordon (*J. Am. Ceram. Soc.*, 56[3], 147, 1973) has shown that the ambipolar diffusion coefficient is given by

$$\tilde{D}_{eff} = \frac{(x+y)\left[D_l^M + \pi\delta D_{gb}^M/d\right]\left[D_l^O + \pi\delta D_{gb}^O/d\right]}{y\left[D_l^M + \pi\delta D_{gb}^M/d\right] + x\left[D_l^O + \pi\delta D_{gb}^O/d\right]} \qquad \text{(Eq. 4)}$$

Comparing Eq. 4 to Eq. 3.73, we see that each of the bracketed terms in Eq. 4 represents the effective diffusion coefficient for one of the species, taking into account both paths of transport. The slowest effective diffusivity is rate-limiting; but within each term, it is the faster of the two paths that dominates, confirming that $\tilde{D}$ is limited by the slowest species along its fastest path.

From the grain size dependence, it is clear that while lattice diffusion limited creep is promoted at the largest grain sizes and boundary diffusion limited creep at the smallest, the actual rate-limiting mechanism depends greatly on the relative values of the lattice and grain boundary diffusion coefficients. At fine grain sizes and low stresses, it is also possible that creep becomes limited by the rate of defect formation at interfaces, in which case the grain size scaling no longer holds. On the other hand, at large grain sizes and high stresses, creep may be accomplished by dislocation plasticity. For magnesia-doped alumina, a well-studied system, the creep behavior over a wide range of grain size and stress is shown in the deformation map in Fig. ST34. This map shows zones in which the different creep mechanisms are dominant (i.e., give the fastest deformation rate). The boundaries between the zones are found by equating the creep rates for different mechanisms, using available transport data for alumina.

3.4 The Electrochemical Potential 251

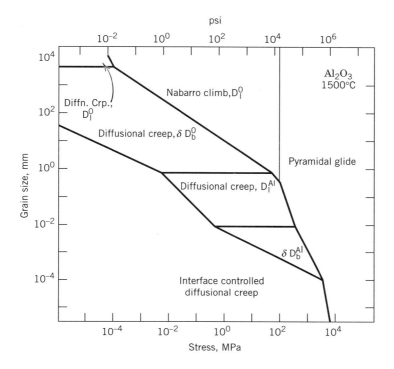

Fig. ST34 Deformation map for MgO-doped Al$_2$O$_3$ at 1500°C, showing regimes of grain size and stress where diffusional creep, interface-limited diffusion creep, and dislocation climb or glide give the fastest rate of deformation. [From A.H. Heuer, N.J. Tighe, and R.M. Cannon, *J. Am. Ceram. Soc.*, 63[1-2], 53 (1980).]

SPECIAL TOPIC 3.3

KINETIC DEMIXING

In discussing the ambipolar diffusion of cations and anions during sintering and diffusional creep, we assumed that the stoichiometry of the compound is largely maintained at both the source and the sink, even while recognizing that gradients in ion (defect) concentrations must be present. Most ceramics are stable enough that differences in transport rate do not lead to decomposition (for example, to gaseous oxygen at one end and metal at the other). On the other hand, in compounds containing multiple cations or anions, differences in the transport rates of like-charged ions can lead to compositional separation during ambipolar diffusion. Differences in *relative* transport rates can allow more of one ion to build up at the sink (or source) than another. *Kinetic demixing* is defined as the compositional separation of an initially homogeneous material, due to differential atomic transport rates under a common driving force. The driving force may be a gradient in oxygen potential, electrical potential, stress, temperature, or others, but we will restrict our discussion to oxygen and stress potential gradients.

The kinetic demixing of multicomponent oxides under an oxygen potential gradient was first described by Schmalzried and coworkers (Z. *Naturforsch.*, 34A[2], 192, 1979, and *Oxid. Metals*, 15[3–4], 339, 1981). Consider a uniform sample of the solid solution $Co_{1-x}Mg_xO$ (rocksalt structure) which is placed under an oxygen potential gradient at high temperature (Fig. ST35). Oxidation of Co^{2+} to Co^{3+} is accompanied by the formation of cation vacancies, and results in a higher steady-state concentration of cation vacancies at the high P_{O_2} end. A gradient in cation vacancy concentration decreasing from right to left in Fig. ST35 is formed, and an opposing flux of cations from left to right occurs. (In fact, from the laboratory reference frame the sample migrates steadily towards the right, Fig ST36.) Since Co diffuses more rapidly than does Mg in this example, an enrichment of Co toward the right is observed. At long times, a steady state is reached in which the concentration profile remains constant; demixing stops when the induced chemical potential (concentration) gradient exactly opposes the applied potential gradient. It is termed *kinetic* demixing since it only exists

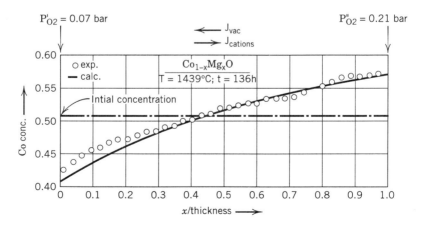

Fig. ST35 Kinetic demixing in a $Co_{1-x}Mg_xO$ solid solution subject to a gradient in oxygen activity. The higher oxygen potential on the right-hand side results in a steady state cation vacancy flux towards the left, with counter-diffusion of Co and Mg. The Co enrichment on the right corresponds to a higher diffusion coefficient for Co than for Mg. [From H. Schmalzried and W. Laqua, *Oxid. Metals*, 15 [3–4], 339 (1981).]

3.4 The Electrochemical Potential

due to differential rates of transport; if the oxygen gradient is removed, the cations simply diffuse down their respective concentration gradients and rehomogenize to a uniform solid solution.

In multicomponent oxides, kinetic demixing of cations can only occur if cation transport is rate-limiting. Thus a necessary condition is that the transport of oxygen be much slower than that of the cations ($D_O \ll D_{cations}$). When this is the case, the oxygen sublattice remains fixed and the cations diffuse in competition with one another (Fig. ST36). A second requirement is that the cation diffusivities differ sufficiently; the exact criterion depends on the cation stoichiometry of the compound. For example, in ABO_3 the flux of A and B must be equal to avoid demixing, whereas in AB_2O_4 the flux of B must be twice that of A.

Because CoO and MgO form a complete solid solution, a continuous variation in composition results (Fig. ST35). However, compounds of more limited solid solution and highly stoichiometric "line" compounds can exhibit kinetic *decomposition* to new phases if the driving force is high enough. Figure ST37 shows the concentration of Ni and Ti across a sample of $NiTiO_3$ (ilmenite structure) which has been subjected to an oxygen potential gradient at high temperature. At the right-hand side the ilmenite has decomposed to the nickel-rich rocksalt solid solution ($Ni_{1-x}Ti_{x/2}O$), demonstrating that $D_{Ni} > D_{Ti}$ in ilmenite, while at the left-hand side Ni-doped rutile TiO_2 has precipitated. Notice that neither phase would be present at equilibrium for this overall composition. Kinetic decomposition occurs when the condition

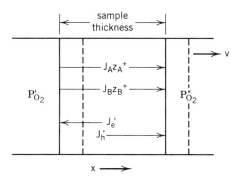

Fig. ST36 During kinetic demixing of an (A, B)O solid in an oxygen potential gradient, charge neutrality is maintained by a flux of electronic species. Cation diffusion must be rate-limiting; and the diffusivities of A and B must differ if demixing is to occur.

254   Chapter 3 / Mass and Electrical Transport

$$\left(\frac{RT}{2}\right)\ln\left(\frac{P_{O_2}{}'}{P_{O_2}{}''}\right) > f(D_A, D_B) \cdot \Delta G_f{}^\circ \qquad \text{(Eq. 1)}$$

is satisfied. Here the left-hand term is the driving force due to the differential in oxygen pressure (units of Joules), $\Delta G_f{}^\circ$ is the standard free energy for decomposition to the end phases, and $f(D_A, D_B)$ is a kinetic factor. The expression for $f(D_A, D_B)$ depends on the cation stoichiometry of the system in question; for a compound ABO_3 it is $(D_A + D_B)/(D_A - 2D_B)$, for AB_2O_4 it is $(2D_A + D_B)/(2D_A - 3D_B)$, and for A_2BO_4 it is $(D_A + 2D_B)/(D_A - 2D_B)$. Equation 1 shows that a critical P_{O_2} differential is necessary to overcome the free energy of decomposition, given a certain difference in cation diffusivities. Conversely, for certain limiting values of the diffusivities (which for $NiTiO_3$ is $D_{Ni} = D_{Ti}$), no demixing can occur; while the greater is the difference in diffusivities, the smaller is the oxygen potential gradient necessary for demixing.

As an example of another potential gradient which can lead to demixing, consider the nonhydrostatic stress generated during diffusional creep or hot forging (Fig. ST38). In contrast to oxygen potential demixing, here anions and cations are transported in the same direction. Since cation diffusion

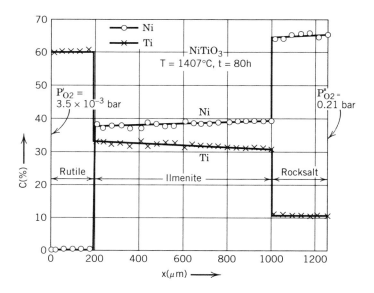

Fig. ST37 Kinetic decomposition of $NiTiO_3$ (ilmenite) to Ti- and Ni-rich phases under an oxygen potential gradient. [From H. Schmalzried and W. Laqua, *Oxid. Metals*, 15 [3–4], 339 (1981).]

3.4 The Electrochemical Potential

must be rate-limiting for demixing to occur, oxygen transport (along whichever path is fastest) must be much faster than cation transport ($D_O \gg D_{cations}$). Stress-induced kinetic demixing has been proposed for ceramics (D. Dimos, J. Wolfenstine, and D. L. Kohlstedt, *Acta Met.*, 36[6], 1543, 1988) but has yet to be conclusively observed. Kinetic decomposition can in principal also occur above a critical stress; the criteria are similar to those given for oxygen potential decomposition except that the driving force (left hand side of Eq. 1) is given by the term ($\Delta\sigma \cdot V_m$), where $\Delta\sigma$ is a critical differential stress and V_m is the molar volume.

Systems that exhibit oxygen potential demixing should not show stress demixing since the necessary conditions concerning relative values of D_O and $D_{cations}$ are exactly the opposite. Furthermore, a comparison of the magnitudes of the driving forces shows that the stress potential gradient is generally much lower in magnitude, being ultimately limited by the failure stress of the material, than the chemical potential gradient which can be achieved through straightforward variations in oxygen pressure. This may explain why oxygen potential demixing and decomposition have been rather widely observed [including the systems (Co, Mg)O, (Ni,Ti)O, Co_2TiO_4, Ni_2SiO_4, and Fe_2SiO_4] while stress-induced demixing has not.

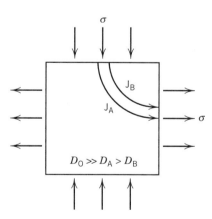

Fig. ST38 Stress-induced kinetic demixing in a solid (A, B)O under nonhydrostatic stress can occur if cation diffusion is rate-limiting and the cation diffusivities differ. Notice that in contrast to oxygen potential demixing (Fig. ST35), D_O must exceed D_A, D_B.

ADDITIONAL READING

Diffusion and Solid State Reactions

P. Shewmon, *Diffusion in Solids*, TMS, Warrendale, PA, 1989.

G. H. Geiger and D. R. Poirier, *Transport Phenomena in Metallurgy*, Addison-Wesley, Reading, MA, 1973.

J. Crank, *The Mathematics of Diffusion*, Oxford University Press, NY, 1975.

H. S. Carslaw and J. C. Jaeger, *Conduction of Heat in Solids*, Clarendon Press, Oxford, 1959.

H. Schmalzried, *Solid State Reactions, 2nd Edition*, Verlag Chemie GmbH, D-6940 Weinheim, and Deerfield Beach, FL, 1981.

Electrical Properties

S. M. Sze, *Physics of Semiconducting Devices, 2nd Edition*, John Wiley and Sons, New York, 1981.

C. Kittel, *Introduction to Solid State Physics, 6th Edition*, John Wiley and Sons, New York, 1986.

G. Burns, *Solid State Physics*, Academic Press, New York, 1985.

A.C. Rose-Innes and E.H. Rhoderick, *Introduction to Superconductivity*, Second Edition, Pergamon Press, Oxford, 1978.

Electronic Ceramics

W. D. Kingery, H. K. Bowen, D. R. Uhlmann, *Introduction to Ceramics, 2nd Edition*, John Wiley and Sons, New York, 1976.

L. L. Hench and J. K. West, *Principles of Electronic Ceramics*, John Wiley and Sons, New York, 1990.

A. J. Moulson and J. M. Herbert, *Electroceramics*, Chapman and Hall, 1990.

Per Kofstad, *Nonstoichiometry, Diffusion, and Electrical Conductivity in Binary Metal Oxides*, Krieger Publications, Malabar, FL, 1983.

O. Toft Sorensen, *Nonstoichiometric Oxides*, Academic Press, New York, 1981.

R. C. Buchanan, Editor, *Ceramic Materials for Electronics, 2nd Edition*, Marcel Dekker, New York, 1991.

PROBLEMS

1. Examine Fig. 3.1 and explain the following observation. In MgO, Al_2O_3, and CoO, the cation self-diffusion coefficient is orders of magnitude greater than the oxygen self-diffusion coefficient. Does your explanation depend on whether the specific sample in question is extrinsic or intrinsic?

Problems   257

2. Consider an oxygen-deficient oxide MO_{2-x} (assume doubly ionized oxygen vacancies), in which oxygen diffusion occurs by a vacancy mechanism. Write an expression for the self-diffusion coefficient of oxygen that explicitly shows the P_{O_2} and temperature dependence, for the undoped material. Then, sketch the expected dependence of $D(O)$ on temperature and P_{O_2} if your material contains a low concentration of solute A_M'. Explain your result fully.

3. NiO is of the rocksalt structure type and has a bandgap of $E_g = 4.2$ eV. We estimate the Schottky defect formation energy to be ~6 eV. Because nickel can be oxidized to the trivalent state, NiO tends to be slightly cation deficient.

   (i) In the hypothetically pure and stoichiometric state, do you expect NiO to be an electronic or ionic conductor at a temperature of 1000K? (Show how you arrive at your answer.)

   (ii) In the nonstoichiometric state, do you think the cation deficiency of $Ni_{1-x}O$ will be accommodated as cation vacancies or as oxygen interstitials, and why? Write a formally correct defect reaction and mass-action equilibrium constant for the oxidation mechanism of your choice.

   (iii) Will the nonstoichiometric material be p- or n-type in its electronic conductivity? Explain. Determine the Brouwer approximation in this regime and find the P_{O_2} dependence of the electronic conductivity.

   (iv) Assume cation vacancies to be the predominant defect accommodating nonstoichiometry. Given a cation vacancy migration energy of 3.2 eV and the lattice parameter for NiO of 0.4177 nm, what concentration of cation vacancies is necessary for the system to be a mixed conductor with an ionic transference number of 0.5? Assume the electron mobility to be 10 cm^2/V·sec and the hole mobility to be 0.1 cm^2/V·sec.

4. In CaO-stabilized cubic ZrO_2, the activation energy for migration of oxygen vacancies is 0.84 eV. Determine the ionic conductivity of a 10% CaO–90% ZrO_2 solid solution at 1000°C. (Assume an ideal solution with no interference from defect association.) What will be the ionic transference number at this temperature? The electronic transference number?

   Calculate the self-diffusion coefficient of oxygen at this temperature for ZrO_2 containing 5% CaO and 10% CaO, respectively.

   (The lattice constant of cubic ZrO_2 may come in handy in this problem; it is 0.51 nm.)

5. Figure 3.18 shows the diffusion coefficient for oxygen in calcia-stabilized zirconia (CSZ) of composition $Zr_{0.85}Ca_{0.15}O_{1.85}$, measured at several temperatures. These data are compared to the diffusion coefficient determined from the electrical conductivity, which has also been measured as a function of temperature.

(a) For CSZ of this composition, do you expect the conductivity to be ionic or electronic, and why? Is the defect structure extrinsic or intrinsic, and why?

(b) In practical use as a solid electrolyte the temperature will be 800°C, and the oxygen pressure may vary from 1 to 10^{-10} atm. However, for these data, the oxygen pressure of the measurement is not given. Does it matter? That is, how many Brouwer regimes do you expect the use conditions to cover? Write the corresponding Brouwer approximations.

(c) Write the relationship(s) between the electrical conductivity and the oxygen self-diffusion coefficient. Then, calculate the electrical conductivity at 800°C.

(d) Determine the activation energy for oxygen diffusion. Is it the same as the activation energy for oxygen vacancy migration? Explain.

Data:

For the reduction of pure ZrO_2

$$O_O^x \Leftrightarrow \frac{1}{2}O_2(g) + V_O^{\cdot\cdot} + 2e'$$

the equilibrium constant is, from Xue and Dieckmann (J. Electrochem. Soc., 138[2], 36C, 1991):

$K_R = 4.78 \times 10^{-19}$ atm$^{1/2}$ (concentrations in mole fraction w.r.t. ZrO_2).

$\rho(ZrO_2) = 5.6$ g/cm^3

Molecular weight of ZrO_2 is 123.22 g/mole.

$E_g = 5.2$ eV

Assume $\mu_e = \mu_h = 1$ cm^2/V sec.

Lattice parameter: $a_0 = 0.51$ nm

6. This problem concerns the transport properties of $Co_{1-x}O$. Data are attached for the nonstoichiometry x and the electrical conductivity of undoped cobalt oxide, measured at 1200°C over a range of oxygen pressures (K. L. Persels, Ph.D. Thesis, Northwestern University, 1990). It is believed that the nonstoichiometry is accommodated by the formation of cobalt vacancies, and that these are predominantly singly ionized: V_{Co}'.

(a) Are the nonstoichiometry data consistent with this explanation? Show using an appropriate defect reaction.

(b) How many Brouwer regimes do the data span? Give the corresponding Brouwer approximations.

(c) Can one determine from the oxygen pressure dependence of conductivity alone whether the conductivity is ionic or electronic? Explain.

(d) A separate tracer diffusion experiment has shown that the activation energy for migration of V_{Co}' is 1.8 eV. Estimate the diffusion coefficient of V'_{Co} at 1200°C.

(e) Determine the ionic conductivity at 1200°C in air (P_{O_2} =0.21 atm). Then, determine whether the conductivity measured is ionic or electronic. What is the approximate value of the electronic transference number?

Data: E_g = 4.0 eV

Schottky energy ~5 eV.

Molecular Weight = 74.93 g/mole

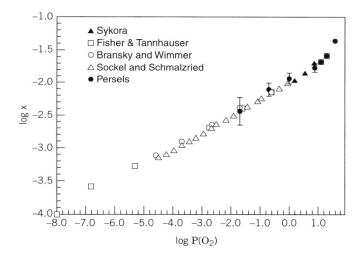

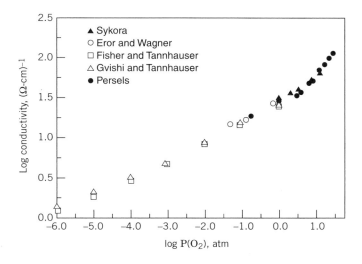

260 Chapter 3 / Mass and Electrical Transport

Density = 6.45 g/cm³

Lattice constant a_o = 0.4331 nm (at 1200°C)

7. In a completely densified, fine-grained aluminum oxide (1 μm average grain diameter), determine the diffusional creep rate at 1400°C under a compressive stress of 200 MPa. At this temperature the following diffusion coefficients have been measured:

$$D_{gb}^{Al} = 5.6 \times 10^{-17} \text{ cm}^3/\text{sec}$$

$$D_{gb}^{O} = 7 \times 10^{-14} \text{ cm}^3/\text{sec}$$

$$D_{lattice}^{Al} = 4 \times 10^{-14} \text{ cm}^2/\text{sec}$$

$$D_{lattice}^{O} = 1 \times 10^{-17} \text{ cm}^2/\text{sec}$$

In another composition, CaO and SiO_2 have been added as liquid phase sintering aids. Upon heating, the liquid dissolves some of the alumina to form, at equilibrium, a CaO-Al_2O_3-SiO_2 liquid of composition 40 wt% Al_2O_3, 40 wt% SiO_2, and 20 wt% CaO. The grain size is larger (10 microns) due to more rapid grain growth. TEM observations show that the glass forms a film of ~2 nm thickness at most of the grain boundaries. Estimate the diffusional creep rate. Is this material suitable for long-term, load-bearing applications at this stress level and temperature?

8. Figure 3.9 shows the diffusion coefficient of Na in NaCl containing a small quantity of $CdCl_2$ solute, plotted against 1/T. Two linear regions are shown, for which the slopes are given (in units of degrees K). Answer the following:
    (a) Write a defect incorporation reaction for $CdCl_2$.
    (b) Explain the existence of intrinsic and extrinsic regimes.
    (c) Write an expression for the diffusion coefficient of sodium in the intrinsic and extrinsic regimes, respectively.
    (d) Determine the activation energy for vacancy migration from the data.
    (e) Determine the Schottky formation enthalpy.
    (f) Using the data, estimate the concentration of Cd in this sample.
    (g) Calculate the ionic conductivity due to sodium vacancies at 550°C.
    (h) NaCl has a bandgap of 7.3 eV. Will this sample be an ionic or electronic conductor over the temperature range shown? Explain briefly.

9. Zinc oxide, which crystallizes in the wurtzite structure, becomes slightly nonstoichiometric at elevated temperatures, forming zinc interstitials. The interstitials are believed to be either singly ionized or doubly ionized.
    (a) Will the electronic conductivity of zinc oxide be *n*-type or *p*-type under these conditions? Explain, using the appropriate defect reactions.

(b) Is zinc oxide likely to be an ionic or electronic conductor under these conditions? Explain briefly.

(c) How (experimentally) could one determine if the interstitials are predominantly singly ionized or doubly ionized? Be explicit.

10. An oxygen ionic conductor ($t_i = 1$) is used as an electrochemical pump to control the oxygen pressure within a closed chamber, which is initially filled with air at 1 atmosphere total pressure. The external atmosphere is air. At 600°C, what voltage must be maintained to keep the oxygen partial pressure within the chamber at 1.5 atm? To hold it at 10^{-6} atm? Indicate the sign of the applied voltage in each case.

11. CeO_{2-x} is easily reduced to an oxygen-deficient stoichiometry, generating oxygen vacancies that have a low migration energy (~0.85 eV). Is reduced CeO_{2-x} suitable for use as the solid electrolyte in a galvanic oxygen sensor? Explain why or why not.

CeO_{2-x} can dissolve >10% of the trivalent solutes La^{3+} and Gd^{3+}, which are charge-compensated by oxygen vacancies. Can this material be used as a galvanic oxygen sensor or fuel cell electrocyte? Explain.

Chapter 4

Phase Equilibria

Ceramic materials are inorganic compounds containing at least two, and often many more, elemental constituents. In technological applications they are used in a variety of morphologies including monoliths, fine powders, thin and thick films, and long or short fibers. Each of these may be single crystalline or polycrystalline in nature; they may include one or many phases. Due to the larger number of components, the number of phases which may exist is generally larger than in metal and semiconductor systems. Ceramics are typically heated to high temperatures during processing in order to speed up chemical reactions and to develop desired microstructures. The number, type, and physical distribution of phases are therefore determined by the phase equilibria and the *thermal history* of the material. In this chapter, we develop a primarily graphical description of the high-temperature behavior of multicomponent systems, emphasizing three-component systems. While it is assumed that the reader is somewhat familiar with one- and two-component phase diagrams, we discuss these first in order to ease the transition into multicomponent systems.

4.1 THERMODYNAMIC EQUILIBRIUM

Phase diagrams are most often intended to describe the condition of *equilibrium*. (In some applications "stability diagrams" are used to show the phases that tend to appear, but may not be at equilibrium.) Equilibrium is the state where all chemical reactions have proceeded to the point where the chemical potential (or *partial molar free energy*) of all components[1], as well as temperature and pressure, are equal throughout the system. In the terminology of reaction rate theory, the rates of all forward chemical reactions are exactly equal to the rates of all reverse reactions, and this balance is maintained for all time, not being "kinetically limited" by the rate of any process.

In the state of equilibrium, there can exist a number of identifiable *phases* or states of matter. A phase is defined as a physically separable and chemically homogeneous arrangement of atoms, distinguishable from all others which may exist for the composition, temperature, and pressure of the system. The different states of matter—vapors, liquids, glasses, and crystals—are all familiar phases. Crystals that differ in both composition and structure are always distinguishable phases (e.g., MgO and Al_2O_3). Different structural forms of a single composition, such as the liquid and vapor states of water, or the monoclinic, tetragonal, and cubic polymorphs of pure zirconia (see p. 29 in Chapter 1), are also considered distinct phases. A particular structure may also exist over a range of compositions. The different compositions may form separate phases that are *immiscible*; examples include two immiscible liquids (oil and water) or two immiscible solids of the same crystal structure (e.g., MgO and CaO), in which case they are separate phases despite being of like structures.

A phase or assemblage of phases need not represent the most thermodynamically stable condition of the system; if so, however, they are considered equilibrium phases. One frequently encounters *metastable* phases which are not the most stable configuration, but cannot transform into the equilibrium phase due to slow atom transport. A glass is one example of a metastable phase, the equilibrium phase for which is the crystal. Water that is cooled in quiescent conditions in a smooth-walled container to a temperature a few degrees below its freezing point can be easily retained as the metastable liquid phase without having the equilibrium phase, ice, crystallize. Another example of a metastable phase is diamond, which is the equilibrium form of carbon only at high temperatures and pressures (see Special Topic 4.1).

The equilibrium state for a composition of matter at any given temperature and pressure may consist of one or more phases. A principal application of the *phase diagram* is to tell us how many phases, and of what specific compositions, co-exist for a given overall composition, temperature, and pressure. Oil

---

[1] Components are chemical entities. While an elemental constituent can always be regarded as a component, the number of components necessary to describe a system can be less than the total number of elements if there are constraints which prevent all elemental concentrations from being independently varied. For instance, in ionic compounds the valences of all elements are frequently fixed under the range of temperature and atmosphere of interest. Charge neutrality then dictates the relative concentrations of the anions and cations, and a degree of freedom is lost. For example a three element system can be represented as a binary system between two compounds. See also the later discussions of Variable Valence Systems and Reciprocal Salt Diagrams.

4.1 Thermodynamic Equilibrium

and water represent two co-existing liquid phases of very different composition, each of which has dissolved in it, at equilibrium, a small amount of the other. An example of multiple solid phases at equilibrium is a geological (rock) formation, in which there are typically several distinguishable crystalline compounds, each of a different composition and crystal structure, co-existing in equilibrium with one another (although arguably they may not be in equilibrium at room temperature, but only at a higher temperature and pressure at which they were formed).

SPECIAL TOPIC 4.1

METASTABILITY IN CARBON: DIAMOND AND DIAMOND-LIKE MATERIALS

The high refractive index of diamond combined with a high dispersion (the variation in refractive index with wavelength, which causes color separation in refracted light) gives this material an extraordinary brilliance and fire, which have made diamonds a popular and valuable gemstone. Because of its extreme hardness, diamond has many important industrial uses as abrasives and cutting tools having a world market of more than a billion dollars. Diamond is not only the hardest material known, but has the highest thermal conductivity at room temperature, is an excellent electrical insulator, is transparent to ultraviolet, infrared, and x-rays as well as visible light. Diamond has the highest elastic modulus of any known material, it is the most incompressible substance known, and has a relatively low thermal expansion coefficient. As a doped semiconductor material, it may have advantages for high temperature electronic applications. With respect to many properties, diamond is superior to all other materials.

There have been efforts to produce synthetic diamonds for more than a century. A Scottish chemist, James Ballantyne Hannay, reported diamond synthesis in 1880 by heating paraffin, bone oil, and lithium in sealed iron tubes. In 1893 the French chemist, Henri Moissan, claimed that he made diamonds from a mixture of carbon and iron heated to a high temperature and then quenched into a water bath. In 1955 F. P. Bundy, H. T. Hall, H. M. Strong and R. H. Wentorf, Jr., announced successful growth of diamonds at high pressure and high temperature at the General Electric Company. The phase equilibrium relationships between diamond and graphite are shown in Fig. ST39. Graphite is the thermodynamic stable form over a wide range of temperature and pressures. Only at high pressures and high temperatures does diamond become the preferred form. The key factor in developing successful diamond synthesis was designing and building high-temperature, high-pressure presses. Graphite in the presence of a liquid metal solvent, such as iron or nickel, is brought to a temperature and pressure range where diamond is the thermodynamic stable phase and at which diamond crystals nucleate and grow. Diamonds are usually formed at pressures of 50–65 kb

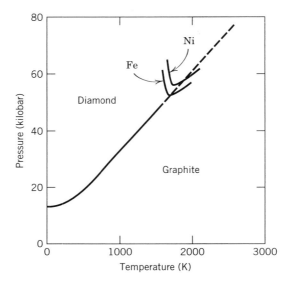

Fig. ST39 Equilibrium phase diagram for the carbon system. The P-T curves for the diamond-iron and diamond-nickel eutectics growth at which a liquid can be produced are also shown.

and at a temperature of about 1800K. More recently, high-temperature, high-pressure cells have been designed with a thermal gradient making it possible for small crystals to dissolve at a higher temperature, diffuse through the liquid, and contribute to the growth of diamond crystals in the lower-temperature part of the reaction cell. Crystals up to 7 mm in diameter are commercially available, and there are reports that crystals as large as 17 mm in diameter and weighing as much as 11 carats have been produced.

The very success of using equilibrium conditions to nucleate and form diamonds turned investigators away from an alternate possibility, metastable diamond synthesis. The likelihood of diamond synthesis in the graphite stable region is suggested by the small free energy difference (500 kcal/mole) between diamond and graphite under ambient conditions. In the late 1950s W. G. Eversole in the United States and B. V. Derjaguin and B. V. Spitsyn in Russia produced metastable diamond growth from a vapor reaction decomposing hydrocarbons. Later, John C. Angus took up the Eversole work and improved the process by using atomic hydrogen to remove graphite that precipitated alongside the new diamond formation; this prepared a graphite-free surface for subsequent diamond growth. The great achievement of the Soviet group looks modest in retrospect. Angus used atomic hydrogen *during* the growth process, which permitted higher growth rates and also the nucleation of new diamond crystallites on non-diamond substrates. Production of atomic hydrogen by the use of hot wire, RF microwave, DC plasma, or thermal plasma generation has been the key to successful diamond growth.

> Initial experimental growth rates were very low, but once the barrier to metastable diamond growth was appreciated, modifications of the technique led to spectacular gains. By use of radio frequency induction heating to produce a thermal plasma, growth rates increased from 1 micrometer per hour to more than 1 micrometer per minute. Diamond synthesis at a high rate has also been found to occur in an oxyacetylene combustion torch when the reducing portion of the flame is targeted on a low-temperature substrate.

4.2 THE GIBBS PHASE RULE

The number of phases that can co-exist in equilibrium is concisely described by the *Gibbs phase rule* (sometimes simply referred to as the phase rule). The phase rule relates the number of equilibrium phases (P) and the number of chemical entities in the system (or *components*, C). The number of phases and number of components together define the total degrees of freedom available to the system (F):

$$F = C - P + 2 \tag{4.1}$$

The degrees of freedom is the number of intensive thermodynamic variables (i.e., those not dependent on the mass of the system) that may be changed without changing the number of phases present in equilibrium. Most often we deal with systems of a known number of components C, whereupon the number of phases P becomes constrained by the intensive variables of temperature, pressure, and overall composition. The form of the Gibbs phase rule in Eq. 4.1 assumes that both temperature and pressure are independently variable. However, in much of materials processing, pressure is kept constant at the ambient value, typically near one atmosphere. Under constant pressure conditions the Gibbs phase rule becomes

$$F = C - P + 1 \tag{4.2}$$

This is sometimes referred to as the phase rule for condensed phases. Even under non-isobaric conditions the assumption of constant pressure is sometimes neglected, but results in little error since small changes in total pressure have only a minor effect on condensed phase equilibria.

Before proceeding to illustrate its utility, we present a simple derivation of the Gibbs phase rule. Each phase contains (to some degree, however dilute) a certain concentration X_i of all of the components; the sum of these concentrations in a given phase equals unity:

$$X_1 + X_2 + X_3 + X_i = 1$$

268   Chapter 4 / Phase Equilibria

The chemical potential of each component i, is a function of the concentration X_i according to:

$$\mu_i = \mu_i^0 + RT \ln \gamma_i X_i$$

where μ_i^0 is the chemical potential of the standard state of component i and γ_i is the activity coefficient. There are thus C chemical potentials to be defined for any phase. However, a constraint is presented by the fact that the sum of all concentrations must be unity, which means that we can only independently vary C–1 chemical potentials. Counting all P phases present in the system, there are P(C–1) chemical potentials in the system that may be independently varied. Including the intensive additional variables of temperature and pressure, we have:

$$\text{Total variables} = P(C-1) + 2$$

Consider next that equilibrium is defined as the state where the chemical potential of each component, μ_i, is the same in all phases that are present. For example, in a three-phase system the chemical potential of component 1 is the same in phase 1 as in phase 2 as in phase 3: $\mu_1^1 = \mu_1^2$ and $\mu_1^2 = \mu_1^3$. For a system of P phases, this results in (P–1) independant equalities for the chemical potential of component 1. Applying this to all C components, the total number of equations defining the condition of equilibrium is

$$\text{Total equations} = C(P-1)$$

As there are fewer equations than the number of variables, some number of variables need to be independently defined in order to solve the system of equations. These are the degrees of freedom, or number of free variables, defined by the Gibbs phase rule:

$$\text{Degrees of Freedom} = \text{Total variables} - \text{Total equations}$$
$$F = P(C-1) + 2 - C(P-1)$$
$$F = C - P + 2$$

Notice that as one adds more components to a system there is greater chemical freedom, and F increases. On the other hand, as more phases are introduced into a system with a fixed number of components, the state of equilibrium becomes increasingly constrained by the need for equality in chemical potential between all phases, and F decreases.

The utility of the Gibbs phase rule can be illustrated using the pressure-temperature phase diagram for the single-component system H_2O in Fig. 4.1. This phase diagram shows the phases present at equilibrium for a broad range of temperature and pressure. At 1 atmosphere pressure, shown by the horizontal dotted line, the familiar phases are ice, water, and steam, encountered in sequence upon increasing the temperature. Let's apply the Gibbs phase rule for liquid water at

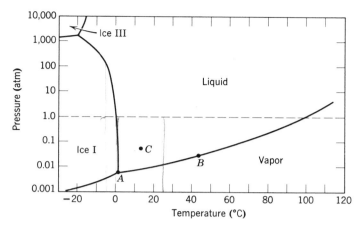

Fig. 4.1 Pressure-temperature phase diagram for water. Behavior at 1 atm pressure is indicated by the horizontal dashed line. The invariant point at which three phases are in equilibrium is labeled "A."

25°C and 1 atmosphere pressure. Since there is one component (C = 1) and one phase (P = 1), the degrees of freedom F is 2. This means that within the bounds of the *phase field* for liquid water, we can freely vary both the temperature and pressure and retain liquid as the equilibrium phase.

If we now consider water at 0°C and 1 atmosphere pressure, this point falls directly on the boundary between liquid water and ice. This is the freezing point of water, at which there are two phases in equilibrium for one component. The Gibbs phase rule (Eq. 4.1) states that there is one degree of freedom. This means that we can preserve the two-phase equilibrium between water and ice by varying *either* temperature or pressure independently, but not both. If we choose to vary one of the variables at will, a two-phase equilibrium can only be maintained if the value of the other is defined by the boundary between the liquid water and ice phase fields in Fig. 4.1. For example, while at 0°C a two-phase equilibrium only exists at a pressure of 1 atmosphere, at −5°C, the ice-water equilibrium exists only at 100 atmospheres. There exists only one combination of temperature and pressure at which all three phases (ice, water, steam) can be in equilibrium, labeled "A". For a one-component system with three phases, according to the Gibbs phase rule there must be zero degrees of freedom. In any phase diagram, a combination of variables at which there are zero degrees of freedom is termed an *invariant point*; for a single-component system such as water this is often referred to as the *triple point* since there are three phases present.

A phase diagram thus provides a graphical depiction of the conditions under which certain phases appear at equilibrium. The determination of a complete phase diagram is, generally speaking, a lot of work. Even in the absence of a detailed phase diagram, however, the Gibbs phase rule is useful for determining the possibility and likelihood that a system of phases is at equilibrium. We illustrate this point by considering the results of a common ceramics processing procedure whereby starting materials are mixed together and fired at high temperature. For example, if a two-component mixture of BaO and TiO_2 powders is fired to high temperature (let's say 1250°C) for a number of hours and quenched to room temperature, preserving the high temperature phase assemblage, a number of phases may result. If four crystalline phases are found: $BaTiO_3$, $BaTi_2O_5$, $BaTi_3O_7$, and TiO_2, is it possible that this system reached equilibrium during firing? (The phase diagram for this binary system appears in Fig. 4.7.) This is a two-component system (C = 2) of constant overall composition, fired at constant temperature, in a constant pressure. Therefore all intensive variables have been fixed, and F = 0. The Gibbs phase rule at constant pressure (Eq. 4.2) tells us that the maximum number of phases (P) that can be present at equilibrium is 3. We conclude that the existence of four crystalline phases cannot constitute a system at equilibrium; most likely the reaction between the starting oxides has not progressed to completion. What if only three of the four crystalline phases ($BaTiO_3$, $BaTi_2O_5$, and $BaTi_3O_7$) are present? Then according to the Gibbs phase rule at constant pressure (Eq. 4.2) this system could, strictly speaking, have been at equilibrium—at an invariant point. However, common sense tells us that the probability that our arbitrarily chosen temperature of 1250°C happened to fall on an invariant point of this system is very low. Thus, while it is thermodynamically *possible* that we reached equilibrium during firing, it is highly unlikely. However, if after quenching we see either two phases or one, according to the phase rule it is certainly possible, and even likely, that the system was at equilibrium.

Let's consider a slightly different experiment where we have fired the sample as before, but then slowly cooled the sample to room temperature. In this instance the overall composition and pressure have both remained constant, but we have independently varied one variable (temperature); hence F = 1. How many phases can be present throughout the cooling process if equilibrium is maintained? The Gibbs phase rule for constant pressure (Eq. 4.2) tells us that at most we can have two phases (P = 2) present in equilibrium. Thus, if either three or four phases are present in the end product, it is clear that equilibrium during firing was not preserved. If only one or two phases are found after cooling, it is then possible (but not proven) that equilibrium was preserved.

Note that the Gibbs phase rule does not tell us the amount of any phase, or the specific phases that will exist; it only tells us the *number* of phases that can coexist at equilibrium. That is, at 0°C and 1 atmosphere pressure, a minute amount of ice floating in water, and a film of water on ice, are both equally in equilibrium. Heating or cooling processes tend to be arrested in temperature at the boundaries between phase fields while the amounts of the respective phases change. Starting with a small amount of ice in water, if we slowly remove heat from the system additional ice will be formed, but the temperature will remain constant at 0°C until all of the water has solidified to ice. The system can then leave the two-phase boundary and enter the single phase ice field, allowing the temperature of the system to decrease (cf. Fig. 4.1). Similarly, during slow heating the temperature remains constant at 0°C while the amount of water increases until all the ice is melted (absorbing the latent heat of fusion), at which point we can leave the phase boundary and enter the single-phase water field. This tendency for systems to hesitate in temperature at phase boundaries upon heating or cooling while phases appear or disappear is especially important in binary and higher-component systems. Frequently, the completeness of such reactions determines the actual phases remaining after a particular thermal history.

4.3 BINARY PHASE DIAGRAMS

Binary phase diagrams are representations of the phase fields present in equilibrium mixtures of two components. For the single-component system in Fig. 4.1, we used temperature and pressure as the coordinates. In a two-component system we have an additional variable, which is the overall composition. (In some cases one plots the activity of one component as the chemical variable, but in the majority of phase diagrams composition is used since it is an easily measured experimental variable.) With these three intensive variables (T, P, and composition) being necessary to define the system completely, a complete binary diagram would be a three-dimensional representation of the phases present for every combination of these variables. For simplicity, and because most studies are concerned with phase equilibria at a constant ambient pressure, binary diagrams most often show a constant pressure slice of the three-dimensional diagram, with temperature and composition as the axes. The implicit assumption of a constant pressure is frequently not mentioned, but should be remembered.

Complete Solid Solution

The complexity of binary phase diagrams for solids depends in large part on the similarity of the endmember components. The simplest case is when the two components have the same crystal structure and the constituent atoms or ions are similar in size and chemical properties. An example is MgO-NiO, shown in Fig. 4.2, in which both components crystallize in the rocksalt structure type, share a common ion (oxygen), and have chemically similar cations. The lattice parameters only differ by ~1 % (0.4213 nm for MgO, 0.4177 nm for NiO). Complete solid solution of MgO in NiO and vice versa is possible, as indicated by the (Ni,Mg)O solid solution (ss) field. At high temperatures above the melting point of either compound, a single phase liquid solution of MgO and NiO corresponding to the overall composition is found. The lens-shaped region separating the single-liquid phase and single-solid-phase fields is a two-phase field, in which a liquid solution enriched in the lower melting component (NiO) is found in equilibrium with a

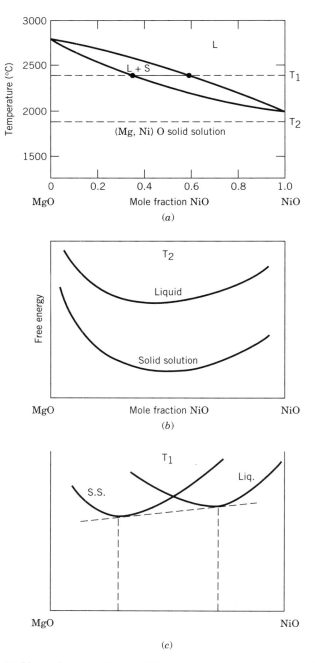

Fig. 4.2 (*a*) Phase diagram of MgO-NiO. Schematic representation of free energy-composition diagram for (*b*) T<2000°C; and (*c*) T=2400°C.

solid solution enriched in the higher melting component (MgO). At any temperature and composition within this two-phase field, the compositions of the liquid and solid are given by the endpoints of a horizontal line, known as the *tie line*, which terminates on the boundaries separating the single-phase and two-phase fields. The boundary curve between the single-phase liquid and two-phase field is the *liquidus*; this defines where a given composition becomes completely liquid (i.e., exits the two-phase field) upon heating. The boundary curve between the single-phase solid solution and the two-phase field is termed the *solidus*; upon cooling this is the temperature at which a given composition completely solidifies.

Limited Solid Solubility

Binary Eutectic System Binary systems with complete solid solutions as in Fig. 4.2 are relatively uncommon in ceramics. Figure 4.3 shows a schematic and an actual example of a slightly more complex binary phase diagram, a binary *eutectic*, that frequently results when the two components are less similar. In this example each solid endmember phase dissolves a limited amount of the other component: The single-phase fields are shown shaded, while the two-phase fields are unshaded. In CaO-MgO the endmembers are of the same structure type (rocksalt), but since Ca^{2+} and Mg^{2+} differ substantially in size they do not form a complete solid solution. (CaO has a lattice parameter, 0.4811 nm, which is 14% larger than that of MgO.) The extent of solid solution varies with temperature, with the maximum solid solubility or *solid solution limit* being given by the boundaries of the single phase fields labeled "MgO ss" and "CaO ss" in Fig. 4.3b. The maximum solid solubility of CaO in MgO is about 8%, while the maximum solubility of MgO in CaO is about 17%, both occuring at 2370°C. Between these two single-phase solid solution fields lies a field containing two solid phases at temperatures below 2370°C, labeled "MgO ss + CaO ss". In this field the solid solution limits of both endmember phases have been exceeded, and the compositions of the two solid phases in equilibrium are given by the endpoints of the tie line shown terminating on the solid solution boundary curves.

Let's apply the condensed phase rule ($F = C - P + 1$) to the various fields of this binary phase diagram. For a single phase field in a binary system, $F = 2$, indicating that it is possible to vary both temperature and composition independently without changing the number of phases. In a two-phase field, the phase rule allows only one degree of freedom; this means that if temperature is varied, so must composition if the same two phases are to be maintained. However, an important point is that it is the *phase composition* that must be varied, not the *average* composition. (In a single-phase field, they are one and the same.) Consider the point A in Fig. 4.3b with an average composition of 40% CaO, at 2100°C. This point is in a two-phase field, where the compositions of the two phases in equilibrium are given by the boundary curves between the one- and two-phase fields. The horizontal tie-line at 2100°C joins these compositions, which are at 5% and 90%

274 Chapter 4 / Phase Equilibria

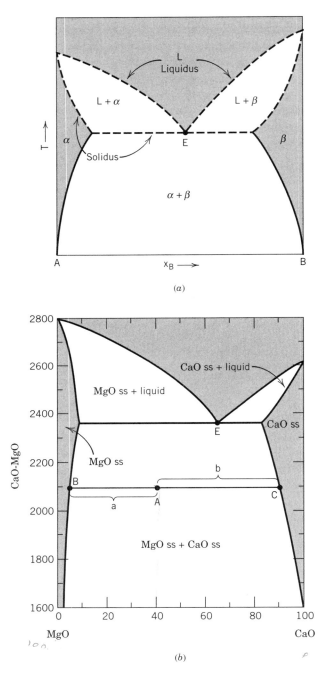

Fig. 4.3 (a) Schematic binary eutectic diagram; (b) CaO-MgO system showing some solid solution of each component in the other endmember phase. Eutectic point is at E. Application of binary lever rule to overall composition A is discussed in text. [adapted from R.C. Doman, J.B. Barr, R.N. McNally, and A.M. Alper, *J. Am. Ceram. Soc.*, **46**[7] 314 (1963).]

CaO, respectively. If temperature is varied up or down the endpoints of the tie-line change accordingly, following the phase boundaries. A change in temperature necessitates a change in the compositions of the individual phases, and therefore the chemical potentials of the constituents, if the two-phase equilibrium is to be maintained. Likewise, an independent variation in the composition of either one of the phases requires a change in temperature if we are to remain on the boundary curves. This demonstrates the one degree of freedom for the two phases in equilibrium. If we instead vary the average composition along the tie-line away from point A, we change the relative amounts of the two phases in equilibrium (according to the binary lever rule discussed later), but not their compositions; those remain at the terminations of the tie-line. Recalling that the Gibbs phase rule does not dictate the relative amounts of phases but only their coexistence, it follows that a change in average composition within the two-phase field does not result in a change in phase equilibrium.

The melting and solidification behavior of a binary eutectic system differs significantly from that in Fig. 4.2. We see that the liquidus curves slope downward toward the center of the diagram, indicating that addition of CaO lowers the melting point of MgO, and vice versa. The two liquidus curves meet at a critical point termed the *eutectic*, labeled E. A liquid of this composition would, upon cooling through the eutectic temperature, solidify directly to two solid phases. At the temperature and composition of the eutectic E, we have three phases of fixed composition co-existing in equilibrium: a liquid of the eutectic composition and the two solid solutions (MgO ss and CaO ss) of compositions corresponding to the solid solution limits at the eutectic temperature (2370°C). For three phases coexisting in equilibrium the Gibbs phase rule yields $F = 0$; hence, the eutectic temperature and composition represent an invariant point where neither temperature nor composition of any of the phases can be varied without causing the disappearance of a phase. In a two-component system, it is not possible to have more than three phases in equilibrium at constant pressure.

Intermediate Compounds

In Fig. 4.3 the endmembers MgO and CaO are of the same structure type, yet are sufficiently mismatched that there is limited solid solution of each in the other. For increasingly dissimilar components, intermediate compounds are often found. Figure 4.4a shows in schematic form a binary system in which the intermediate compound AB forms. The three solid phases: α, β, and ab, have each been drawn as vertical lines to indicate that these are *line compounds* with no visible solid solution range. This diagram has two eutectic points, one between L, A and AB and one between L, AB and B. Typically these will occur at different temperatures. ab is a *congruently melting* compound, which means that it melts to form a liquid of the same composition. For example, the Czochralski technique, widely used in the growth of silicon crystals for semiconductor devices, can only be used for congruently melting compounds. As discussed later, whether or not a com-

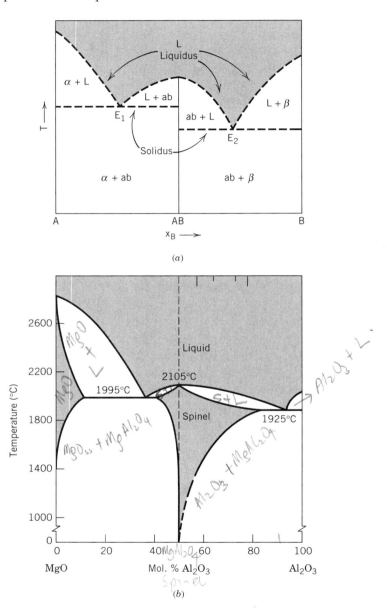

Fig. 4.4 (a) Schematic binary diagram with one intermediate compound ab forming a eutectic with each of the endmembers A and B; all solid phases are line compounds. (b) MgO-Al$_2$O$_3$ binary diagram showing two eutectics and an intermediate compound MgAl$_2$O$_4$ (spinel). Some range of solid solution exists for periclase and spinel, while corundum dissolves negligible MgO.

pound melts congruently determines the methods that can be used for growth of single crystals.

An example of this kind of phase diagram is shown in Fig. 4.4b for the system MgO - Al_2O_3, in which the intermediate compound is $MgAl_2O_4$ (spinel). Unlike the schematic in Fig. 4.4a, not all compounds in the system are line compounds. There is an extensive solid solution of Al_2O_3 in spinel that reaches a maximum of about 85 mole% Al_2O_3 at 1860°C, and a more limited solid solution of MgO in spinel reaching 62% Al_2O_3 at 1995°C. The solid solution field of the spinel phase appears as the wide shaded region in Fig. 4.4. There also exists a substantial solid solution of Al_2O_3 in the MgO endmember at elevated temperatures. On the other hand, at the Al_2O_3 endmember no solid solution field is shown for corundum since the solubility of MgO in corundum is at most a few hundred parts-per-million. We may view Fig. 4.4b as either a binary diagram between the components MgO and Al_2O_3 with an intermediate compound, or as two binary diagrams between the components MgO-$MgAl_2O_4$ and $MgAl_2O_4$ - Al_2O_3.

SPECIAL TOPIC 4.2

FREE ENERGY CURVES AND THE COMMON TANGENT CONSTRUCTION

The fields and boundaries on a phase diagram depict the phase or combination of phases which represent the minimum free energy for a material of a certain composition and at a given temperature. The origin of these features can be visualized using curves of free energy versus composition for each phase. Figure ST40 shows the free energy for a simple binary solution of two components A and B at a constant temperature. If there is no atomic mixing between the two endmembers, that is, the composition of the whole

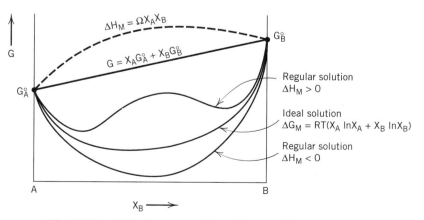

Fig. ST40 Gibbs free energy for a binary solution of A and B.

278 Chapter 4 / Phase Equilibria

is arrived at by a simple physical mixture of the two phases consisting of pure A and pure B, then the free energy on a per-mole basis is a linear combination of the free energies of each pure phase

$$G = G_A^o X_A + G_B^o X_B \qquad \text{(Eq. 1)}$$

where G_A^o and G_B^o are the free energies of the phases corresponding to pure A and pure B, and X_A and X_B are the mole fractions of A and B (where $X_A + X_B = 1$). The variation in free energy with composition is given by the straight line in Fig. ST40. Notice that it is necessary to specify the "phase" of A and B for which free energies are given rather than just the composition, since even a pure component can exist in different phases (liquid, vapor, and various crystalline polymorphs).

If A and B are thoroughly mixed at the atomic level, the free energy will now deviate from this simple linear relationship in Eq. 1 by an amount ΔG_m, which can be separated into an enthalpy (heat) of mixing and an entropy of mixing

$$\Delta G_m = \Delta H_m - T\Delta S_m. \qquad \text{(Eq. 2)}$$

The values of ΔH_m and ΔS_m depends on the nature of the interactions between atoms of A and B in the solution. Several limiting cases can be used to illustrate the origin of phase fields in a phase diagram. An *ideal solution* has no heat of mixing ($\Delta H_m = 0$), and has a configurational entropy due solely to random mixing, which gives a free energy of mixing

$$\Delta G_m \text{ (ideal solution)} = -T\Delta S_m = +RT(X_A \ln X_A + X_B \ln X_B) \qquad \text{(Eq. 3)}$$

This change in free energy is always negative, as illustrated in Fig. ST40. Any composition of the ideal solution is lower in free energy than the same composition present as separate phases, due to the increased entropy upon mixing.

A *regular solution* approximates the heat of mixing as a function which is parabolic in composition, and treats the entropy of mixing as ideal, giving a free energy of mixing

$$\Delta G_m \text{ (reg. solution)} = \Omega X_A X_B + RT(X_A \ln X_A + X_B \ln X_B) \qquad \text{(Eq. 4)}$$

Ω is an interaction coefficient, which characterizes whether the interaction between A and B atoms is mutually attractive or repulsive. A negative value of Ω indicates an attractive interaction that lowers the free energy of mixing beyond that of the ideal solution by the amount ΔH_m (as shown in Fig. ST40), whereas if Ω is positive the interaction is repulsive, and the free energy curve is raised by the amount ΔH_m. When this positive contribution is sufficiently large, an upward inflection develops in the free energy curve. This is now a situation in which phase separation can lower the free energy of the solution. Upon drawing a common tangent between the energy minima as shown in Fig. ST41, we see that the free energy of the homogeneous solution (given by the heavy curve) lies above the tangent between compositions X_B' and X_B''. Any solution of a composition X lying between these two limits can lower its free energy by phase separating into a physical mixture of the compositions X_B' and X_B''. The change in free energy upon phase separation is depicted by the arrow in Fig. ST41; the free energy after phase separation is given by a linear combination of the free energies of phases X_B' and X_B''. Notice also that the common tangent between X_B' and X_B'' identifies the two compositions for which the chemical potentials of A and B are equal, since the slope of the free energy curve is the chemical potential,

$$\mu = \left(\frac{\partial G}{\partial X}\right)_{T,P} \tag{Eq. 5}$$

Thus the common tangent automatically satisfies the definition of equilibrium between phases. Other solution models in addition to the ideal and regular solutions can be used to approximate the free energy of a phase as a function of its composition and temperature. In the most general case the free energy is a complex function that is best determined from experimental data or detailed atomistic computations.

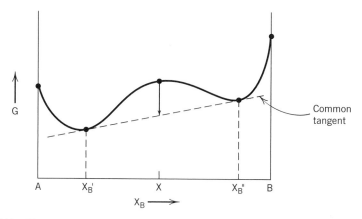

Fig. ST41 Tangent to free energy curve gives the compositions of the equilibrium phases of compositions X_B' and X_B''.

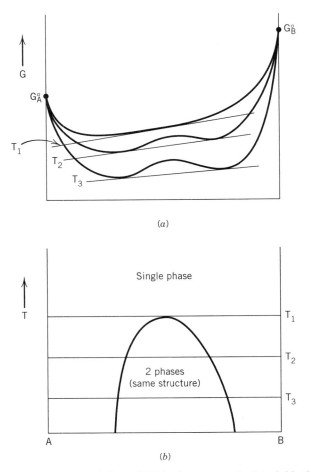

Fig. ST42 Temperature variation of Gibbs free energy in (a) yields the immiscibility region shown in (b).

The temperature dependence of demixing is shown in Fig. ST42a as a series of free energy curves for which the phase separation region broadens with decreasing temperature. A region of immiscibility (also called a "miscibility gap") is formed within the phase diagram (Fig. ST42b). The temperature of the peak is termed the *consolute* temperature and corresponds to where an inflection first develops in the free energy curve. Above that temperature, the single phase has the the lowest free energy at every composition whereas below, the phase-separated mixture has lower free energy for all compositions within the gap. In this particular phase diagram the two phases are different compositions of a single solution being represented by a single free energy curve and are implicitly of the same structure.

Free energy curves can be drawn for every possible phase of a material. Figure 4.2b shows the free energy curves for the MgO-NiO system at a temperature T_2 where the solid solution is the equilibrium phase. The liquid solution is therefore of higher energy at this temperature. At temperatures above 2750°C, where only liquid is present, the liquid solution curve is lower. At a temperature T_1, however, the two curves overlap, and the common tangent between the two curves gives us the compositions of the liquid and solid phases which are in equilibrium.

Metastable phases are those with a higher free energy, but which for kinetic reasons may exist in preference to an equilibrium phase. In Fig. 4.2*b* the liquid has higher free energy—yet if it were quenched rapidly, it might be preserved for a time as a metastable phase at temperatures where the solid is the most stable. In Fig. ST43 a binary phase diagram containing an intermediate compound AB, which forms a simple eutectic with each

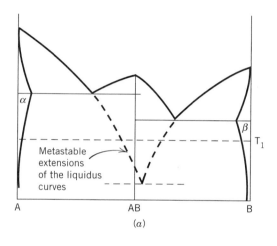

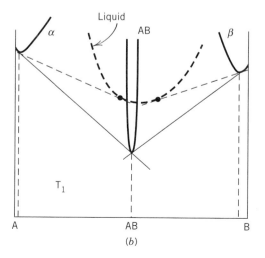

Fig. ST43 (*a*) Equilibrium binary system with intermediate compound AB, and metastable phase diagram which can result when AB is kinetically prevented from crystallizing. (*b*) Corresponding Gibbs free energy curves.

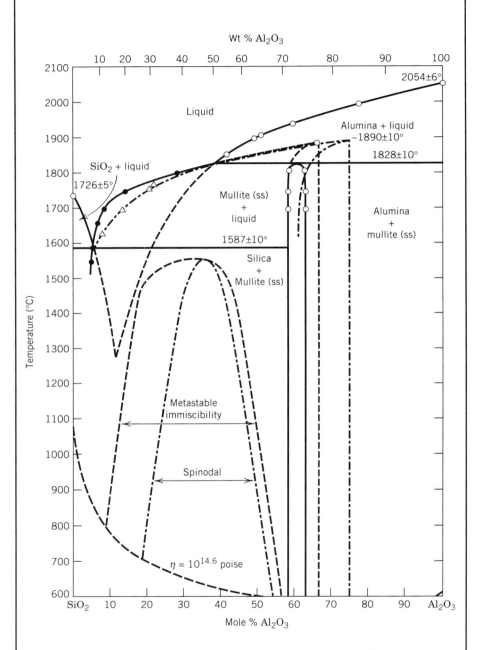

Fig. ST44 SiO$_2$-Al$_2$O$_3$ system showing metastable liquidus and solidus extensions and metastable immiscibility region at lower temperature. From S.H. Risbud and J.A. Pask., *J. Mat. Sci.*, 13 [11], 2449 (1978).

endmember A and B is shown. At the temperature T_1 the phases at equilibrium are the solid phases α, AB, and β. The tangents to the free energy curves (solid lines) give the limits of solid solution of α and β (Fig. ST43b). The very limited solid solution range of AB corresponds to a free energy curve that is narrow and steep. The higher free energy curve for the liquid shows that it is metastable, but if the phase AB is difficult to crystallize, the liquid may continue to exist upon cooling as a metastable phase. In this case the common tangents between the liquid free energy curve and the solid phases yield the *metastable extensions* of the liquidus curves, as shown in Fig. ST43a. The phase relations shown by metastable diagrams of this type are most commonly seen in compositions where the liquid phase is viscous and from which crystallization is difficult, such as silicate glasses, or where a liquid has been rapidly quenched. Figure 44 shows a more complex example with several metastable features that have been proposed for the SiO_2-Al_2O_3 system. Another form of metastable phase formation is *solid state amorphization*, a phenomenon where two solid phases placed into contact form a metastable glassy solid rather than an equilibrium crystalline intermediate compound.

Peritectic Diagrams and Incongruent Melting

Incongruent melting, whereupon a solid phase melts to form a liquid of a different composition, is the hallmark of *peritectic* diagrams. Figure 4.5 shows a schematic peritectic phase diagram for a two-component system with one intermediate compound, and in which all compounds are line compounds. The compound ab melts incongruently; with increasing temperature, this line compound intersects the L + α two phase field at its melting point. The melting reaction is therefore ab $\rightarrow$ L + α. At its melting point, ab is in equilibrium with α and liquid of composition P, constituting a three-phase equilibrium and therefore an invariant point. The peritectic *temperature* is the melting point of ab, and by convention, the peritectic *point* "P" is the composition of the liquid that is in equilibrium with the two solids at this temperature. A eutectic also appears in Fig. 4.5, between the compounds ab and β.

We may think of the eutectic and peritectic as being in a sense inverted with respect to one another; the transition from a single-phase to two-phase region occurs upon cooling for the eutectic point, and upon heating for the peritectic point. Peritectic solidification is characterized by the formation of a new solid from the reaction between a pre-existing solid and liquid. In the case of the compound ab, the peritectic solidification reaction is L + α $\rightarrow$ ab. Figure 4.6 shows a peritectic phase diagram for a binary system with no intermediate compound, and in which both solid phases have extensive solid solubility. In this instance, the phase β decomposes incongruently upon heating into a mixture of liquid and α.

284 Chapter 4 / Phase Equilibria

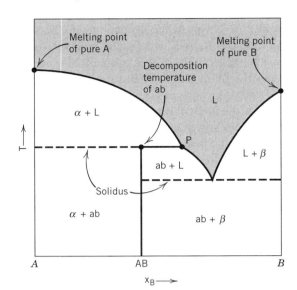

Fig. 4.5 Binary diagram containing an incongruently melting compound ab, and containing a peritectic (P) and a eutectic invariant point. No solid solution is shown for the solid phases.

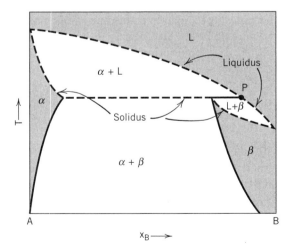

Fig. 4.6 Binary peritectic diagram in which both solid phases have extensive solid solubility. The phase β melts incongruently.

4.3 Binary Phase Diagrams

An example of a binary system with numerous intermediate compounds and several eutectic and peritectic invariant points is the technologically important system $BaO\text{-}TiO_2$, shown in Fig. 4.7. $BaTiO_3$ is an important ferroelectric material, as discussed in Chapter 1, and $BaTi_4O_9$ and $Ba_2Ti_9O_{20}$ are useful dielectric materials for microwave frequency communications. Most of the intermediate compounds in this system exhibit very limited solid solution. The exceptions are hexagonal and cubic $BaTiO_3$, both of which appear to have some solid solubility for excess TiO_2 at elevated temperatures. It is left as an exercise to the reader to identify the two eutectic and four peritectic points in this diagram.

Subsolidus Phase Equilibria

Figure 4.8 introduces a new level of complexity, wherein a pure endmember exists as more than one solid phase. This diagram shows the region between ZrO_2 and the intermediate compound $ZrCaO_3$ in the binary system ZrO_2–CaO. ZrO_2

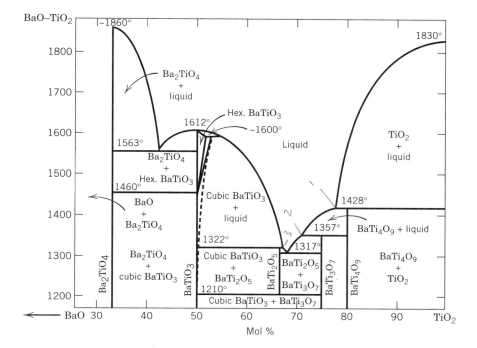

Fig. 4.7 $BaO\text{-}TiO_2$ binary system, showing several intermediate compounds of both congruently and incongruently melting character. (Adapted from Phase Diagrams for Ceramists, The American Ceramic Society, Columbus, Ohio, 1975.)

286 Chapter 4 / Phase Equilibria

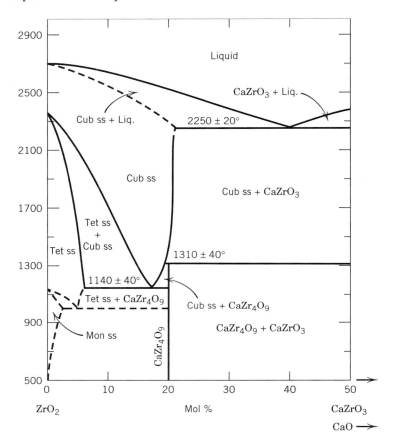

Fig. 4.8 Portion of the ZrO_2-CaO binary system, showing subsolidus phase equilibria. {From V.S. Stubican and J.R. Hellman, *Adv. Ceram.*, 3, 25 (1981), and V.S. Stubican and S.P. Ray, *J. Am. Ceram. Soc.*, 60[11–12], 534 (1977).]

can exist as tetragonal (*t*), cubic (*c*), and monoclinic (*m*) polymorphs (see p. 29), and the relative stability of each polymorph is affected by dopants and temperature. Additives known as "stabilizers" such as CaO, MgO, Y_2O_3, and Gd_2O_3 extend the equilibrium single phase fields of the tetragonal and cubic phases to lower temperatures from where they can frequently be preserved to room temperature as metastable phases. Figure 4.8 shows that there exists a conventional eutectic between cubic ZrO_2 and $ZrCaO_3$ at 2250°C. At lower temperature there is a region of *subsolidus* phase equilibria where features like those of the eutectic appear between solid phases. At approximately 1140°C and 17% CaO, there is an invariant point termed a *eutectoid* (to indicate only solid phases in equilibrium) where cubic ZrO_2, tetragonal ZrO_2, and $CaZr_4O_9$ are in equilibrium. The cubic phase decomposes to the latter two upon cooling if the kinetics are fast enough. The solid-phase equivalent of the peritectic is the *peritectoid* and can be treated in an analogous manner to these examples. In Fig 4.7 there also exists a eutectoid at 1210°C between the line compounds $BaTi_2O_5$, $BaTiO_3$ and $BaTi_3O_7$. The eutectoid decomposition reaction is $BaTi_2O_5 \rightarrow BaTiO_3 + BaTi_3O_7$.

Solidus and Liquidus Temperatures

We referred earlier with respect to Fig. 4.2 to the *liquidus*, defined as the lowest temperature at which a particular composition is completely liquid. This is also, upon heating, the highest temperature at which any remnant of solid exists. The *solidus* is conversely the highest temperature at which a particular composition is completely solid, or the lowest temperature at which any liquid exists. The solidus is an important temperature in the firing process for ceramics, since the presence of a small amount of liquid can dramatically increase diffusional transport, and along with it the rates of chemical reaction, sintering, and grain growth. It is important to emphasize that it is not a single characteristic temperature for a system, but instead varies with the specific composition.

Figure 4.3a highlights the liquidus and solidus curves in a simple binary eutectic. To the left of the eutectic composition, a single-phase liquid will, upon cooling, first begin to crystallize phase α when the liquidus is reached. To the right side of the eutectic, the liquid will first solidify phase β. In this diagram the solidus temperature is constant across the wide central region where two solid phases coexist. Upon heating, all two-phase mixtures with an overall composition within this range exhibit the first appearance of a liquid at the same temperature: that of the eutectic. Depending on which side of the eutectic composition we are on, the solid phase that remains in equilibrium with the liquid is either α or β. However, for compositions within the single-phase solid solution fields of α or β, the solidus is not flat but a curve which varies with temperature, as shown in Fig. 4.3a. The corresponding liquidus and solidus temperatures for a peritectic system are highlighted in Figs. 4.5 and 4.6. In all cases, the liquidus and solidus lines bound the two-phase (solid + liquid) fields.

In eutectic and peritectic systems containing intermediate compounds, such as Figs. 4.4a and 4.5, we have again highlighted the locations of the solidus lines to show that they vary with composition. Using Fig. 4.5 as an example, to the right of the compound ab, the two equilibrium solids ab and β will first react upon heating at the eutectic temperature to form a liquid of the eutectic composition. The temperature of first melting is considerably higher to the left of ab, and occurs at the peritectic temperature.

In the processing of ceramics, starting materials are often those of convenience and economy, rather than solid phases in thermodynamic equilibrium. The critical firing temperature is often the lowest solidus temperature in the system. Let us illustrate by referring again to Fig. 4.5. Consider the starting materials to be the solids α and β, present as finely mixed powders. Upon heating, at microscopic contacts between the powder particles the local composition can span the entire range between pure α and pure β. The first liquid that forms in this material will therefore be of eutectic composition, and will form at the eutectic temperature regardless of whether the overall composition lies to the left or right of the compound ab. Depending on the overall composition this liquid may remain or disappear as the reaction between α and β proceeds toward equilibrium. As another example, consider the BaO-TiO_2 phase diagram in Fig. 4.7. If we heat a mixture of the powders BaO and TiO_2, the first temperature at which a liquid will form, however small the actual amount, is the eutectic at 1317°C. Only after some extent of compositional mixing will equilibrium compositions and phases be achieved. The relation between phase equilibria and processing and microstructure is discussed further in Chapter 5.

Variable Valence Systems: Example in Fe-O

Care must be taken in selecting the correct number of components for a system. In the binary examples Figs. 4.2 through 4.8, we have chosen the respective oxides as the components. In fact there are three elements involved in each instance (two metals and oxygen). Why is this not a three component system? Notice that if the valences of the cations are fixed, the concentration of oxygen in any mixture is dictated by charge balance. This removes one compositional variable-there are only two independently variable chemical potentials. Presuming that the oxygen activity is high enough that the metals remain fully oxidized, it is then convenient and valid to treat these as binary systems between the endmember oxides.

Nonetheless, situations arise in which the metal has more than one stable valence state, as with the transition metal oxides. In these cases oxygen must be considered as one of the components. An important example of a variable valence system is iron-oxygen, the binary diagram for which is shown in Fig. 4.9. The stable valences for iron in oxides are 2+ and 3+. One way to depict the relevant phase equilibria is as a binary diagram between the reduced phase wüstite and the oxidized phase hematite. As the diagram shows, wüstite does not exist at the ideal stoichiometry of FeO that is required of divalent iron, but is always somewhat cation deficient ($Fe_{1-x}O$) due to a fraction being trivalent. The hematite phase does exist in the ideal stoichiometry Fe_2O_3 corresponding to all iron being in the Fe^{3+} state. There is also an "intermediate compound" Fe_3O_4, magnetite, occurring at the composition $FeO \cdot Fe_2O_3$, which is narrow in stoichiometry

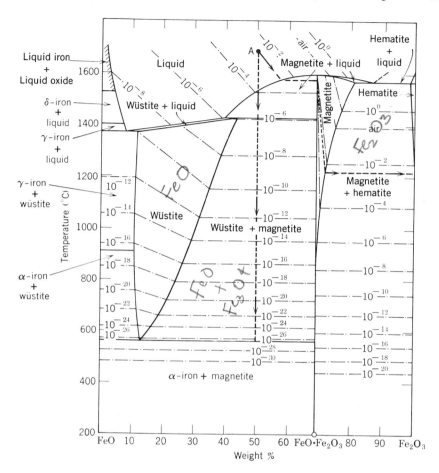

Fig. 4.9 Fe-O diagram, represented as a binary system between FeO and Fe_2O_3 endmembers. Contours of constant oxygen activity in equilibrium with the equilibrium phases are shown.

at temperatures below 1000°C but has a range of solid solution above. The single- and two-phase fields between these three solids and the liquid can be identified by close inspection of Fig. 4.9.

Since the average valence state of the iron is determined by both temperature and oxygen activity, an important aspect of this diagram is the equilibrium between the condensed and vapor phases. The *total pressure* represented in Fig. 4.9 is implied to be one atmosphere, as in the earlier binary diagrams. However, the oxygen *partial pressure* (P_{O_2}) co-existing in equilibrium with the condensed phases varies throughout. The dot-dash contours represent the partial pressure of oxygen that is in equilibrium with the condensed phases at the given temperature. Notice that in the two-phase fields these lines are horizontal, indicating a fixed oxygen partial pressure in equilibrium with the two-phase mixture. This is consistent with our earlier result, required by the Gibbs phase rule, that variations in average composition within a two-phase field change only the relative amounts of the two phases, and not their individual phase compositions. In the single-phase fields, however, the equilibrium oxygen partial pressure must change with both temperature and phase composition, so that the P_{O_2} contours slant across these regions.

If the oxygen activity were to be fixed, then one compositional degree of freedom would be lost and the system must behave as a single component. To confirm that this is the case, let us trace the solidification behavior of the liquid composition labeled A, at a constant oxygen activity. Following the 10^{-2} atm isobar down in temperature, one finds that single phases can exist over a range of temperature, but any two phases coexist only at a single temperature - just like any other single component system. If the composition A is instead cooled without constraining oxygen activity (i.e., following the vertical dashed line), the behavior is that of a binary system.

Binary Lever Rule

We have utilized the fact that for a binary system in two phase-equilibrium, the overall composition does not change the composition of the individual phases, but only their relative amounts. From the equilibrium phase diagram, we can calculate these relative amounts using the *binary lever rule*, a simple statement of mass conservation. We use Fig. 4.3b for illustration. At an average or overall composition of 40% CaO and a temperature of 2100°C, the two phases present at equilibrium are MgO containing about 5% CaO and CaO containing about 10% MgO. These two solid phases must be present in amounts that yield an average composition of 40% CaO.

Treating the tie-line that joins the two phase compositions as a lever, and the average composition at point A as a fulcrum, the relative amounts of the two phases must balance the lever to give us the overall composition. The lever arms are given by the composition difference a = A − B, and b = C − A, with the total lever length being a + b. Thus, the amount of the MgO solid solution is given by b/(a+b), and the amount of the CaO solid solution is a/(a+b). Measuring with a ruler, from Fig. 4.3 we find that there is 41% of the CaO solid solution and 59% of the MgO solid solution at equilibrium for an overall composition of 40% CaO. (These percentages are in the same mole percent units given on the composition scale of the diagram.) The binary lever rule can be applied to any two-phase field of a binary phase diagram.

4.3 Binary Phase Diagrams 291

___SPECIAL TOPIC 4.3___

CRYSTAL GROWTH AND PHASE EQUILIBRIA

Many different optical or electronics applications require the growth of relatively large single crystals. These crystals are grown by a variety of means, depending on the material's melting temperature, the desired composition and crystal structure, and other practical considerations. The characteristics of the phase diagram often determine the crystal growth methods that can be used.

The most important consideration for crystal growth from the melt is whether the material melts congruently, incongruently, or decomposes without melting altogether. This determines the source material with which the crystal is in equilibrium at its highest temperature of stability. The ideal circumstance is when a compound melts congruently (the liquid is the same composition as the compound), for then the crystals may be grown directly by solidification of the melt. This can be relatively fast since no long-range transport of material either toward or away from the crystal interface is required (with the possible exception of dopants). On the other hand, if the compound melts incongruently, then the crystal and the melt are clearly of different compositions, and diffusion and compositional mixing must be carefully induced to achieve high-quality single-crystal material of the desired composition; the rate of growth is usually much slower. If a crystal decomposes without melting, it is usually necessary to grow it either from a vapor phase, or by finding a lower melting liquid of another composition to act as a solvent from which the crystal can be grown.

The most widely used technique for single-crystal growth is the Czochralski (CZ) method, which can only be used for congruently melting materials. A "seed" crystal is pulled upward from the surface of the melt while being rotated about its axis, in a furnace with a carefully controlled temperature profile. Single crystal silicon for the electronics industry is widely grown by the Czochralski method, in cylindrical ingots ("boules") as large as 12" in diameter. An important ceramic material that is grown using this method is lithium niobate. The binary phase diagram for this system, Fig. ST45a, shows that $LiNbO_3$ melts congruently. However, notice that single-phase $LiNbO_3$ exists over a range of compositions. The congruent melting point is not exactly at the stoichiometric composition, but is in fact slightly niobium-rich. Solidifying a composition of the congruent composition can give a high-quality single crystal. But if crystal growth is performed using melt compositions on either side of the congruent point, then rejection of one of the components is necessary as solidification of the congruent crystal proceeds, resulting in poorer crystal quality and lower yield. Grow-

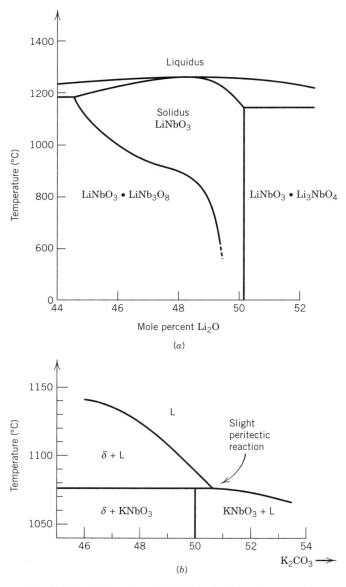

Fig. ST45 (a) Portion of Li_2O-Nb_2O_5 phase diagram showing that $LiNbO_3$ melts congruently, but at a slightly Nb-rich composition. (b) Peritectic melting of $KNbO_3$ occurs only a fraction of a percent away from the stoichiometric composition in the Nb_2O_5-K_2CO_3 system.

ing off-congruency can also lead to composition variation along the length of the crystal boule as it is pulled.

Another interesting optical material that does not melt congruently is $KNbO_3$. This compound is a perovskite-structure electro-optic material which is of interest for frequency conversion applications. Figure ST45*b* shows a portion of the binary diagram very close to the ideal composition. Even though the peritectic point differs only slightly in composition from the phase composition, crystal growth from the melt is not feasible. High-quality crystals must be grown using some sort of solution or flux growth technique. Solution growth refers to the growth of a crystal from a liquid containing the same components, but which is quite different in composition (e.g., $BaTiO_3$ can be grown from a TiO_2-rich melt). Flux growth refers to growth from a foreign liquid solvent in which the constituents of the crystal are soluble. Preferred fluxes are usually low melting liquids (e.g., halides, borates, or vanadates) in order to lower the temperature of growth. These must have a low solubility in the crystal which is to be grown in order to avoid contamination. Since in both solution and flux growth the liquid differs greatly in composition from the crystal, growth must be performed slowly if crystals of high perfection (e.g., of optical quality) are to be obtained.

4.4 FEATURES OF TERNARY PHASE DIAGRAMS

As with binary phase diagrams, ternary phase diagrams are a graphical representation of the phases in equilibrium at given compositions and temperatures. These diagrams are most often presented in a triangular coordinate system, where each corner of the triangle represents one of the three components. In a three-dimensional depiction this triangle is the base (where for the binary diagram we had only a line) of a volume for which temperature is the vertical axis. For condensed phases a constant pressure is usually assumed or specified, in which case there are three independent variables: temperature and the concentrations of two of the three components (the third being determined by difference). While the features of the ternary diagram are best seen in a three-dimensional model, Fig. 4.10, these are cumbersome to present graphically, so typically the temperature axis is not explicitly shown. Instead, a view along the temperature axis is used, with the composition axes given as a equilateral triangle. The three-dimensionality of features is often shown as constant-temperature contours, which are analogous to the constant-elevation contours on a topographical map. For example, the liquidus, which appears as a line in a binary diagram, is a *surface* in a ternary diagram, the shape of which can be shown using constant temperature contours.

294 Chapter 4 / Phase Equilibria

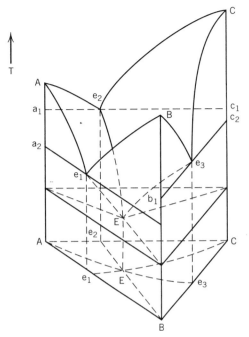

Fig. 4.10 Ternary eutectic phase diagram with limited solid solution for all three endmembers and no intermediate compounds.

Reading Composition

Compositions are read from a ternary diagram by constructing lines parallel to each side through the point of interest. Figure 4.11 shows a triangular grid where each apex is the pure component (A, B, and C), and the composition of interest is indicated by the dot. The composition of this point is 27% A, 46% B, and 27% C. Line A is parallel to the base of the triangle (at which the concentration of A is zero) opposite from the apex A (where the concentration is 100%). We read the amount of A to be 27% from the scale on the right side of the diagram. Line B is parallel to the side of the triangle opposite from apex B, and we read the concentration from the scale on the base of the triangle to be 46% B. Similarly, line C is parallel to the side opposite C and the concentration is read as 27% C. Alternatively, we can determine the concentration of C by difference since the sum of all three must equal 100%

Note that the composition can also be read from just one side of the triangle. For example, the right side of the triangle is a "join" between A and B, and is intersected by lines A and B. The fraction of A is represented by the segment "a". The remainder must then be made up of "b" and "c". The fraction of B can be obtained by running the scale in the reverse direction, and is represented by seg-

4.4 Features of Ternary Phase Diagrams

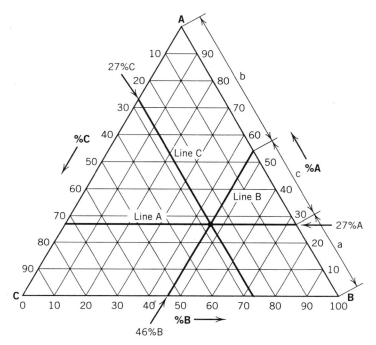

Fig. 4.11 Reading composition from a ternary diagram. The lines parallel to the three sides show concentration at 10% intervals. The composition of the point is read from the intersection of lines parallel to the three sides.

ment "b". The fraction of C is the remainder, given by line segment "c". These segments give the fraction of each component in the overall composition. Upon closer examination, the three concentrations are also represented by the line segments a, b, and c, and sum to a total of 100% along the right side of the triangle. This procedure works because of the mathematical identity of similar triangles, and is not limited to equilateral triangles, as we shall see later with respect to the ternary lever rule.

Example: SiO_2-Al_2O_3-"FeO"

Let us now identify compositions in a real ternary phase diagram for the system SiO_2-Al_3O_3-"FeO" (the latter indicating "FeO" to be an approximate composition due to the nonstoichiometry of wüstite) shown in Fig. 4.12. Ternary diagrams are available from various sources; a quite complete compilation is the series "Phase Diagrams for Ceramists," published by The American Ceramic Society in volumes which are updated regularly. Fig. 4.12 illustrates a typical representation of a ternary diagram from this source. First we can locate specific stoichiometric compounds (solid phases) that form in this system. These are listed in the upper-

296 Chapter 4 / Phase Equilibria

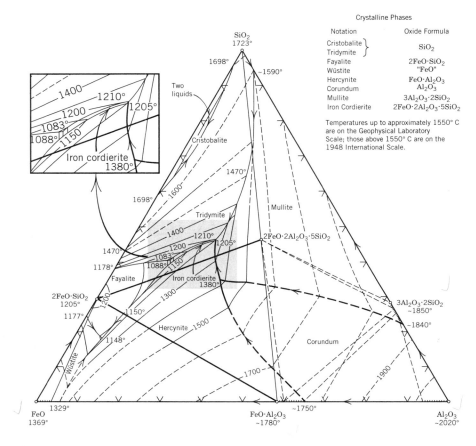

Fig. 4.12 SiO_2-Al_2O_3-"FeO" ternary phase diagram. (Fig. 696 from *Phase Diagrams for Ceramists*, the American Ceramic Society, Columbus, Ohio, 1975.)

right-hand corner of the diagram, along with their mineral names. Each of these compounds can be found marked with a small open dot "o" in the diagram. There are three compounds that are the pure endmember components: SiO_2, Al_2O_3, and "FeO." The sides of the triangle represent compositions between any two of the three endmembers, and all binary intermediate compounds appear here (these are the intermediate compounds that would appear in the corresponding binary diagrams). Ternary phases (of which there is only one in this system, iron cordierite, $2FeO \cdot 2Al_2O_3 \cdot 5SiO_2$) appear within the triangle and their compositions are read as previously described. Some of these compounds also have a temperature shown near the dot. This indicates the melting point of the compound, and furthermore, if given, indicates that the compound melts congruently. (If a temperature is not given, it does not necessarily mean that the compound melts incongruently!)

Some of the solid phases that are present in this diagram have extensive ranges of solid solubility. These can be identified by hatchmarks near the compounds emanating from the ideal composition indicated by the "o." In this diagram the compounds with noticeable solid solubility are corundum (Al_2O_3), mullite ($3Al_2O_3 \cdot 2SiO_2$), hercynite ($FeO \cdot Al_2O_3$), and wüstite (FeO). In general, solid solubility is a function of temperature and increases at higher temperature. However, in the two-dimensional projection of the ternary diagram this is difficult to show, and the hatched regions are drawn to show the *maximum* solid solubility of the compound, which typically occurs at the solidus temperature for a congruently melting compound. (See for example the binary compounds in Figs. 4.3 and 4.4.)

Primary Phase Fields

The *primary phase* in any multicomponent system is the first solid to crystallize from the liquid upon cooling. The *primary phase field* of a compound is the region of composition over which it is the first to crystallize. In a binary diagram, this composition region is spanned by a segment of the liquidus curve, lying over a liquid+solid, two-phase field. Referring to the binary diagram in Fig. 4.5, notice that there are three solid compounds (α, ab, and β) and three liquid + solid two-phase fields. There are also three corresponding liquidus curves. Any liquid composition, upon cooling, can cross only a single liquidus curve and therefore crystallize one primary phase. Thus, the compositions spanned by the three liquidus curves in Fig. 4.5 are the primary phase fields for the compounds α, ab, and β, respectively.

In a ternary diagram, the primary phase fields are portions of the liquidus *surface*, and are typically labeled with the name of the crystallizing compound. In the simplified ternary diagrams in Fig. 4.13, these are labeled as α, β, γ, ab, and ac, representing the phases of pure composition A, B, C, AB and AC respectively. In Fig. 4.12, the primary phase fields are bounded by heavy lines, some of which are solid and some dashed, depending on the level of confidence in the data. For example, the primary phase field for corundum appears in the lower right-hand corner of the diagram; upon cooling of a liquid of any composition within this field, corundum will be the first phase to crystallize. There is a large primary phase field for mullite ($3Al_2O_3 \cdot 2SiO_2$) along the right side of the diagram, within which any liquid first crystallizes mullite upon cooling. The primary phase field for iron cordierite is a very small slice of composition space, indicated by the label "iron cordierite."

Congruently and Incongruently Melting Compounds

A congruently melting compound has a composition that lies *within* its own primary phase field (including lying on the boundary of the field), whereas an incongruently melting compound lies *outside* its primary phase field. We can illustrate this point by again referring to a binary diagram, Fig. 4.5, in which the compounds α and β are congruently melting and ab is incongruently melting. Notice that the segment of the liquidus curve lying over the L + ab two-phase field does not span the composition ab itself. The primary phase field of ab therefore does not include its stoichiometric composition. In contrast, segments of the liquidus defining the primary phase fields of the α and β phases do include the phase compositions.

In the schematic ternary diagrams in Fig. 4.13a and 4.13b, all compounds melt congruently, being within their primary phase fields. In Fig. 4.13c, the intermediate compound ac along the A-C join melts incongruently. Its composition in fact lies in the γ phase field, indicating that upon cooling, a liquid of composition given by AC first solidifies the phase γ. In the real ternary system in Fig. 4.12, notice that the primary phase field for iron cordierite does not include the iron cordierite composition. This tells us that the compound melts incongruently, and

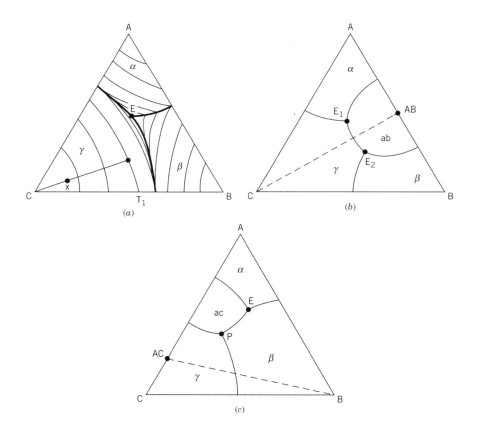

Fig. 4.13 Schematics of simple ternary diagrams. (*a*) Ternary eutectic; (*b*) Ternary double eutectic with one congruently melting binary compound; (*c*) Ternary with one eutectic and one peritectic and an incongruently melting binary compound (ac).

4.4 Features of Ternary Phase Diagrams

is consistent with the absence of a labeled melting point for iron cordierite. The iron cordierite composition lies in the mullite phase field, indicating that a liquid of the exact iron-cordierite composition first crystallizes mullite upon cooling through the liquidus surface. With the exception of iron cordierite, all of the compounds in Fig. 4.12 melt congruently.

Boundary Curves and Temperature Contours

In a binary phase diagram, a three-phase equilibrium between one liquid and two solids exists only at a eutectic or peritectic point where there are zero degrees of freedom. In a ternary phase diagram, the additional compositional variable allows a three-phase equilibrium to exist along a line of varying temperature and composition where there is one degree of freedom. These lines are the *boundary curves* that separate primary phase fields, along which one liquid is in equilibrium with two solid phases defined by the adjoining primary phase fields (Figs. 4.12, 4.13). The point at which a boundary curve intersects the side of a ternary phase diagram is where the system becomes reduced to only two components. The presence of three phases requires that this intersection be a critical point in the binary system (either a eutectic or peritectic). The temperature of this point is usually labeled on a ternary diagram.

Temperature contours are lines of constant temperature that show us the shape of the liquidus surfaces. Recalling that ternary diagrams are viewed along the temperature axis, these temperature contours are exactly analogous to the constant altitude contours that show the lay of the land on a topographical map. In Fig. 4.12 these are the curved lines labeled with temperature at regular intervals. At a boundary curve where two primary phase fields meet, the temperature contours must join, often showing a cusp representing a sharp "valley" between the two primary phase fields (cf. Fig. 4.10). Boundary curves are often labeled with arrows to indicate the direction of decreasing temperature. When a boundary curve is short, these arrows can be useful for determining the shape of the liquidus surfaces. They are also important for determining crystallization paths, as discussed later.

Beneath a liquidus surface in a ternary diagram, but above the solidus temperature, a two-phase, liquid-solid equilibrium exists. For a binary system the compositions of the liquid and the solid in a two-phase field are given by the constant temperature tie line (cf. Fig. 4.2). In a ternary system we can visualize the tie line connecting the compositions of the liquid and the solid at any temperature by using the temperature contours. For example, consider the composition labeled x in Fig. 4.13a. This composition lies in the primary phase field of γ, so at temperatures below the liquidus surface (but above the solidus), γ is in equilibrium with a liquid. We can obtain the composition of this liquid by noting that mass balance requires that the liquid and γ together have the average composition x. Thus, the tie line must run through point x from the endmember *c* to a point on the liquidus surface that marks the composition of the liquid. The tie line shown is for a temperature labeled T_1. Note that we can apply the binary lever rule to this tie line to obtain the relative amounts of γ and liquid at this temperature.

Ternary Invariant Points

The intersection of three boundary curves in a ternary diagram defines a ternary invariant point (critical point). These points are the intersections of three primary phase fields, at which the three primary phases are in equilibrium with a single liquid, the latter having a composition given by the invariant point. Examination of the Gibbs phase rule for a three-component system at constant pressure shows that a four-phase equilibrium represents a condition with zero degrees of freedom (F = 0). Thus, at a ternary critical point, neither temperature nor composition can be varied without the disappearance of one or more phases. The temperatures of these invariant points are usually labeled on a ternary phase diagram.

Ternary invariant points can be eutectics or peritectics, and there are two simple methods for deciding which they are. First, if the decreasing temperature arrows on the boundary curves all point toward the critical point then it is a ternary eutectic; if one or more of the arrows point away, it is a ternary peritectic. A second and more reliable method (since the arrows and temperature contours are not always given) is to consider the location of the critical point with respect to the three solid phases in equilibrium with the liquid. If the critical point lies *inside* a triangle connecting the three solid phases at equilibrium, then it is a eutectic. If the critical point lies *outside* this triangle, then it is a peritectic. We illustrate by analogy with the binary system in Fig. 4.5. Notice that the eutectic between AB and B lies within the line joining the two compounds, whereas the peritectic between A and AB lies outside the line between them. Notice also that the liquidus curves adjacent to the eutectic between AB and B descend in temperature towards the eutectic, whereas one of those adjacent to the peritectic between A and AB descends away in temperature.

The single invariant point depicted in Figs. 4.13*a* and the two in 4.13*b* are eutectics, while Fig. 4.13*c* contains one eutectic and one peritectic. Another example of a ternary eutectic appears in the lower-left-hand corner of Fig. 4.12, labeled 1148°(C). This point represents an equilibrium between liquid of the eutectic composition, hercynite, wustite, and fayalite, and lies inside the triangle connecting the three solid phases. An example of a ternary peritectic is near the center of the same diagram, labeled 1380°. This equilibrium is between liquid of the peritectic composition, corundum, hercynite, and mullite. The three equilibrium solid phases can be found by reading directly the primary phase fields surrounding the critical point. Since the critical point lies outside the triangle formed by these three phases, it is a peritectic.

Compatibility Triangles[2]

The triangles that join three solid phases in equilibrium at a critical point have an additional use. They are known as *compatibility triangles*, and define the three solid phases that are in equilibrium after cooling below the solidus temperature. We find these triangles by: (1) locating all of the ternary critical points; (2) identifying the primary phase fields surrounding each critical point; and (3) joining the compositions of the primary phases. In Fig. 4.13a, there is only a single critical point, thus the compatibility triangle is the entire diagram. In Figs. 4.13b and 4.13c, there are two critical points and two compatibility triangles. In the complex diagram in Fig. 4.12, all compatibility triangles have been drawn, with both solid and dashed lines. The dashed lines are used to indicate a compatibility triangle for a phase with some extent of solid solubility, for instance between $3Al_2O_3 \cdot 2SiO_2$ and $FeO \cdot Al_2O_3$. These may represent the maximum extent of solid solution or merely that observed after cooling to room temperature, and is frequently not intended to show strict thermodynamic equilibrium (in which case the compatibility triangle would be a function of temperature) but rather the approximate compositions of coexisting phases.

Solidus Temperatures

The solidus temperature in ternary diagrams is read in a somewhat different way than in binary diagrams. Referring first to binary diagrams containing compounds with limited solid solution, Figs. 4.4a and 4.5, notice that two-phase compositions lying between A and AB, or AB and B, first melt upon heating to the critical temperature, either a eutectic or peritectic. In a ternary system the same is true for the three solids in equilibrium, only now the solidus is a triangular plane that we view face-on. For compositions lying within a compatibility line (in a binary) or a compatibility triangle (in a ternary), *the solidus temperature is the same as the critical temperature of the compatibility*. Thus we find the solidus temperature for a given composition simply by identifying 1) the compatibility triangle, and 2) its critical point.

Identifying the melting point of compositions within the solid solution field of a compound in a ternary diagram presents a more difficult problem. Referring to binary diagrams with extensive solid solution as in Figs. 4.3, 4.4b, and 4.6, we see that the solidus temperature for compositions within the single phase field of a compound varies with composition. The same is true of the solidus *surface* for ternary compounds with extensive solid solutions. The details of such solidus surfaces are typically not shown in a ternary diagram.

---

[2]Also known as compatibility fields.

Liquid-Liquid Immiscibility

One final feature of importance is the region of liquid–liquid immiscibility. This is a range of compositions over which two liquids coexist in equilibrium (and perhaps also with one or two solid phases). A liquid–liquid phase separation region can be found lying against the upper–left-hand side of the diagram in Fig. 4.12. In this region the label "two liquids" points to a straight line that is the tie line joining the equilibrium liquid compositions at what appears to be a temperature of 1700°C. This tie-line runs approximately parallel to the side of the diagram, indicating a tendency for liquids in this composition region to phase-separate upon cooling into SiO_2-rich (about 90% SiO_2, 2% Al_2O_3, and 8% FeO) and FeO-rich liquids (about 65% SiO_2, 2% Al_2O_3, and 33% FeO). Notice that tie-lines at any particular temperature must terminate on the isothermal contour. In Fig. 4.12 the labels "cristobalite" and "tridymite" indicate the primary phase fields of these crystalline polymorphs of pure SiO_2. A boundary curve surrounds the liquid–liquid immiscibility region and is again a line representing three phases in equilibrium. However, unlike the boundary curves discussed above, this curve represents two-liquid phases in equilibrium with one solid phase (cristobalite), rather than one liquid and two solids.

A liquid–liquid immiscibility region can also appear as a dome, as shown for the Na_2O-B_2O_3-SiO_2 system in Fig. 5.86. These are the three-dimensional counterparts of the binary immiscibility region in Fig. ST42. Phase separation in this system plays an important role in the microstructures and processing of sodium borosilicate glass products typified by Pyrex™ and Vycor™. Direct examination of microstructures has shown that the tie lines join equilibrium liquids rich in SiO_2 and rich in B_2O_3, respectively.

SPECIAL TOPIC 4.4

RECIPROCAL SALT DIAGRAMS

Reciprocal salt,[*] or *equivalence* diagrams, are a special form of ternary phase diagram useful for showing phase relations in systems where the cation and anion components simultaneously vary. In the 1970s interest in this type of representation was revived[**] with the development of new oxynitride ceramics. We will use the sialon system, Si-Al-O-N, discussed in Chapter 1, as an example. The reciprocal salt diagram for this system is shown in Fig. ST46. It is a square-planar representation in which each corner is an oxide or nitride. The concentration units along each axis, however, are in terms of *equivalents*, which have the following meaning. If we proceed along the top edge from Si_3O_6 (i.e., 3 SiO_2) to Al_4O_6 (i.e., 2 Al_2O_3), the oxygen content is fixed while the cation concentrations vary. We have added an

4.4 Features of Ternary Phase Diagrams

amount of Al^{3+}, *which is in terms of valence equivalent to 3 Si^{4+}*. The same substitution takes place proceeding along the bottom edge from Si_3N_4 to Al_4N_4 (i.e., 4 AlN). Proceeding down the left edge from Si_3O_6 to Si_3N_4 or the right edge from Al_4O_6 to Al_4N_4, one substitutes $4N^{3-}$ for $6O^{2-}$, maintaining a constant total anion valence. That is, six oxygen ions are, in terms of valence, equivalent to four nitrogen ions. In Fig. ST47, we can see that the equivalence diagram is in fact a planar section of the quaternary system Si-Al-O-N. Notice that as long as the valences of all components remain fixed, the compositions of all condensed phases must appear in this plane. Thus all phase information can be contained within the square-planar representation, which is more convenient to use than the three-dimensional diagram in Fig. ST47.

It is important to realize that by constraining the valences of all four components, we have removed a composition variable and lost one degree of freedom. The reciprocal salt diagram is a quasi-ternary diagram, in which the phase fields are similar to those in the isothermal section of a conventional ternary. In Fig. ST46, single-phase fields appear as an area (when there is a range of solution) or as a point (for a line compound). Two-phase fields appear as a line if both phases are line compounds, and as an area (with implicit tie lines) if there is a range of solution. Three-phase compatibility triangles exist in regions of composition where the temperature of the diagram is below the solidus temperature, and four phases can coexist only at a single composition in the isothermal section.

We have in fact already carried out this type of reduction in the number of components in discussing binary and ternary phase diagrams. For example, the three-component system Mg-Ca-O (Fig. 4.3) can be treated as a binary system between the components MgO and CaO only when the valences of all ions are fixed; the requirement of charge neutrality then allows only two of the three elemental concentrations to be independently varied. It is this assumption of fixed valence, which is sometimes not explicitly stated, that permits the representation of compounds rather than elements as the components. For the Fe-O system in Fig. 4.9, several important solid phases with variable Fe valence are readily accessible depending on the oxygen activity and temperature. It is necessary to treat Fe-O as a two-component system to show these phases.

While one identifies phase fields in reciprocal salt diagrams in a similar fashion to those in conventional ternary diagrams, the concentration units often need to be converted in practical usage. The concentration scale in

---

*The name is derived from fused salt phase equilibria, in which a simple reciprocal salt is one containing two cations and two anions, for example Na-K-Cl-F, in which at least the compounds NaCl, NaF, KCl, KF would appear. See J.E. Ricci, The Phase Rule and Heterogeneous Equilibrium, D. Van Nostrand Company, NY, 1951, for further discussion.

**L.J. Gauckler and J. Huseby, *J. Am. Ceram. Soc.*, 58, 346 (1975), and 58, 377 (1975).

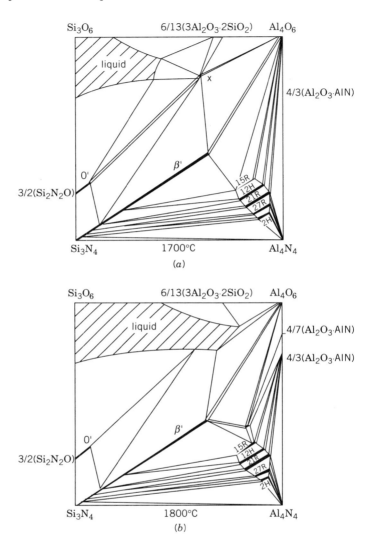

Fig. ST46 Equivalence diagrams for the Si-Al-O-N system at 1700°C and 1800°C. Courtesy K.H. Jack.

Fig. ST46 is in equivalent percent, whereas one may wish to know the concentration of the elemental constituents in atomic or weight percent. The following conversion factors apply for Si-Al-O-N (the numerical factors are the valences of respective ions and are adjusted appropriately for other systems):

4.4 Features of Ternary Phase Diagrams

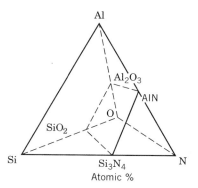

Fig. ST47 Constant-valence plane of the Si-Al-O-N system in which the equivalence diagram lies. (From L.J. Gauckler and G. Petzow, *Nitrogen Ceramics*, F.L. Riley, ed., Noordhoff Pub., Leyden, 1977.)

Atomic Percent To Equivalent Percent

$$Eq\% \, Al = 100 \left[\frac{at\% \, Al \cdot 3}{at\% \, Al \cdot 3 + at\% \, Si \cdot 4} \right]; \quad Eq\% \, Si = 100 - Eq\% \, Al$$

$$Eq\% \, O = 100 \left[\frac{at\% \, O \cdot 2}{at\% \, O \cdot 2 + at\% \, N \cdot 3} \right]; \quad Eq\% \, N = 100 - Eq\% \, O$$

Equivalent Percent To Atomic Percent

$$at\% \, Al = 100 \left[\frac{Eq.\% \, Al \cdot 4}{Eq\% \, Si \cdot 3 + Eq\% \, Al \cdot 4 + Eq\% \, O \cdot 6 + Eq\% \, N \cdot 4} \right]$$

$$at\% \, Si = 100 \left[\frac{Eq\% \, Si \cdot 3}{Eq\% \, Si \cdot 3 + Eq\% \, Al \cdot 4 + Eq\% \, O \cdot 6 + Eq\% \, N \cdot 4} \right]$$

$$at\% \, O = 100 \left[\frac{Eq\% \, O \cdot 6}{Eq\% \, Si \cdot 3 + Eq\% \, Al \cdot 4 + Eq\% \, O \cdot 6 + Eq\% \, N \cdot 4} \right]$$

$$at\% \, N = 100 \left[\frac{Eq\% \, N \cdot 4}{Eq\% \, Si \cdot 3 + Eq\% \, Al \cdot 4 + Eq\% \, O \cdot 6 + Eq\% \, N \cdot 4} \right]$$

Returning to the Si-Al-O-N phase diagram (Fig. ST46), we see that a large solid solution region extends from Si_3N_4 toward Al_3O_3N, and a lesser one from Si_2N_2O toward Al_2O_3. As discussed in Chapter 1, extensive sialon

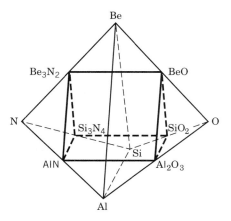

Fig. ST48 Triangular prism representation of the Be-Si-Al-O-N system. From (From L.J. Gauckler and G. Petzow, *Nitrogen Ceramics*, F.L. Riley, ed., Noordhoff Pub., Leyden, 1977.)

solid solutions are able to form when Al^{3+} substitutes for Si^{4+} in the parent phase with additional nitrogen or oxygen providing charge compensation. As an exercise in processing, consider firing a mixture that is 1/3 Si_3N_4 and 2/3 Al_2O_3 by mole fraction, in a closed system such as a hot-pressing die. The phase diagram indicates that this composition, which is half-way between Si_3N_4 and Al_4O_6, lies in a two-phase field at 1800°C containing β'

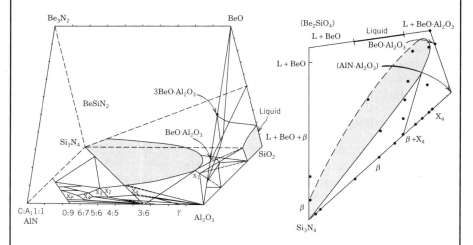

Fig. ST49 Phase equilibria in the Be-Si-Al-O-N system, illustrating plane of constant metal:anion ratio of 3:4, within which solid solutions of β'-sialon are contained. (From L.J. Gauckler and G. Petzow, *Nitrogen Ceramics*, F.L. Riley, ed., Noordhoff Pub., Leyden, 1977.)

sialon and liquid, with the tie line (not shown) running from the liquid field to the β´ solid solution. At 1700°C, the liquid has solidified and is replaced by the X-phase sialon. In order to obtain a single-phase β´ sialon, it is necessary to adjust the composition towards the nitrogen-rich side, for instance, by using a mixture of Si_3N_4, Al_2O_3, and AlN.

Sialon systems are often prepared using more than one oxide additive, such as the composition M-Si-Al-O-N, where M may be Be, Mg, Ca, or Y. The next step of increasing complexity in equivalence diagrams is the use of a triangular prism as a quasi-quaternary representation of a five-component system, as shown in Fig. ST48 for Be-sialon. Each of the three square faces is an equivalence diagram like that discussed above, in which only two cations appear. The two triangles at the ends are conventional ternary diagrams, for the pure oxides and nitrides. If all valences in the five-component system are fixed, all possible phase compositions must fall within this triangular prism. As an example of the use of this type of diagram, Fig. ST49 shows phase equilibria that have been determined in a study of possible β´ sialon compositions. It was found that an extensive range of solid solution is possible among the cations Be, Al, and Si, but that the overall cation/anion ratio remains very close to 3:4 as in the parent Si_3N_4, for charge compensation reasons. Therefore the β´ sialon compositions fall in the plane defined by Si_3N_4, Be_2SiO_4, $BeAl_2O_4$, and Al_3O_3N.

4.5 OPERATIONS USING TERNARY DIAGRAMS

Having identified the basic features of ternary diagrams, we now illustrate some frequently used operations. First, we will show how a complete binary diagram may be constructed from the temperature contours in a ternary diagram. Secondly, we will construct an *isothermal section* of a ternary diagram, which simplifies the ternary into a plane of constant temperature from which phases present at equilibrium can be easily read. Then, the *ternary lever rule* will be applied; it is the tool by which relative amounts of phases present at equilibrium are determined. In the following section, the changes of phases and compositions during heating and cooling of ternary compositions are discussed, using detailed examples to illustrate the effects of thermal history on phase constitution and microstructure. Throughout, a graphical approach will be used with the emphasis on facilitating practical use of ternary diagrams.

Constructing Binary Diagrams from Ternary Diagrams

If all phases found along the line or "join" between two compounds in a ternary system can be expressed in terms of the endmembers, the join represents a "true" two-component system. In Fig. 4.12, the three sides of the ternary as well as the

308 Chapter 4 / Phase Equilibria

fayalite-hercynite join are true binary systems; the other joins are not. For those that are, we can reconstruct the binary diagram as follows:

1. Along the join between the two components read the melting points of compounds, and the temperatures at which isothermal contour lines and boundary curves intersect that join. Mark these on your binary diagram.
2. Connect these temperatures in order to form the liquidus curves of the binary.
3. At each point where a boundary curve intersects the join, determine the three phases in equilibrium. Identify this binary invariant point (either a eutectic or peritectic).
4. At each invariant point, connect the two solid phases are in equilibrium with a horizontal tie line.
5. Draw vertical lines representing all compounds, assumed for the moment to be line compounds. These should continue upward until a tie line is reached.
6. If the compound has significant, solid solubility, the maximum solubility may be shown and if so can be marked at the invariant point. Normally there is insufficient information to determine the temperature contours of the solidus line.
7. Enter any liquid–liquid immiscibility information that can be read directly from the ternary diagram.

Example: "FeO"-SiO_2 Binary Diagram

To illustrate the procedure, we will construct the "FeO"-SiO_2 diagram from the ternary phase diagram in Fig. 4.12. Reading along the join from FeO toward SiO_2, we find the following melting points and intersections:

1369°C	Melting point: wüstite.
1300	Isothermal contour: wüstite liquidus surface.
1200	Isothermal contour: wüstite liquidus surface.
1177	Boundary curve: wüstite and fayalite.
1200	Isothermal contour: fayalite liquidus surface.
1205	Melting point: fayalite.
1200	Isothermal contour: fayalite liquidus surface.
1178	Boundary curve: fayalite and tridymite.
1200	Isothermal contour: tridymite liquidus surface.
1300	Isothermal contour: tridymite liquidus surface.
1400	Isothermal contour: tridymite liquidus surface.
1470	Boundary curve: tridymite and cristobalite.
1500	Isothermal contour: cristobalite liquidus.
1600	Isothermal contour: cristobalite liquidus.
1698	Boundary curve between cristobalite and two liquid region.
1700	Isothermal contour: cristobalite liquidus surface.
1723	Melting point: cristobalite.

4.5 Operations Using Ternary Diagrams   309

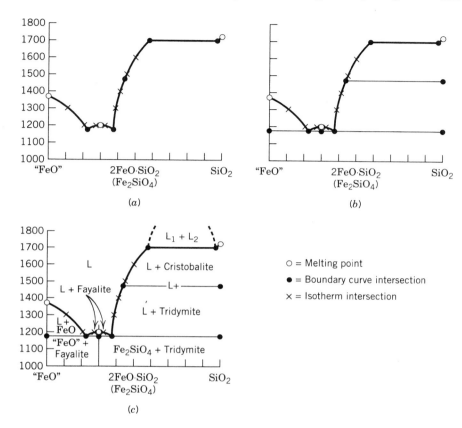

Fig. 4.14 (a) Liquidus contour of the "FeO"-SiO$_2$ binary system, constructed from the liquidus features along the FeO-SiO$_2$ join in Fig. 4.12. (b) Binary critical points added, defining solidus features and polymorphic transformation for SiO$_2$. (c) Completed binary phase diagram, with intermediate compound (fayalite) and liquid-liquid immiscibility added and phase fields labeled.

Upon connecting these points, we have constructed the liquidus curve of the FeO-SiO$_2$ system, shown in Fig. 4.14a.

There are five boundary curve intersections with the FeO-SiO$_2$ join. These represent invariant points in the binary diagram, with the following three-phase equilibria:

1177°C	Wüstite + fayalite + liquid
1178	Fayalite + tridymite + liquid
1470	Tridymite + cristobalite + liquid
1698	Cristobalite + liquid (1) + liquid (2)
1698	Cristobalite + liquid (1) + liquid (2)

The two points at 1698°C are compositions at which two different liquids are in mutual equilibrium with cristobalite.

Horizontal lines are drawn at each of these temperatures to connect the phases in equilibrium in Fig. 4.14b. Only two of these lines, at 1177 and 1178°C, represent the solidus. Both are binary eutectic temperatures. The other three critical points represent three-phase equilibria but do not define conditions for first melting during a heating process. The line at 1470°C represents a polymorphic phase transformation from tridymite to cristobalite without change in the crystal composition. Since above and below the transformation temperature, the liquid is in equilibrium with a different polymorph, the liquidus curve should show an inflection at the critical temperature. This inflection is barely detectable in Fig. 4.14b. At 1698°C the line represents the bottom of the liquid–liquid immiscibility region. Upon heating through this temperature an additional liquid forms. Since this is not the first liquid to form, this temperature does not define a solidus.

Figure 4.14c shows the completed binary phase diagram. Vertical lines for the intermediate compounds (in this case only fayalite) have been added, and since Fig. 4.12 shows all compounds in this binary to be line compounds, we need not account for solid solubility. A schematic representation of the liquid–liquid immiscibility region has been drawn as a dashed line in Fig. 4.14c since this particular ternary diagram (Fig. 4.12) does not provide sufficiently detailed isothermal contours for this region.

Note that *subsolidus* phase equilibria in the binary system, such as eutectoid and peritectoid reactions, cannot in most cases be determined from a ternary diagram. Since the typical ternary diagram emphasizes the liquidus surface, often the best that we can do is to determine binary phase equilibria at higher temperatures where some liquid is present.

Constructing Isothermal Sections

An important use of the ternary diagram is the determination of phases in equilibrium at any given temperature or composition. This can be plotted for the entire composition range as an *isothermal section*, a constant-temperature slice of the three-dimensional ternary diagram. We will illustrate with several examples based on Fig. 4.12.

Consider the isothermal section at a temperature of 1600°C. The first step is to locate and draw in the 1600°C isotherm wherever it appears within a primary phase field. Figure 4.15a shows these contours. Working clockwise from the upper left, the different arcs represent liquid in equilibrium, respectively, with the solid phases cristobalite, mullite, corundum, and hercynite.

Since each segment of the 1600°C isotherm represents a two-phase equilibrium between a liquid of the composition at the isotherm and the primary solid phase, the tie lines in the two-phase field must naturally join the arc and the solid phase composition. For example, all points on the cristobalite isotherm are liquids in equilibrium with SiO$_2$. The tie lines therefore radiate from SiO$_2$ out to points along the isothermal contour. Figure 4.15b shows a number of these tie lines drawn in for each of the primary phase fields. These regions are the liquid–solid, two-phase fields at 1600°C.

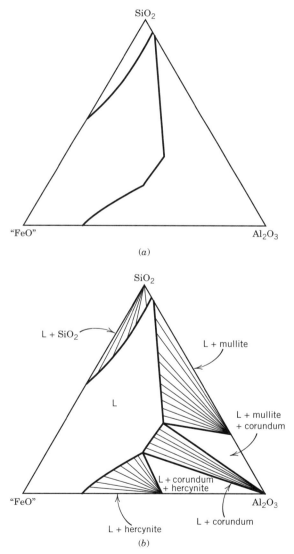

Fig. 4.15 Isothermal section at 1600°C in the SiO$_2$-Al$_2$O$_3$- "FeO" system. (*a*) Isothermal contours only. (*b*) Tie-lines drawn to show the liquid compositions in equilibrium with solid phases. One, two, and three phase fields are labeled.

Now we can identify one- and three-phase fields as well. The central region bounded by the isothermal contours is a single-phase liquid field. Referring back to Fig. 4.12, it can be seen that the isotherms in this region are all higher in temperature than 1600°C. Three-phase fields are identified as the triangular regions bounded by two-phase fields. These are regions where the two solids of the adjacent primary phase fields are in equilibrium with one liquid. The corresponding boundary curves between the primary phase fields will cross these regions. In Fig. 4.15b two such fields exist: liquid + corundum + hercynite, and liquid + mullite + corundum. As discussed later compositions lying in these regions will have precipitated two solid phases upon cooling to 1600°C.

Figure 4.15b shows that 1600°C is a high enough temperature that, no matter what composition we choose, there is always some liquid present. That is, we are above all solidus temperatures in this system. Using the ternary critical points, we can also construct isothermal sections for subsolidus regions. Figure 4.16a shows the first stage in developing an isothermal section for a temperature of 1300°C, in which the isothermal contours have been drawn. Figure 4.16b shows the addition of tie lines for the two-phase fields and the location of two three-phase fields. The lower-right-hand corner of the diagram contains subsolidus information. Notice, that the liquid composition labeled A lies at the intersection of two two-phase fields, and represents a liquid in equilibrium with both hercynite and mullite. This tells us that hercynite and mullite must also be in equilibrium with each other at this temperature, and we can draw a tie line between the two. The triangle between liquid of composition A, hercynite, and mullite, is therefore a three-phase field. The remaining triangular region between hercynite, mullite, and corundum is also a three-phase field, which we recognize as a compatibility triangle between the three solid phases. The temperature of 1300°C must then lie below the critical temperature for this compatibility triangle. Referring back to Fig. 4.12, we can see that hercynite, mullite, and corundum have a ternary peritectic temperature at 1380°C.

Without considering detailed liquidus information, we can also identify regions of three solid-phase equilibria simply by inspecting the ternary critical points. Those critical points with a temperature *higher* than the temperature of interest indicate that there must be a compatibility triangle between the three solid phases in existence at the temperature of the isothermal section. In Fig. 4.12, the critical points and respective phases in equilibrium are:

1083°C	Liquid + tridymite + fayalite + iron cordierite
1088	Liquid + fayalite + iron cordierite + hercynite
1148	Liquid + wustite + fayalite + hercynite
1210	Liquid + tridymite + mullite + iron cordierite
1205	Liquid + mullite + iron cordierite + hercynite
1380	Liquid + mullite + hercynite + corundum

4.5 Operations Using Ternary Diagrams

It is left as an exercise to the reader to construct isothermal sections at temperatures above and below various critical points in this system.

The limiting case of the isothermal section occurs at temperatures low enough that we are below *all* of the critical temperatures. This isothermal section is simply composed of all of the compatibility triangles in the system. For Fig. 4.12 we

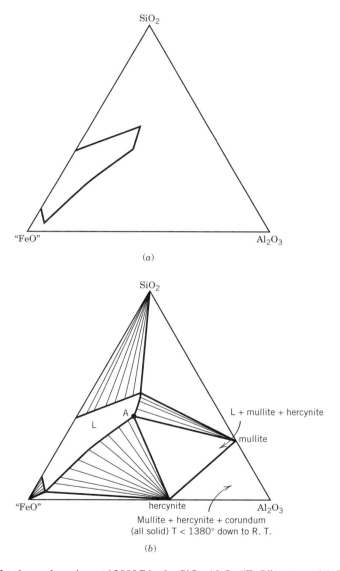

Fig. 4.16 Isothermal section at 1300°C in the SiO_2-Al_2O_3-"FeO" system. (*a*) Isothermal contours only. (*b*) Completed isothermal section with phase fields labeled. Notice three solid phase compatibility at the Al_2O_3-rich corner.

locate these from the solid phases in equilibrium at each of the critical points listed above. (Note that some are eutectic and some are peritectic points.) The result is shown in Fig. 4.17. As a caution, we point out that while an isothermal section in principle represents thermodynamic equilibrium, for many refractory systems the experimentally determined room temperature phase compatibilities may not represent true equilibrium due to the freezing out of kinetic processes during cooling.

A final feature we discuss is the liquid–liquid immiscibility region. In a ternary diagram these appear as domes, with isothermal contours sometimes given to show the shape of the dome. Tie lines may also be given to show the compositions of the two liquids in equilibrium at a given temperature. For example, Fig. 4.14c shows a liquid–liquid immiscibility region near the SiO_2-rich end of the FeO-SiO_2 binary. In the ternary diagram (Fig. 4.12) this region extends outward slightly from the join, and one tie line is shown at a temperature that appears to be about 1650°C. Note that the boundary curve for the two-liquid region, like other boundary curves, is a projection along which temperature varies. This particular boundary curve is shown as a dotted line in Fig. 4.18. It intersects the FeO-SiO_2 join at 1698°C at both ends, but dips slightly in temperature as it extends outward. In this admittedly complicated depiction, one must try to visualize the presence of a surface separating the two-liquid dome from the underlying cristobalite primary phase field. This surface is not hori-

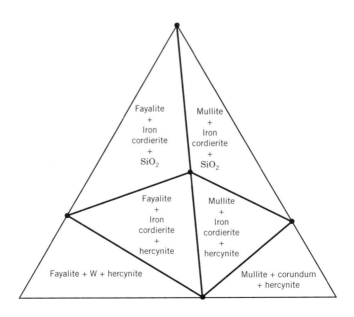

Fig. 4.17 Compatibility triangles at room temperature for the SiO_2-Al_2O_3-"FeO" system.

zontal (i.e., at constant temperature in composition space), but descends from 1698°C at the FeO-SiO$_2$ join down to a minimum temperature at the outermost part of the boundary curve that appears to be about 1630°C. The 1650°C isothermal contour cuts across this surface as shown in Fig. 4.18.

The Ternary Lever Rule

In two-phase regions of a ternary diagram, we apply the binary lever rule to determine the relative amounts of the two phases. In three-phase regions, the ternary lever rule must be used. The compositions of the three phases will form a triangle that is typically not equilateral, but scalene. Just as the binary lever is a graphical technique using the overall composition as a fulcrum to determine relative amounts of the two phases, the average composition in a three-phase field acts as a fulcrum by which the relative amounts of the three phases forming the vertices of the triangular plane can be "weighed."

Figure 4.19 illustrates this method. Compositions A, B, and C are those of the three phases in mutual equilibrium (although equilibrium between phases is not

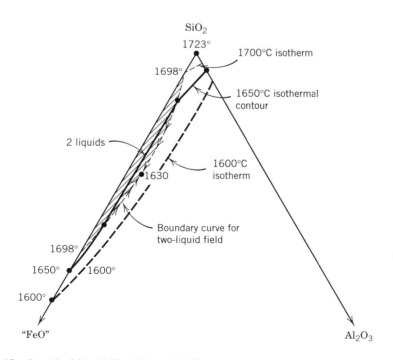

Fig. 4.18 Detail of liquid-liquid immiscibility region at the SiO$_2$-rich end of the SiO$_2$-Al$_2$O$_3$-"FeO" system (cf. Fig. 4.12). The 1650°C isothermal contour is shown.

316 Chapter 4 / Phase Equilibria

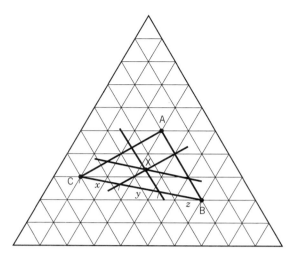

Fig. 4.19 Application of ternary lever rule to a scalene triangle representing a three-phase equilibrium between A, B and C. Lines drawn through the overall composition at X and parallel to the sides of the triangle are used to determine the relative amounts of A, B and C (see text).

specifically required to apply the lever rule; we can apply it any time we know the endpoint and overall compositions). A, B, and C may represent any combination of liquid and solid phases; it does not matter whether we are above or below the solidus. The overall or average composition is labeled X and must of course lie within the triangle. Through this point X, three lines have been drawn, each of which is parallel to one side of the triangle formed by A, B, and C. We determine the fraction of each phase, which is unity at any apex, by a method similar to that by which compositions are read. To illustrate, notice that the line through X and parallel to the side A–C lies a fraction of the total distance to the apex B. This fraction is the amount of B. We can perhaps measure it more easily from the intersection along the side C–B. The line parallel to A–C intersects this side at a distance x. The sum of x, y, and z is the length of side C–B. The fraction of phase B is given by the relative distance between point B and the side C–A:

$$\text{Fraction B} = x / (x+y+z)$$

Similarly the segment z represents the fractional distance between C and the opposing side A-B. The fraction of C is thus

$$\text{Fraction C} = z / (x+y+z)$$

This leaves only A undetermined, and since the sum of fractions A, B, and C must equal unity, the fraction of A is

4.5 Operations Using Ternary Diagrams

$$\text{Fraction A} = y / (x+y+z)$$

If only one phase fraction needs to be determined, then a still simpler construction illustrated in Fig. 4.20 can be used. For an overall composition of X, a line from A through X to the opposing side of the triangle gives the fraction of A using the two line segments x and y:

$$\text{Fraction A} = y / (x+y)$$

The remaining phases are given by:

$$\text{Fraction (B + C)} = x / (x+y)$$

For completeness, notice that the relative amounts of B and C can be determined from the intersection of the line with the side of the triangle opposite A. This intersection divides the side into segments z and w, from which we obtain the following fractions *relative to one another*:

$$\text{Fraction of B} = z / (z+w)$$

$$\text{Fraction of C} = w / (z+w)$$

It is then a simple matter to determine the actual fractions of B and C.

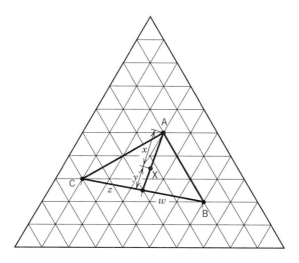

Fig. 4.20 Alternative geometry for applying ternary lever rule (see text).

4.6 REACTIONS UPON HEATING AND COOLING

Reactions upon heating tend to be of greater importance for the processing of ceramics than for most other materials. This is because relatively few ceramics are processed directly from the melt (due to their high melting points), despite notable exceptions such as glasses and glass ceramics, and single crystals for optical and electronic applications. Most polycrystalline ceramics are prepared from powder mixtures fired to high temperatures, frequently using a small amount of liquid phase to facilitate reaction and densification. We may contrast this with metals processing, which almost always begins with a melt, and in which solidification reactions upon cooling are the most important. Thus it is important to emphasize reactions upon heating as well as cooling; together, they determine the final microstructure and properties of polyphase ceramics. In this section we discuss basic ternary eutectic and peritectic reactions and then give several detailed examples in real systems.

Ternary Eutectic Reaction

Eutectic solidification is characterized by the crystallization of a single liquid into multiple solid phases, often forming a finely divided or lamellar microstructure. The last liquid to solidify (i.e., lowest melting temperature liquid) has the eutectic composition, and solidifies upon cooling through the solidus (eutectic) temperature to give, for a binary system

$$L \text{ (eutectic composition)} \rightarrow \text{solid A} + \text{solid B}$$

In a ternary system the solidification reaction is

$$L \text{ (eutectic composition)} \rightarrow \text{solid A} + \text{solid B} + \text{solid C}$$

If no other solid is present prior to eutectic solidification, one progresses directly from a single-phase liquid to three solids. This occurs only for an overall composition that is exactly that of the eutectic. More commonly the overall composition lies away from that of the eutectic and one or two solid phases precipitate prior to the eutectic solidification.

A simple ternary eutectic crystallization path is illustrated in Fig. 4.21. Beginning with a single-phase liquid of overall composition x, which lies within the primary phase field of α, α is the first solid to appear upon cooling through the liquidus surface. With further cooling, more α precipitates, and the liquid composition moves directly away from the α composition along the path labeled 1. This continues until the boundary curve between α and γ is reached, whereupon γ begins to precipitate. The liquid composition then follows the boundary curve, along the path 2, solidifying both α and γ simultaneously and maintaining a three-phase equilibrium. Upon reaching the eutectic temperature, the last remaining liquid, of composition E, solidifies directly into three phases in a finely divided mixture: α, β, and γ.

4.6 Reactions upon Heating and Cooling   319

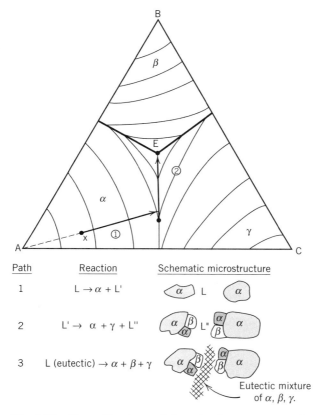

Fig. 4.21 Crystallization path in a simple ternary eutectic.

The heating reaction is exactly the reverse of this sequence. Thus for three solid phases at equilibrium in a compatibility triangle characterized by a eutectic, the first liquid forms at the eutectic temperature and is of the eutectic composition.

Ternary Peritectic Reaction

Peritectic solidification is characterized by the reaction between a liquid and a solid to form a new solid. Often, a pre-existing solid is consumed or *resorbed* in the process. Upon heating, a peritectic reaction is characterized by incongruent melting: a solid phase melts to form a new solid phase and a liquid. We first illustrate using the binary system in Fig. 4.22, which contains the incongruently melting compound ab. The composition x is exactly that of ab. Upon solidification, the peritectic reaction is

$$L \text{ (peritectic composition)} + \beta \rightarrow ab$$

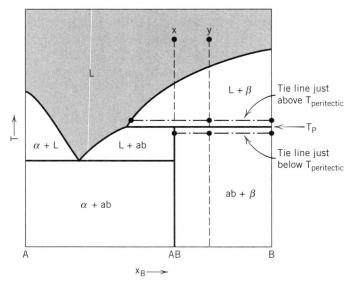

Fig. 4.22 Crystallization in a binary peritectic. Solidification to form AB causes partial or complete resorption of the primary phase of composition B.

which is simply the reverse of the incongruent melting of ab. Consider now the composition y, which lies over the primary phase field of β, and for which the tie lines just above and just below the peritectic temperature are shown. Just above the peritectic temperature we have liquid of peritectic composition and solid phase β; just below, the equilibrium phases are α and β. Thus the solidification reaction is

$$L \text{ (peritectic composition)} + \beta \rightarrow ab + \beta$$

a process in which β phase is *partially resorbed*.

The ternary counterpart to Fig. 4.22 is shown in Fig. 4.23. The composition x lies withing the α phase field, and therefore α crystallizes while cooling along path 1, while α and γ crystallize along the boundary curve, path 2, much as discussed above. Upon reaching the peritectic temperature, the liquid of composition P is in equilibrium with the solid phases α and γ, and the new phase that must appear is ab. Notice that the new phase has a composition that lies outside the triangle formed by P, A, and C. In order for ab to form, the reaction must be:

$$L \text{ (peritectic)} + \alpha \rightarrow ab + \gamma$$

Thus in this solidification reaction α is partially resorbed, while additional γ is formed, in the process of crystallizing ab. The heating reaction, conversely, actually forms new α phase in the process of melting.

4.6 Reactions upon Heating and Cooling

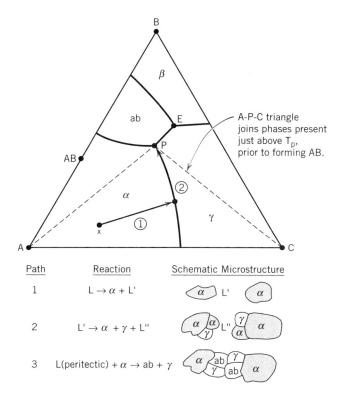

Fig. 4.23 Crystallization in a ternary peritectic.

Reactions Upon Heating

Now let us consider in heating reactions in a real and somewhat complex ternary system, $MgO\text{-}Al_2O_3\text{-}SiO_2$, shown in Fig. 4.24. For simplicity, the following constraints are assumed:

(i) The starting solids form an equilibrium phase assemblage.
(ii) The heating rate is slow enough that reactions are allowed to go to completion, maintaining equilibrium throughout.

We again caution that these conditions are often not satisfied in ceramics processing, since one may begin with convenient materials rather than those at thermodynamic equilibrium, and it may not be practical or even desirable to allow reactions to proceed to completion since the brief appearance of a nonequilibrium liquid may be advantageous to processing. The overall compositions labeled A and B in Fig. 4.24 will be used to illustrate ternary eutectic and ternary peritectic reactions, respectively, in considerable detail.

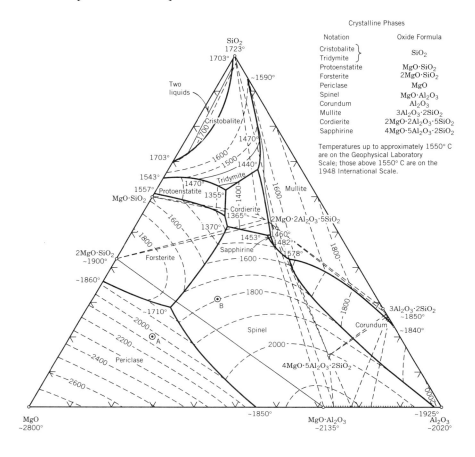

Fig. 4.24 MgO-Al$_2$O$_3$-SiO$_2$ system. (Fig. 712 from *Phase Diagrams for Ceramists*, the American Ceramic Society, Columbus, Ohio, 1975.)

Heating Through a Ternary Eutectic

The composition A in Fig. 4.24 lies at 20% Al$_2$O$_3$, 20% SiO$_2$, and 60% MgO. Beginning at room temperature, we identify the phases present from the compatibility triangle within which the composition lies. From the room temperature isothermal section for this system shown in Fig. 4.25 (i.e. collection of compatibility triangles), we find that composition A lies within the periclase + forsterite + spinel triangle. The critical point for this phase assemblage, labeled ~1710°, falls within the compatibility triangle and identifies the critical point as a ternary eutectic. The amount of each phase in the starting assemblage can be found by applying the ternary lever rule to the compatibility triangle using composition A as the fulcrum, with the following results:

4.6 Reactions upon Heating and Cooling

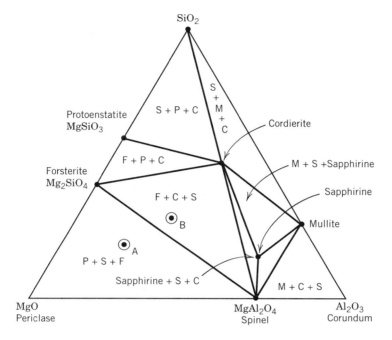

Fig. 4.25 Room temperature isothermal section for the MgO-Al$_2$O$_3$-SiO$_2$ system.

28% Spinel (MgAl$_2$O$_4$)
47% Forsterite (Mg$_2$SiO$_4$)
25% Periclase (MgO).

The compatibility triangles in Fig. 4.25 assume no solid solution for any compound. Since each of these three solids is in fact highly stoichiometric, this computation is reasonably accurate over a range of temperatures below the solidus.

Upon heating, little change in composition of any of the three phases takes place until the solidus is reached. At that critical temperature, the phase rule indicates that four phases will be in equilibrium: the three solid phases plus a liquid of the critical point composition. Reading from the position of the eutectic point in Fig. 4.24, this liquid has the composition:

20% Al$_2$O$_3$
28% SiO$_2$
52% MgO

The eutectic reaction upon heating is thus the formation of a liquid from the pre-existing solid phases:

$$MgO + Mg_2SiO_4 + MgAl_2O_4 \rightarrow liquid$$

Since the eutectic liquid composition lies within the compatibility triangle, a positive amount of all three solid phases is consumed in this reaction.

We may now ask, when is this reaction complete, and how much of each solid phase will be consumed in the reaction? At heating rates slow enough to preserve equilibrium, the temperature of the system will be arrested at the critical temperature while this melting reaction proceeds. Heat added to the system is absorbed as the latent heat of fusion of the eutectic liquid until the reaction is complete, whereupon the system temperature will rise again.

The Gibbs phase rule tells us that above the eutectic temperature, there can be at most three phases in equilibrium while we vary temperature, so that if we begin with three solid phases and form a liquid, then at least one solid phase must disappear. At most, all of the solid phases may disappear in a eutectic reaction, leaving behind a single-phase liquid. We can determine the phases that remain or disappear in a couple of different ways.

One way is to examine the proportion of each phase consumed by the eutectic reaction, for it must cease when any one of the reactant phases is completely depleted. The relative amount of each solid consumed in the formation of the eutectic liquid can be found by applying the ternary lever rule to the compatibility triangle, using the *eutectic composition* as the fulcrum. Reading from Fig. 4.26, note that the amounts consumed are

28% Spinel
67% Forsterite
5% Periclase

Since these fractions differ from the starting fractions of the three phases (obtained from the same compatibility triangle using the overall composition as the fulcrum), solid phases will remain after the reaction is complete. Upon comparing the ratio of starting phase fraction to the fraction consumed:

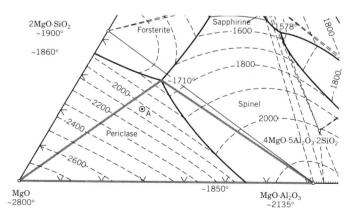

Fig. 4.26 Phases present just above the eutectic temperature (at 1710°C) for an overall composition at A.

4.6 Reactions upon Heating and Cooling

Phase	Starting Fraction	Consumed Fraction	Ratio
Spinel	0.274	0.280	0.979
Forsterite	0.471	0.670	0.703
Periclase	0.255	0.050	5.100

it becomes apparent that forsterite will be the first to be depleted, leaving behind a three-phase mixture of spinel, periclase, and liquid after the reaction at the critical temperature is complete. Clearly, if the eutectic composition and overall composition are one and the same, then all three solid phases would disappear in the eutectic reaction, forming one liquid. It is also possible to have one solid and the liquid left after the reaction. If so, the remaining solid is the primary phase for the overall composition.

Another, perhaps simpler way to determine both the phases and amounts of phases present after heating through the eutectic temperature is to construct an isothermal section just above the eutectic temperature. Application of the ternary lever rule using the overall composition as the fulcrum yields the amount of each phase. How do we identify the phases present just above the eutectic? Certainly one phase will be the eutectic liquid we just discussed. A second phase will be the primary phase for the field within which the overall composition lies. In this case composition A lies in the periclase primary phase field. Then, we find the third phase simply by noting that the phase compositions must surround the overall composition (i.e., positive quantities of each phase are required). From Fig. 4.26, we see that the third phase is spinel if the overall composition is A. The amount of each phase left after eutectic melting, obtained by applying the ternary lever rule to the triangle outlined in Fig. 4.26 with A as the fulcrum, yields

7.7%	Spinel
0%	Forsterite
23.6%	Periclase
68.7%	Liquid

Let's now consider what happens with further heating. A series of isothermal sections constructed at successively higher temperatures shows that the liquid composition leaves the eutectic point and tracks the boundary curve between the two solid phases, spinel and periclase. This is simply the reverse of the cooling path. The equilibrium composition of the liquid at any temperature is given by the position of the boundary curve. The reaction taking place is

$$\text{spinel} + \text{periclase} \rightarrow \text{liquid}$$

Furthermore, the *tangent* to the boundary curve provides useful information regarding the relative amounts of the two solids being consumed in this reaction, as it gives the instantaneous trajectory of the liquid composition. As shown in Fig. 4.27, this tangent intersects the side of the diagram between the compositions of the spinel and periclase phases, and divides the join between them

326   Chapter 4 / Phase Equilibria

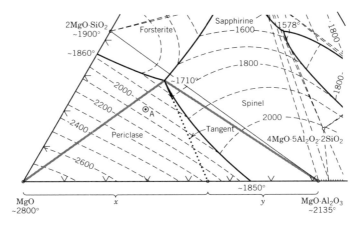

Fig. 4.27 A three-phase equilibrium exists for composition A above the eutectic temperature. The arrow shows the direction in which the liquid composition changes upon heating; it proceeds up the boundary curve between MgO and MgO·Al$_2$O$_3$ phase fields. The tangent to the boundary curve intersects the Alkemade line between the two solid phases.

into the segments marked x and y. Application of the binary lever rule to these segments yields the relative amounts of periclase and spinel being consumed during liquid formation:

Fraction MgO $= y/(x+y) = 37\%$
Fraction MgAl$_2$O$_4$ $= x/(x+y) = 63\%$

For the short segment of the boundary curve between 1710°C and 1800°C, these amounts are relatively constant.

If we follow the boundary curve beyond about 1800°C, the triangle formed between the liquid and the two solid compositions no longer surrounds the overall composition A. This is an indication that the three-phase equilibrium is no longer maintained. All of the spinel phase will have been consumed, leaving us with a *two-phase* equilibrium between periclase and liquid. A line joining periclase and the liquid composition must pass through the overall composition A, as shown in Fig. 4.28. Above 1800°C, the liquid composition tracks along the liquidus surface of the periclase primary phase field toward the composition A, becoming enriched in periclase along the way. The reaction that takes place is

$$\text{MgO} + \text{liquid} \rightarrow \text{liquid}$$

Finally, when the liquid composition reaches composition A at about 2050°C, all of the periclase will have been consumed, leaving a single-phase liquid at all higher temperatures.

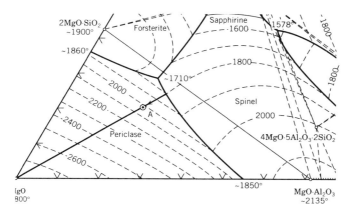

Fig. 4.28 Above 1800°C and below the liquidus at 2050°C, composition A exists as a two phase equilibrium between liquid and periclase. The arrow shows the changing composition of the liquid upon heating in this temperature range.

We may summarize the sequence of reactions upon heating and the temperatures at which they occur as follows:

Phases Present	Applicable Reaction	Range of Temperature
P + F + S	three-phase equilibrium	RT → 1710°C
P + F + S + L	P + F + S → L	1710°C exactly (invariant point)
P + S + L	P + S → L	1710–1800°C
P + L	P + L → L	1800–2050°C
L	single-phase only	>2050°C

where

L = liquid, of varying composition
P = MgO (periclase)
F = Mg_2SiO_4 (forsterite)
S = $MgAl_2O_4$ (spinel).

Heating Through a Ternary Peritectic

Let's now consider composition B in Fig. 4.29, which lies in the forsterite + cordierite + spinel compatibility triangle (cf. Fig. 4.25) and has the composition 30% Al_2O_3, 30% SiO_2, and 40% MgO. The primary phase fields of these three phases intersect at a critical point at 1370°C, which lies outside the compatibility tri-

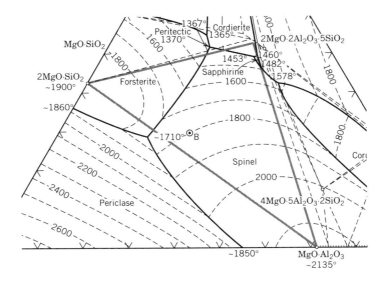

Fig. 4.29 Compatibility triangle and peritectic invariant point for composition B.

angle, thereby identifying it as a ternary peritectic. Forsterite and spinel are congruently melting phases, while cordierite melts incongruently. A new detail is that since cordierite exhibits some solid solution, the observed compatibility triangle does not terminate exactly at the stoichiometric composition. Applying the ternary lever rule to this compatibility triangle with B as the fulcrum, the equilibrium amounts of the solid phases at subsolidus temperatures are found to be:

34% Spinel
13% Cordierite
53% Forsterite

Upon heating, melting firsts takes place at the peritectic temperature of 1370°C. The new phase is a liquid of this critical point composition. Since the peritectic composition lies *outside* the compatibility triangle, in order to maintain mass balance the melting reaction must be of the type

$$\text{solid 1} + \text{solid 2} \rightarrow \text{solid 3} + \text{liquid}$$

The amount of one of the solid phases must increase. Starting with the phases forsterite, cordierite, and spinel, one can form liquid of the peritectic composition only by a reaction of the type:

4.6 Reactions upon Heating and Cooling

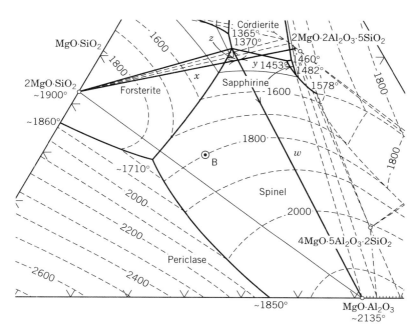

Fig. 4.30 Peritectic melting of composition B: Cordierite and forsterite react to form spinel and liquid of the peritectic composition.

$$\text{forsterite} + \text{cordierite} \rightarrow \text{spinel} + \text{liquid}$$

$$(\text{Mg}_2\text{SiO}_4 + \text{Mg}_2\text{Al}_4\text{Si}_5\text{O}_{18} \rightarrow \text{MgAl}_2\text{O}_4 + \text{L})$$

This reaction is illustrated by the arrows in Fig. 4.30. To draw a parallel with incongruent melting in a binary peritectic, Fig. 4.22, the incongruently melting compound AB melts at the peritectic temperature to form a liquid richer in A than itself, thereby requiring simultaneous formation of solid B in order to preserve mass balance.

The amounts of the phases consumed and formed during a ternary peritectic reaction are determined by a construction involving two linear levers. These join the *reactant phases* and the *product phases,* respectively, as illustrated in Fig. 4.30. The intersection of these two lines serves as a fulcrum for both levers and divides the levers into the segments x, y, z, and w. The levers then define the relative fractions reacting and forming as follows:

$$\text{forsterite} + \text{cordierite} \rightarrow \text{spinel} + \text{liquid}$$

$$y/(x+y) + x/(x+y) \rightarrow z/(z+w) + w/(z+w)$$

$$28\% + 72\% \rightarrow 5\% + 95\%$$

Notice that the product of the reaction is almost all liquid and relatively little spinel, since the peritectic point is close to the fulcrum.

This construction gives the *relative* amounts of phases consumed and formed by the peritectic reaction. As in the case of the ternary eutectic, the extent to which the reaction actually proceeds depends on how much of the reactant phases are present in the starting assemblage. When either forsterite or cordierite have been completely consumed, the peritectic reaction is complete, leaving at most two solids and a liquid.

Since composition B lies within the spinel primary phase field, just above the peritectic temperature, two of the three phases present must be spinel and the peritectic liquid. The location of the overall composition indicates that the third must be forsterite. These three compositions form a compatibility triangle just above the peritectic temperature, for which phase fractions are obtained by applying the lever rule with B as the fulcrum:

43% Forsterite
35% Spinel
22% Liquid, of peritectic composition

With further heating, the liquid composition will travel up the boundary curve between the forsterite and spinel primary phase fields, Fig. 4.31. The reaction taking place during this path is

$$\text{forsterite} + \text{spinel} \rightarrow \text{liquid}$$

The amount of forsterite and spinel being consumed at any temperature is obtained by drawing a tangent to the boundary curve and applying the linear lever rule where this tangent intersects the join between forsterite and spinel. One of the two phases will have been completely reacted away when the triangle formed by the liquid composition, forsterite, and spinel borders on the overall composition. From Fig. 4.31, this will occur at a temperature of about 1680°C. Thereafter only two phases are present, spinel and liquid. Reactions upon further heating are straightforward; the liquid composition leaves the boundary curve and travels up the spinel liquidus surface until composition B is reached. Along this path, the amounts of liquid and spinel are obtained by applying a binary lever to the line joining the liquid and spinel. Above 1800°C, there remains only a single-phase liquid.

4.6 Reactions upon Heating and Cooling

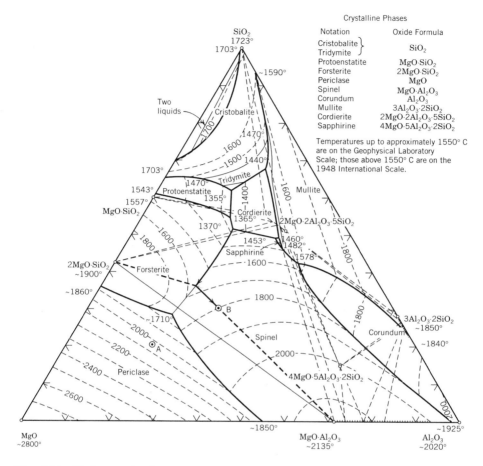

Fig. 4.31 Trajectory for liquid composition with further heating above the peritectic for an overall composition at B.

To summarize the heating path, the following reactions and temperatures apply:

Phases Present	Reaction	Range of Temperature
F + S + C	(none)	RT → 1370°C
F + S + C + L	F + C → S + L	1370°C exactly
F + S + L	F + S → L	1370 - 1680°C
S + L	S → L	1680 - 1800°C
L	(none)	>1800°C

where F, S, and C are forsterite, spinel and cordierite, and L is liquid.

Equilibrium Crystallization Paths

Eutectic Solidification We next use the CaO-Al$_2$O$_3$-SiO$_2$ system, shown in its entirety in Fig. 4.32, to give detailed examples of equilibrium crystallization reactions. As the first example, consider an overall composition at 26% CaO, 26% Al$_2$O$_3$, and 48% SiO$_2$, labeled A in Fig. 4.33*a*. This composition lies within the compatibility triangle formed by tridymite, anorthite, and pseudowollastonite. The invariant point at 1170°C where the three primary phase fields meet lies within the compatibility triangle, hence it is a eutectic. Let's examine the crystallization reactions upon cooling a single-phase liquid of this composition from a temperature above the liquidus (~1450°C). Since A lies in the anorthite primary phase field (CaO·Al$_2$O$_3$·2SiO$_2$), the first solid to form upon cooling is an-

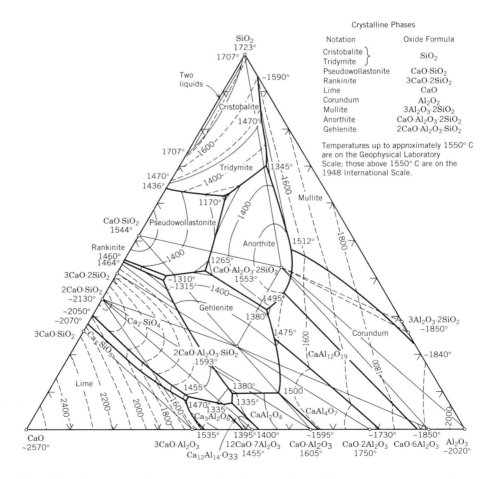

Fig. 4.32 CaO-Al$_2$O$_3$-SiO$_2$ system. (Fig. 630 from *Phase Diagrams for Ceramists*, The American Ceramic Society, Columbus, Ohio, 1975.)

4.6 Reactions upon Heating and Cooling

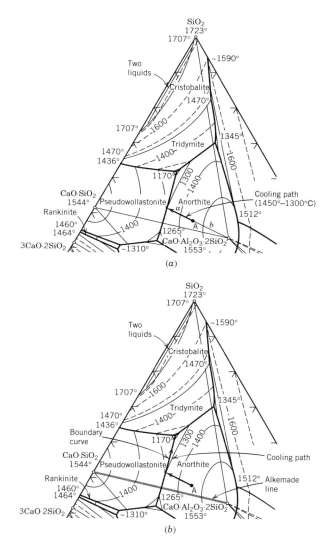

Fig. 4.33 (a) Crystallization of a liquid of composition A between 1450°C and 1300°C yields anorthite and liquid of varying composition given by the arrow. (b) Below 1300°C, anorthite and pseudowollastonite crystallize simultaneously. The liquid composition follows the boundary curve.

orthite, crystallizing when the liquid reaches 1450°C. Below the liquidus, anorthite continues to solidify from the liquid and the composition of the remaining liquid moves directly away from that of anorthite (Fig. 4.33a). This continues until the liquid composition reaches the boundary curve between anorthite and pseudowollastonite ($CaO \cdot SiO_2$) at ~1300°C. The reaction taking place between 1450°C and 1300°C is

$$L \rightarrow \text{anorthite} + L'$$

where L´ indicates the remaining liquid depleted in anorthite. The fractions of liquid and anorthite at any point along this trajectory can be calculated using a binary lever with A as the fulcrum between the anorthite and liquid compositions; at 1300°C, the fractions of liquid and anorthite are given by $b/(a + b)$ and $a/(a+b)$, respectively.

Figure 4.33b shows the trajectory of the liquid composition with further cooling. Below 1300°C, we have a three-phase equilibrium between liquid, anorthite, and pseudowollastonite. The liquid composition travels down the boundary curve between the anorthite and pseudowollastonite phase fields, away from the *Alkemade line*[3] joining anorthite ($CaO \cdot Al_2O_3 \cdot 2SiO_2$) and pseudowollastonite ($CaO \cdot SiO_2$). At any point along the boundary curve, we may draw a tangent back to the Alkemade line to quantify the solidification reaction. In this instance the tangent extends back to intersect the Alkemade line (Fig. 4.33b), indicating that a positive amount of each phase is being simultaneously precipitated (in the next example we discuss what happens when the extension lies outside the Alkemade line). The solidification reaction along the boundary curve is

$$L \rightarrow \text{anorthite} + \text{pseudowollastonite} + L'$$

and the relative fractions at which the two solid phases are being precipitated at any instant are obtained by applying the linear lever to the Alkemade line, using as the fulcrum the intersection of the tangent to the boundary curve (Fig. 4.33c). Thus, at 1250°C the fraction of anorthite being crystallized is $x/(x + y)$ and of pseudowollastonite is $y/(x + y)$.

The cooling reaction just described continues until the liquid composition reaches the eutectic point at 1170°C. The new primary phase field we reach is that of tridymite (one of the polymorphs of SiO_2). The eutectic reaction is

$$L \rightarrow \text{anorthite} + \text{pseudowollastonite} + \text{tridymite}.$$

The relative fractions of the three solid phases being crystallized in this reaction are obtained from the ternary lever applied to the compatibility triangle using the *eutectic composition* as the fulcrum (Fig. 4.33d). However, the amounts of each solid phase left in the sample as a whole after the eutectic reaction is complete are obtained using the *overall composition A* as the fulcrum (Fig. 4.33e).

---

[3]The Alkemade line is a useful reference for cooling and heating paths along boundary curves. It is defined as the line joining the two compounds whose primary phase fields meet at the boundary curve. The *Alkemade theorem* states that the temperature of decreasing temperature on a boundary curve is always away from the Alkemade line or its extension (if the boundary curve projects back to an intersection outside the Alkemade line).

Crystallization with Partial Resorption

As a second example, let's consider the cooling path for a bulk composition of 80% CaO, 10% Al_2O_3, and 10% SiO_2, labeled B in Fig. 4.34a. The composition lies within the compatibility triangle formed by CaO, $3CaO \cdot SiO_2$, and $3CaO \cdot Al_2O_3$, which has its invariant point (at 1470°C) outside the compatibility triangle. Equilibrium solidification will cease at this invariant point, a ternary peritectic. Since the overall composition lies in the primary phase field labeled Lime (CaO), that is

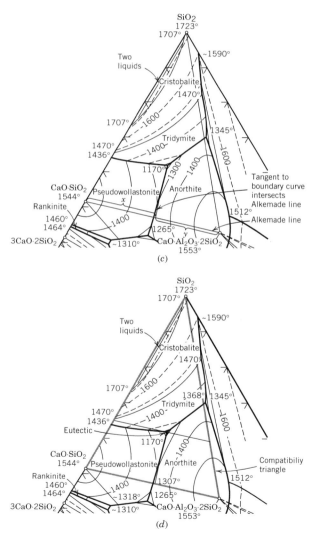

Fig. 4.33 (c) Relative amounts of anorthite and pseudowollastonite being crystallized from the liquid are obtained from the intersection of the tangent to the boundary curve with the Alkemade line (see text). (d) The eutectic liquid crystallizes three solid phases, in relative proportions given by the ternary lever rule.

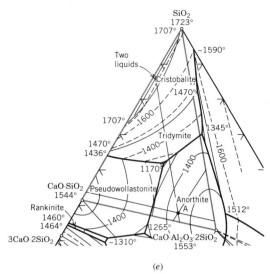

(e)

Fig. 4.33 (*e*) The final amounts of the three solid phases are obtained by applying the ternary lever rule using the overall composition A.

therefore the first solid phase to crystallize upon cooling. We read from the isothermal contours that this will first occur at about 2250°C. Below the liquidus temperature, CaO continues to solidify, leaving the liquid increasingly enriched in Al_2O_3 and SiO_2. This continues until the liquid composition reaches the boundary curve between the CaO and Ca_3SiO_5 primary phase fields, at 1850°C. The path the liquid composition has taken between 2250°C and 1850°C corresponds to the reaction

$$L \rightarrow CaO + L'$$

where L' is the CaO-deficient liquid. (As in previous examples, the fractions of liquid and CaO present at any point along this trajectory can be calculated using a binary lever rule, with B as the fulcrum between CaO and the liquid composition. When the liquid has just reached the boundary curve, there is about 42% CaO and 58% liquid.)

Figure 4.34*a* shows the liquid composition trajectory upon further cooling. We now have a three-phase equilibrium between CaO, Ca_3SiO_5, and liquid, with the liquid composition following the boundary curve away from the Alkemade line connecting the two solid phases in equilibrium, CaO and Ca_3SiO_5. However, in this instance the tangent to the boundary curve extends back to an intersection *outside* the Alkemade line. The intersection is on the SiO_2-rich side of the 3:1 compound, which indicates that some CaO must be consumed to form Ca_3SiO_5.

4.6 Reactions upon Heating and Cooling

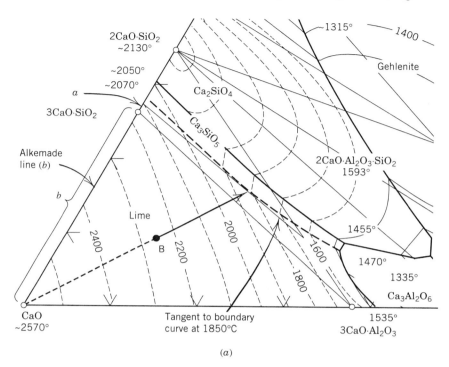

Fig. 4.34 (a) Crystallization of a liquid of composition B results first in the solidification of lime in equilibrium with a liquid increasingly depleted in CaO as temperature decreases. Upon reaching the boundary curve, the liquid composition traces the boundary curve with a trajectory that intersects outside the Alkemade line between CaO and Ca_3SiO_5, necessitating the partial resorption of CaO as Ca_3SiO_5 is formed.

The reaction is

$$CaO + L \rightarrow Ca_3SiO_5 + L'$$

We apply the lever rule at the Alkemade line to determine the relative amounts of CaO and liquid being consumed: $a/(a+b)$ is the fraction of CaO reacting, and $b/(a+b)$ is the fraction of L reacting (Fig. 4.34a). This ratio varies with temperature as the intersection shifts. At about 1700°C, the tangent to the boundary curve intersects the Alkemade line at the 3:1 compound rather than on its extension. The reaction is at this point the direct solidification

$$L \rightarrow Ca_3SiO_5 + L'$$

Figure 4.34b shows the ternary lever used to find the fractions of phases present at this temperature. We have CaO, Ca_3SiO_5, and liquid of a composition corresponding to the boundary curve at 1700°C. Lines have been drawn through B parallel to the sides of the triangle joining these phases. Using the ternary lever rule, the fraction of CaO is $a/(a + b + c)$, the fraction of 3:1 is $c/(a + b + c)$, and the fraction of liquid is $b/(a + b + c)$.

At temperatures below 1700°C and above the critical point temperature, the tangent to the boundary curve intersects the Alkemade line between CaO and Ca_3SiO_5; therefore, both of these solid phases are simultaneously crystallizing from the liquid:

$$L \rightarrow CaO + Ca_3SiO_5 + L'$$

This reaction will continue until the liquid composition reaches the peritectic at 1470°C. Figure 4.34c shows the ternary lever used to determine the phases present just above 1470°C, where we still have a three-phase equilibrium. We find that the liquid fraction has decreased significantly, while the fraction of Ca_3SiO_5 has increased considerably since the liquid trajectory is almost directly away from this composition.

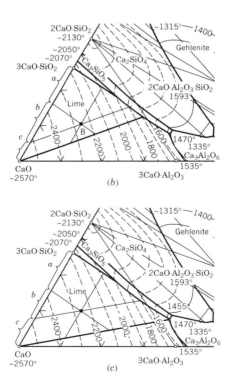

Fig. 4.34 (b) Ternary lever at 1700°C, from which the amounts of CaO, 3CaO·SiO$_2$, and liquid are calculated. (c) Ternary lever just above the peritectic temperature of 1470°C.

4.6 Reactions upon Heating and Cooling

At the 1470°C peritectic point, the liquid is in equilibrium with three solids: CaO, Ca_3SiO_5, and $Ca_3Al_2O_6$. From the Gibbs phase rule one phase must disappear upon further cooling. The peritectic solidification reaction, depicted by the arrows in Fig. 4.34d, consumes some of one solid in order to form the other two:

$$L + CaO \rightarrow Ca_3SiO_5 + Ca_3Al_2O_6$$

The relative amounts of phases consumed and produced are defined by the two linear levers in Fig. 4.34d, using their intersection as the fulcrum. The fraction of liquid consumed is $b/(a + b)$ and of CaO consumed is $a/(a + b)$, while the fraction of $Ca_3Al_2O_6$ produced is $d/(c + d)$ and of Ca_3SiO_5 produced is $c/(c + d)$. We may think of the peritectic reaction as a two-step reaction between the peritectic liquid and CaO (inward arrows), followed by a decomposition to the products (outward arrows). For equilibrium cooling, the temperature will remain at 1470°C until all of the liquid is consumed.

As in previous examples, these phase fractions are the relative amounts involved in the reaction and not the total remaining after the reaction is complete.

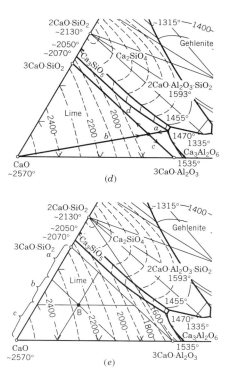

Fig. 4.34 (d) Peritectic reaction. (e) Ternary lever from which final phase fractions are calculated.

340 Chapter 4 / Phase Equilibria

To obtain that, we apply the ternary lever rule to the compatibility triangle using the overall composition B as the fulcrum, as shown in Fig. 4.34e. The fraction of CaO is $a/(a+b+c)$, of Ca_3SiO_5 is $c/(a+b+c)$, and of $Ca_3Al_2O_6$ is $b/(a+b+c)$. Since no solid solution is indicated for any of these three solids, we assume that they are line compounds. This compatibility triangle then applies for all temperatures below 1470°C.

Nonequilibrium Crystallization

Equilibrium phase diagrams provide useful insight even when equilibrium is not maintained. One example we have mentioned is when starting materials do not constitute an equilibrium phase assemblage. In the MgO-SiO_2-Al_2O_3 system, for example (Fig. 4.24), the lowest critical point is a ternary eutectic at 1365°C. Were we to start with a powder compact of MgO, SiO_2, and Al_2O_3, and to heat to just above 1365°C, it is quite likely that during subsequent reaction and interdiffusion, there will exist somewhere within the sample the composition of this lowest solidus. That region of the sample will melt; and liquid may be detected regardless of the overall composition of the sample. Still more complex reactions can occur when starting materials decompose during heating, as is common for metal salts such as carbonates, nitrates, sulfates, or hydroxides.

Another common type of nonequilibrium reaction takes place during crystallization due to physical isolation of a pre-existing phase. This is particularly likely during resorption reactions, and causes departures from the equilibrium cooling path, as the isolation of phases causes the "apparent" composition of the system to deviate from the actual composition. The phenomenon is less likely during heating reactions where the amount of liquid is increasing and pre-existing solid phases are being dissolved.

To illustrate this phenomenon, consider composition C in Fig. 4.35a again in the CaO primary phase field. As discussed previously, when the tangent to the boundary curve lies on an extension of the Alkemade line, rather than on the line itself, resorption of the previously crystallized solid is required in order to form the new solid: $CaO + L \rightarrow Ca_3SiO_5$. Physical isolation of the primary phase can occur when the reaction takes place at the liquid–solid interface and particularly if it forms a continuous reaction layer. This is shown schematically in Fig. 4.35b. The liquid is unable to react with the primary phase since transport through the reaction product is slow. The "system" of concern becomes Ca_3SiO_5 and the liquid alone. With CaO behaving as an inert phase due to its physical isolation, we have an *apparent two-phase equilibrium* between liquid and Ca_3SiO_5. The apparent composition is deficient in CaO compared to the overall composition B, and as a result, the crystallization path need not adhere to the boundary curve in the absence of a true three-phase equilibrium. The new cooling path will be away from Ca_3SiO_5 since that is the single solidifying phase, as shown in Fig. 4.35a.

4.6 Reactions upon Heating and Cooling

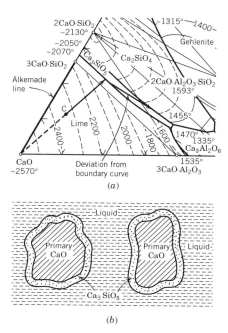

Fig. 4.35 (a) Nonequilibrium crystallization of composition C causes a deviation from the boundary curve between CaO and Ca$_3$SiO$_5$ when (b) the product phase Ca$_3$SiO$_5$ causes the primary phase CaO to become physically isolated. Final crystallization occurs at the 1455°C peritectic rather than at the equilibrium 1470°C peritectic.

Upon reaching the boundary curve between Ca$_3$SiO$_5$ and Ca$_2$SiO$_4$, a new three-phase local equilibrium is established. Since the tangent to this boundary curve does intersect the Alkemade line joining Ca$_3$SiO$_5$ and Ca$_2$SiO$_4$, it is likely that the liquid will follow the new boundary curve, the "equilibrium" reaction being L → Ca$_3$SiO$_5$ + Ca$_2$SiO$_4$. The liquid composition may follow this boundary curve until the invariant point at 1455°C is reached. This is also a peritectic, lying outside the compatibility triangle Ca$_3$SiO$_5$-Ca$_2$SiO$_4$-Ca$_2$Al$_2$O$_6$. The new solid, Ca$_3$Al$_2$O$_6$, is formed via the reaction:

$$L + Ca_3SiO_5 \rightarrow Ca_2SiO_4 + Ca_2Al_2O_6$$

If this nonequilibrium reaction takes place as described, we are left with four solid phases: CaO, Ca$_3$SiO$_5$, Ca$_2$SiO$_4$, and Ca$_2$Al$_2$O$_6$. The presence of four phases while temperature is freely varied violates the Gibbs phase rule, and is certainly one sign that the system is not at equilibrium.

This scenario is one relatively simple example of a nonequilibrium reaction. More complex situations where more than one nonequilibrium reaction occurs in sequence, and causes additional deviations from the equilibrium cooling path, are easy to imagine. Additional solid phases result in the final product. Another cause of nonequilibrium phase assemblages is the slow nucleation and growth of equilibrium crystals from the liquid. This is especially common for viscous silicate liquids, and reaches an extreme case in the formation of glasses, where no crystalline phases form upon cooling the liquid.

SPECIAL TOPIC 4.3

PORCELAIN

Porcelain is defined as a strong, white, impermeable, sonorous, translucent ceramic with less than 0.5% residual porosity. Its strength and sonorous qualities result from the low porosity (hold a piece of porcelain on your fingertips and tap it to demonstrate its tone, which only slowly is dampened; compare it with a cheaper, more porous ordinary dinnerware). Its white color results from using pure ingredients which form a microstructure consisting of quartz particles and tiny mullite crystals immersed in a viscous silicate glass (Fig. ST50). Porcelain was first made in China. By the T'ang Dynasty (618–906 A.D.) porcelain was a major export product. It became a staple of the china trade from the sixteenth century onward. There were many extensive European efforts to produce something equivalent.

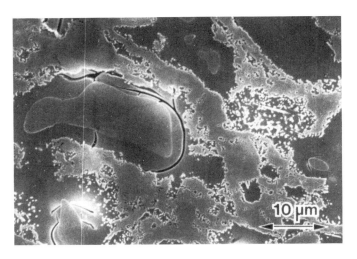

Fig. ST50 K'ang Hsi porcelain made in China at the beginning of the eighteenth century is dense, hard and translucent. The microstructure consists of partly dissolved rounded quartz particles in a matrix of silicate glass which contains fine mullite crystals. The circumferential crack around the quartz grain is enhanced by light etching with 1% hydrofluoric acid.

4.6 Reactions upon Heating and Cooling   343

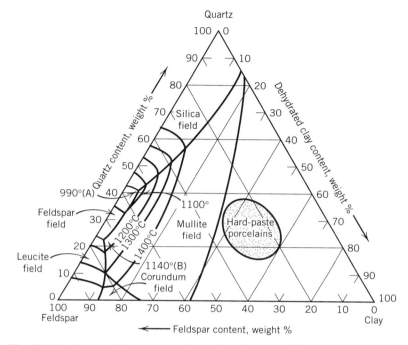

Fig. ST51 Quartz-feldspar-dehydrated clay phase equilibrium diagram shows the region of typical hard-paste porcelain. The lowest temperature eutectic for these compositions is about 990°C.

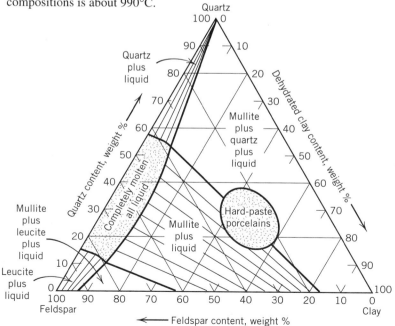

Fig. ST52 An isothermal cut through the phase diagram illustrates the phases present at the firing temperature of hard paste porcelains.

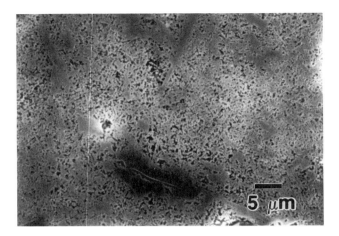

Fig. ST53 The microstructure of a Böttger porcelain sample consists of mullite crystals in a glass matrix. This composition did not include quartz, which is present in most porcelains now being made.

By the seventeenth century, Chinese porcelain was being made from a mixture of quartz, feldspar, and kaolinite (China clay), as shown in the phase diagram of Fig. ST51. This is part of the larger K_2O-Al_2O_3-SiO_2 diagram. On heating, kaolinite $(Al_2Si_2O_5)(OH_4)$ decomposes to form dehydrated amorphous "meta-kaolin," a metastable phase, at 450–500°C. Feldspar grains react with excess silica and the clay to form a viscous liquid at about 990°C. As heating continues, mullite crystallizes from the metakaolin, quartz dissolves and potassia diffuses throughout the system with the further precipitation of mullite and more rapid solution of quartz. (Question: What do you think limits the quartz solution rate?) At a typical firing temperature of 1300°C, the phases present are mullite, silica (almost always), and a potassium alumino silicate glass as shown in Fig. ST52. When quartz is present, its large volume contraction at the 573°C β–α transformation leads to circumferential cracks forming in the surrounding glass during cooling (Fig. ST50).

A happy circumstance of available raw materials and high-temperature kilns dug into the insulating clayey hillside soil gave rise to Chinese porcelain. Efforts to make a European equivalent were only partially successful until a modern research effort was mounted by Augustus the Strong, Elector of Saxony in the opening years of the eighteenth century. Johann Böttger was the principal investigator and Count Ehrenfried von Tschirnhaus was in overall charge of the project. Hundreds of compositions were tested with a solar furnace based on a meter-diameter lens. Lime was developed as the principle flux and a critical success was designing a high-temperature furnace with multiple fire boxes able to reach the necessary temperature, which was above the 1345°C eutectic (CaO-Al_2O_3-SiO_2 diagram, Fig. 4.32). The project was successful in developing "Böttger" porcelain in 1708 (Fig. ST53). This material consisted of mullite crystallites in a lime alumina silicate glass and had extraordinary properties. Subsequently, the formula was modified to contain more K_2O which made it possible to fire at a lower temperature.

ADDITIONAL READING

D. V. Ragone, *Thermodynamics of Materials*, Vols. 1 and 2, John Wiley and Sons, New York, 1995.

C. H. P. Lupis, *Chemical Thermodynamics of Materials*, Elsevier Science Publishing Co., New York, 1983.

E. M. Levin, C. R. Robbins, H. F. McMurdie, *Phase Diagrams for Ceramists*, pp. 5–31, The American Ceramic Society, Columbus, OH, 1964. (Later volumes in this continuing series give periodic updates with many new phase diagrams.)

C. G. Bergeron and S. H. Risbud, *Introduction to Phase Equilibria in Ceramics*, The American Ceramic Society, Columbus, OH, 1984.

A. Muan and E. F. Osborn, *Phase Equilibria Among Oxides in Steelmaking*, Addison-Wesley, , Inc., Reading, MA, 1965.

PROBLEMS

1. Consider the SiO_2-Al_2O_3 binary phase diagram shown in Fig. ST44. The dotted lines for metastable equilibria may be neglected for the purposes of this question.

   (i) Identify the phases, their compositions, and amounts present for an overall composition of 45 mole% Al_2O_3 at 1500° and 1700°C, respectively, and for an 80% Al_2O_3 composition at 1750°C.

   (ii) Notice that mullite, which melts incongruently, has a range of solid solution at temperatures below the peritectic but narrows to a single composition at the peritectic. In contrast, eutectic systems usually show a maxiumum in the solid solubility of the endmember phases at the eutectic temperature. Explain, using the Gibbs phase rule, why a range of mullite compositions is not permissible at the peritectic.

2. For the BaO-TiO_2 phase diagram in Fig. 4.7, show, using sketches and reactions, the heating path (beginning at room temperature) and cooling path (beginning with a single-phase liquid) for the following compositions:

(i) 38 mole% TiO$_2$
(ii) 79 mole% TiO$_2$
(iii) 73 mole% TiO$_2$

In the case of cooling, also sketch what the microstructure might look like after reaching room temperature. Then, choose one of the three compositions above for which you think nonequilibrium cooling could take place, and discuss how that path would differ from the equilibrium path. How would you determine from either: (a) the distribution of phases in the microstructure; and/or (b) the phase content (e.g., by x-ray diffraction) if nonequilibrium cooling had occurred?

3. The KF-NaF-MgF$_2$ ternary diagram is shown below. Do the following:
   (a) Identify the intermediate compounds and whether they melt congruently or incongruently.
   (b) Identify the ternary critical points and whether they are ternary eutectic or peritectic points.
   (c) Show the compatibility triangles for this system.

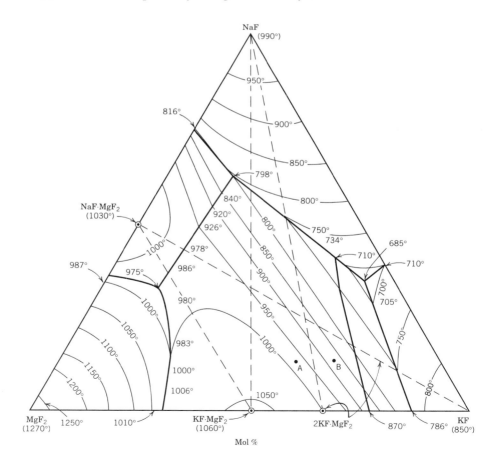

(d) If a composition of 10 mole % KF, 10 mole % NaF, 80 mole % MgF_2 is heated, at what temperature does it form a single-phase liquid?

(e) If a powder mixture of the three pure endmembers is prepared, what is the *minimum* temperature at which a liquid will form? Assume that a very sensitive calorimeter is used, which can detect the exotherm for forming very small amounts of liquid.

(f) Explain the crystallization path for a single-phase liquid of the composition marked A. How does it differ from that marked B? Is there an opportunity for nonequilibrium cooling in either case? Explain.

(g) For the composition A, determine the phases present and the fractions of each phase at a temperatures of: 800°C, 725°C, and room temperature.

4. Imagine that you have received a sample of an unknown ceramic material. The supplier of the material says that it was fired in ambient air (pressure constant). Upon analyzing the overall composition, you determine that it belongs to a four-oxide-component system, AO-BO-CO-DO. You now carry out an x-ray diffraction phase analysis of the sample and find five crystalline phases to be present.

(i) Is it *possible* that this sample was heat-treated such that equilibrium was achieved? Is it *likely*? Why or why not?

(ii) Answer the same questions for the case where *six* crystalline phases are present. What if *three* crystalline phases are found?

5. The Y_2O_3-SiO_2-Al_2O_3 phase diagram is shown below. By answering the following questions we will explore characteristic features of ternary diagrams, and how phase equilibria influence processing of polyphase materials.

(a) Identify the ternary critical points, making a table showing the three solid phases in equilibrium with the liquid at each critical point. Identify whether the critical point is a peritectic or eutectic. In a separate diagram, show the compatibility triangles in this system.

(b) Which compounds in this system melt incongruently?

(c) Yttrium aluminum garnet (YAG, or $Y_3Al_5O_{12}$) is of interest as a laser host (when doped with rare earths such as Nd) and as an IR-transparent window material. What (approximately) is the melting point of YAG? Can single crystals of YAG in principal be grown by pulling a seed in contact with the melt (Czochralski method)? Why or why not?

(d) A small amount of SiO_2 powder (5 wt %) is added to a fine YAG powder, in hopes of densifying the powder compact into a transparent polycrystalline YAG (suitable for IR window material) by forming a minor amount of liquid phase during firing. After mixing uniformly, the powder mixture is pressed into a pellet and rapidly heated in a furnace. What is the minimum temperature (approximately) at which a liquid phase might form? What is the corresponding liquid composition?

After firing the pellet at 1800°C for many hours, the sample is cooled to room temperature at a rate slow enough to maintain equilibrium. Upon reheating, at what temperature would a liquid first form in this sample? What would be the liquid composition? (Explain how you arrived at your answer.)

(e) In a separate study, fine Y_2O_3 and SiO_2 powder are added to Al_2O_3 powder as liquid-phase sintering aids. The overall composition is 80 wt % Al_2O_3, 15 % SiO_2, and 5 % Y_2O_3. If the sample is equilibrated at 1700°C, what phases are present, and in what amounts?

(f) Toward the SiO_2-rich end of the SiO_2-Y_2O_3 join, there exists a region of liquid immiscibility where two liquid phases are present at equilibrium. Can you determine from this phase diagram the exact compositions of the two liquids? The approximate composition? Explain.

(g) What is the most refractory (highest melting) three-phase composite one can prepare in this ternary system?

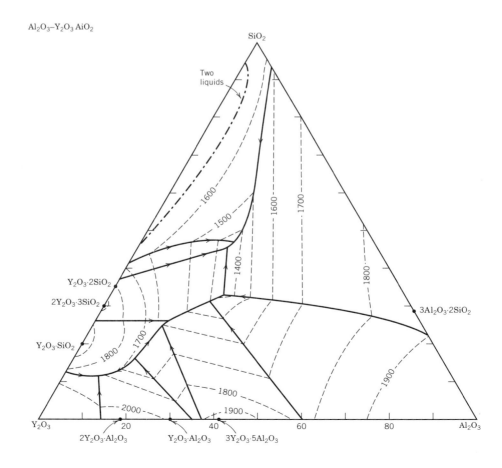

6. The CaO-SiO$_2$-TiO$_2$ ternary phase diagram shown below contains 10 compounds and 9 invariant points. Rankinite is $3CaO \cdot 2SiO_2$, pseudowollastonite is $CaO \cdot SiO_2$, sphene is $CaO \cdot TiO_2 \cdot SiO_2$, and $CaO \cdot TiO_2$ is perovskite ($CaTiO_3$). Answer the following:

(i) What is the solidus temperature (invariant point) for the sphene-perovskite-pseudowollastonite compatibility triangle? Is it a eutectic or a peritectic? Show (in words or on the diagram) the equilibrium crystallization path for a composition 40% TiO$_2$, 20% SiO$_2$, and 40% CaO.

(ii) For a composition 20% TiO$_2$, 25% SiO$_2$, 55% CaO, identify the compatibility triangle and invariant point. Is this one a peritectic or eutectic? Show or describe the equilibrium crystallization path.

(iii) Now let's say that you have added 5% SiO$_2$ (ground quartz) and 2.5% CaO to a powder compact of CaTiO$_3$ because you wish to form a small amount of a liquid phase to help the sintering process. If during microstructural analysis of the finished ceramic product you find a small amount of crystalline SiO$_2$, would you conclude that this was a result of incomplete reaction or non-equilbrium cooling? (Either way, explain your answer.)

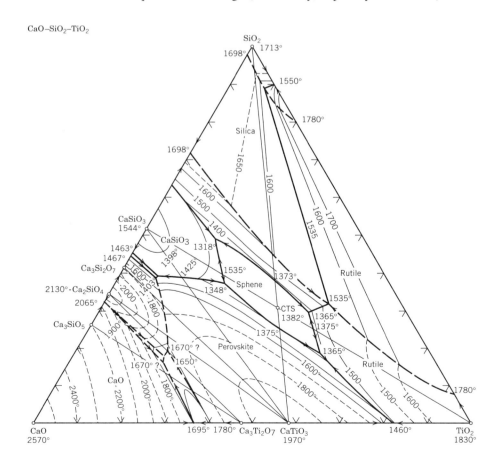

7. From the YN·Si$_3$N$_4$·Y$_2$O$_3$·SiO$_2$ system, a composite was made with fine Si$_3$N$_4$ containing 10 mole% SiO$_2$ and 2 mol% YN.

   (a) What would be the compositions and amounts of the phases present at a sintering temperature of (a) 1500°C? (b) 1550°C?

   (b) Which would be a better choice as a sintering temperature? Why?

   (c) What would be the phases present at room temperature?

   (d) What is the approximate solidus temperature for this composition?

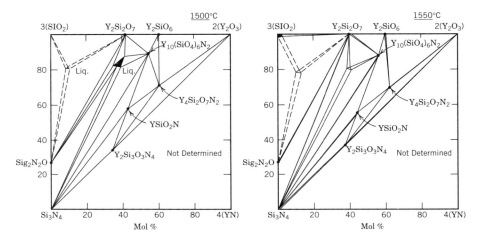

8. Use the CaO-SiO$_2$-Al$_2$O$_3$ phase diagram (Fig. 4.32) to answer the following questions:

   (a) The compound rankinite, which has chemical formula 3CaO·2SiO$_2$, melts incongruently. At what temperature does it melt, and into what phases? Identify the compositions of the phases formed.

   (b) Draw the binary phase diagram between anorthite and pseudowollastonite.

   (c) A mixture containing 20 wt% quartz, 20 wt% pseudowollastonite, and 60 wt% anorthite, each in the form of fine powders, is mixed and pressed into a pellet, then fired. During heating, at what temperature will melting first occur in the sample? If you are observing the pellet through a window in the furnace, do you expect it to retain its shape and perhaps densify when this temperature is reached, or instead to slump into a puddle? Explain.

   (d) Consider a polycrystalline aluminum oxide that contains minor amounts (percent levels or less) of both CaO and SiO$_2$. The relative amounts of the two impurities vary somewhat from batch to batch of the alumina powder, but the total amount is always less than 4 wt%. What is the highest temperature at which you could expect a furnace tube sintered from this material to be usable in a long-term application (months), under a moderate stress of ~100 MPa? Explain your reasoning.

Chapter 5

MICROSTRUCTURE

Although single crystals have many applications in special ceramics, and single phase glasses consitute a significant industry with many important applications, most of the advanced ceramics on which we focus, as well as traditional clay-based silicate ceramics and cements, are a mixture of crystals, glass, grain boundaries, and porosity. The paradigm of Materials Science is that processing of synthetic or natural materials gives rise to an internal structure that determines properties and consequently the performance of the ceramic products. This is the ultimate aim of ceramic engineering. The constituents of ceramic microstructures are the crystals and glasses discussed in Chapter 1. Fundamental processes affecting the formation of these constituents are the topics of Chapters 3 and 4. In this chapter we discuss the capillary forces that direct much of microstructure development, the processes of microstructure change, and then a few consequences of microstructure development on properties.

5.1 CAPILLARITY

Capillary forces are *surface forces* exerted by the external and internal surfaces of condensed phases. A simple experimental demonstration that a force is required to extend a liquid surface is depicted in Fig. 5.1. A soap film on a wire loop is attached at the edges and the stable configuration is a flat sheet of minimum area. If a gentle pressure is applied by blowing on one side, spherical soap bubbles can be formed. With just the right pressure, we can blow out the film to form a hemisphere; when we stop blowing, the extended soap bubble pops back into position as a flat plate (Fig. 5.1). A disturbed surface will thus tend toward its center of curvature to become planar with a smaller area. In stretching the surface, we perform work against the *surface tension* (or *interfacial tension* in the case of interfaces between condensed phases), γ, defined as the reversible work, w_r, required to increase the surface by a unit area:

$$\gamma = \frac{dw_r}{dA} \tag{5.1}$$

The most familiar manifestation of surface tension is observed in the tendency of liquids to form a configuration having minimum surface area. Soap bubbles, falling liquid droplets, and oil droplets suspended in water always tend toward a spherical shape.

There is also an excess energy to a surface due to differences in the surroundings of atoms at the surface and in the interior, as shown in Fig. 5.2. Since each atom on the surface has only a partial compliment of its internal coordination number, there are broken or distorted bonds resulting in an increase in energy. Surface energy is the increase in free energy per unit area of new surface formed:

$$\gamma_e = \left(\frac{\partial G}{\partial A}\right)_{T,P,n_i} \tag{5.2}$$

For a liquid the surface tension and surface energy are one and the same—the same value is obtained by either reversibly stretching an existing surface or by

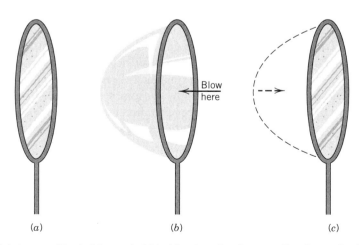

Fig. 5.1 (*a*) A soap film held on a bubble-blowing ring forms a flat sheet of minimum surface area. (*b*) Steady, gentle blowing extends the film to become a hemisphere. (*c*) If we stop blowing, the half-bubble springs back towards its concave side. All curved surfaces tend to move towards their center of curvature in the absence of external forces.

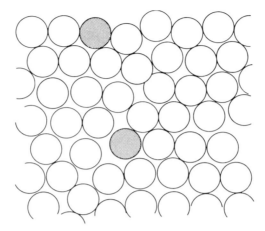

Fig. 5.2 Cross-section of a liquid illustrating the difference in surroundings of a surface atom and an interior atom.

reversibly creating new surface. Atoms move to the surface and adopt the same structural configuration in either case. However, for solid surfaces, particularly at low temperatures and for small displacements, the surface can be stretched without changing atomic arrangements. A solid surface can also support deformations out of the plane of the surface, since a solid can support shearing stresses while a liquid cannot. Thus for a solid the surface tension is replaced by the *surface stress*, a tensor quantity, which is not in general equal to the surface energy. For high-temperature processes where atom transport occurs fast enough that solid surfaces are free to reconfigure, the distinction between surface stress or surface tension and the surface energy is often unimportant.

An important difference between liquid and solid surfaces is that of surface energy anisotropy. Whereas liquid droplets always tend toward a spherical shape, indicating that the surface energy is *isotropic*, crystals often grow with faceted surfaces. In general, different crystallographic faces of crystals have different surface energies, due to differences in bonding or atom density. The planes of lowest energy and consequently the most stable ones are often those planes with the most dense atomic packing. The equilibrium shape of a crystal corresponds to a minimization of the total surface energy (product of specific surface energy and surface area, $\gamma \cdot A$), and not the surface area alone. The *Wulff plot* is a construction that allows us to determine the equilibrium crystal shape from the variation of surface energy with crystal orientation. By drawing a vector normal to each surface orientation of length equal to the magnitude of its surface energy (shown as a polar plot for two dimensions in Fig. 5.3), an envelope representing the surface energy is obtained. An inner envelope can then be constructed, by drawing lines normal to each surface energy vector at its endpoint. This inner envelope is the equilibrium crystal shape. Notice that if the surface energy is the same for all orientations, the

354 Chapter 5 / Microstructure

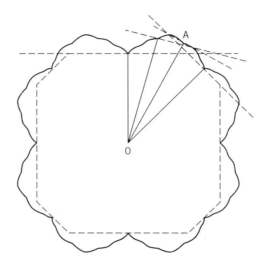

Fig. 5.3 The Wulff construction. A two-dimensional section of a polar plot of surface tension for a cubic crystal, in which the direction of the vector from the origin to any point on the outer envelope (solid line) represents the direction of the surface normal, and its magnitude is the surface tension for that particular plane. The equilibrium shape of a crystal is given by the inner envelope (dotted line) representing a minimum total surface energy (specific surface tension x area). Surfaces with orientations such as A are higher in energy with respect to the equilibrium shape and may facet to form lower energy surfaces. (After C. Herring in *Structure and Properties of Solid Surfaces*, ed. R. Gomer and C.S. Smith, University of Chicago Press, 1953, Chapter 1.)

envelopes are reduced to circles (or, in three dimensions, spheres), and we obtain the equilibrium shape of liquid droplets.

The preferred, low energy surfaces of crystalline grains can be observed when surfaces are changed by precipitation, vaporization, surface diffusion, volume diffusion, or grain growth. An example is shown in Fig. 5.4 for the surfaces of pores within a crystal of UO_2. It is important to note that the observed shape or "growth habit" of a crystal or grain can result from both surface energy anisotropy and *growth rate anisotropy*, the latter occurring when a particular crystal face affords easy atom attachment and grows rapidly. In practice the two effects can be difficult to separate, and are often simultaneously present. Growth rate anisotropies can result in shapes which exaggerate actual surface energy anisotropies. Examples of highly anisotropic crystallite shapes in several ceramic systems are shown in Figs. ST59, 5.32, and 5.33.

Pressure Due to Curved Surfaces

The snapping back into a planar form of a curved soap film illustrated in Fig. 5.1 results from the fact that the surface energy causes a pressure difference across a

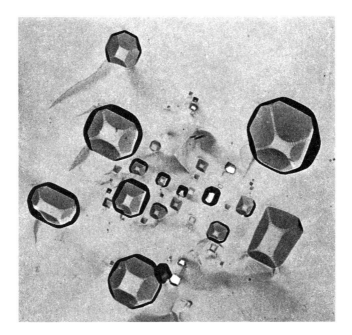

Fig. 5.4 Pores which have formed "negative" crystals of Wulff shape in UO_2; (100) planes are parallel to surface (18,000×). (Courtesy T.R. Padden.)

curved surface. This is the basis for methods of measuring surface tension of liquids by the height of rise in a capillary, or by the pressure required to blow a spherical bubble at the end of a submerged capillary. As illustrated in Fig. 5.5, the work of expanding a spherical surface $\Delta P\, dv$ must equal the increase in surface energy $\gamma\, dA$. For a sphere, dv and dA are given by

$$\Delta P dv = \gamma dA$$

$$dv = 4\pi r^2 dr \qquad dA = 8\pi r dr$$

and so the pressure exerted by the spherical surface is

$$\Delta P = \gamma \frac{dA}{dv} = \gamma \frac{8\pi r dr}{4\pi r^2 dr} = \gamma \left(\frac{2}{r}\right) \tag{5.3}$$

In general when the surface is not spherical, similar analysis gives

$$\Delta P = \gamma \left(\frac{1}{r_1} + \frac{1}{r_2}\right) \frac{1}{2} \tag{5.4}$$

where r_1 and r_2 are the principal radii of curvature.

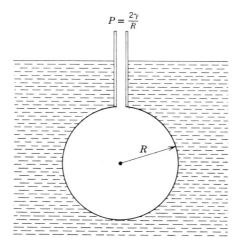

Fig. 5.5 Determination of the pressure at equilibrium to maintain a spherical surface of radius R.

Chemical Potential Changes at Curved Surfaces

One result of the pressure difference across a curved surface is a change in solubility or vapor pressure as compared to a planar surface. The pressure applied to the liquid or solid by the curved surface increases the chemical potential of its constituents and the pressure of the vapor phase in equilibrium with it. As illustrated in Fig. 5.6, a surface of convex curvature (positive r) has a greater equilibrium vapor pressure than a planar surface (infinite r), which in turn has a greater vapor pressure than the concave surface (negative r). The amount of this increase can be derived by considering the transfer of one mole of material from the flat surface, through a liquid or vapor, to the spherical surface. With temperature, external pressure, and overall composition held constant, the work done is equal to the change in chemical potential ($\mu = \mu^\circ + RT \ln a$, where μ° is the standard chemical potential and a the activity). Assuming a constant activity coefficient, the chemical potential difference is given by

$$\Delta\mu = (RT \ln c - RT \ln c_o) \text{ or } \Delta\mu = (RT \ln p - RT \ln p_o)$$

where c is the solubility and p the vapor pressure, and c_o and p_o are the equilibrium solubility and vapor pressure over a flat surface. If done reversibly this work is equal to the change in surface energy, γdA.

$$RT \ln \frac{c}{c_o} = RT \ln \frac{p}{p_o} = \gamma \, dA = \gamma \, 8\pi \, r^2 dr$$

Since the change in volume is $dv = 4\pi r^2 \, dr$, the radius change for a one-mole

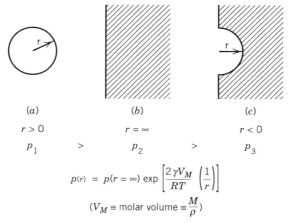

Fig. 5.6 The equilibrium vapor pressure over (a) a surface of positive curvature exceeds that over (b) a flat surface, which exceeds that over (c) a surface of negative curvature.

transfer is $dr = V_m/4\pi r^2$ (where V_m is the molar volume) and

$$c = c_o \exp\left[\frac{\gamma V_m}{RT}\frac{2}{r}\right] \quad (5.5)$$

For a non-spherical surface the solubility is

$$c = c_o \exp\left[\frac{\gamma V_m}{RT}\left(\frac{1}{r_1}+\frac{1}{r_2}\right)\right] \quad (5.6)$$

The argument of the exponential term is usually small, and since $e^x \sim 1+x$ for small x, upon rearranging Eq. 5.6 we find that the change in solubility or vapor pressure is approximately inversely proportional to r. The increase in solubility (Table 5.1) can be substantial for particles of micron size and smaller. The relations in Eqs. 5.5 and 5.6 are known as the Thompson–Freundlich equation. The effect of particle diameter in these relations is an important reason for using fine particle-size materials. It is common in ceramic processing to synthesize powders of submicron particle size, which give rise to capillary forces of appreciable magnitude.

Not only do surfaces form configurations that minimize surface energy, in systems of two or more components there is also a strong tendency for the distribution of constituents to occur in a way that minimizes surface energy. If a small amount of a low surface tension component is present, it concentrates in the surface layer; and the surface energy is sharply decreased with small additions. In contrast, when a high surface tension component is added to one of low surface

Table 5.1 Effect of Surface Curvature on Pressure Difference and Relative Vapor Pressure across a Curved Surface

Material	Surface Diameter (microns)	Pressure Difference (psi)	Relative Vapor Pressure (p/p_o)	Gradient $\Delta p/r$ Units of p_o/cm
Silica glass	0.1	1750	1.02	4000
(1700°C)	1.0	175	1.002	40
$\gamma = 300$ ergs/cm^2	10.0	17.5	1.0002	0.40
Liquid cobalt	0.1	9750	1.02	4000
(1450°C)	1.0	975	1.002	40
$\gamma = 1700$ ergs/cm^2	10.0	97.5	1.0002	0.40
Liquid water	0.1	418	1.02	4000
(25°C)	1.0	41.8	1.002	40
$\gamma = 72$ ergs/cm^2	10.0	4.18	1.0002	0.40
Solid Al$_2$O$_3$	0.1	5250	1.02	4000
(1850°C)	1.0	525	1.002	40
$\gamma = 905$ ergs/cm^2	10.0	52.5	1.0002	0.40

energy, it tends to be less concentrated in the surface layer than in the bulk and has only a slight influence on the surface tension. As a result the surface energy does not change linearly with composition between endmembers, but rather as shown in Fig. 5.7, where B is the surface energy lowering component when added to A, and A increases the surface energy when added to B. The Gibbs adsorption isotherm relates the surface excess Γ_2 of a component, in units of moles per square centimeter, to the surface energy and bulk concentration c_2:

$$\Gamma_2 = -d\gamma / RT \, d\ln c_2 \quad (5.7)$$

This interrelationship shows that the greater the reduction in γ caused by a certain addition c_2, the more component 2 will concentrate in the surface layer. By plotting γ versus $\ln c_2$ and taking the slope, it is possible to determine the surface excess. For many materials having a high surface activity, this slope is constant over a considerable composition range, nearly up to values of Γ_2 corresponding to the adsorption of a monolayer at the surface. For high surface energy materials, the influence of surface-active additives is very large. Even at small concentrations, oxygen tends to concentrate in the surface layers on metals, carbides, and nitrides. Values for the surface energies of a number of materials are shown in Table 5.2.

Just as free surfaces have an associated surface energy, boundaries between phases— solid–solid, liquid–liquid, liquid–solid—are characterized by an interface

Table 5.2 Measured Surface Energies of Various Materials in Vacuum or Inert Atmospheres

	Temperature (°C)	Surface Energy (ergs/cm^2)
Liquids		
Water	25	72
Lead	350	442
Copper	1120	1270
Silver	1000	920
Iron	1535	1880
Platinum	1770	1865
Sodium chloride	801	114
Sodium sulfate	884	196
Sodium phosphate, NaPO$_3$	620	209
Sodium silicate	1000	250
B$_2$O$_3$	900	80
FeO	1420	585
Al$_2$O$_3$	2080	700
0.13 Na$_2$O–0.13 CaO–0.74 SiO$_2$	1350	350
Solids		
Copper	1080	1430
Silver	750	1140
Iron (γ phase)	1350	2100
Platinum	1300	2200
Al$_2$O$_3$	1850	905
TiC	1100	1190
0.20 Na$_2$O–0.80 SiO$_2$	1350	380
NaCl (100)	25	300
LiF (100)	−196 (77 K)	340
CaF$_2$ (111)	−196 (77 K)	500
BaF$_2$ (111)	−196 (77 K)	280
CaCO$_3$ (1010)	25	230
MgO (100)	−196 (77 K)	1500
Al$_2$O$_3$ (10$\bar{1}$2)	25	6000
(10$\bar{1}$0)	25	7300
(0001)	25	>40,000
MgAl$_2$O$_4$ (100)	25	3000
(111)	25	5000
SiC (11$\bar{2}$0)	25	20,000
Si (111)	—	1230
Ge (111)	—	1060

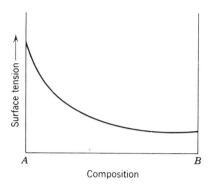

Fig. 5.7 Surface tension in a two-component system.

energy corresponding to the energy required to form a unit area of new interface. This interfacial energy is always less than the sum of the separate surface energies, for otherwise the interface would spontaneously separate to lower its energy. It may be almost any value less than this sum. Occasionally, when two miscible phases are mixed, the surface is observed to extend itself, to increase its area as the first stage of complete mixing, corresponding to a nonequilibrium "negative" surface energy. Some interfacial energies for mutually saturated systems are given in Table 5.3.

Wetting and Dihedral Angles

When we have a vapor–liquid–solid at equilibrium, the shape conforms to a minimum total interface energy for the boundaries present. When the solid–liquid in-

Table 5.3 Interfacial Energies

System	Temperature (°C)	Interface Energy (ergs/cm^2)
$Al_2O_3(s)$–silicate glaze(l)	1000	<700
$Al_2O_3(s)$–Pb(l)	400	1440
$Al_2O_3(s)$–Ag(l)	1000	1770
$Al_2O_3(s)$–Fe(l)	1570	2300
SiO_2(glass)–sodium silicate(l)	1000	<25
SiO_2(glass)–Cu(l)	1120	1370
Ag(s)–Na_2SiO_3(l)	900	1040
Cu(s)–Na_2SiO_3(l)	900	1500
Cu(s)–Cu_2S(l)	1131	90
TiC(s)–Cu(l)	1200	1225
MgO(s)–Ag(l)	1300	850
MgO(s)–Fe(l)	1725	1600

terface energy (γ_{SL}) is high, the liquid forms a ball having a small solid–liquid interface area as shown in Fig. 5.8. With an incompressible solid a balance of forces determines the contact angle θ,

$$\gamma_{LV} \cos\theta = \gamma_{SV} - \gamma_{SL}$$

$$\cos\theta = \frac{\gamma_{SV} - \gamma_{SL}}{\gamma_{LV}} \tag{5.8}$$

where γ_{SV}, γ_{SL}, and γ_{LV} are the interface energies between the phases present in the system at the time of measurement, usually not pure surfaces. A value θ = 90° is often defined as the boundary between "nonwetting" and wetting behavior. *Spreading* occurs when the liquid completely covers the solid surface, that is, θ = 0°. The spreading tendency is sometimes defined by a spreading coefficient S:

$$S_{LS} = \gamma_{SV} - (\gamma_{LV} + \gamma_{SL}) \tag{5.9}$$

Spreading occurs when S is positive.

Just as the liquid droplet on a solid surface reaches an equilibrium configuration based on the interface energies, solid systems and solid–liquid systems reach an equilibrium defined by the surface energies. If a polycrystalline solid is immersed in a liquid or vapor phase, grooves are formed where grain boundaries intersect the surface. The angle of etching, or *dihedral angle*, is determined by ratio of grain boundary energy to surface-liquid or surface-vapor interface energy, as shown in Fig. 5.9. If we have a pore, solid inclusion, or liquid inclusion internal to a polycrystalline material, the configuration is also determined by the relative interface energies as shown in Fig. 5.10. The dihedral angle φ is given by the relationships:

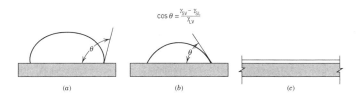

Fig. 5.8 Illustration of (a) nonwetting (θ>90°), (b) wetting (θ<90°), and (c) spreading (θ=0°) of a liquid on a solid.

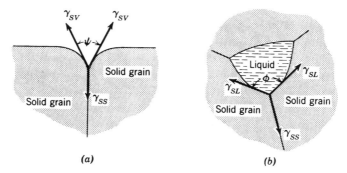

Fig. 5.9 Dihedral angle at equilibrium where a grain boundary meets a surface.

$$\gamma_{SS} = 2\gamma_{SV} \cos\frac{\phi}{2}$$

$$\gamma_{SS} = 2\gamma_{SL} \cos\frac{\phi}{2}$$

$$\gamma_{SS} = 2\gamma_{SS}' \cos\frac{\phi}{2}$$

This angle may not be the same everywhere, since interfacial energies of solids vary for the same reasons that surface energies are anisotropic. The grain boundary energy is a function of the grain misorientation, and solid–liquid and solid–vapor interface energies vary with crystal orientation. All are susceptible to change upon adsorption of dilute impurities, and the adsorption itself can be anisotropic. Ceramic surface energies tend to be more anisotropic than those in metals, and careful measurements of thermal grooves show that a rather wide distribution of dihedral angles exists, as shown in Fig. 5.11 for MgO, and Al_2O_3. The distribution for undoped Al_2O_3 is particularly broad due to surface faceting, but becomes considerably narrower upon adding a small amount of MgO, a result of importance for the sintering of alumina. Notice also that many of the

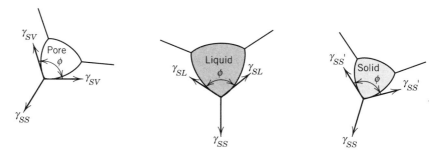

Fig. 5.10 Dihedral angles at inclusions within a polycrystal.

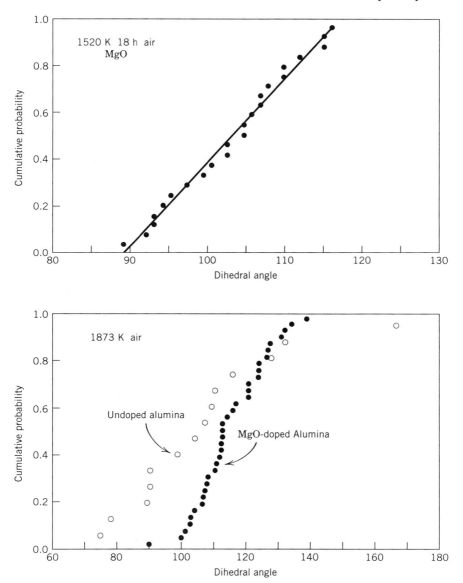

Fig. 5.11 Cumulative distributions of dihedral angles where grain boundaries meet the free surface in MgO and Al_2O_3. [From C.A. Handwerker, J.M. Dynys, R.M. Cannon, and R.L. Coble, *J. Am. Ceram. Soc.*, 73 [5] 1371 (1990).]

angles are less than 120°, corresponding to a surface energy that is *less* than the grain boundary energy ($\gamma_{sv} < \gamma_{gb}$). For MgO, the entire distribution lies below 120°. This feature is unique to ceramics and has not been observed in metals. While it may be intuitively surprising to find $\gamma_{sv} < \gamma_{gb}$ since the free surface has more broken bonds than the grain boundary and might well be expected to have

364 Chapter 5 / Microstructure

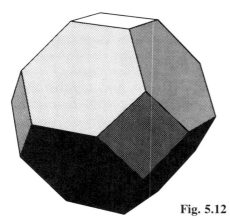

Fig. 5.12 Truncated octahedron (tetrakaidecahedron).

the higher energy, other effects such as bond distortion at grain boundaries or differences in impurity adsorption between surfaces and grain boundaries may explain this result.

For a polycrystalline single–phase material, if all grain boundary energies are the same, three grains meet along a line (the *three grain junction*) with a dihedral angle of 120°. In two dimensions an idealized polycrystalline network consists of honeycomb array of space-filling hexagons. In three dimensions there are no regular geometric polyhedra for which all angles between faces at three grain junctions

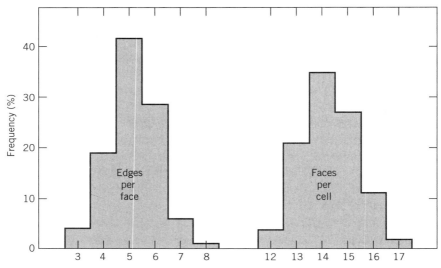

Fig. 5.13 Distributions of edges per face and faces per cell for random close-packing of nearly-identical sized deformable spheres, measured by J.L. Finney [*Proc. Roy. Soc. (London) A*, **319**, 479 (1970).]

5.1 Capillarity **365**

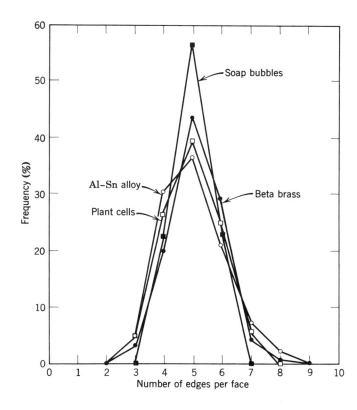

Fig. 5.14 Experimentally observed frequency distribution of the number of edges per face in various "polycrystalline" systems. [From C.S. Smith, "The Shape of Things," *Sci. Am.*, **190**, 58 (January, 1954).]

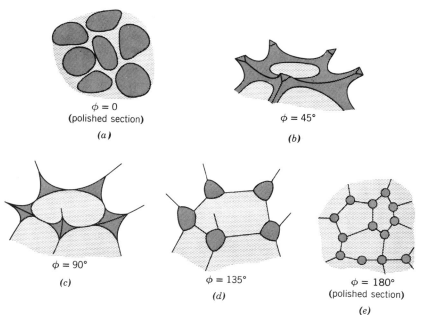

Fig. 5.15 Second-phase distribution for different values of the dihedral angle.

are 120° (in a plane taken normal to the junction line). The truncated octahedron (tetrakaidecahedron) (Fig. 5.12) comes close to this relationship. This shape has six square faces and eight hexagonal faces with 24 vertices, each having two angles of 120° and one of 90°. In an ordered packing array these can be distorted to give a lower total surface energy by having all of the edges meet at a four grain junction with an angle of 109° (the tetrahedral angle) and all the faces meet at an edge with 120° angle. However, real polycrystalline grains are neither packed in such orderly arrays nor are they all of identical size. In the next level of approximation,

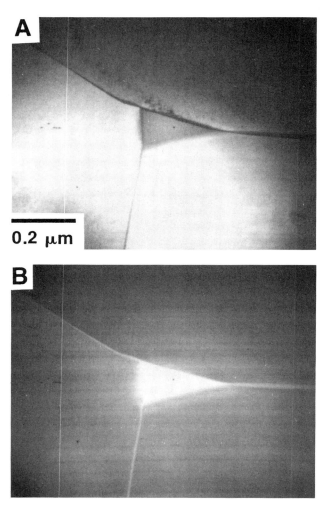

Fig. 5.16 Cross-section of a silicate secondary phase in strontium titanate which is continuous along a three-grain junction. (*a*) bright field transmission electron microscope image, (*b*) dark field image. [From C.J. Peng and Y.-M. Chiang, *J. Mater. Res.*, **5**[6], 1237-45 (1990).]

5.1 Capillarity   367

consider spheres of identical size that are randomly packed and compressed to remove interstitial space. This is the arrangement found in foamed polystyrene (Styrofoam™), for example. Distributions compiled by J. L. Finney [*Proc. Royal Soc. (London) A*, 319, 479 (1970)] from a model of nearly 8000 like-sized polyhedra in the "random-close-packed" arrangements (Fig. 5.13) showed that the number of faces per grain varies from 12 to 17 and is peaked near 14, the tetrakaidecahedral value, and the number of edges per face ranges from 3 to 8 and is peaked near 5. Pentagonal faces are also characteristic of soap bubbles, plant

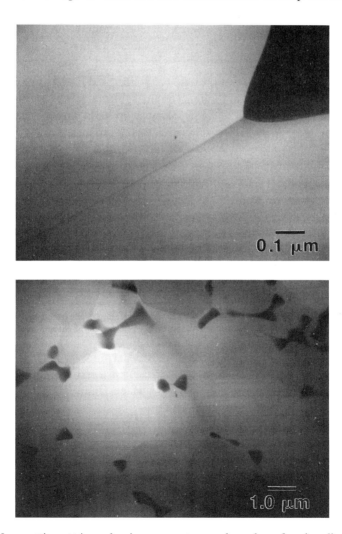

Fig. 5.17 Non-wetting yttrium-aluminum garnet secondary phase forming discrete particles at grain junctions in a high thermal conductivity aluminum nitride due to a high dihedral angle (approximately 120°). (From L. Weisenbach, J.A.S. Ikeda, Y-M Chiang, *Advances in Ceramics*, Vol. 26, The American Ceramic Society, Westerville, Ohio, 1989.)

cells, and polycrystalline materials (Fig. 5.14). The number of faces observed on realistic three-dimensional grain structures ranges from 9 to 18.

When a small amount of a second phase is present in a polycrystalline microstructure, its distribution amongst the primary grains is critically dependent on the dihedral angle, as illustrated in Fig. 5.15. When γ_{SS}/γ_{SL} is equal to or greater than 2, ϕ is 0, and at equilibrium the faces of all grains are penetrated by the second phase. This is often the case when a small amount of liquid phase is present to facilitate reactive-liquid sintering. When the ratio γ_{SS}/γ_{SL} is between 2 and $\sqrt{3}$, ϕ is in the range between 0° and 60° and the liquid forms a continuous skeleton along the grain edges forming approximately triangular prisms at the intersections of three grains. The cross section of a silicate secondary phase with this kind of configuration in strontium titanate is shown in Fig. 5.16. When the ratio γ_{SS}/γ_{SL} is between $\sqrt{3}$ and 1, ϕ is between 60° and 120° and the second phase partially penetrates along the three grain junction lines, but no longer forms a continuous network, illustrated in Fig. 5.17 for yttrium aluminum garnet phase in polycrystalline aluminum nitride. At still higher ratios the second phase remains discrete and tends to spheroidize. Pores as well as crystalline secondary phases are often distributed in this way, as shown in Fig. 5.27c for an incompletely densified alumina and Fig. 5.31 for a zirconia-alumina two phase composite. Differences in the distribution of minor phases can have an enormous effect on processing and on properties.

SPECIAL TOPIC 5.1

RAYLEIGH INSTABILITY AND MICROSTRUCTURE

The Rayleigh instability can be observed in streams of water flowing from faucets: The water leaves the tap as a continuous stream, but then breaks up into individual droplets as the stream narrows (Fig. ST54). This is a spontaneous process driven by the surface tension of the water, and is due to the fact that a sphere has a lower surface area than a cylinder of the same volume. Since mass must be redistributed during the break-up, the fluid stream does not progress directly from a long cylinder to a single sphere, but instead breaks up into multiple spheres, limited in number by the consideration that there exists a maximum number of spheres with surface area less than a cylinder of equal volume. Analysing in detail, for infinitesimal sinusoidal fluctuations in the cross-sectional area of a cylinder of radius R, the surface area decreases only for fluctuations of a wavelength greater than the circumference of the cylinder: $\gamma_{min} = 2\pi R$. Lord Rayleigh [*Proc. London Math. Soc.*, 10, 4-13 (1879)] further showed that the fastest-growing wavelength which comes to dominate the periodicity of the liquid break-up is larger than λ_{min}, being given by $\lambda = 9.02R$. This is

5.1 Capillarity 369

Fig. ST54 A thin stream of water from a faucet undergoes Rayleigh break-up into discrete droplets in order to lower surface energy.

the fluctuation which is ultimately amplified and becomes the characteristic wavelength of the stream, resulting in spherical droplets of identical size, Fig. ST55.

This kind of surface energy instability can also occur for a solid cylinder, or a cylindrical pore channel [analysed by F. A. Nichols and W. W. Mullins, *J. Appl. Phys.* 36[6] 1826–35, (1965), and *Trans. AIME*, 233[10] 1840–48 (1965)]. In other geometric configurations such as a thin film which breaks up into islands or continuous phases in a phase-separated glass which break up into droplets, the basic principle causing breakup is the same although the details differ. Additional complexities are introduced if the surface en-

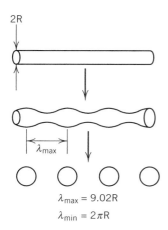

Fig. ST55 Rayleigh instability is characterized by the selective amplification of fluctuations in cross-section until break-up occurs. For a cylinder with isotropic surface energy the minimum wavelength at which the fluctuation is unstable is $\lambda_{min} = 2\pi R$, while the wavelength which grows fastest is $\lambda_{max} = 9.02R$.

370 Chapter 5 / Microstructure

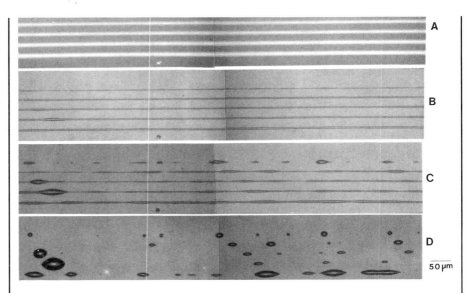

Fig. ST56 Model experiment in which pore channels have been lithographically patterned between sheets of Corning 7056 glass and subsequently annealed at 650°C, showing the fluctuation and eventual breakup of the pore channels into spheres. (A) as prepared; (B) 3 h.; (C) 6 h.; and (D) 13 h. [From H.D. Ackler, M.S. Thesis, Univ. California at Berkley, 1992. (Lawrence Berkeley Laboratory Report LBL-33488, UC-404).]

ergy is anisotropic, so that the minimum energy shape is not a sphere. Figure ST56 shows the results from a model experiment in which grooves have been lithographically patterned [after J. Roedel and A. M. Glaeser, *J. Am. Ceram. Soc.*, 70, C172 (1987)] between two plates of glass, and subsequently annealed at high temperature. The straight grooves eventually develop perturbations and spheroidize.

In polycrystalline microstructures, cylinder-like phases often occur at the junction between three grains. If the dihedral angle between the second phase and the grain boundary is less than 90° (see Fig. 5.15), a continuous three-grain junction phase is possible. The second phase may be a liquid which is aggregating during a liquid phase sintering process, a solid second phase, or porosity. A second phase channel may undergo break-up into discrete particles, just like a liquid stream. Further analysis has shown that the stability of an intergranular second-phase channel depends strongly on the dihedral angle between the second phase and the grain boundaries [W. C. Carter and A. M. Glaeser, *Acta Metall.*, 35[1] 237–45 (1987)]. At large dihedral angles, the three-grain-junction phase is close to a true cylinder in shape and undergoes Rayleigh instability to form droplets as discussed above. However, at smaller dihedral angles, the second-phase channel is not cylindrical, having a triangular cross section with sides that go from convex outward to concave outward as the dihedral angle decreases. It is found that as the sides become increasingly concave, the second-phase channel becomes stabilized and tends to remain as a completely interconnected feature of the microstructure.

The physical distribution of secondary phases has important consequences for the properties of multiphase composites. A continuous three-dimensionally interconnected second phase can be a short-circuit transport path for electrical conduction, fluid flow, or gas permeability through the polycrystal. Certain processes tend to generate microstructures with high or low dihedral angle phases. In solid-state sintering, a smaller dihedral angle increases the capillary pressure at particle contacts early in the process, and helps to remove large multisided pores later in sintering. A high dihedral angle allows Rayleigh break-up of the pore channels, and as a result solid-state sintered microstructures tend to develop closed porosity. (*Containerless* hot-isostatic pressing requires closed internal porosity so that applied gas pressure can exerted on the body as a whole.) On the other hand, liquid phase sintering requires a low or zero dihedral angle between the liquid and the grain boundary. Liquid phase sintered materials tend to form microstructures in which the second phase is completely interconnected.

5.2 GRAIN GROWTH AND COARSENING

When a polycrystalline aggregate is heated, the grain size may increase and the grain size distribution may change as a result of several possible processes. In metals that have been plastically deformed, a new generation of strainfree grains occurs by a process of nucleation and growth called recrystallization. This process is also seen with some soft ceramics such as the alkali halides but is not of much general importance for ceramics. The primary driving force for growth of strain-free grains is reduction in the interfacial energy per unit volume. In *normal* grain growth the distribution of grain sizes is relatively unchanged and shifts to larger size (Fig. 5.18). *Discontinuous* grain growth (or *abnormal* growth) occurs when a few large grains grow faster than the surrounding fine grained matrix. *Coarsening*, or Ostwald ripening, occurs when smaller particles go into solution or vaporize and larger particles grow. For all these processes, the basic thermodynamic and kinetic factors are well understood. Impurity effects and crystalline anisotropy can influence secondary variables such as interface energies, diffusion coefficients and boundary mobilities giving rise to a good deal of complexity and making specific predictions difficult.

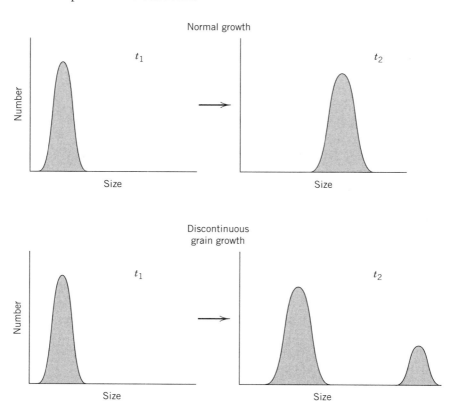

Fig. 5.18 Beginning with a single-modal distribution of grain sizes, normal grain growth is characterized by an increase in average size while retaining similar size distribution, while discontinuous growth results in a broadened or bimodal distribution with some grains growing much faster than the average.

Grain Boundary Migration and Grain Growth

Grain growth is driven by the reduction in boundary energy as the total grain boundary area of a polycrystal is decreased

$$\Delta G = \int_{A_1}^{A_2} \gamma \, dA = \gamma(A_2 - A_1)$$

A demonstration of the excess energy of grain boundaries is seen in the thermal signature during coarsening of a very fine-grained titanium dioxide (Fig. 5.19). When a polycrystal of about 20 nm starting grain size is heated in a differential scanning calorimeter, and grows to a final grain size of about 250 nm, an exotherm is seen which corresponds to the release of the excess grain boundary enthalpy. (That is, the excess free energy of grain boundaries is given by $G_{gb} = H_{gb} - TS_{gb}$;

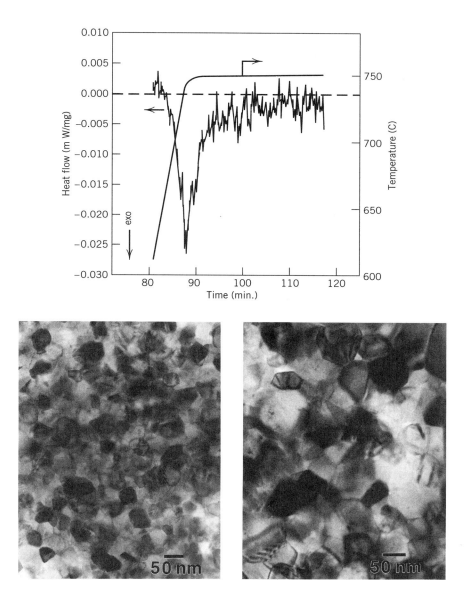

Fig. 5.19 Calorimetric measurement of the enthalpy released during grain growth. When a dense TiO_2 with initial average grain size of 20 nm is heated in a differential scanning calorimeter, an exotherm is observed as the grains coarsen to a final grain size of about 250 nm. (From C.D. Terwilliger, Ph. D. Thesis, MIT, 1993.)

and the exotherm in Fig. 5.19 corresponds to the release of H_{gb} only. There is also a release of grain boundary entropy, which is seen as a change in heat capacity with grain size, since $\Delta S_{gb} = \int_{T_1}^{T_2} C_p / T \, dT$.)

Since the average grain size increases during grain growth, the total number of grains must decrease in order to conserve volume. An equivalent way of looking at grain growth is to evaluate the rate of grain disappearance. The change in chemical potential of atoms across a curved grain boundary has been given in Eq. 5.5. This difference is the driving force that makes the boundary move toward its center of curvature. As discussed previously, grain boundaries which are equal in energy meet at three-grain junctions to form angles of 120°. Using the two-dimensional example in Fig. 5.20 for illustrative purposes, we see that if all boundaries are required to meet with an angle of 120°, grain boundaries without curvature only occur for six-sided grains. Grains with fewer sides have boundaries that are concave when observed from the center of the grain. These are the grains that shrink and eventually disappear as grain boundaries migrate toward their center of curvature. Grains with more than six sides have convex boundaries that migrate outward and tend to grow larger. In three dimensions, the net curvature ($1/r_1 + 1/r_2$) determines the direction of migration. By observing a soap froth it can be seen that grains with fewer sides shrink while those with more sides grow.

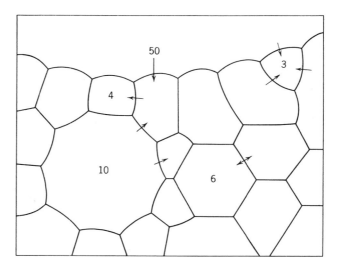

Fig. 5.20 Schematic drawing of two-dimensional polycrystalline specimen. If all boundaries meet at 120°, the sign of curvature of the boundaries changes as the number of sides increases from less than six to more than six. The radius of curvature becomes less, the more the number of sides deviates from six. Arrows indicate the directions in which boundaries migrate. (From J.E. Burke.)

5.2 Grain Growth and Coarsening

The rate of boundary motion (velocity) depends on the applied driving force from curvature and the grain boundary mobility, M_b:

$$V = M_b F \qquad (5.10)$$

where g is the grain diameter ($g = 2r$). As with other mobilities discussed in Chapter 3, the grain boundary mobility is a velocity per unit driving force. If we consider the driving force to be the interfacial energy per unit volume (J/m³), M_b has units of (m/sec)/(J/m³) = m⁴/J sec. In normal grain growth, for which the grain size distribution remains narrow and the average size increases (Fig. 5.18), the mobility is relatively constant from boundary to boundary. To determine the rate of grain growth we first evaluate the driving force. Taking into account that each boundary is shared by two grains, in a unit volume of polycrystal in which the average grain diameter is g, the driving force available from interfacial energy is

$$F = \gamma S_v \approx \gamma (3/g) \qquad (5.10a)$$

where S_v is the boundary area per unit volume, and we have made the simplifying assumption of spherical grains. However, grain growth is not driven by the excess energy alone but rather by the grain boundary curvature (i.e., a two-dimensional honeycomb grain structure has only straight grain boundaries, and therefore does not undergo grain growth even though there is an excess of boundary energy.) In a dense polycrystal the average boundary curvature is about a factor of 6 less than it is for a equivalent sized sphere (M. Hillert, *Acta Met.*, 13, 227, 1965). The rate of growth of the grain of average size is

$$dg/dt = 2V = 2M_b F = M_b \gamma/g \qquad (5.11)$$

Upon integrating Eq. 5.11 we have

$$g^2 - g_0^2 = 2M_b \gamma t \qquad (5.12)$$

where g_0 is the grain size at zero time. Grain growth that obeys this relationship is termed parabolic growth, so-named for the dependence of g on the square root of time. Not all instances of normal grain growth exhibit parabolic time kinetics. Experimentally, it is found that when log g is plotted versus log t, a straight line is obtained but frequently the slope is smaller than 1/2, often being about 1/3. This occurs for a number of reasons: g_0 is often not a large amount smaller than g; or the mobility is not constant with grain size. The latter can be due to the coalescence of inclusions and pores, increase in solute segregation, or thickening of a liquid film at grain boundaries as the grain size grows. In thin films and fibers, grain growth rates also tend to decrease as the grain size approaches the sample dimensions.

Grain boundaries move as a consequence of many individual atomic jumps across the plane of the boundary (Fig. 5.21). While this process is always necessary, it is not always the slowest or rate-limiting step. When it is, the mobility is termed *intrinsic*, and is given by the mobility of atoms moving across the boundary (D_b/kT, see Chapter 3) divided by the number of atoms involved per unit area of grain boundary (δ/Ω)

$$M_b = \frac{D_b \Omega}{\delta kT} \tag{5.13}$$

where Ω is the atomic volume and δ is the boundary thickness (~0.5 nm). For an ionic solid in which both cations and anions must diffuse, the limiting diffusivity D_b is the slower of the two. However, rarely in ceramics is the mobility as high as Eq. 5.13 would indicate, since segregated solutes, pores and inclusions, and second phase films can also exert a drag force on the boundary (Fig. 5.22). The mobility of a grain boundary with a liquid film, when diffusion through the liquid is rate-limiting, is given by

$$M_b \text{ (liquid film)} = \frac{D_l C_l \Omega}{\omega kT} \tag{5.14}$$

in which ω is the thickness of the film, C_l the solubility of the grain in the liquid, and D_l is the rate-limiting diffusivity in the liquid. Since boundary and liquid diffusivities can be comparable in magnitude, for very thin liquid films this mobility may approach the intrinsic limit, but it is typically lower. The slow step in atomic transport across a boundary film may also be the rate of atom detachment

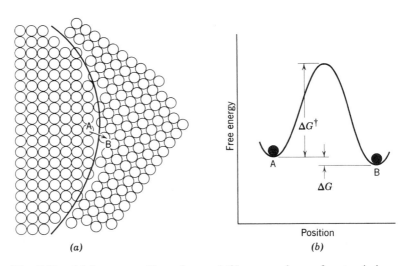

Fig. 5.21 (*a*) Structure of boundary and (*b*) energy change for atomic jump.

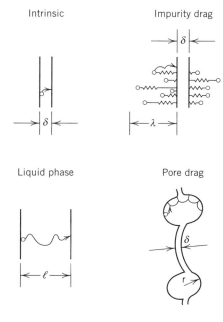

Fig. 5.22 Schematic representation of grain boundary motion limited by intrinsic drag, impurity drag, liquid phase diffusion, and pore drag mechanisms. (From M.F. Yan, R.M. Cannon, and H.K. Bower, pp. 276–307 in *Ceramic Microstrucures '76*, R.M. Fulrath and J.A. Pask, eds., Westview Press, Boulder, CO, 1977.)

or adsorption at the liquid-crystal interface rather than diffusion. If so, the mobility is of form

$$M_b \text{ (liquid film)} = \frac{K\Omega}{kT} \tag{5.15}$$

where K is an interface reaction rate constant.

Even ceramics free of liquid phase films are rarely pure enough for intrinsic grain boundary migration to take place. Some form of impurity segregation nearly always occurs. MgO in Al_2O_3, $CaCl_2$ in KCl, ThO_2 in Y_2O_3, and Nb in $BaTiO_3$ are all effective grain growth inhibitors in amounts well below the solubility limits. *Solute drag* occurs when a strong interaction between segregated impurities and the grain boundary requires the solute distribution to be carried along with the moving boundary. Solutes may segregrate right at the disordered "core" of the grain boundary or in an adjacent space-charge cloud as discussed in Chapter 2. Solute diffusion in the near boundary region is usually slower than the intrinsic diffusion of host atoms across the boundary plane, and becomes rate-limiting. Even solutes that are depleted from a grain boundary exert drag since the depletion zone must be carried along; for an equivalent amount of depletion, the drag is the same as it is for segregation. From the solute drag theory of J. W. Cahn [*Acta Met.*, 10, 789 (1962)], in the limit where the intrinsic drag of the boundary is negligible compared to the solute drag, the mobility is given by the following integral over distance from the boundary, x,

$$M_b \text{ (solute drag)} = \left[C_s\, 4 N_v kT \int_{-\infty}^{+\infty} \frac{\sinh^2\left(\dfrac{U(x)}{2kT}\right)}{D(x)} dx \right]^{-1} \quad (5.16)$$

where C_s is the bulk concentration, N_v is the density of atoms per unit volume, $D(x)$ is the spatially-varying solute diffusivity, and $U(x)$ is the spatially varying interaction energy between the solute and the boundary plane (at $x = 0$). With the simplifying assumptions that segregation occurs uniformly over a width 2δ in which the solute diffusivity is D_s and its concentration is given by $C_s \exp(U/RT)$, an approximate expression is

$$M_b \text{ (solute drag)} \approx \frac{D_s \Omega}{2 \delta kT \left(C_s e^{U/RT} \right)} \quad (5.17)$$

We see that the grain boundary mobility is directly proportional to the rate-limiting diffusivity and inversely proportional to the segregated concentration of solute.

Over some range of driving force (e.g., grain size) the solute drag mobility is a constant —the velocity increases in direct proportion to the driving force. However, the amount of segregation carried by the boundary is not constant with velocity. Cahn and also Lücke and Stüwe (pp. 171–210 in *Recovery and Recrystallization of Metals*, L. Himmel, Editor, Gordon and Breach, New York, 1962) showed that as the boundary increases in velocity it continually sheds solute (Fig. 5.23). For a sufficiently high driving force, it is able to break away from the solute cloud altogether and migrate at the intrinsic velocity. The theoretical relationship between velocity and driving force is shown in Fig. 5.24. The slope is the inverse mobility, which has both intrinsic and solute drag components

$$M_b = \frac{V}{F} = \left[\frac{1}{M_b \text{ (intrinsic)}} + \frac{1}{M_b \text{ (solute drag)}} \right]^{-1} \quad (5.18)$$

At low velocity and low driving force the boundary mobility is solute-drag limited and the mobility is given by Eqs. 5.16 or 5.17. With increasing velocity there is a sudden transition ("breakaway") to the intrinsic mobility. The dotted line of negative slope in between these two branches is a range of driving force over the intermediate mobility that is unstable; the boundary either slows down to the solute-limited mobility or increases to the intrinsic limit. Direct observations [A. M. Glaeser, H. K. Bowen, R. M. Cannon, *J. Am. Ceram. Soc.*, 69[2], 119–25 (1986)] have shown that grain boundary motion is often not uniform, but starts and stops

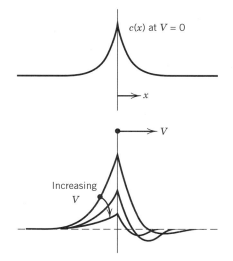

Fig. 5.23 Decreasing solute segregation and the development of an asymmetric concentration profile as the velocity of a moving boundary increases. (After J.W. Cahn, *Acta Met.*, **10**, 789, 1962.)

fitfully as if making transitions between solute-drag and intrinsic regimes or between different solute-drag regimes.

Pores and second-phase inclusions that intersect the boundary also present a barrier to boundary migration. C. Zener (reported in C. S. Smith, Trans. AIME, 1948) considered the situation where the second phase inclusions are of a constant size and are immobile. As a grain boundary which intersects an inclusion

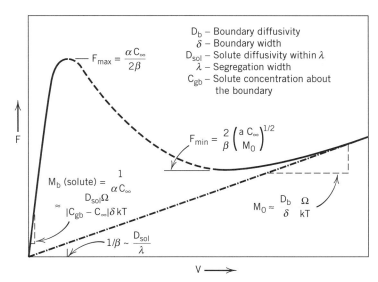

Fig. 5.24 Force-velocity relationship for boundary migration both in the presence of solute drag and in the intrinsic regime. The slope represents the reciprocal grain boundary mobility, M_b^{-1}. (From M.F. Yan, R.M. Cannon, and H.K. Bower, pp. 276–307 in *Ceramic Microstrucures '76*, R.M. Fulrath and J.A. Pask, eds., Westview Press, Boulder, CO, 1977.)

migrates toward its center of curvature (Fig. 5.25) under a driving force $F_d = \gamma/R$ where R^{-1} is the net grain curvature, the boundary area is extended. The inclusion exerts a restraining force

$$F_r = 2\pi\gamma r \sin\phi \cos\phi \qquad (5.19)$$

where r is the inclusion radius and ϕ is the angle depicted in Fig. 5.25. Since the function $\sin\phi \cos\phi$ has a maximum value of 1/2 at $\phi = 45$, the maximum restraining force per inclusion is $\pi\gamma r$. For a boundary with n_s inclusions per unit area, the total restraining force per unit area is $F_T = n_s \pi\gamma r$. If the driving force exceeds the drag force, $F_d > F_T$, then the inclusions are left behind. This may be the situation in the early stages of grain growth. However, as grain growth progresses the average R increases; thus F_d decreases while F_T remains constant (in this idealized case where the inclusions neither coarsen nor move). At a limiting grain size R_1 we reach an equilibrium condition where the drag force is exactly opposed by the driving force, and grain growth must stop altogether. From $F_d = F_T$, this condition is

$$n_s \pi r R_1 = 1$$

Since the area density and volume density of precipitates are related by $n_s \approx n_v r$, where n_v is the number of inclusions per unit volume, and the volume fraction of inclusions is given by $V_f = n_v (4/3)\pi r^3$, we have the following relationship for the limiting grain size R_1:

$$\frac{R_1}{r} = \frac{4}{3V_f} \qquad (5.20)$$

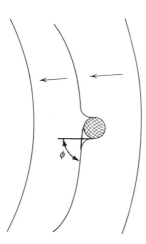

Fig. 5.25 Changing configuration of a boundary while passing an inclusion.

5.2 Grain Growth and Coarsening

For example, if there is 1% by volume of inclusions, the grains can only grow to approximately 100 times the inclusion size; for 10% by volume of inclusions, the grains can grow to 10 times the inclusion size. This simple and approximate relationship shows that the effectiveness of inclusions in limiting grain growth increases as the particle size is decreased and the volume fraction increased.

In polycrystalline ceramics the most common inclusions are pores, which are generally neither immobile, spherical, constant in size, nor of constant volume fraction (due to densification during firing). Grain growth in porous ceramics is complex, but a few limiting cases can be discussed. The pore space in a densifying powder compact is initially located at the grain boundaries, is interconnected, and is a large fraction of the grain diameter in size. As densification proceeds, the pores shrink, and first become completely isolated when the pore volume fraction is a few percent. From the Zener type of analysis (Eq. 5.20) we expect that the grains cannot grow to more than a few times their initial size during this stage. Grains remain pinned by porosity and can only grow at a rate determined by the rate of decrease in pore volume fraction and average pore size. It is experimentally observed that the rate of grain growth in porous compacts increases as the material densifies, and frequently does not begin in earnest until only a few percent porosity is left.

It is an oversimplification to consider pores to be immobile as assumed in the Zener analysis. If there is material transport across the pore by surface diffusion, vapor transport, or lattice diffusion, the pores may have a significant mobility. Solid or liquid inclusions may also move by atom diffusion along the interface or through the body of the inclusion. The pore mobility is given by an equation of form

$$M_p = \frac{A \exp\left(\frac{-Q}{RT}\right)}{r^n} \tag{5.21}$$

where A is a constant, Q is the activation energy for the rate-limiting process, and the exponent n is 3 for lattice diffusion, 4 for surface diffusion, and 2 or 3 for vapor transport. Thus smaller pores in a given material always have a higher mobility, regardless of the migration mechanism.

The effect of mobile pores on grain boundary mobility has been discussed by R. J. Brook [*J. Am. Ceram. Soc.*, 52, 59 (1969)]. Pores that remain attached to the grain boundary each exert a drag force F_p, resulting in a net drag force from N pores per unit area of NF_p. This drag force reduces the net driving force on the boundary accordingly:

$$V_b = M_b(F - NF_p) \tag{5.22}$$

where M_b is the boundary mobility in the absence of pores and F is the driving force due to boundary curvature alone. Since "attachment" by definition means the boundary and pore velocities are equal, $V_b = V_p = M_p F_p$. After rearranging Eq. 5.22 the net mobility of the boundary with attached pores is given by

$$M_{net} = \frac{V_b}{F} = \frac{M_b M_p}{(NM_b + M_p)} \tag{5.23}$$

In the limit where the boundary mobility is high relative to pore mobility or the pore density is low, that is, $NM_b \gg M_p$, we have a mobility of $M_{net} = M_p/N$. Boundary migration is entirely pore-limited and the mobility decreases as the areal density of pores N increases. When there are few pores or they are of high mobility, i.e., $NM_b \ll M_p$, there is negligible pore drag and the boundary migrates as if the pores were absent, $M_{net} = M_b$. Otherwise, the grain boundary mobility lies between these limits as given by Eq. 5.23.

These results pertain only to pores that remain attached to the grain boundaries. Since the maximum drag force a pore can exert on a grain boundary is $\pi \gamma r$, the maximum pore velocity is $V_p(\max) = M_p F_p(\max) = M_p \pi \gamma r$. When the boundary moves at a velocity greater than this, the pore separates from the boundary. From the inverse dependence of M_p on r^n (Eq. 5.21), we can see that larger pores are less mobile; however, they also exert more boundary drag (Eq. 5.19). Whether or not pore-boundary separation occurs depends on the driving force for boundary migration, that is, the grain size, as well as the pore size and pore mobility. Figure 5.26 is a type of map due to R.J. Brook (here using MgO as an example) that has as its axes grain size and pore size, and on which three fields representing limiting cases for boundary migration in the presence of pores are shown. There is region bounded by the x-axis where the grain size is fine (driving force high) and the pore size is large. Here pores exert enough boundary drag to remain attached, and the boundary mobility is pore-drag limited. A second field bounded by the y-axis is where pores are small enough, and therefore mobile enough, to remain attached to the boundaries. Here the pore mobility is high enough that the boundary migrates as if pore-free. These are the two limiting cases discussed above, in both of which the pores remain attached. The third region between these two is where pore-boundary separation occurs, and in which the boundary again migrates unimpeded by pores. The pore volume fraction and impurity content have a large influence on this behavior. The results in Fig. 5.26 are specific to one composition of Fe-doped MgO in which boundary and pore mobilities are relatively well known, and other materials will have in detail different maps. In practice a quantitatively accurate map can only be constructed for a few well-studied materials, but nonetheless these general features are useful in understanding grain growth in porous materials, and especially in understanding simultaneous densification and grain growth during sintering.

For several reasons, the sizes of pores and inclusions tend not to remain static during grain growth, and this can lead to transitions between different limiting

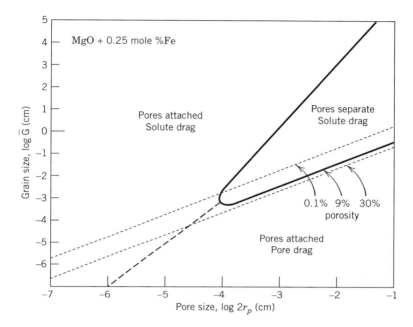

Fig. 5.26 Grain size – pore size map showing regions where pores remain attached to boundaries and where pore-boundary separation occurs. When the pores are attached, the boundary mobility may be limited by solute drag or pore drag. When pores are separated from boundaries, solute drag is limiting. (Adapted from C.A. Handwerker, R.M. Cannon, and R.L. Coble, pp. 619–643 in *Advances in Ceramics*, Vol 10, W.D. Kingery, ed., The American Ceramic Society, Columbus, Ohio, 1984.)

mechanism and effective grain boundary mobilities. Inclusions that remain attached to grain boundaries tend to become concentrated at boundary intersections and agglomerate into larger particles as grain growth proceeds, as illustrated in Fig. 5.27. Coarsening of pores and inclusions by Ostwald ripening can also occur during firing, and competes with pore shrinkage by densification processes (Section 5.3). Attention to whether or not pores separate from grain boundaries as grain growth proceeds is critical to attaining high-fired densities, and is discussed in Section 5.3 and the related article on sintering of magnesia-doped alumina.

Discontinuous grain growth occurs when some small fraction of the grains grow much faster than the average (Fig. 5.28). A large grain growing at the expense of a uniform, fine-grained matrix which is static often will exhibit a growth rate proportional to $1/d_m$, where d_m is the matrix grain size. It was previously believed that if the boundary mobility is the same everywhere, discontinuous grains would grow at a faster *relative* rate than the matrix grains, and a broadening of the

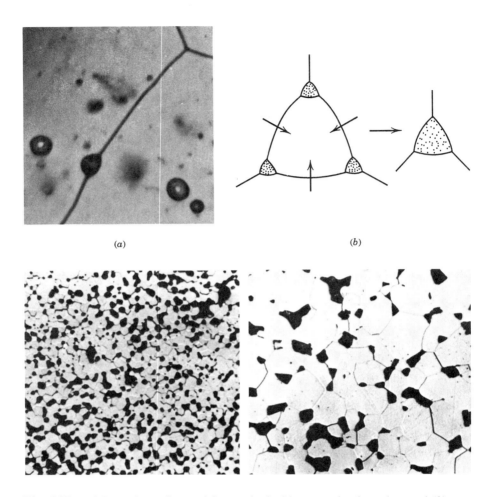

Fig. 5.27 (*a*) Pore shape distorted from spherical by a moving boundary and (*b*) pore agglomeration during grain growth. (*c*) Simultaneous grain growth and pore growth in sample of UO$_2$ after 2 min., 91.5% dense, and 5 h., 91.9% dense, at 1600°C. (From Francois and Kingery.)

size distribution would occur. More recent analyses [C. V. Thompson, H. J. Frost and F. Spaepen, *Acta Metall.*, 35[4] 887 (1987), and D. J. Srolovitz, G. S. Grest, and M. P. Anderson, *Acta Metall.*, 33, 2233 (1985)] indicate that the relative growth rates of large and small grains are not sufficiently different to cause a broadening of the size distribution if the mobility of all boundaries is the same. One concludes that other causes of local variations in growth rate such as nonuniformities in impurity content, liquid phases, or porosity must be invoked to explain discontinuous growth. Inhomogeneous packing in the starting powder compact can lead to local variations in densification rate, with denser regions subsequently exhibit-

5.2 Grain Growth and Coarsening **385**

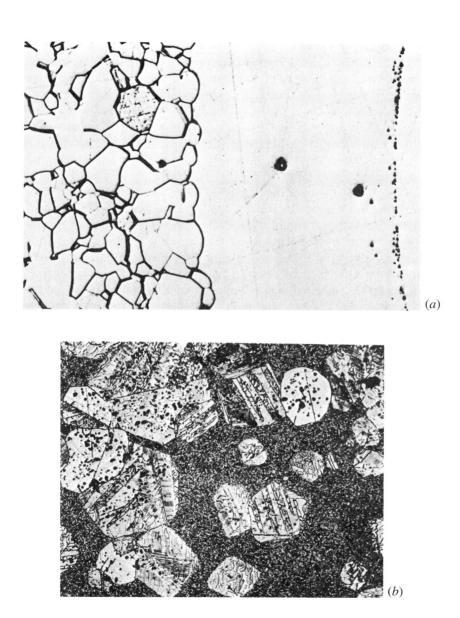

Fig. 5.28 Discontinuous grain growth. (*a*) A large Al_2O_3 crystal into a matrix of uniformly sized grains (495×) (courtesy R.L. Coble) and (*b*) Large grains of barium titanate growing in a fine-grained matrix (250×) (Courtesy R.C. DeVries).

ing more rapid grain growth (i.e., less pore-pinning). Inhomogeneities such as these formed earlier in processing are generally irreversible. Anisotropies in grain boundary energy and mobility can also cause abnormal growth. Boundaries of large discontinuously growing grains often show preferential faceting on low index crystallographic orientations, as illustrated in Fig. 5.29. Close examination frequently shows that the growth planes are those of low surface energy, and that the transfer of material from the matrix grains occurs through a thin layer of intergranular material, often liquid, in which the diffusion coefficient is relatively high. Alternatively, low energy boundaries have a lower mobility than other types, even when wet with a thin liquid film (see related sidebar on magnesia-doped alumina). Growth that is slow normal to and rapid parallel to a low energy surface leads to platelike grains (Figs. ST59, 5.33). Occasionally it is seen that large discontinuously growing grains have impinged after consuming a fine-grained matrix, resulting in a narrow size distribution of coarse grains. It is also possible to stimulate this result by intentionally seeding a fine-grained starting material with uniform dispersion of coarse grains, as has been done for $BaTiO_3$ and Al_2O_3.

A useful polycrystalline microstructure may consist of a fine-grained dense material of narrow size distribution and equiaxed grain shape, such as the translucent magnesia-doped alumina lamp tube material in Fig. ST58, the sintered MnZn ferrite in Fig. 5.30 (where grain size controls the magnetic domain configuration and eddy current losses), and the transformation-toughened zirconia-alumina composite in Fig. 5.31. In other applications the desired microstructure may consist of a random array of highly anisotropic grains, shown in Fig. 5.32 for a sintered Si_3N_4 where the interlocking platelet grains give a high fracture toughness. As another example, the ceramic superconductor shown in Fig. 5.33

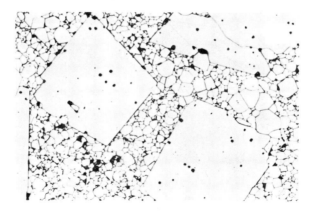

Fig. 5.29 Large faceted grains in a polycrystalline spinel. The flat faces of the large grains suggest a crystallographic orientation of low energy, possibly with a wetting liquid (350x). Courtesy R.L. Coble.

5.2 Grain Growth and Coarsening 387

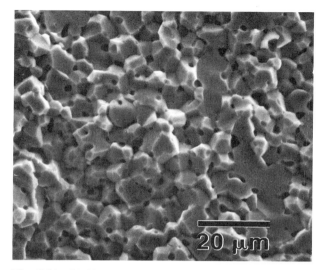

Fig. 5.30 Uniform equiaxed grains in a manganese-zinc ferrite (fracture surface).

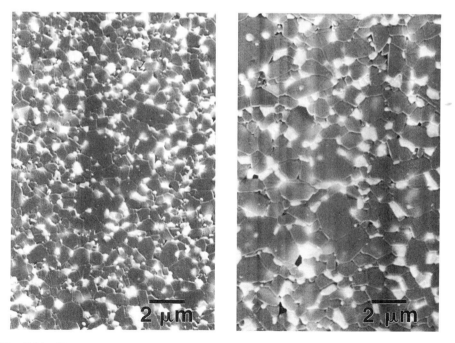

Fig. 5.31 Two-phase microstructure of an Al_2O_3-15 vol% tetragonal ZrO_2 microstructure in a transformation toughened, high strength composite. (Courtesy of N. Claussen)

388 Chapter 5 / Microstructure

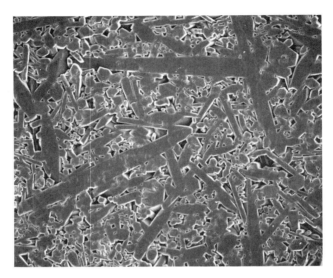

Fig. 5.32 Elongated grain microstructure in a high fracture toughness gas-pressure-sintered silicon nitride, composition 90% Si_3N_4, 10% $Y_3Al_5O_{12}$. (Courtesy I.-W. Chen and D. Jacobs, University of Michigan.)

has undergone a grain nucleation and growth process resulting in highly aligned platelet grains, which are able to carry a high superconducting current. Aligned microstructures are also important for ceramic thin films. In many instances the ability to control the morphology and rate of grain growth is the key to obtaining useful properties.

Particle Coarsening (Ostwald Ripening)

Coarsening is a process that, like grain growth, occurs due to the difference in free energy between curved surfaces. In a system of dispersed particles having a range of sizes in a medium, which may be solid, liquid, or vapor, if there is appreciable solubility or vapor pressure the smaller particles dissolve and the larger ones grow. The Thompson–Freundlich equation discussed previously gives the increased solubility of a small particle,

$$c_a = c_o \exp\left[\frac{\gamma M}{\rho RT} \frac{2}{a}\right] \tag{5.24}$$

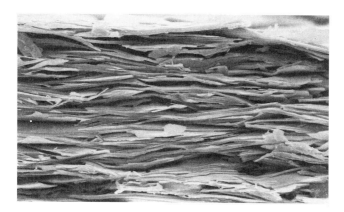

Fig. 5.33 Plate-shaped grains with a common crystallographic orientation (texture) in a high current density oxide superconductor of the Bi-Sr-Ca-Cu-O family. (Courtesy of M.W. Rupich and G.N. Riley, American Superconductor Corporation.)

in which c_o is the equilibrium solubility (planar interface) and a is the particle radius. Since the argument of the exponential term is usually small, $2M\gamma/a\rho RT \ll 1$, then $c_a \approx c_o(1 + 2M\gamma/a\rho RT)$. If particles dissolve and grow readily without being limited by the rate of interfacial reactions, the growth rate of particles is likely to be limited by diffusion through the surrounding medium. As discussed by G. W. Greenwood [*Acta Met.*, 4, 243 (1956)], a particle of radius a_1 surrounded by a diffusion field of radius r (Fig. 5.34) grows at a rate given by

$$4\pi a^2 \frac{da}{dt} = D(4\pi r^2)\frac{dc}{dr} \qquad (5.25)$$

where D is the diffusivity in the medium and dc/dr is the gradient in concentration at distance r. For well-dispersed particles, at a distance $r \gg a$ the solubility is that of the average particle size, $\bar{a}$, while the solubility at the particle surface ($r = a$) is given by Eq. 5.24. Upon integrating the right-hand side of Eq. 5.24 and evaluating at $r = a$, the growth rate of any particle is given by

$$\frac{da}{dt} = -\frac{D}{a}[c(a)-c(\bar{a})] = -\frac{D}{a}\left(\frac{c_o \, 2\gamma M}{\rho RT}\right)\left(\frac{1}{a}-\frac{1}{\bar{a}}\right) \qquad (5.26)$$

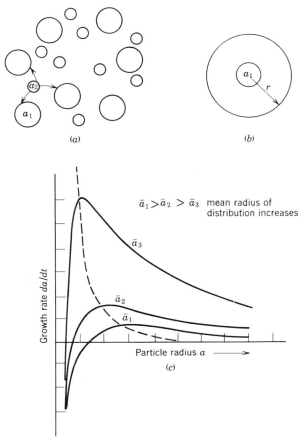

Fig. 5.34 (a) Coarsening of particles in a two-size particle system; (b) growth of particles a_1 in a diffusion field of radius r; (c) variation in the particle growth rate with particle radius.

There are several notable features to the growth rate predicted by Eq. 5.26, drawn in Fig. 5.34c as curves of growth rate versus particle radius. Each of the three solid curves represents a material of a certain average particle size, $\bar{a}$. First, where the growth rate curves cross zero a particle has exactly the average size $\bar{a}$. In this instance no concentration gradient exists. Particles smaller than $\bar{a}$ have a negative growth rate; they shrink and disappear, and the finer the particle, the faster it shrinks. While all particles of radius greater than $\bar{a}$ will grow, notice that the curves go through a maximum, with the fastest growing particles being those that are twice as large as the average. Thus the size distribution of particles becomes self-regulating, since those that are small disappear rapidly and those that are largest grow slowly. More detailed analyses confirm that the size distribution remains constant as the average size increases. If growth is dominated by the fastest-growing particles, which remain twice the size of the average, then

$d\bar{a}/dt = 1/2\, da_{max}/dt$. Upon substituting $a_{max} = 2\bar{a}$ into Eq. 5.26 and integrating, the rate of growth of the average size is

$$\bar{a}^3 - \bar{a}_o^3 = \frac{3Dc_o\gamma M}{4\rho RT}\cdot t \qquad (5.27)$$

More rigorous analyses give essentially the same result [J. M. Lifshitz and V. V. Slyozov, *J. Physics Chem. Solids*, 19, 35 (1961); C. Wagner, *Z. Electrochem.* 65, 581–591 (1961); G. W. Greenwood, *Acta Met.* 4, 243–248 (1956)]. The diffusion-limited growth of precipitates in a solid matrix and of grains in a liquid phase are observed to have the $t^{1/3}$ time dependence given by Eq. 5.27 as illustrated in Fig. 5.35.

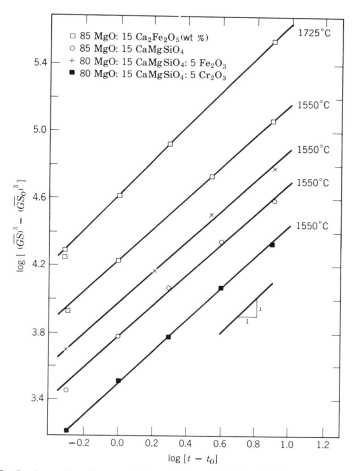

Fig. 5.35 Isothermal grain growth in systems containing MgO and liquid (GS = grain size in microns, t = time in hours) $GS^3 \propto t$. (From J. White in *Materials Sci. Research*, Vol. 6, Plenum Publishing Corporation, New York, 1973, p. 81.)

5.3 SINGLE-PHASE SINTERING

From a broad perspective the vast majority of useful ceramics are multiphase, and most are densified with the aid of a small amount of reactive liquid phase or applied pressure (see Table 5.4). Nonetheless, for many applications purely single-phase polycrystalline ceramics are essential. *Sintering* refers to the process of firing and consolidating a body shaped from powder particles. Single-phase sintering has also served an historically important role as a model scientific problem, through which the understanding of ceramics processing has been greatly advanced. Magnesia-doped alumina, as discussed in Special Topic 5.2, was the first single-phase crystalline ceramic to be sintered to complete density without either a liquid phase or applied pressure. This achievement enabled the use of sintered alumina as an envelope for sodium-vapor lamps, an application requiring tranparency and extreme corrosion resistance at high temperatures. For high temperature structural applications, single-phase polycrystalline ceramics such as alumina, silicon carbide, or silicon nitride have high temperature strength and creep resistance that is superior to that of their liquid-phase-sintered counterparts. While a reactive liquid phase can allow densification at lower temperatures, the maximum use temperature of the material may then be defined by the melting point of the liquid. This limitation is important for structural ceramics used at high temperature but less so for electronic ceramics which are used near room temperature. Hence, ceramics varistors, capacitors, thermistors, and ceramic substrates for electronic components are often liquid-phase sintered.

A powder compact has an excess of surface energy which is reduced during sintering. The excess surface energy per gram of powder of particle radius r is given by $3\gamma/r\rho$, and for typical surface energies of $\gamma = 0.3-1$ J/m^2, and material densities of $\rho = 3-7$ g/cm^3, the excess energy is about 4 J/g at $r = 0.1$ micron and 0.04 J/g at $r = 10$ micron. That is to say, these are small driving forces compared to chemical driving forces: For example, the heat of fusion of water is 334 J/g, and the standard free energy of the decomposition reaction $CaCO_3 \rightarrow CaO + CO_2$ is 3.31 kJ/g. At the microscopic level, the driving force for sintering is the capillary pressure associated with surface curvature where particles come into contact (Fig. 5.36). The neck formed between particles has a saddle curvature characterized by two radii of curvature, x, the diameter of the neck, and ρ, which is small and negative. The capillary pressure is $\Delta P = \gamma(1/\rho + 1/x)$ and is negative. Summed over a powder compact, this capillary pressure gives a modest *sintering pressure,* which is typically 0.1–1 MPa (14.5–145 psi) and which can be measured in a zero-creep experiment such as that illustrated in Fig. 5.37. That is, merely by having a fine powder the compact is subjected to a net compressive pressure which is often sufficient to densify the material at elevated temperatures.

Viscous Sintering

The simplest case of single-phase sintering occurs when an aggregate of glass particles is heated. Two simple conditions have been analysed: an initial state

Table 5.4 Selected Polycrystalline Ceramics and How They Are Densified

System	Application	Densification Method
Al_2O_3	sodium vapor arc lamp tubes	Solid-state sintered with MgO additive.
	furnace tubes, refractories	Liquid-phase sintered with a small amount of modified silicate glass.
SiC	high-strength structural ceramics	Solid-state sintered with boron or aluminum additive plus carbon; also liquid phase sintered or hot-pressed using oxide additives.
Si_3N_4	high-strength and high fracture toughness structural ceramics for low and high temperatures (to ~1500°C)	Liquid-phase sintered with oxide additives (e.g., Y_2O_3, CaO, Al_2O_3, MgO) under elevated nitrogen pressure; also hot-pressed and hot-isostatically pressed.
ZnO	varistors	Liquid-phase sintered using Bi_2O_3 and other oxide additives, which also impart electrical doping functions.
$BaTiO_3$	capacitor dielectrics, positive-temperature-coefficient (PTC) thermistors	Liquid-phase sintered using titania-rich eutectic liquid, sometimes with silica and other oxide additives.
$SrTiO_3$	capacitor dielectrics	Liquid-phase sintered with silica and other oxide additives.
Mn-Zn and Ni-Zn based ferrites	soft ferrites for ac magnetic applications	Solid-state-sintered under controlled firing atmosphere (PO_2).
$Pb(Zr,Ti)O_3$ (PZT) and $(Pb,La)(Zr,T)O_3$ (PLZT)	piezoelectric actuators and electro-optic devices	Sintered or hot-pressed, possibly assisted by lead-rich liquid phase.
ZrO_2	transformation-toughened structural ceramics, solid electrolytes and gas sensors	Sintered, hot-pressed, and hot-isostatically pressed. Minor silica impurity frequently results in a small amount of liquid phase.

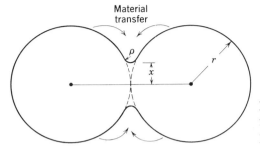

Fig. 5.36 The neck between particles forms a miniscus which exerts capillary pressure drawing particles together.

represented by two particles in contact (Fig. 5.38) and a later situation with closed pores in a continuous matrix (Fig. 5.39). In the initial stage, shrinkage, densification, and pore elimination occur at a rate determined by the intitial particle size, the value of the surface energy and the glass viscosity. As the spheres begin to coalesce (Fig. 5.38) the neck has a radius of curvature that remains small compared with the surface curvature of the particles, resulting in a negative pressure that causes viscous flow of material toward the interparticle region. In a classic paper, J. Frenkel [*J. Phys. (USSR)*, 9, 385 (1945)] found that the increase of the contact radius, x, is proportional to $t^{1/2}$. The rate of initial neck growth is given by

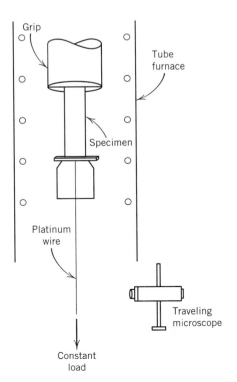

Fig. 5.37 The sintering pressure exerted by a powder compact can be measured in a zero-creep experiment. A specimen consisting of compacted powder spontaneously shrinks when heated to high temperature. By applying a restraining load, the stress at which no creep occurs can be determined, and is a measure of the sintering pressure. [From T. Cheng and R. Raj, *J. Am. Ceram. Soc.*, **71**[4], 276-80 (1988).]

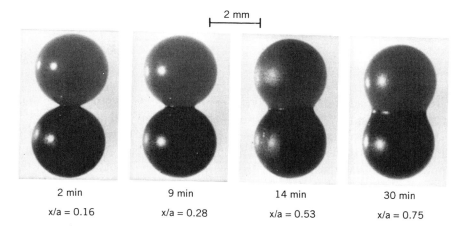

| 2 min | 9 min | 14 min | 30 min |
| x/a = 0.16 | x/a = 0.28 | x/a = 0.53 | x/a = 0.75 |

Fig. 5.38 Two-particle model made of glass spheres (3 mm diameter) sintered at 1000°C. x and r are, respectively, the radius of the neck and the sphere. (From H.E. Exner, "Solid State Sintering," *Reviews on Metallurgy and Physical Ceramics*, Vol. 1, No. 1–4, pp. 1–251, Freund Publishing House, Tel Aviv, Israel, 1979.)

$$\frac{x}{r} = \left(\frac{3\gamma}{2\eta r}\right)^{1/2} t^{1/2} \tag{5.28}$$

The macroscopic result of this coalescence is a shrinkage of the powder compact. The linear shrinkage is determined by the approach of the particles' centers toward one another. For the geometry of overlapping spheres (Fig. 5.40) the shrinkage, y, is given by

$$y \approx \rho \approx \frac{x^2}{4r} \tag{5.29}$$

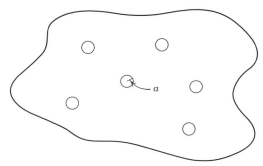

Fig. 5.39 Compact with isolated spherical pores of radius a near the end of the sintering process.

396 Chapter 5 / Microstructure

from which the relative change in length, y/r, is

$$\frac{y}{r} = \frac{\Delta L}{L_o} = \frac{\Delta V}{3V_o} = \frac{3\gamma}{8\eta r}t \tag{5.30}$$

Thus the initial rate of shrinkage is linear in time, and directly proportional to the surface tension and inversely proportional to the viscosity and particle size.

After long times when the structure can be represented as shown in Fig. 5.39, there is a negative pressure equal to $2\gamma/a$ in each pore, which is equivalent to the application of an external positive pressure. J. K. Mackenzie and R. Shuttleworth [*Proc. Phys. Soc. London*, B62, 833 (1949)] derived a relation for the shrinkage rate after deducing the properties of the porous material from the porosity and viscosity of the matrix and obtained the equation

$$\frac{d\rho'}{dt} = \frac{2}{3}\left(\frac{4\pi}{3}\right)^{1/3} n^{1/3} \frac{\gamma}{\eta}(1-\rho')^{2/3} \rho'^{1/3} \tag{5.31}$$

where ρ' is the relative density (fraction of the theoretical density) and n is the number of pores per unit volume of the matrix. Since the number of pores is in turn related to the pore size and the relative density by

$$n\frac{4\pi}{3}a^3 = \frac{\text{Pore volume}}{\text{Solid volume}} = \frac{1-\rho'}{\rho'} \tag{5.32}$$

and

$$n^{1/3} = \left(\frac{1-\rho'}{\rho'}\right)^{1/3}\left(\frac{3}{4\pi}\right)^{1/3}\frac{1}{a} \tag{5.33}$$

by combining with Eq. 5.31, we have

$$\frac{d\rho'}{dt} = \frac{3\gamma}{2a_o\eta}(1-\rho') \tag{5.36}$$

where a_o is the initial radius of the pores. Thus the rate of densification slows as full density ($\rho'=1$) is approached. G. W. Scherer [*J. Am. Ceram. Soc.*, 60, 236 (1977)] has developed a latticework model for the intermediate stage of sintering

5.3 Single-Phase Sintering   397

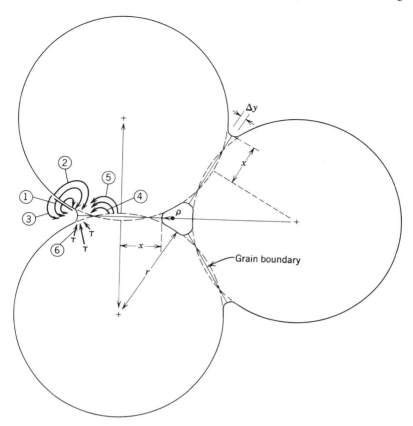

Fig. 5.40 Alternate paths for matter transport during the initial stages of sintering. (Courtesy M.A. Ashby.)

Alternate Paths for Matter Transport During the Initial Stages of Sintering

Mechanism Number	Transport Path	Source of Matter	Sink of Matter
1	Surface diffusion	Surface	Neck
2	Lattice diffusion	Surface	Neck
3	Vapor transport	Surface	Neck
4	Boundary diffusion	Grain boundary	Neck
5	Lattice diffusion	Grain boundary	Neck
6	Lattice diffusion	Dislocations	Neck

when the pores are interconnected and open to the atmosphere, which gives results comparable to the Mackenzie-Shuttleworth model. For complete densification the time required is approximately

$$t_s \sim \frac{1.5 a_o \eta}{\gamma} \qquad (5.35)$$

For densification of a soda lime silica glass the models developed are in good agreement with experimental results giving us confidence in applying them to vitrification processes.

When pressure is applied to the outside of the sintering body, the shrinkage rate is increased. P. Murray, E. P. Rogers, and A. E. Williams [*Trans. Brit. Ceram. Soc.*, 53, 474 (1954)] showed that when the applied pressure P_a is large, as in hot-pressing, the capillary pressure is negligible in comparison, and the Mackenzie–Shuttleworth relationship simplifies to

$$\ln[1-\rho(t)] = \ln(1-p_c) + \frac{3 P_a t}{4 \eta} \qquad (5.36)$$

In fact, the capillary pressure p_c becomes infinite when the pore radius approaches zero so that this relationship cannot apply for the very final period of densification.

Thus far we have assumed that there is no internal gas pressure developed in the pores; that is, the gas initially present, if any, diffused out more rapidly than pore shrinkage took place. If the gas is insoluble, the bubble shrinks until the pressure of the gas equals the capillary pressure $p_c = 2\gamma/a$ where a is the pore radius. If the radius of the bubble is a_o when the pore first becomes a closed sphere and the pressure of the gas is initially p_o, then the equilibrium size of the bubble will be

$$a_{eq}^2 = \frac{p_o a_o^3}{2 \gamma} \qquad (5.37)$$

This pressure will prevent sintered bodies from reaching full density.

Crystalline Ceramics

For crystalline ceramics the driving forces are essentially the same as for viscous sintering, although there are added complexities due to anisotropies of surface energies and the formation of thermal grooves at the contacts between particles determined by the relative values of the surface and grain boundary tensions. (Powder particles are also rarely spherical as we have depicted, but we shall continue with this simplifiying assumption in order to focus on the basic thermodynamics and kinetics of sintering.) More importantly, there are a number of competing paths for material transport to the neck area in the initial stage. Some of these lead to *densification*, which refers to the shrinkage process and requires that the centers of particles approach one another (Fig. 5.40). Other transport mechanisms lead to *coarsening*, which is a growth of the neck between particles leading to reduction of the specific surface area without shrinkage. Generally densification is desired, which then is largely a matter of promoting those mechanisms that lead to shrinkage over those that do not.

Two possible paths for material transport are by *evaporation–condensation* or by *surface diffusion* as shown in Fig. 5.40. These are coarsening mechanisms which move material from the particle surface to the neck and so reduce the surface energy; neither process causes densification since no mass is removed from the plane between the particles. For vapor transport, we can calculate the rate at which the bonding area between particles grows, from the rate of material transfer to the lenticular region between the spheres. As shown in Fig. 5.6, the vapor pressure over a neck of small negative radius of curvature is lowered with respect to the large positive radius curvature of the particle surface. If ρ is much less than the particle radius r, from the Thompson–Freundlich equation the difference in vapor pressure is given to good approximation by

$$\Delta p = \frac{p_o \gamma V_m}{RT}\left(\frac{1}{\rho}\right) \tag{5.38}$$

where V_m is the molar volume (given by molecular weight/density) and p_o is the equilibrium vapor pressure over a flat surface. The rate of condensation is given by the Langmuir equation as

$$m = \alpha \Delta p \left(\frac{M}{2\pi RT}\right)^{1/2} \quad (g/cm^2/sec) \tag{5.39}$$

where α is a sticking coefficient approximately equal to 1. The rate of condensation should be equal to the volume increase of the neck region,

$$\frac{mA}{d} = \frac{dv}{dt} \quad (cm^3/sec) \tag{5.40}$$

where A is the area of condensation and d is density. From the geometry of contacting spheres in Fig. 5.36, the radius of curvature at the contact point is approximately $\rho = x^2/2r$ when x/r is small.[1] The outer area of the lenticular volume onto which mass is condensing is approximately $\pi x^3/r$ and the volume contained in the lens is approximately $\pi x^4/2r$. Substituting these values into Eq. 5.40 and integrating, a relationship is obtained for the rate of growth of the neck between particles

$$\left(\frac{x}{r}\right)^3 = \left[\frac{3\sqrt{\pi}\gamma M^{3/2} p_o}{\sqrt{2}R^{3/2}T^{3/2}d^2}\right]\frac{t}{r^2} \tag{5.41}$$

---

[1] A dihedral angle of 180° is assumed. If the dihedral angle is less, the effective radius of curvature at the neck is reduced, lowering the capillary driving force.

Experimental data has confirmed the general form of this relationship, which indicates that the neck diameter increases as $t^{1/3}$, $r^{-2/3}$, and $p_o^{1/3}$. Since the vapor pressure increases exponentially with temperature, the process of vapor phase coarsening is strongly temperature dependent. If surface diffusion from the particle surface to the neck (Fig. 5.40) is the fastest mechanism of neck growth, solutions which have been obtained for diffusion to the cylindrical neck are of the form

$$\left(\frac{x}{r}\right)^n = \left[\frac{A \delta D_s V_m \gamma_{sv}}{RT r^4}\right] t \tag{5.42}$$

where A is a numerical constant, δ is the surface "thickness" (a few angstroms), D_s is the surface diffusion coefficient, and n has values of 5 to 7 depending on how the neck shape is treated in the model. (See W. S. Coblenz, J. M. Dynys, R. M. Cannon, and R. L. Coble, *Proceedings of the Fifth International Conference on Sintering and Related Phenomena*, edited by G. C. Kuczynski, Plenum Press, NY, 1983, for a detailed discussion of the different surface diffusion models.) Thus there is a somewhat weaker time dependence $t^{1/n}$ for surface diffusion, than for evaporation-condensation.

Densification mechanisms (in the absence of viscous flow and plastic deformation) are those in which the grain boundary plane serves as the "source" for diffusional transport (Fig. 5.40), and the neck as the "sink" or repository for atoms. This kind of transport can take place either along the grain boundary or through the lattice. In both cases the driving force for diffusion is the gradient in stress between the surface of the neck and the grain boundary plane. The net curvature of the neck causes a tensile pressure of

$$P = -\gamma \left(\frac{1}{x} - \frac{1}{\rho}\right) \approx \frac{\gamma}{\rho}$$

at its surface, and applies a compressive stress of equal magnitude to the grain boundary plane. As discussed for the process of diffusional creep in Chapter 3, under a stress σ the equilibrium concentration of vacancies changes as

$$C_v(\sigma) = C_v(\sigma = 0) \exp\left[\frac{\sigma \Omega}{kT}\right] \tag{5.43}$$

where Ω is the vacancy volume. Since usually $\frac{\sigma \Omega}{kT} \ll 1$, this can be simplified to

$$C_v(\sigma) \approx C_v(\sigma = 0)\left[1 + \frac{\sigma \Omega}{kT}\right]$$

At the tensile stress at the surface of the neck, the vacancy concentration is increased, and at the compressive stress at the grain boundary, the vacancy concentration is decreased. Vacancy diffusion occurs down its concentration gradient from the neck to the grain boundary, which necessitates an equivalent diffusion of atoms in the opposite direction, from the grain boundary to the neck—this is the essential step which removes mass from the plane between particles. (While we use a vacancy mechanism as the example, this result is general and does not depend on the diffusion mechanism.) While it is difficult to calculate accurately the concentration gradients or precisely the exact paths along which material transport to the neck area occurs, number of models have been derived which are based on reasonable assumptions. Following the same line of reasoning as for the viscous case, W. D. Kingery and M. Berg [*J. Appl. Phys.*, 26, 1205 (1955)] calculated the rate of neck growth by volume diffusion from the grain boundary to the neck surface to be given by

$$\left(\frac{x}{r}\right)^5 = \left(\frac{80\gamma\Omega D_l}{kT}\right)\frac{t}{r^3} \tag{5.44}$$

in which D_l is the lattice self-diffusion coefficient for the rate-limiting ion. From the initial stage sintering geometry, the relative shrinkage and the ratio x/r are related by

$$\frac{\Delta V}{V} = \frac{3\Delta L}{L} = \frac{3x^2}{4r^2} \tag{5.45}$$

or,

$$\frac{\Delta L}{L} = 1.4\left(\frac{\gamma\Omega D_l}{kT}\right)^{2/5}\frac{t^{2/5}}{r^{6/5}} \tag{5.46}$$

If grain boundary diffusion is rate-limiting, solving for the radial diffusion outward from the grain boundary to the neck through a boundary of thickness δ [R. L. Coble, *J. Am. Ceram. Soc.*, 41, 55 (1958)] yields

$$\left(\frac{x}{r}\right)^6 = \left[\frac{192\,\delta D_b\,\gamma\Omega}{kT}\right]\frac{t}{r^4} \tag{5.47}$$

and

$$\frac{\Delta L}{L} = \left[\frac{3\,\delta D_b\,\gamma\Omega}{kT}\right]^{1/3}\frac{t^{1/3}}{r^{4/3}} \tag{5.48}$$

The time dependence for densification of a powder compact by solid state diffusion is weaker than the linear time dependence in viscous sintering (Eq. 5.30).

The geometric models on which the results in Eqs. 5.41, 5.42, and 5.44-5.48 are based apply only to the *initial stage* of sintering, generally considered to be when the ratio x/r is less than 0.3. Evaporation–condensation, surface diffusion, grain boundary diffusion, and lattice diffusion are competitive paths of transport; upon firing, the mechanism that gives the fastest rate of neck growth will dominate, and either cause coarsening or densification. In addition to the availability of multiple paths, however, there are multiple species to be considered. It is to be emphasized that solid-state sintering is an *ambipolar* diffusion process in which both cations and anions (or, for covalent compounds, all constituents) must be transported from the source to the sink. As in the case of diffusional creep (see Special Topic 3.2), the rate-limiting mechanism will be *the slowest species along its fastest path*. For example, sintering of Al_2O_3 may in principle be rate-limited by any one of six diffusion mechanisms (surface, grain boundary, or lattice diffusion of Al or O) in addition to vapor phase transport. Since most compositions of alumina do densify when fired at high temperature, we may safely presume that grain boundary or lattice diffusion of either Al or O is rate-limiting. Given the small difference in time-dependence and particle size dependence of shrinkage between the grain boundary diffusion and lattice diffusion mechanisms (Eqs. 5.44 and 5.47), and the approximate nature of the geometric models which lead to these results, one cannot easily distinguish between densification mechanisms on this basis alone. Additional information regarding the lattice and boundary diffusion coefficients of the anion and cation is necessary. Since oxygen grain boundary diffusion is much faster than oxygen lattice diffusion, as it is in many close-packed oxides, we may narrow down the rate-limiting possibilities to grain boundary diffusion of oxygen, grain boundary diffusion of aluminum, or lattice diffusion of aluminum. Studies of solute effects on sintering and diffusional creep in Al_2O_3 suggest that one of the latter two is the rate-limiting mechanism. Similarly, in a material used as an oxygen electrolyte such as stabilized zirconia, lattice diffusion of oxygen is extremely rapid, suggesting that sintering will be rate-limited by cation diffusion, in either the lattice or grain boundary.

From the dependence of the sintering equations on diffusivities, one can see that solid solution additives which influence diffusion can have an enormous impact on the rate of sintering and on whether or not densification occurs. We already discussed in Chapters 2 and 3 how defects introduced to compensate aliovalent solutes can greatly influence lattice diffusivities. Solute additions also influence grain boundary and surface diffusivities, although the effects are less predictable and the mechanisms generally unknown. In nonstoichiometric ceramics, where the defect structure is determined by oxidation and reduction (one example being spinel ferrites), the control of firing atmosphere can also be important.

The importance of particle size in determining sintering rate can be seen in the inverse dependencies of the sintering equations on particle radius r. For a fixed temperature and composition, the rate equations for densification and coarsening (Eqs. 5.41, 5.42, 5.44, 5.47) are of form

$$\left(\frac{x}{r}\right)^n = \frac{\text{Constant}}{r^p} t \quad (5.49)$$

where the constant contains the surface tension and the relevant diffusion coefficient, n is between 2 and 6, and p is between 2 and 4 (except for viscous sintering, where it is 1). For a particular mechanism, the time to reach a particular shape change (i.e., x/r ratio) for two particle sizes r_1 and r_2 is, upon rearranging Eq. 5.49,

$$\frac{t(r_2)}{t(r_1)} = \left(\frac{r_2}{r_1}\right)^p \quad (5.50)$$

This relationship is known as *Herring's scaling law* [C. Herring, *J. Appl. Phys.*, 21, 201 (1950)]. For example, applying Eq. 5.50 to the densification mechanisms of grain boundary and lattice diffusion where $p = 3$ or 4, we find that the time required to reach a particular x/r is 10^3 to 10^4 times shorter for 0.1 μm particles than it is for 1 μm particles! The recognition that fine particle sizes promote faster sintering and at lower temperatures has motivated an entire field of study in the synthesis of fine ceramic powders.

The importance of temperature in promoting densification can be seen in the dependence of the rate equations on diffusivity. The activation energies for solid-state diffusion coefficients ($D = D_0 \exp[-Q/RT]$) in any given material typically scale as

$$Q_s < Q_{gb} < Q_l$$

As illustrated in Fig. 5.41, grain boundary and lattice diffusion have a steeper temperature dependence and tend to dominate at higher temperatures, thereby promoting densification. Surface diffusion and evaporation–condensation have weaker temperature dependencies and often dominate at low temperatures where lattice and grain boundary transport is frozen out. The principle behind *fast-firing* [R. J. Brook, *Proc. Brit. Ceram. Soc.*, 32, 7, (1982)] is that, by rapid heating to bypass the coarsening regime, one can promote densification. Conversely, firing at lower temperature results in coarsening without significant densification and can be useful for fabricating porous materials. Of course, if the activation energy for a particular coarsening process is higher than for the densification mechanism, firing at higher temperature promotes coarsening rather than densification.

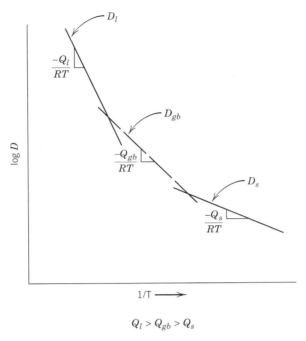

Fig. 5.41 Lattice diffusion typically occurs with an activation energy greater than that for grain boundary diffusion, which is in turn greater than that for surface diffusion. As a result, lattice and boundary diffusion mechanisms of sintering which lead to densification tend to be promoted at higher temperatures.

Later-Stage Sintering

If we presume that densification has occurred in the initial stage of sintering, the microstructure of a powder compact then enters an *intermediate stage*, where voids form continuous pore channels along three grain junctions (Fig. 5.42). These pore channels become narrower as densification proceeds and eventually undergo Rayleigh breakup (see Special Topic 5.1) forming discrete pores when a few percent porosity is left. In the *final stage* of sintering these isolated pores are eliminated by mass transport from the grain boundary to the pore, as illustrated in Fig. 5.43. Ambipolar diffusion involving all constituent ions is also required in these processes. For equivalent volume fraction porosity, samples with smaller pores have a higher curvature driving force and a smaller mean diffusion distance between pores and the boundary and will tend to densify faster. The diffusion path or paths depend on where the pore is located; pores on grain boundaries can be eliminated by grain boundary or lattice diffusion, while pores within grains can only be removed by lattice diffusion. The location of the pores becomes a critical issue for continued densification, for in most ceramics the lattice diffusivity of the

5.3 Single-Phase Sintering 405

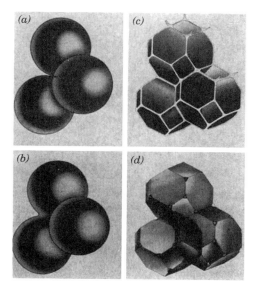

Fig. 5.42 Stages of sintering. (*a*) Initial stage; model structure represented by spheres in tangential contact. (*b*) Near end of initial stage. Spheres have begun to coalesce. The neck growth illustrated is for center-center shrinkage of 4%. (*c*) Intermediate stage; dark grains have adoped shape of tetrakaidecahedron, enclosing white pore channels at grain edges. (*d*) Final stage; pores are tetrahedral inclusions at corners where four tetrakaidecahedra meet. [From R.L. Coble, *J. Appl. Phys.*, 32 [5] 787 (1961).]

slower, rate-limiting constituent is too low for effective annihilation of pores that have become trapped within grains. If high-sintered density is to be achieved, it is important for pores to remain attached to grain boundaries throughout the later stages of sintering. Thus we return to the grain-size–pore-size map, Fig. 5.26, in which it is shown that pore-boundary separation will occur only for certain ratios of the grain size to pore size: Pores that are large relative to the grain size remain attached and slow down the boundary, and pores that are small relative to the grain size stay attached to the boundary as well but do not exert much drag. In between these limits of behavior is a range of grain size to pore size where separation occurs, to the detriment of densification.

It is common for polycrystalline microstructures to simultaneously undergo grain growth and densification during the firing process. Since densification is accompanied by a decrease in the size (and number) of pores, the ratio of grain size to pore size increases continuously during densification, as shown by the trajectory in Fig. 5.44. R. J. Brook (*Treatise on Materials Science and Technology*, Vol. 9, edited by F. F. Y. Wang, Academic Press, NY, 1976) pointed out that the key to avoiding pore entrapment is the avoidance of the pore separation region by this trajectory. A plot of grain size against density can be used to show these

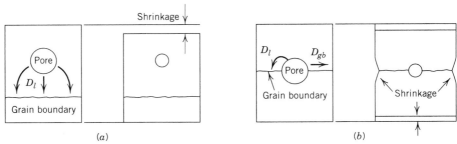

Fig. 5.43 (a) Pores located within a grain can shrink by diffusion of vacancies from the surface of the pore to the grain boundary, where they are annhilated. Lattice diffusion is rate-limiting. (b) Pores located along grain boundaries can shrink by diffusion of vacancies from the pore surface to the grain boundary by grain boundary diffusion or by lattice diffusion.

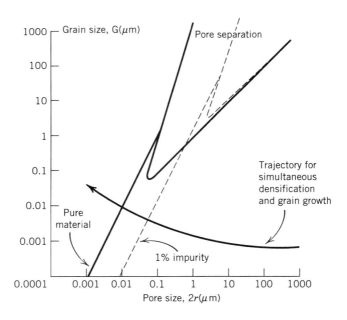

Fig. 5.44 On a grain-size – pore-size map (see also Fig. 5.25), simultaneous densification and grain growth is represented by the trajectory shown. If the pore separation field is not interersted, complete densification is possible. [Adapted from R.J. Brook, *J. Am. Ceram. Soc.* 52, 56 (1969).]

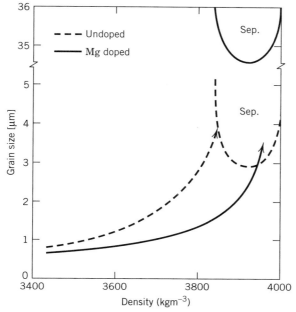

Fig. 5.45 Grain size – density map for Al_2O_3. Depicted are the effects of raising surface diffusivity by a factor of 4; lowering lattice diffusivity by a factor of 2; and lowering grain boundary mobility by a factor of 34. This raises pore separation region and lowers grain size trajectory (dotted lines), making it possible to sinter to complete density. [From S.J. Bennison and M.P. Harmer, *J.Am. Ceram. Soc.*, 73[4], 833 (1990).]

trajectories, Fig. 5.45. The influence of temperature, solutes, and other variables can then be discussed in terms of their effects on the position of the pore separation field, or on the grain size-density trajectory. Figure 5.45 illustrates that the pore separation field is moved to higher grain sizes by adding an impurity which lowers grain boundary mobility, making it easier for the trajectory to bypass the separation field. Flattening the grain size to pore size trajectory, that is, increasing the densification rate relative to the grain growth rate, is also beneficial and can be done for example by adding a solute that increases the rate of densification or lowers the boundary mobility. The grain-size–density trajectory can also be flattened by altering time–temperature profile in a firing cycle to rapidly reach a temperature regime where densification is dominant. The results in Fig. 5.46 show that fast-firing has this effect.

Notice that the pore-separation region we are trying to avoid is in fact a moving target, since its position is also dependent on the fraction of porosity, which decreases continuously during densification. An additional complication is that a range of pore and grain sizes usually exists in any given microstructure, and exaggerated grain growth may occur with only local pore separation. Generally speak-

408 Chapter 5 / Microstructure

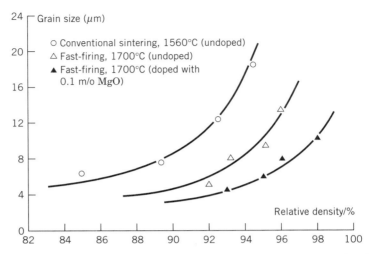

Fig. 5.46 Actual grain size – density trajectories measured for Al_2O_3. Notice the effect of MgO addition is to lower the rate of grain growth relative to density. (From S. Wu, E. Gilbart and R.J. Brook, pp. 574–582 in *Advances in Ceramics,* Vol. 10, W.D. Kingery, ed., The American Ceramic Society, Columbus, Ohio, 1984.)

ing, however, additives that (1) decrease the rate of coarsening in earlier stages of sintering, (2) increase the rate of densification, (3) decrease the rate of grain growth, or (4) increase pore mobility such that pore-grain boundary separation does not occur, are beneficial to achieving high density. Solutes that lower the dihedral angle where a pore meets the boundary can in principal be beneficial, since, as shown in Fig. 5.47, the effective area of grain boundary intersected by a pore of constant volume increases as the dihedral angle decreases, resulting in greater pore drag and less likelihood of separation. (Decreasing dihedral angle does lower the adjacent surface curvature and thereby the driving force for pore elimination

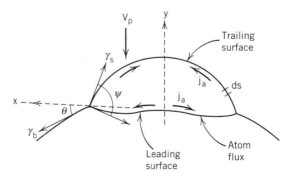

Fig. 5.47 Schematic of a moving pore indicating the atom flux from leading to trailing surfaces. [From M. Sakarcan, C.H. Hsueh, and A.G. Evans, *J. Am. Ceram. Soc.*, 66[6], 456 (1983).]

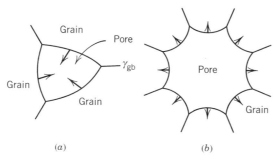

Fig. 5.48 (a) Pores with few neighboring grains tend to shrink, while (b) pores with many sides tend to grow.

as well as initial stage sintering, however.) These are desirable objectives that may not always be achievable. Materials having high vapor pressure and low solid state diffusivities such as covalent solids (Si, SiC, Si_3N_4) often cannot be sintered without the addition of dopants or second-phase constituents.

Packing, Agglomeration, and Pore Growth

The surface curvature of internal pores located at grain junctions depend on the number of surrounding grains as well as the dihedral angle (Fig. 5.48). If a pore is surrounded by many grains, the curvature of the pore surface is negative and the pore has a thermodynamic tendency to grow rather than shrink. In two dimensions, a pore surrounded by six grains has flat sides and is stable when the dihedral angle is 120° In three dimensions, if the dihedral angle is 120°, twelve grains

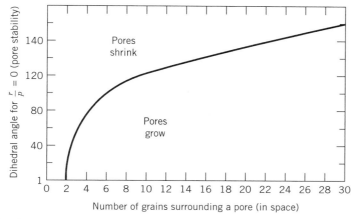

Fig. 5.49 Conditions for pore stability in three dimensions as a function of pore coordination number.

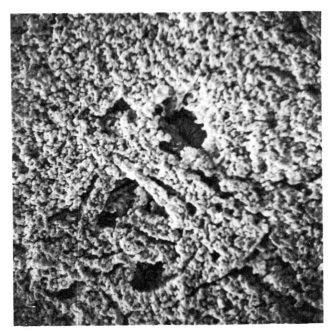

Fig. 5.50 Large voids formed by bridging of agglomerates in fine Al_2O_3 powder (2000x). (Courtesy C. Greskovich.)

surrounding a pore will be stable. As the dihedral angle varies, pores tend to grow or shrink depending on the number of surrounding grains, as illustrated in Fig. 5.49.

From Figs. 5.48 and 5.49 we can appreciate that packing errors made earlier in powder processing will tend to persist throughout sintering. Particle packing is rarely perfect, and fine powders are often agglomerated. The bridging of agglomerates leaves large voids after pressing (Fig. 5.50) which are unstable against growth. In addition to agglomerates and variations in packing, nonuniform porosity in the later stages of sintering results from particle size variations in the starting material, from green density variations caused by die-wall friction, from incompletely removed organic binders, and from rapid elimination of porosity near surfaces caused by temperature gradients during heating (Fig. 5.51). Nonuniform pore concentrations causes different parts of a sample to shrink at different rates; the more slowly shrinking parts are restrained by the pore-free parts, and the effective diffusion distance is no longer from the pore to an adjacent grain boundary but a pore–pore or pore–surface distance many orders of magnitude larger. A comparison of a pore-free sintered sample of yttria compared with one having residual pores clusters resulting from improper powder processing is shown in Figs. 5.52 and 5.53. One approach striving for perfection in the powder handling stage is illustrated in Fig. 5.54, in which chemically precipitated spheres of TiO_2 have been settled into a densely packed array, and can then be fired to high densities at reduced temperatures. However, such high

5.3 Single-Phase Sintering 411

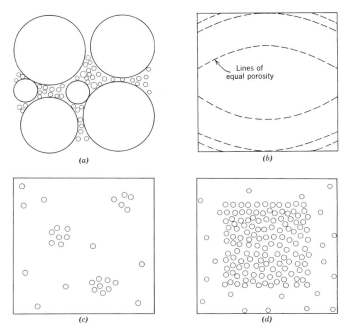

Fig. 5.51 Pore-concentration variations resulting from (*a*) a variation in grain sizes, (*b*) die friction, (*c*) local packing and agglomeration differences, and (*d*) more rapid pore elimination near surfaces.

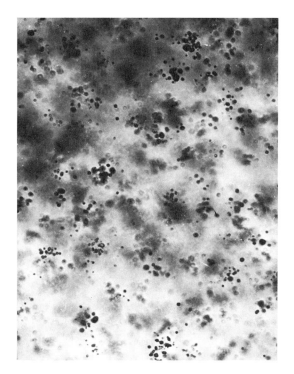

Fig. 5.52 Residual pore clusters resulting from improper powder processing in a sample of 90 mole % Y_2O_3-10 mole % ThO_2. Transmitted light, 110×. (Courtesy C. Greskovich and K.N. Woods.)

412 Chapter 5 / Microstructure

Fig. 5.53 Polished section of Y_2O_3 + 10 mole % ThO_2 sintered to pore-free state. 80×. (Courtesy C. Greskovich and K.N. Woods.)

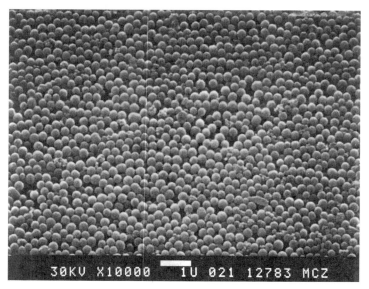

Fig. 5.54 Top surface of a compact of monosized TiO_2 (0.2% BaO and 0.5% Nb_2O_5). (From E.A. Barringer, N. Jubb, B. Fegley, R.L. Pober, and H.K. Bowen, pp. 315–333 in *Ultrastructure Processing of Ceramics, Glasses, and Composites,* L.L. Hench and D.R. Ulrich, eds., J. Wiley and Sons, New York, 1984.)

5.3 Single-Phase Sintering

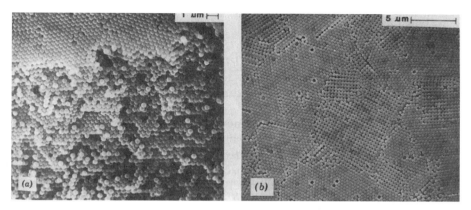

Fig. 5.55 (*a*) Fracture surface and (*b*) top surface of a compact formed from monodisperse boron-doped SiO_2. (From E.A. Barringer, N. Jubb, B. Fegley, R.L. Pober, and H.K. Bowen, pp. 315–333 in *Ultrastructure Processing of Ceramics, Glasses, and Composites*, L.L. Hench and D.R. Ulrich, eds., J. Wiley and Sons, New York, 1984.)

perfection is not easily achievable; and even for monosized powders domains of different packing density can occur in this kind of settling experiment (Fig. 5.55), which may undergo differing rates of local densification upon firing.

SPECIAL TOPIC 5.2

MAGNESIA-DOPED ALUMINA

In the late 1950s, researchers at the General Electric Company seeking a suitable material for containment of the corrosive plasma in high pressure sodium vapor lamps developed the first single phase polycrystalline ceramic which could be sintered to complete density — a polycrystalline alumina (corundum phase) which came to be named Lucalox™ (for transLUCent ALuminum OXide). The sodium vapor arc lamp is a two-envelope device (Fig. ST57) which can be seen in widespread use in lighting fixtures along our nation's highways. It consists of an inner envelope, partially filled with a halogen gas, within which a sodium vapor of 3700°C plasma temperature is struck. The alumina tube itself operates at a wall temperature of about 1200°C and is enclosed by an evacuated outer envelope of glass. Without a highly refractory yet transparent containment tube to contain the plasma, which can furthermore be economically manufactured (compared with, say, single crystal sapphire), this advance in lighting technology could not have been achieved.

The key to Lucalox™ was a discovery by Robert L. Coble (U.S. Patent 3,026,210) that magnesia was a critical additive which allowed alumina to be sintered to theoretical density. P. D. S. St. Pierre and A. Gatti at General Electric had developed a firing process (U.S. Patent 3,026,177), which re-

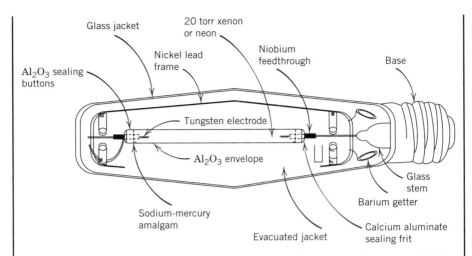

Fig. ST57 Construction of typical high-pressure sodium lamp. (Courtesy of W.H. Rhodes, Osram Sylvania, Inc.)

sulted in translucent material, but it was not known until Coble's work that magnesia, accidentally supplied by the furnace refractories, was the key ingredient. Magnesia-doped alumina came to be not only a landmark in ceramics processing, but a vehicle for the testing and development of sintering science over the next three decades. Its long life as a topic of scientific interest has been largely due to the elusiveness of an adequate explanation for the effect of magnesia; according to S. J. Bennison and M. P. Harmer (pp. 13*ff* in *Ceramic Transactions*, Vol. 7, 1990, The American Ceramic Society) by 1989 sixty papers related to the sintering of this one system had been published.

The dramatic influence of magnesia on the microstructure of sintered alumina can be seen in Fig. ST58. In the absence of magnesia, pores become entrapped within the alumina grains as abnormal grain growth takes place during sintering. These pores are impossible to remove in a reasonable firing time since the lattice transport required is extremely slow; the pores scatter light and render the alumina opaque. Coble showed that by using about 0.25 weight % magnesia, and firing at ~1900°C in hydrogen atmosphere, a completely dense alumina with no entrapped pores could be obtained. (Lucalox™ is actually not completely tranparent, but somewhat translucent since the refractive index of corundum is anisotropic (birefringent), and some light scattering takes place in the randomly oriented polycrystal even if it is fully dense.) It was later shown that firing in vacuum or a soluble gas such as hydrogen or oxygen yields similar results, while firing in an insoluble gas such as nitrogen, air (which is mostly nitrogen), helium or argon prevents full densification due to internal gas pressure building to an equilibrium with the capillary pressure of the pore.

5.3 Single-Phase Sintering 415

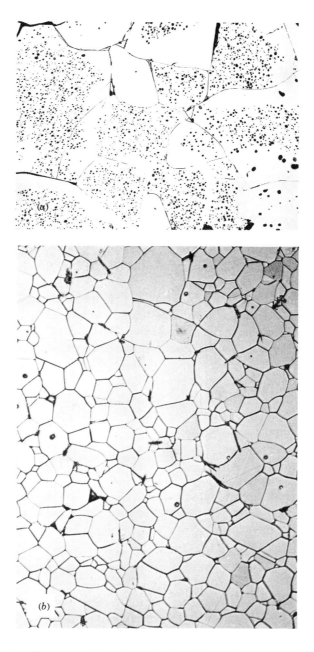

Fig. ST58 (*a*) Sintered Al_2O_3 without MgO addition illustrating elimination of porosity adjacent to grain boundaries with residual porosity remaining at grain centers. (Courtesy J.E. Burke.) (*b*) Al_2O_3 sintered with MgO additions, showing nearly porefree microstructure with only a few pores located within grains (500×). (Courtesy C. Greskovich and K.W. Lay.)

While the *effect* of magnesia was easily demonstrated, understanding the *mechanism* by which it acts took much longer. As discussed in the section on single phase sintering, densification occurs when pores located at grain boundaries are removed by lattice or grain boundary diffusional processes. Once a pore becomes entrapped within the grain, however, lattice diffusion (in corundum as well as most other ceramics) is prohibitively slow for much further densification. Thus the key to achieving transparency is the prevention of pore-grain boundary separation. Coble (*J. Appl. Phys.*, 32[5], 793, 1961) outlined several specific mechanisms by which this could be accomplished:

1. Second phase particles of $MgAl_2O_4$ spinel, resulting from an excess of magnesia beyond the solid solution limit, pin grain boundaries and prevent abnormal grain growth.
2. Magnesia in solution segregates to grain boundaries and lowers grain boundary mobility by solid solution-drag. The pores then remain attached to boundaries and can be removed by the usual densification processes.
3. Magnesia changes the equilibrium pore shape by changing the relative values of the surface energy and grain boundary energy. For a pore of constant volume, a lowering of the dihedral angle causes a greater area of the grain boundary to be intersected by the pore, and results in a larger drag force.
4. The rate of densification is increased relative to the rate of grain growth by magnesia in solid solution. Coble believed that the lattice diffusion of aluminum was rate-limiting, while oxygen was more rapidly transported along the grain boundaries.

A. H. Heuer [*J. Am. Ceram. Soc.*, 62[5–6], 317, 1979, and 63[3–4], 230, (1980)] later added a fifth possibility:

5. Magnesia increases the rate of surface diffusion in alumina, thereby increasing the mobility of pores and allowing them to keep up with migrating boundaries.

It is of course possible for more than one of these mechanisms to be acting at the same time. During the 1970s and 1980s, much effort was expended in model experiments and measuring fundamental parameters necessary to support or exclude particular mechanisms. In the case of both grain boundary segregation of magnesia (necessary for solid solution drag) and the enhancement of surface diffusion by magnesia, opinions were reversed as new studies appeared. [The history of these events can be read in the comprehensive review by Bennison and Harmer (1990)]. Only mechanism 1 was completely ruled out; this by an ingenious experiment by W. C. Johnson and R. L. Coble [*J. Am. Ceram. Soc.*, 61[3–4], 110 (1978)] in which a two-phase mixture of $MgAl_2O_4$ and Al_2O_3 was used as the source of magnesia vapor to dope an undoped alumina powder compact. By using a two-phase equilibrium mixture (see also the phase diagram in Fig. 4.4), the thermodynamic activity of MgO is pinned at a constant value (at constant firing temperature). The undoped alumina, held at the same temperature, may be doped by the magnesia vapor up to, but not in excess of, the solid solution concentration limit. Thus the *supersaturation* of magnesia necessary to precipitate spinel particles cannot occur in this experiment. Johnson and Coble nonetheless observed the same dense, equiaxed microstructure characteristic of Lucalox in the surface of their alumina sample, while the undoped interior showed the usual abnormal grain growth and entrapped porosity, and therefore proved that a second phase was not necessary to achieve theoretical densities.

The current understanding is that mechanisms 2–5 are *all* to some degree influenced by magnesia additions. The single most affected parameter seems to be the grain boundary mobility [S. J. Bennison and M. P. Harmer, *J. Am. Ceram. Soc.*, 68[1], C22 (1985)]. From considerations of simultaneous densification and grain growth, it appears that while the measured changes in surface diffusivity or densification rate are by themselves not sufficient for the avoidance of pore-boundary separation, the combined effects of a slight increase in the densification rate and pore mobility (factors of 3–4) and a substantial decrease in grain boundary mobility (factor of 25 or more) are adequate [K. A. Berry and M. P. Harmer, *J. Am. Ceram. Soc.*, 69[2], 143 (1986)].

It is now also recognized that an important role of magnesia is to lessen anisotropies in surface and grain boundary energies and mobilities. Pore entrapment does not require the separation of pores uniformly from *all* boundaries; a few high mobility grain boundaries can lead to local pore separation and discontinuous grain growth. Magnesia has been found to narrow the distribution of dihedral angles at the alumina free surface [C. A. Handwerker, J. M. Dynys, R. M. Cannon, R. L. Coble, *J. Am. Ceram. Soc.*, 73, 1371 (1990)], reflecting a homogenization of surface energies, and/or grain boundary energies, which should reduce the local variation in pore and grain boundary velocity. In Figs. ST59 and ST60(a) it can be seen that grain shapes tend to be tabular in the absence of magnesia. The long, straight boundaries are those in which one grain has a basal plane (0001) orientation. Most aluminas contain a small amount of silica, which together with other impurities forms a small amount of glass, which tends to wet preferentially along these boundaries. The prevalence of this type of boundary suggests that it is of low energy. The addition of magnesia suppresses both the average grain size and the pronounced faceting, indicating a reduction in the grain boundary energy and/or mobility, especially of those boundaries *not* oriented along a basal facet (which must grow at a faster relative rate to obtain tabular grains). An experiment where single-crystal sapphire spheres are used as "seeds" to represent discontinuous grains (Fig. ST60) further illustrates these

418   Chapter 5 / Microstructure

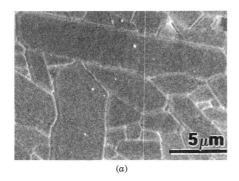

(a)

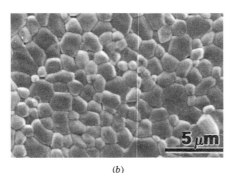

(b)

Fig. ST59 (a) Dense, hot-pressed alumina without MgO addition shows larger grains with tabular shape indicating grain boundary energy anisotropy, while (b) a similar sample with MgO addition shows more isotropic grain morphology. [From S.J. Bennison and M.P. Harmer, *J. Am. Ceram. Soc.*, **66**[5] C90-C92 (1983).]

points. For undoped alumina with a typical small amount of silica impurity, (a), the matrix grains show rapid growth and faceting, and any anisotropy in the growth of the seed is hard to detect since many of the surrounding boundaries are already faceted. With 0.1% magnesia, (b), the matrix grains are finer and equiaxed, and the seed grows discontinuously but in a relatively isotropic manner. With 0.9% added anorthite glass, (c), the seed exhibits pronounced faceting along basal planes (top and bottom surfaces). At this fairly high level of added glass, however, 0.1% magnesia is not able to prevent faceting (d).

While the last word on this subject has not been written, magnesia-doped alumina illustrates many of the paradigms of ceramics processing and is an example of the success of single-phase sintering theories. The lamp envelopes themselves continue to be widely produced by a number of manufacturers worldwide at an estimated rate of 16 million per year (W. H. Rhodes and G. C. Wei, pp. 273–76 in *Concise Encyclopedia of Advanced Ceramic Materials*, edited by R. J. Brook, Pergamon Press and The MIT Press, NY, 1991). Magnesia-doped alumina has even been used as the basis for art ceramics, with a pleasing translucency (Fig. ST61).

5.3 Single-Phase Sintering **419**

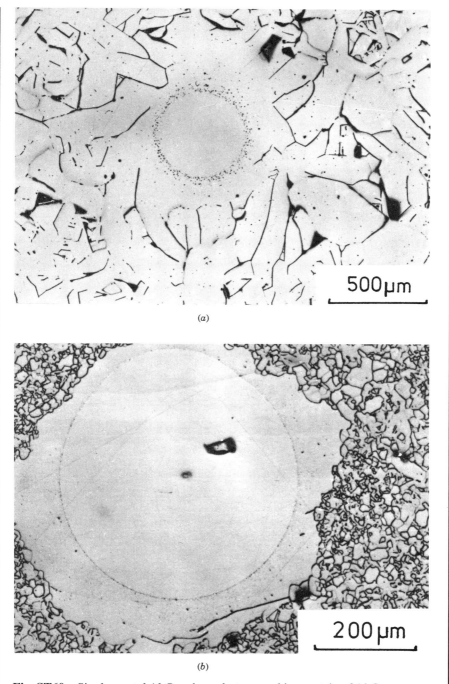

Fig. ST60 Single crystal Al_2O_3 spheres hot-pressed in a matrix of Al_2O_3 powder at 1800°C. (a) Undoped Al_2O_3 matrix, 60 min. (b) 0.1 wt % MgO, 60 min.

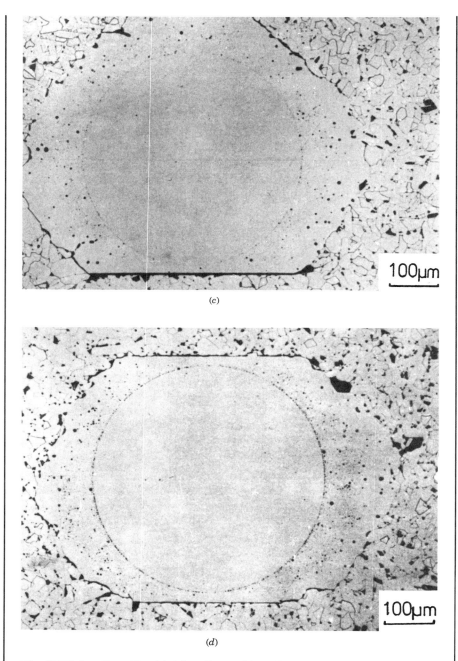

Fig. ST60 (continued) (c) 1.0 wt% anorthite, 10 min.. (d) 0.9 wt % anorthite and 0.1 wt % MgO, 10 min. [From W.A. Kaysser, M. Sprissler, C.A. Handwerker, and J.E. Blendell, *J. Am. Ceram. Soc.*, **70**[5] 339-43 (1987).]

Fig. ST61 Magnesia-doped alumina sintered art ceramics exhibit aesthetically pleasing translucency. (Courtesy of Kyocera Corporation.)

5.4 REACTIVE ADDITIVE SINTERING

While single-phase sintering processes have been the most studied from a basic viewpoint, the majority of commercial ceramics have a small amount of reactive liquid which markedly accelerates the densification rate as compared with the pure system (see Table 5.4). A simple type of reactive liquid is a eutectic liquid in which the primary phase is partly soluble. Examples include $BaTiO_3$, in which a slight excess of titania in the overall composition forms a Ti-rich eutectic liquid at temperatures above 1320°C (see the phase diagram in Fig. 4.7), and ZnO varistors with Bi_2O_3 additive, in which a Bi-rich liquid phase forms above the eutectic temperature of 740°C (Fig. 5.56). Zinc oxide that has been liquid-phase sintered using a small amount of bismuth oxide and quenched to room temperature exhibits a wetting Bi-rich liquid phase at the three-grain junctions, Fig. 5.57. Silica and other oxide additives that react with the parent solid to form a modified siliceous glass are also frequently used for liquid phase sintering. Since silica is a nearly ubiquitous impurity in oxide ceramics, small amounts of glass are sometimes unavoidable. Many electronic ceramics, which may be co-fired with metals at reduced temperatures, are formulated with low melting glasses as densification aids. Covalent structural ceramics such as SiC and Si_3N_4 are difficult to sinter as a single-solid phase even at high temperatures, and are often liquid-phase sintered using oxide additives that form an silicon oxynitride or oxycarbide glass. Tungsten carbide used for cutting tool applications is liquid-phase sintered with a small amount of Ni or Co metal. A wide variety of results can occur depending on the rate of reaction processes, amount of liquid formed at the sintering temperature, relative solubilities of the additive and principle constituent, and the wetting behavior.

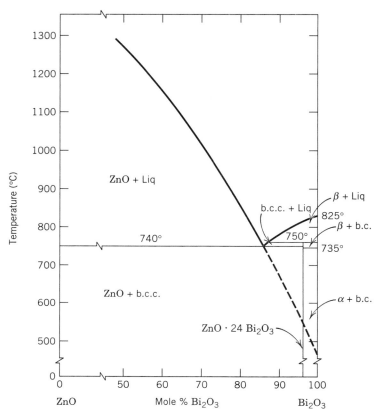

Fig. 5.56 ZnO-Bi_2O_3 phase diagram, showing limited solid solubility of bismuth in zinc oxide and formation of eutectic liquid containing 86% Bi_2O_3 at 740°C. [From G. M. Safronov, V. N. Batog, T. V. Stepanyuk, and P. M. Fedorov, *Russ. J. Inorg. Chem.*, **16**, 460 (1971).]

When a small volume fraction of additive is present, its composition can be quickly saturated by dissolving a small amount of the major phase. During heating and cooling, the composition of the additive phase may easily follow the liquidus curve of the phase diagram (Fig. 5.56). In contrast, there is little change of composition for the major phase during a relatively rapid heating process.

The essential requirements for a successful liquid phase additive are an appreciable solubility for the principle constituent, a reasonably low viscosity for rapid diffusion kinetics, and a contact angle that allows it to wet and penetrate between the principal constituent particles,

$$\phi = 0; \quad 2\gamma_{sl} \leq \gamma_{ss} \tag{5.51}$$

The wetting liquid concentrates at the particle contacts and forms a meniscus (Fig.

5.4 Reactive Additive Sintering 423

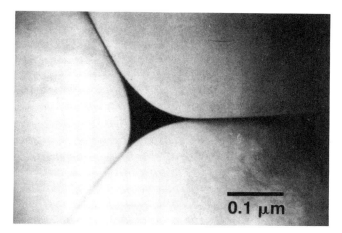

Fig. 5.57 Bismuth-rich phase at a three-grain junction in zinc oxide exhibiting a zero dihedral angle.

5.58) similar to the neck between sintering single-phase particles and exerts an effective compressive pressure on the compact

$$\Delta P \approx \frac{-2\delta_{lv}}{\rho} \tag{5.52}$$

where ρ is the meniscus radius. Hot-stage microscopy has shown that there is an initial, rapid rearrangement of particles into a higher density packing configuration when this occurs. Often the greatest amount of densification takes place here. (In fact, with enough liquid the composite can fully densify at this stage, whereupon further microstructure development becomes strictly a matter of grain-coarsening and reshaping.) After initial rearrangement, further densification takes place as particle contacts flatten under the compressive stress applied to the point contacts by capillary pressure. The compressive stress causes an increased solubility of the solid in the liquid given by

$$c = c_o \exp\left[\frac{KV_m \, 2\gamma_{lv}}{RT\rho}\right] \tag{5.53}$$

where K is a constant that depends on the geometry of the system. Dissolution at the particle contacts and transport of the material towards stress-free interfaces results in *contact flattening*. This is a densification process similar to solid-state sintering by grain boundary diffusion; the thin liquid film takes on the role of the grain boundary as a path of rapid transport. The shrinkage during this process was analyzed by W. D. Kingery [*J. Appl. Phys.*, 30, 301 (1959)]:

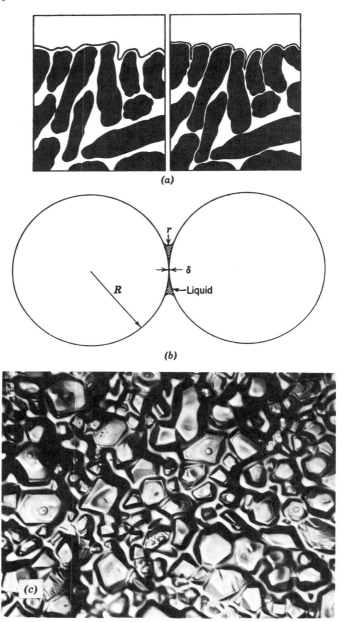

Fig. 5.58 (*a*) Surface of solid-liquid composite with varying amounts of liquid phase. (*b*) Drop of liquid between two solid spheres exerts pressure to pull them together. (*c*) Surface of forsterite ceramic showing liquid capillary depression between crystals.

5.4 Reactive Additive Sintering

$$\frac{\Delta L}{L_o} = \left[\frac{12\,\delta D_{liq}\,C_o\,\gamma_{LV}\,V_m}{RT}\right]^{1/3}\frac{t^{1/3}}{r^{4/3}} \tag{5.54}$$

This result is similar to Eq. 5.48 for grain boundary diffusion; here δ is the width of the liquid film, which may change with time, and C_o is the solubility of the solid in the liquid. Gessinger et al. [G. H. Gessinger, H. F. Fischmeister, and H. L. Lukas, *Acta Met.*, 21, 715 (1973)] have analyzed the situation where the liquid wets the surface sufficiently to form a meniscus but does fully penetrate the grain boundary (Fig. 5.59), and find that the kinetics are of similar form to Eq. 5.54, except that the applicable diffusion coefficient is that along the grain boundary (the driving force is also slightly reduced, depending on the relative values of the dihedral angle between the grain boundary and the liquid and the contact angle where the liquid meets the surface).

After some densification the grains may be mostly surrounded by liquid. Capillary pressure is diminished but particle coarsening can still take place simultaneously with densification. Since the solubility of the smallest particles is greatest, that is,

$$c = c_o \exp\left[\frac{\gamma_{sl}V_m}{RT}\left(\frac{1}{r_1}+\frac{1}{r_2}\right)\right] \tag{5.55}$$

where r_1 and r_2 are the principle radii of curvature of the solid particles, dissolving grains reprecipitate onto larger grains in a way that leads to flattened boundaries

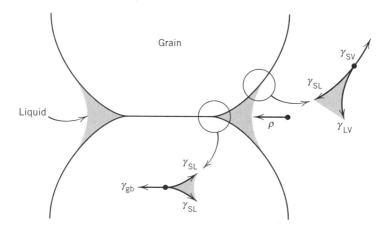

Fig. 5.59 Liquid phase sintering configuration with non-penetrating liquid phase.

426 Chapter 5 / Microstructure

and a more densely packed array of grains (Fig. 5.60). While these processes of rearrangement, densification, and solution-precipitation have been modeled, the changing compositions and overlap of processes occurring as sintering proceeds have not allowed the development of precise time dependencies well confirmed by experiment.

An example of a liquid-phase-sintered ceramic microstructure in which the solidified liquid penetrates fully between grains is shown in Fig. 5.61. In other systems, significant changes in the microstructure occur during cooling. Figure 5.57 shows that in a quenched $ZnO-Bi_2O_3$ sample viewed at high magnification indeed has the zero dihedral angle assumed in the theory. However, during cooling the liquid and solid compositions change, frequently with precipitation of the

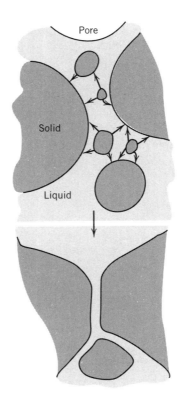

Fig. 5.60 Solution-reprecipitation from smaller particles and highly curved surfaces leads to flattened boundaries and more densely packed grains.

5.4 Reactive Additive Sintering

principal phase. This can give rise to a substantial change of the observed dihedral angle in the microstructure (Fig. 5.62, 5.63). As the sample cools below the solidus temperature, crystallization of a liquid phase can cause a discontinuous increase in the dihedral angle to a nonwetting configuration, since solid–solid interfacial tensions are usually greater than solid–liquid tensions. For silicate liquids that are slower to crystallize, such as the oxynitride glass phase in sintered silicon nitride or the magnesia-glass material in Fig. 5.61, leisurely cooling is often sufficient to preserve the high-temperature phase distribution.

Data for the tungsten-nickel system and the CaF_2-NaF system show that during heating in the solid state, there is already substantially enhanced transport in the boundary regions and preferential dissolution of the smallest particles of the principle phase. In these and several other systems, large amounts of densification may be observed upon heating even before the eutectic temperature is reached, indicating enhanced transport along boundary regions prior to actual melting. This kind of effect is often termed *activated sintering* and is a process that is not well understood. In some cases the presumption of no liquid may be in error due to impurities which lower the solidus temperature below the *apparent* eutectic of the system.

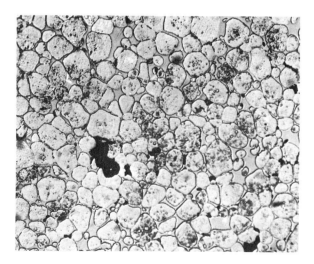

Fig. 5.61 Microstructure of magnesia - 2% kaolin body resulting from reactive-liquid sintering (245×).

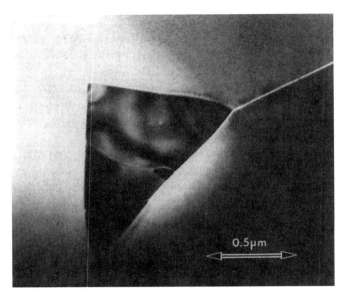

Fig. 5.62 ZnO-Bi$_2$O$_3$ varistor composition, annealed 43.5h. at 610°C resulting in nonwetting secondary phase. [From J.P. Gambino, Ph.D. thesis, Massachusetts Institute of Technology, 1984.]

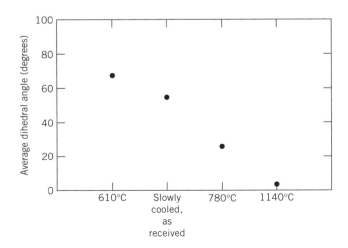

Fig. 5.63 The average dihedral angle observed in ZnO-Bi$_2$O$_3$ for different post-sintering heat treatment conditions.

5.5 HOT-PRESSING

The compressive sintering pressure due to capillarity is modest, on the order of a few megapascals. In most systems there is a limited amount of control over materials parameters such as diffusion and vapor transport rates, interfacial energies, and phase compositions, which may or may not be compatible with densification by solid state and liquid-phase sintering. Hot-pressing is an effective approach to densification which supplements the capillary driving force with an externally applied pressure. Figure 5.64 shows that an applied pressure P_a is magnified at the contact area between particles. The pressure at the grain boundary is [R. L. Coble, J. Appl. Phys., 41, 4798 (1970)]:

$$\frac{\text{force/particle}}{\pi x^2} \approx \frac{P_a (2r)^2}{\pi x^2} \quad (5.56)$$

This pressure is added to the capillary pressure to give a total pressure on the neck of

$$P = \frac{\gamma}{\rho} + \frac{P_a \, 4r^2}{\pi x^2} \approx \frac{\gamma}{\rho} + \frac{P_a r}{\pi \rho}$$

using the relationship $\rho \sim x^2/4r$. Earlier results for densification by lattice and grain boundary diffusion (Eqs. 5.46 and 5.48) remain applicable, except that γ is re-

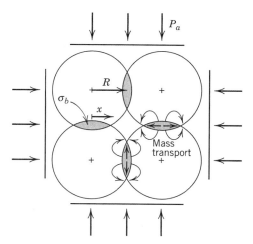

Fig. 5.64 Hot-pressing causes the applied pressure to be magnified at particle contacts and increases the rate of densification.

placed by the term $(\gamma + P_a r/\pi)$. For example, shrinkage by grain boundary diffusion during hot pressing is given by

$$\frac{\Delta L}{L} = \left(\frac{1.5 \delta D_{gb} \Omega}{r^4 kT}\right)^{1/3} \left(\gamma + \frac{P_a r}{\pi}\right)^{1/3} t^{1/3} \qquad (5.57)$$

Since the surface tension γ is ~ 1 J/m², in a compact consisting of 1-µm-radius particles, an applied pressure of 35–70 MPa (5,000–10,000 psi) is sufficient to increase the driving force by a factor of 10 or 20. This applied pressure increases the driving force for densification but not that for coarsening, since the pressure is applied to the grain boundary and does not directly affect the neck curvature which drives coarsening processes. Therefore, hot-pressing promotes densification over coarsening, by increasing the rate of densification. Examination of Eq. 5.57 shows that a factor of 10 increase in the driving force effectively doubles the rate of shrinkage. For simple shapes, hot-pressing of ceramics is typically done under uniaxial loading in a die, often made of graphite to withstand high temperatures, and sometimes lined with a thin protective layer of boron nitride. For more complex shapes, *hot-isostatic pressing* can be used. Two techniques have been used to produce high-density parts using HIP. The first is to enclose the green body, which has a density of 50–80% of theoretical, within a metal capsule or a glass capsule, which is evacuated and sealed. The assembly is then heated in a high-pressure gas atmosphere (typically argon or nitrogen), which compresses and densifies the part uniformly. After cooling, the metal or glass coating is chemically removed. The sealing techniques are complicated and not well adapted to mass production. A second method that is increasingly becoming standard for the production of uniform high-quality, high-density complex ceramics is to use sintering-plus-HIP. If sintering is carried out to 93–95% of theoretical density, the residual pores are closed and the sample surface is gas tight. Hot isostatic pressing can then be accomplished without encapsulation by heating the part to a high temperature in a high-pressure gas atmosphere, achieving complete densification and porosity elimination.

5.6 GLASSES AND GLASS-CERAMICS

Although the transparency and uniform color of many glass objects suggest a uniform material, many glasses contain a microstructure that is inhomogenous at levels varying from a few nanometers to a few microns. Many commercially processed glasses contain microscopic defects due to incomplete mixing, undissolved impurities, or gas bubbles. For glasses which are initially homogeneous, phase separation and crystallization upon cooling can result in varying levels of inhomogeneity. Rather than being a liability, these processes are carefully controlled in many engineered glass products in order to achieve properties not available in a homogeneous glass.

5.6 Glasses and Glass-Ceramics

We discussed in Chapter 1 the structure of glasses and the characteristics of the glass transition. A liquid that is supercooled sufficiently far below the melting temperature without the onset of crystallization must eventually undergo the transition to a glass. The glass is a metastable state; the equilibrium form is crystalline. Most useful glasses are based on complex multicomponent systems that have multiple crystalline phases at equilibrium. *Glass-ceramics* is the term for a broad class of materials that are melted and formed as glasses because of the ease of processing, but are subsequently crystallized to take advantage of properties of the crystalline phase. A highly crystalline ceramic is thus achieved without the shrinkage associated with sintering; and the formability of glass is used to great advantage. The most common glass-ceramics are cookware products such as Corning's Pyroceram™ and Visions™, which after final processing may contain more than 95% by volume of crystalline phases in the Li_2O-Al_2O_3-SiO_2 family. The two crystalline phases of primary interest, β-spodumene and β-quartz solid solutions, lend a low thermal expansion coefficient and high degree of thermal shock resistance to the product. Similar compositions have been used in aerospace applications as thermal-shock-resistant nosecones and radomes, or as fiber-reinforced composites in which the glass ceramic constitutes a matrix of low thermal expansion relative to the fiber, causing useful residual compressive stresses to be developed in the matrix after cooling. "Machinable" glass ceramics are another type of crystallized glass containing a high fraction of a micaceous phase (e.g., potassium phlogopite), which has planes of easy cleavage. When the phase is dispersed as randomly oriented polycrystalline grains in a glass matrix, local fracture occurs easily without the catastrophic propagation of cracks through the material. These glass-ceramics can be drilled and shaped with ordinary machine tools. Crystallized glasses are also potential bioceramic materials for uses in dental implants and prostheses, often containing hydroxyapatite, a biocompatible phase to which bone can knit.

Many glasses also phase separate while in the liquid state into individual phases each of which may crystallize or remain glassy. A variety of interesting and useful microstructures can be obtained by controlling phase separation. Corning's Corelle™ is a strengthened glass product based on a central layer of opaque white glass, which derives its appearance from phase separation, laminated between two borosilicate glass layers. The lower thermal expansion coefficient of the borosilicate causes surface compressive stresses to develop upon cooling to room temperature. Corning's Vycor™ is a nearly pure silica glass that is made at lower temperatures than an equivalent fused silica, by a process involving phase-separation of a sodium-borosilicate glass, chemical leaching of the high alkali phase, and viscous sintering of the remaining porous high-silica network.

Crystallization and Glass Formation

Glasses and glass-ceramics are two sides of the same coin; in forming glasses we seek to *prevent* crystallization, whereas in the processing of glass-ceramics we *promote* crystallization of desired phases and crystal morphologies. Both types of materials can therefore be discussed from the common viewpoint of crystallization. It is convenient to separate the phenomenon of crystallization into two distinct physical processes: (1) nucleation of crystallites; and (2) growth of crystallites. The suppression of either may be sufficient to render a supercooled liquid glassy. In the following, we will aim to understand the combined effects of nucleation and growth on the extent of crystallization after a homogeneous liquid has been cooled.

D. R. Uhlmann [*J. Noncryst. Solids*, 7, 337 (1972)] proposed the use of time-temperature-transformation (TTT) diagrams (Fig. 5.65) to depict the cooling rates necessary to prevent "detectable" crystallization. What is "detectable" depends to some extent upon how closely we scrutinize the material, and its intended purpose, but a generally sufficient definition is a crystal volume fraction, V_c/V_{total}, of $\sim 10^{-6}$ or so. In a TTT diagram (which is widely used to depict other types of phase transformations as well), a characteristic knee-shaped curve (Fig. 5.65) shows the time necessary to reach a certain crystalline fraction at a temperature of interest. The shape of the curve arises from the fact that the overall rate of crystallization is reduced at both high and low temperatures. Specifically, the rate of crystallite *nucleation* is low close to the melting point due to reduced driving force for crystallization. At temperatures below the "knee" in the curve, crystallization is again slow, because crystal *growth rates* are low. At large undercoolings below the melting point, the nucleation rate may be very high, but in the absence of growth the overall crystallization rate remains low. The knee therefore represents the combination of nucleation rate and growth rate which gives the maximum rate of crystallization. Liquids that are cooled faster than a certain critical rate, R_c, avoid intersecting the transformation curve and can form a glass. The critical rate is given by

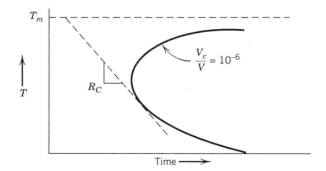

Fig. 5.65 Schematic time-temperature-transformation (TTT) diagram showing the time necessary to crystallize a certain volume fraction of the material (V_c/V) upon cooling to a temperature below the melting point T_m.

5.6 Glasses and Glass-Ceramics

$$R_c = \frac{T_m - T_n}{t_n} \tag{5.58}$$

where T_m is the melting point and T_n and t_n are the temperature and time of the nose.

Proceeding further, these features of the TTT diagram can be understood from the thermodynamics and kinetics of crystal nucleation and growth. A stable crystallite will only form when the liquid is sufficiently undercooled for the volume free energy decrease upon solidification to overcome the surface energy increase. For the homogeneous nucleation of spherical crystallites, the Gibbs free energy change for a nucleus of radius r is

$$\Delta G(r) = 4\pi r^2 \gamma_{SL} + \frac{4}{3}\pi r^3 \Delta G_v \tag{5.59}$$

where γ_{SL} is the solid–liquid interfacial energy and ΔG_v is the volume free energy change of crystallization. ΔG_v is zero at the melting point, and becomes negative in sign and increases in magnitude with undercooling below T_m. The individual terms in Eq. 5.59 and their sum are shown against the crystal radius r in Fig. 5.66. There is a maximum in $\Delta G(r)$ at a critical size r^*, beyond which growth is favored since the free energy decreases with increasing r. Below r^* dissolution of the nucleus is favored since ΔG decreases as r decreases. Taking the derivative of Eq. 5.59 with respect to r and equating to zero at $r = r^*$, we obtain $r^* = -2\gamma_{SL}/\Delta G_v$.

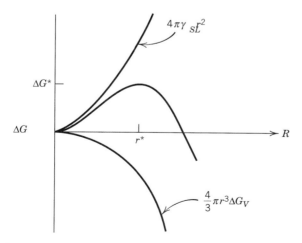

Fig. 5.66 The free energy change for growth of a crystal of radius r at temperatures below the melting point includes a positive surface energy and a negative volume free energy contribution.

Chapter 5 / Microstructure

Substituting back into Eq. 5.59 gives us the free energy change for the formation of supercritical nuclei

$$\Delta G^* = \frac{16 \pi \gamma_{SL}^3}{3(\Delta G_V)^2} \tag{5.60}$$

Treating the formation of nuclei as an activated process, the steady-state number density (per unit volume) of critical nuclei is given by

$$n^* = N_v \exp\left(\frac{-\Delta G^*}{kT}\right) \tag{5.61}$$

where N_v is the number of molecules per unit volume. Since the addition of one additional molecule to a critical nucleus is in principle sufficient to make it supercritical, the rate at which supercritical nuclei are formed (in terms of number per unit volume per unit time) is given by:

$$I_v = \upsilon N_v \exp\left(\frac{-\Delta G^*}{kT}\right) \tag{5.62}$$

where υ is the rate of molecular attachment. In a fluid liquid the rate of attachment is on the order of 10^{11} sec^{-1}, and since $N_v \sim 10^{22}$–10^{23} cm^{-3}, the nucleation frequency is

$$I_v \approx 10^{33} \exp\left(\frac{-\Delta G^*}{kT}\right) \quad (\text{no./cm}^3/\text{sec}) \tag{5.63}$$

In a viscous melt the attachment frequency is lower and is strongly dependent on viscosity, η. Using the Stokes–Einstein relation to estimate the atom diffusivity in a viscous melt,

$$D = \frac{kT}{3\pi a_0 \eta} = \upsilon a_o^2 \tag{5.64}$$

where a_0 is the atomic jump distance (~0.3 nm), we may expect that υ will be inversely proportional to viscosity. Experimental data suggests that υ is generally greater than what Eq. 5.63 predicts, by approximately a factor of forty [D. R. Uhlmann, et al., *J. Phys.*, 43, C9–175 (1983)]. Combining Eqs. 5.64 and 5.62 we have for the nucleation rate

$$I_v = \frac{40 N_v kT}{3\pi a_o^3 \eta} \exp\left(\frac{-\Delta G^*}{kT}\right) \tag{5.65}$$

Since the temperature dependence of other quantities in the pre-exponential term is small compared to that of the viscosity, Eq. 5.65 can be further approximated as

$$I_v \approx \frac{K}{\eta} \exp\left(\frac{-\Delta G^*}{kT}\right) \tag{5.66}$$

where K is a constant, of value $\sim 10^{30}$ cm^{-3}sec^{-1}poise for oxide glassformers.

Before proceeding further, we evaluate the volume free energy of crystallization, ΔG_v, as this is the driving force that determines ΔG^* (Eq. 5.60). (The volume free energy is related to the molar free energy by $\Delta G_v = \Delta G_m/V_m$, where V_m is the molar volume.) ΔG_m is zero at the melting point, where the liquid and the crystal are in equilibrium with one another, and increases in magnitude with decreasing temperature. A number of approximations have been developed for the dependence of ΔG_m on undercooling ($\Delta T = T - T_m$). The simplest approximation is by Turnbull [*J. Appl. Phys.*, 21, 1022 (1950)], who considered that for small undercoolings the molar heat and entropy of fusion, ΔH_m and ΔS_m, are not greatly changed from their values at the melting point. At the melting point we have the relationship

$$\Delta G_m = \Delta H_m - T_m \Delta S_m = 0 \tag{5.67}$$

which upon rearranging,

$$\Delta S_m = \frac{\Delta H_m}{T_m} \tag{5.68}$$

and substituting back into 5.67, gives the result

$$\Delta G_m \approx \frac{\Delta H_m \Delta T}{T_m} \tag{5.69}$$

Notice that this allows the free energy of fusion to be approximated from the heat of fusion and melting point, known for many materials and easily measured by calorimetric methods. While Eq. 5.69 is a reasonable approximation at small undercooling, at larger undercoolings it fails since ΔH_m and ΔS_m cannot be assumed to remain independent of temperature. Other approximations for ΔG_m account for the temperature dependence of ΔH_m and ΔS_m in various ways. J. D. Hoffman [*J. Chem. Phys.*, 29, 1192 (1958)] assumed that the difference in heat capacity between the liquid and the crystal, ΔC_p, remains constant with undercooling and obtained the approximation

$$\Delta G_m \approx \frac{\Delta H_m \Delta T \cdot T}{T_m^2} \qquad (5.70)$$

C. V. Thompson and F. Spaepen [*Acta Met.*, 27, 1855 (1979)] pointed out that Hoffman's result is applicable when ΔC_p is large (as it is for oxide glasses and polymers) but not when it is small (as for liquid metals). They obtained an alternative approximation suitable for materials of small ΔC_p

$$\Delta G_m \approx \frac{\Delta H_m \Delta T}{T_m} \left(\frac{2T}{T_m + T} \right) \qquad (5.71)$$

Using the free energy approximations in Eqs. 5.69–5.71, the nucleation frequency can be determined from Eqs. 5.63, 5.65, or 5.66. For example, the substitution of Turnbull's approximation into Eq. 5.63 gives as the nucleation rate for fluid systems at small undercoolings

$$I_v \approx 10^{33} \exp\left[\frac{-16\pi V_m^2 \gamma_{SL}^3}{3kT\left(\Delta H_m^2 \Delta T^2 / T_m^2\right)} \right] \qquad (5.72)$$

The temperature dependence of nucleation in this relationship (or the equivalent expression obtained using Eq. 5.71) is sharp, so much so that homogeneous nucleation can be thought of as occurring at a characteristic temperature for a material. In very fluid melts the onset of nucleation is easily detected since subsequent crystal growth is very fast. Table 5.5 shows that homogeneous nucleation typically takes place at a temperature that is 75%–85% of the absolute melting point. Notice that the nucleation rate expressed in Eq. 5.72 is also extremely sensitive to surface energy $[I_v \propto \exp(\gamma_{SL}^3)]$; in fact, working backwards, the temperature of homogeneous nucleation can be used to estimate γ_{SL}.

There is no upper limit to the nucleation frequency expressed in Eq. 5.72, which results from the simplifying assumption, incorrect in detail, that the atom attachment rate is independent of temperature. In the viscous melts typical of many oxide glassforming systems, viscosity can increase sharply with decreasing temperature, and undercoolings are frequently large. The nucleation frequency is then better approximated by substituting Hoffman's result into Eq. 5.66

$$I_v \approx \frac{K}{\eta} \exp\left[\frac{-16\pi}{3kT} \frac{V_m^2 \gamma_{SL}^3}{\Delta H_m^2 \left(\Delta T \cdot T / T_m^2\right)^2} \right] \qquad (5.73)$$

5.6 Glasses and Glass-Ceramics

Table 5.5 Experimental Nucleation Temperatures

	T_m (K)	T^* (K)	$\Delta T^*/T_m$
Mercury	234.3	176.3	0.247
Tin	505.7	400.7	0.208
Lead	600.7	520.7	0.133
Aluminum	931.7	801.7	0.140
Germanium	1231.7	1004.7	0.184
Silver	1233.7	1006.7	0.184
Gold	1336	1106	0.172
Copper	1356	1120	0.174
Iron	1803	1508	0.164
Platinum	2043	1673	0.181
Boron trifluoride	144.5	126.7	0.123
Sulfur dioxide	197.6	164.6	0.167
CCl_4	250.2	200.2±2	0.202
H_2O	273.2	232.7±1	0.148
C_6H_6	278.4	208.2±2	0.252
Naphthalene	353.1	258.7±1	0.267
LiF	1121	889	0.21
NaF	1265	984	0.22
NaCl	1074	905	0.16
KCl	1045	874	0.16
KBr	1013	845	0.17
KI	958	799	0.15
RbCl	988	832	0.16
CsCl	918	766	0.17

T_m (K) is the melting point, T^* (K) the lowest temperature to which the liquid could be supercooled. $\Delta T^*/T_m$ is the maximum supercooling in reduced temperature units.
Source: K. A. Jackson in *Nucleation Phenomena*, American Chemical Society, Washington, 1965.

In this case an increase in the driving force with decreasing temperature, represented by the exponential term, can be offset by an increase in the viscosity η. A maximum in the nucleation frequency with decreasing temperature occurs, below which the increase in viscosity freezes out further nucleation.

Once a nucleus is formed, its rate of growth is determined by the rate of atom attachment at the crystal–liquid interface. K. A. Jackson (*Progress in Solid State Chemistry*, Vol. 3, Pergamon Press, NY, 1967) related the atomic morphology of the crystal–liquid interface during growth to the entropy change during crystallization, ΔS_m, and deduced that when a small entropy change is involved in crystallization ($\Delta S_m < 2R$, where R is the gas constant), the growing interface is rough on an atomic scale (often called *normal* or *Wilson-Frenkel* growth) and the growth

438 Chapter 5 / Microstructure

rate is relatively isotropic (Fig. 5.67). On the other hand, if the entropy change is large ($\Delta S > 4R$), an atomically smooth interface forms on the densely packed crystal faces and a rough interface on the less densely packed faces. The growth rate anisotropy tends to be large under these conditions, and faceted crystals result. (Notice that since crystal growth from the vapor phase or dilute liquid solutions generally involves a large ΔS, faceted crystals are the norm for these processes.)

The growth rate of an atomically rough interface is given by reaction-rate theory as

$$u = va_o \left[1 - \exp\left(\frac{-\Delta G_m}{RT} \right) \right] \tag{5.74}$$

where v is the frequency with which atoms are transported across the interface, a_o is the distance the crystal advances for a layer of attached molecules, and ΔG_m is the free energy of crystallization. For small undercooling ΔG_m is given by Eq. 5.69; and when $\Delta G_m \ll RT$, Eq. 5.74 is approximately

$$u \approx va_o \left(\frac{\Delta H_m \, \Delta T}{RT T_m} \right) \tag{5.75}$$

Thus for small undercoolings, the growth rate is proportional to the degree of undercooling and the heat of fusion. At the other extreme of very large undercooling ($\Delta G_m \gg RT$), Eq. 5.74 reaches a limiting value

$$u \approx va_o \tag{5.76}$$

indicating that growth is restricted only by the rate of atom transport across the interface. Since a_o is a few angstroms and v has a maximum value equal to the

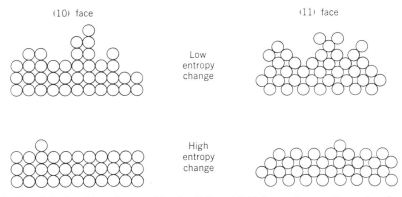

Fig. 5.67 Calculated interface profiles for (10) and (11) faces of a two-dimensional crystal, showing the effects of entropy change on interface structure. (Courtesy of K.A. Jackson.)

atomic vibrational frequency, the maximum crystal growth velocity is ~10^5 cm/sec. For viscous fluids, v can be approximated by Stokes' law

$$v = \frac{kT}{3\pi a_o^3 \eta} \quad (5.77)$$

and the growth rate obtained by substituting Eq. 5.77 in Eq. 5.74, using an appropriate approximation for ΔG_m. The net impact of temperature on growth rate can be seen in Fig. 5.68 for GeO_2 crystals, which undergo normal growth. The growth rate is slow at low undercooling due to a small driving force, reaches a maximum with further undercooling, and slows again at large undercooling due to the influence of temperature on viscosity (Eq. 5.77).

The growth of atomically smooth interfaces follows dependencies similar to that of rough interfaces, with the exception that the density of surface sites at which atoms can be attached is lower. The growth rate is expressed by relations like Eqs. 5.74–5.75, except that a multiplying factor f, which represents the fraction of the total sites available for atom attachment, must be included. This fraction may be much less than 1. If screw dislocations are present, they often provide

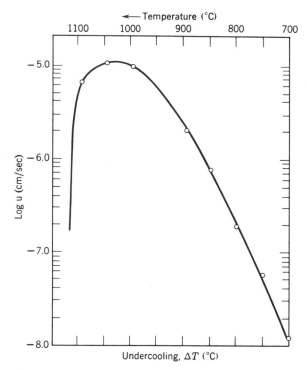

Fig. 5.68 Growth rate versus undercooling relation for GeO_2. [From V.J. Vergano and D.R. Uhlmann, *Phys. Chem. Glasses*, **11**, 30 (1970).]

a ledge of preferred growth sites where they exit the crystal (Fig. 5.69). Growth spirals like that illustrated in Fig. 5.70 are characteristic of this type of growth. The screw dislocation is a heterogeneous site for the growth of a two-dimensional nucleus; a tighter spiral provides more growth sites (larger f) at the expense of an increase in surface energy. It is found that the density of sites increases in proportion to the undercooling [W. B. Hillig and D. Turnbull, *J. Chem. Phys.*, 24, 914, (1956)]

$$f \approx \frac{3 \Delta H_m a_o \Delta T}{4 \pi V_m \gamma_{SL} T_m} \approx \frac{\Delta T}{2 \pi T_m} \tag{5.78}$$

In the absence of screw dislocations, an atomically smooth crystal still can grow by nucleating two-dimensional steps on the surface, which then propagate laterally, but this "homogeneous" nucleation process has a high activation barrier and requires a greater driving force (greater undercooling) before it becomes significant compared to spiral growth.

These growth processes are applicable when growth is limited by atom attachment at the interface. When the growing crystal differs in composition from the liquid, solute must be rejected as the crystal grows, and if the interface attachment process is comparatively fast, diffusion of the rejected solute away from the crystal interface will limit growth. In this case the growth rate decreases with time as the solute diffusion layer thickens. If the morphology of growth is such that the

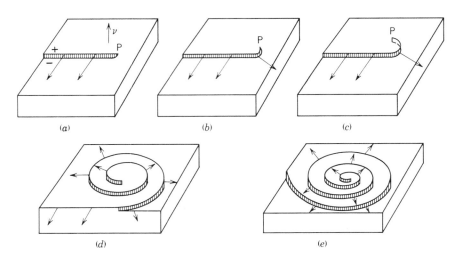

Fig. 5.69 (a) Step attached to the emergence point P of a dislocation with a Burgers vector which is not parallel to the surface. The step height $h = \bar{b} \cdot \bar{v}$ where $\bar{v}$ is the unit normal on the surface. (b) to (e): Under the influence of supersaturation the step in (a) winds up into a spiral centered on P.

5.6 Glasses and Glass-Ceramics

diffusion length is small compared to the crystal dimension, for example when rodlike dendrites grow lengthwise by rejecting solute radially, the growth rate may be constant in time, but remains diffusion-limited.

Combining the time and temperature dependencies of the nucleation and growth rates, the TTT diagram can be determined. For simplicity, consider a spherical nucleus that reaches its critical size at time t' and then grows outward at rate u.

Fig. 5.70 Growth spiral on cadmium iodide crystals growing from water solution. Interference contrast micrograph at 1025×. (Courtesy K.A. Jackson.)

From the theory of phase transformations the volume fraction crystallized at a time t is given by

$$\frac{V_c}{V} = 1 - \exp\left[-\int_0^t I_v \frac{4}{3}\pi \left(\int_{t'}^t u\, d\tau\right)^3 dt'\right], \qquad (5.79)$$

and if both I_v and u are assumed to be constant in time, Eq. 5.79 reduces to

$$\frac{V_c}{V} = 1 - \exp\left[-\frac{\pi}{3} I_v u^3 t^4\right] \qquad (5.80)$$

which is known as the Johnson–Mehl equation. Since we are concerned with small values of V_c/V, the argument of the exponential term must be small, in which case Eq. 5.80 can be further simplified to

$$\frac{V_c}{V} \approx \frac{\pi}{3} I_v u^3 t^4 \qquad (5.81)$$

Equations 5.79–5.81 are useful results for determining the TTT diagram. Upon inserting the nucleation rate and growth rate from earlier relationships, the time dependence of the crystallized fraction at a given temperature can be calculated. Notice that since the crystallization time is proportional to $(V_c/V)^{1/4}$, it is not especially critical what volume fraction we choose as the "detectable" limit for glass formation. In a few glassforming systems sufficient data exists for a reasonable test of the theory. Figure 5.71a shows TTT curves calculated for a crystal fraction $V_c/V \sim 10^{-3}$ in sodium disilicate ($Na_2O \cdot 2SiO_2$) and anorthite ($CaO \cdot Al_2O_3 \cdot 2SiO_2$), which agree well with the onset of detectable crystallinity determined by x-ray diffraction (Fig. 5.71b) [D. Cranmer, R. Salomaa, H. Yinnon, and D. R. Uhlmann, *J. Non-Cryst. Solids*, 45, 127 (1981)].

Notice, however, that there is somewhat of an inconsistency in the way we have defined the critical cooling rate (Fig. 5.65) and the way in which we have calculated the TTT diagram; the cooling rate is a *continuous* variation of temperature, whereas the TTT diagram was calculated from *isothermal* crystallization. A better estimate of the critical cooling rate is achieved by taking into account the crystallization during each incremental step of cooling. Figure 5.72 compares TTT curves calculated for isothermal crystallization ($V_c/V=10^{-6}$) and for crystallization during continuous cooling (CT curve) of a lunar glass composition. The difference in critical cooling rate between the two is about an order of magnitude. In Fig. 5.72 the transformation curve for substantially complete crystallization, $V_c/V=0.5–0.9$, is also shown. This occurs at a continuous cooling rate approximately another order of magnitude slower.

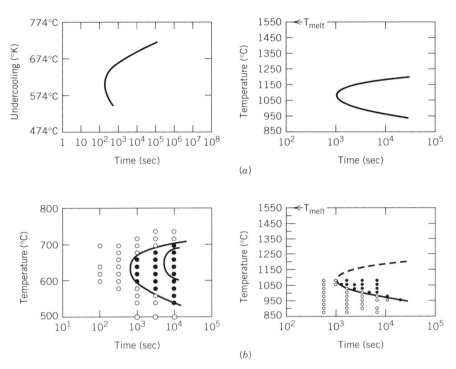

Fig. 5.71 (a) Calculated TTT curves for $Na_2O \cdot 2SiO_2$ at $V_c/V = 10^{-6}$ and $CaO \cdot Al_2O_3 \cdot 2SiO_2$ (anorthite) at $V_c/V = 10^{-3}$. (b) Experimental observations of crystallized (closed symbols) and glassy (open symbols) samples. (From G.S. Meiling and D.R. Uhlmann, *Phys. Chem. Glasses*, **8**, 62 (1967), D. Cranmer, R. Salomaa, H. Yinnon, and D.R. Uhlmann, *J. Non-Cryst. Solids*, **45**, 127, 1981, and H. Yinnon and D.R. Uhlmann, in *Glass: Science and Technology*, Vol. 1, D.R. Uhlmann and N.J. Kreidl, Editors, Academic Press, 1983.)

In this discussion we have considered nucleation to be homogeneous. The presence of external and internal surfaces and other defects, which act as heterogeneous nucleation sites, can dramatically lower the supersaturation (undercooling) necessary for crystallization. A nucleus that forms in the shape of a spherical cap (Fig. 5.73) incurs less of an increase in surface energy than does the homogeneous nucleation of a sphere of equivalent volume. The spherical cap introduces two new interfaces of energies γ_1 and γ_3 where there initially was one (γ_2), with an increase in surface energy that depends on the shape of the cap, that is, the relative values of the interfacial energies,

$$\gamma_2 = \gamma_1 \cos\theta + \gamma_3 \tag{5.81a}$$

where θ is the contact angle. As θ decreases to zero, the excess surface energy of the nucleus vanishes (since $\gamma_2 = \gamma_1 + \gamma_3$) and there is no barrier to nucleation. For

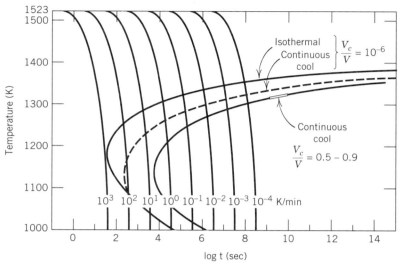

Fig. 5.72 Isothermal and constant-rate cooling TTT curves corresponding to $V_c/V = 10^{-6}$ for lunar glass. (From H. Yinnon and D.R. Uhlmann, in *Glass: Science and Technology*, Vol. 1, D.R. Uhlmann and N.J. Kreidl, Editors, Academic Press, 1983.)

nonzero θ the nucleation barrier is given by

$$\Delta G^*_{heterog} = \Delta G^*_{homog} f(\theta) \tag{5.81b}$$

where

$$f(\theta) = \frac{(2+\cos\theta)(1-\cos\theta)^2}{4} \tag{5.82}$$

and ΔG^*_{homog} is given by Eq. 5.59. Figure 5.74 compares TTT curves calculated for heterogeneous nucleation in $Na_2O \cdot 2SiO_2$ at various contact angles with that for homogeneous nucleation, and shows that the critical cooling rate necessary to form a glass can increase by several orders of magnitude when the contact angle is small.

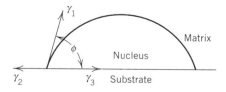

Fig. 5.73 Spherical cap model of heterogeneous nucleation.

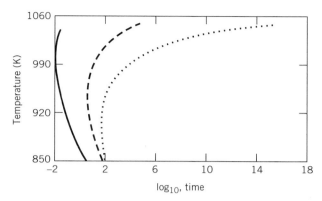

Fig. 5.74 TTT curves for $Na_2O\text{-}2SiO_2$ showing effects of nucleating heterogeneities, at $V_c/V = 10^{-6}$. Solid line, contact angle = 40°; dashed line, contact angle = 80°, dotted line, homogeneous nucleation or contact angle = 120° and 160°. (From P. Onorato and D.R. Uhlmann.)

At an atomistic level, the increase in interfacial energy when one crystal nucleates upon another is dependent on the lattice mismatch. Closely matching crystals can grow *epitaxially* upon one another, forming a coherent interface of low energy in which lattice planes maintain their continuity across the interface. Seed crystals of this kind provide low energy heterogeneous nucleation sites. In glass-ceramics, crystallization is stimulated by *nucleating agents* believed to act in this way, and also by phase separation in the starting glass, which provides internal surfaces for heterogeneous nucleation.

Often, the data necessary to perform detailed calculations of TTT diagrams are not available. Uhlmann et al. (1979) and Onorato et al. (1981) have presented a simplified model by which the critical cooling rates necessary to form a glass can be determined to order-of-magnitude accuracy using limited kinetic data. From TTT curves for a wide variety of materials, they noticed that the nose in the curves is approximately at 0.77 of the melting point. Using this approximation, the time to crystallize a given fraction can be calculated for just one temperature, and the critical cooling rate is that necessary to avoid the nose in the curve. A second useful approximation is of the nucleation barrier ΔG^* (Eq. 5.60). From collected data for homogeneous nucleation in a wide range of materials (of large entropy of fusion), they found the nucleation barrier to be well approximated by

$$\Delta G^* \approx 12.6 \frac{\Delta S_m}{R} kT^* = BkT^* \qquad (5.83)$$

where $T^* \approx 0.8 T_m$ and the constant B is of order 50–60. Using these approximations, an expression which gives the critical cooling rate for glass formation to order-of-magnitude accuracy is

$$R_c \approx \frac{AT_m^2}{\eta} \exp(-0.212\,B) \left[1 - \exp\left(-\frac{0.3\,\Delta H_m}{RT_m}\right)\right]^{3/4} \qquad (5.84)$$

where $A = 4 \times 10^5$ erg/cm^3 K and η is understood to be the viscosity at the temperature of the nose, $0.77\,T_m$.

The critical cooling rates for glass formation vary greatly between materials, as shown in Table 5.6. As is our experience when observing a glassblower at work, many silicate glasses can be cooled at quite leisurely rates without crystallization. On the other hand, metallic glasses can only be formed when the cooling rate is on the order of 10^5-10^7 K/sec. This is achieved with procedures such as "splat" quenching or melt-spinning, in which the melt is flattened into a thin film or continuous ribbon against a spinning metal wheel for a high rate of heat transfer. Some materials cannot be prepared as a glass from the melt at currently achievable cooling rates. Vapor phase deposition onto a cold substrate and sol-gel chemical synthesis are alternative low temperature methods of preparing glassy materials.

To summarize these many considerations involved in glass formation, materials that are good glass formers tend to be those which possess the intrinsic properties of a high viscosity at the melting point, a large liquid–crystal interfacial energy, and a low value of the ratio $\Delta H_m/T_m$ (Since this characterizes the bulk free energy of crystallization, see Eq. 5.69). It is also helpful if the system is one in which there is a large difference in composition between the crystallizing phase and the matrix, which slows down crystal growth due to solute rejection. Finally, it is important to avoid nucleating heterogeneities if glass formation is to be promoted.

Controlled Crystallization in Glass-Ceramics

The exemplar of controllably crystallized glasses is the lithia-alumina-silica family, first developed at the Corning Company in the late 1950s by S. D. Stookey

Table 5.6 Estimated Cooling Rates for Glass Formation

Material	dT/dt (°K/sec) Homogeneous nucleus $\Delta G^* = 50kT$ $T_r = 0.2$	dT/dt (°K/sec) Heterogeneous nucleus $\theta = 80°$ $\Delta G^* = 50kT$ $T_r = 0.2$	dT/dt (°K/sec) Homogeneous nucleus $\Delta G^* = 60kT$ $T_r = 0.2$
Na$_2$O·2SiO$_2$	4.8	46	0.6
GeO$_2$	1.2	4.3	0.2
SiO$_2$	7×10^{-4}	6×10^{-3}	9×10^{-5}
Salol	14	220	1.7
Metal	1×10^{10}	2×10^{10}	2×10^9
H$_2$O	1×10^7	3×10^7	2×10^6

$T_r = \Delta T/T_m$

(U.S. Patent No. 2,920,971, January 12, 1960).[2] These materials derive their utility from a crystalline phase of low thermal expansion coefficient. One such phase is the β-quartz solid solution, in which Al^{3+} substitutes for Si^{4+} with interstitial Li^{1+} maintaining local charge neutrality (analogous to the structural role of aluminum in silicate glasses when paired with alkali ions, discussed in Chapter 1). Extensive substitution into the quartz structure is possible, up to the stoichiometric composition representing β-eucryptite, [Li Al]SiO_4. The volume thermal expansion coefficient of β-quartz decreases with increasing substitution by Al and Li, reaching a value of zero at a Li_2O:Al_2O_3:SiO_2 ratio of about 1:1:3.5, and becoming in fact *negative* at higher substitutions and at the stoichiometric β-eucryptite composition. The lithia-alumina-silica formulations typical of Pyroceram™ contain β-spodumene (LiAl[Si_2O_6]) as the major crystalline phase, which has a weakly positive thermal expansion coefficient (~10^{-6} per degree). The ultimate thermal expansion coefficient is determined by the specific compositions and relative amounts of β-eucryptite, β-spodumene, residual glass phase, and other minor crystalline phases; a typical value is 0.5×10^{-6} per degree. (The resulting thermal shock resistance is discussed in Special Topic 5.3.) The size of the β-eucryptite and β-spodumene crystals in Pyroceram™ is on the order of a micrometer, which is large enough for light scattering that gives the characteristic white appearance of these products. In comparison, transparent glass ceramics (e.g., Corning's Visions™, Schott Glass's Zerodur™, and Nippon Electric's Narumi™ line of products) contain β-quartz as the principal crystalline phase in the form of finer crystallites less than a tenth of a micrometer in dimension. Crystallite sizes below the wavelength of visible light, combined with a good match in refractive index between the crystal and residual glass (achieved by small adjustments in the glass composition) results in a high transparency material. Various rare-earth or transition metal additives are used when a tinted color is desired. In addition to possibly improved aesthetics as cookware, transparent glass ceramics are used for oven windows and rangetops.

The crystallization of the desired phases at controlled grain sizes is achieved by adding heterogeneous nucleating agents such as TiO_2 and ZrO_2, and by carrying out a two-step heat treatment sequence causing first the copious nucleation of crystallites, followed by crystal growth (Fig. 5.75). The object is formed from the glass at a working-range viscosity of about 10^4 poise, corresponding to a temperature range near the liquidus at 1250°C. The glass object is then held at a moderate temperature of 800–850°C for several hours, during which an extremely high density of heterogeneous nuclei [10^{12}–10^{15} nuclei per cubic centimeter] form. In opaque glass ceramics nucleated with TiO_2, the nucleation sites are believed to be crystallites of rutile which act as heteroepitaxial seeds for crystallization of β-eucryptite and β-spodumene due to a reasonably close lattice match between the crystals (within 15% for certain crystal planes). The crystallization of insoluble TiO_2 early

---

[2] See S.D. Stookey, *Journey to the Center of the Crystal Ball*, The American Ceramic Society, 1985, for a personal history of the development of glass-ceramic materials and a list of relevant Corning patents.

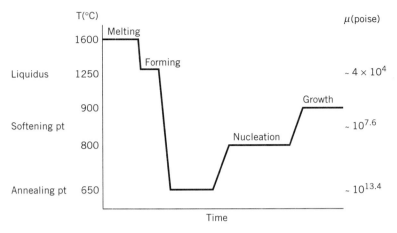

Fig. 5.75 Typical processing cycle for Li_2O-Al_2O_3-SiO_2 glass ceramics.

in the process is aided by the phase separation of a low viscosity glass phase enriched in Li, Al, Ti, and other minor constituents. Following the nucleation heat treatment, a higher temperature treatment at 875–900°C for several hours grows the crystalline phases until the grains impingement upon one another, as shown in Fig. 5.76. The final grain size is largely determined by the mean separation between nuclei. Transparent glass ceramics have a higher density of nuclei, achieved using the dual nucleating agents TiO_2 and ZrO_2, to precipitate $ZrTiO_4$ nucleation sites. The importance of a high nucleation density is illustrated by the microstructures in Fig. 5.76, showing a large difference in grain size depending on whether or not a lower-temperature nucleation heat treatment was used.

A variation on the Pyroceram™ theme is carried out in the chemically strengthened Corning material Chemcor™, a glass-ceramic based on sodium/potassium aluminosilicate rather than lithium aluminosilicate. When the base glass is ion-exchanged with a lithium salt, a surface layer of lithia aluminosilicate is formed which can be further processed into a low-thermal expansion glass-ceramic layer. Upon cooling from heat treatment temperatures large compressive stresses develop in the surface layer due to its thermal expansion mismatch with the core glass. The resulting material have a bending strength of about 600–700 MPa, as compared to a strength of about 40 MPa in conventional Pyroceram.

A wide variety of glass-ceramic materials in many compositional families have been developed since the original inventions of Stookey, including those in which crystalline phases are tailored for high mechanical strength, high temperature capability, photosensitivity, low dielectric constant for electronic packaging, dielectric-breakdown resistance, and biological compatibility.[3] One interesting family of materials is based on the crystallization of micaceous phases which, due to easy cleav-

---

[3] Many of these are discussed in a recent review by G.H. Beall, "Design and Properties of Glass-Ceramics," *Ann. Rev. Mater. Sci.*, 91–119, 1992.

5.6 Glasses and Glass-Ceramics

age along the basal plane, results in machinable glass ceramics (Fig. 5.77). These materials are generally derivatives of the K_2O-MgO-Al_2O_3-SiO_2 system containing some fluorine, in which the important crystalline phase is potassium phlogopite, $KMg_3(AlSi_3O_{10}F_2)$. Machinable glass ceramics are relatively soft ceramics, but some possess strengths (200–300 MPa) comparable to many sintered ceramics.

Phase Separation

Single-phase glasses can undergo phase separation upon cooling into fields of immiscibility defined by the equilibrium or metastable phase relations, discussed earlier in Chapter 4. Phase equilibria dictates the type, number, and composition of phases that are formed. The microstructural size scale and morphology of phase separated glasses are determined by kinetic processes as well. J. W. Gibbs (*Scientific Papers*, Vol. 1, Dover Publications, New York, 1961) first pointed out that phase separation could take place in a manner that is *large in degree but small in spatial extent, or small in degree and large in spatial extent*. Nucleation and growth, which we discussed above with respect to crystallization of glasses, may be con-

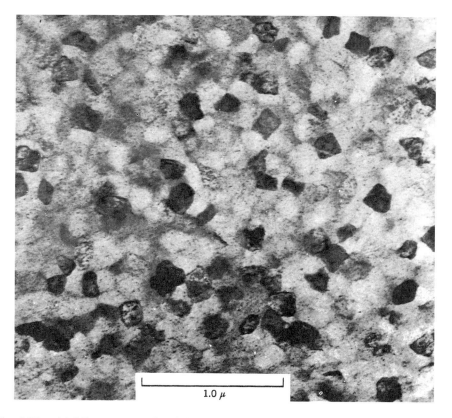

Fig. **5.76** (*a*) Microstructure in Li_2O-Al_2O_3-SiO_2 glass ceramic held at 775°C for 2 h. before heating to 975°C for 2 min.

450 Chapter 5 / Microstructure

sidered the former: The nucleus is a small region at which there is a large discontinuous change in composition which grows in size (Fig. 5.78b). *Spinodal decomposition* is a process of phase separation corresponding to the latter, in which a small fluctuation in composition over a large length scale amplifies until the phases reach their equilibrium compositions (Fig. 5.78a). These two processes often result in different microstructures, especially at the earlier stages of phase separation.

Detailed theoretical descriptions of spinodal decomposition were developed during the 1960s [see J. W. Cahn *Acta Met.*, 9, 795, 1961, and *Trans. Met. Soc.*

Fig. 5.76 (b) Identical composition heated rapidly to 875°C and held for 25 min. (From P.E. Doherty in R.M. Fulrath and J.A. Pask, Eds., *Ceramic Microstructures*, John Wiley and Sons, New York, 1968, pp. 161-185.)

5.6 Glasses and Glass-Ceramics **451**

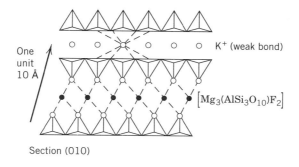

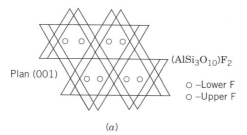

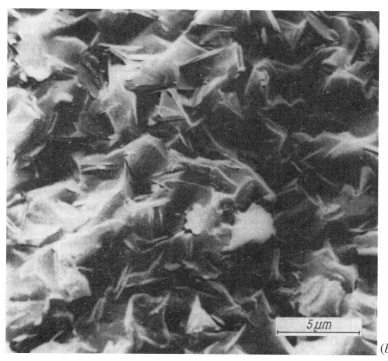

Fig. 5.77 (*a*) Structure of fluorophlogopite (from G. Beall). (*b*) Micaceous grains of sodium phlogopite in a machinable glass-ceramic. (From W. Vogel, *Chemistry of Glass*, The American Ceramic Society, Columbus, Ohio, 1985.)

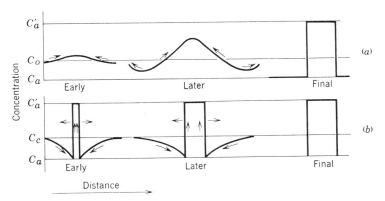

Fig. 5.78 Schematic evaluation of concentration profiles for (*a*) spinodal decomposition and (*b*) nucleation and growth. [From J.W. Cahn, *Trans. Met. Soc. AIME*, **242**, 166 (1968).]

AIME, 242, 166, (1968)]. It is a process that has been found to occur in a variety of materials including glasses, crystalline ceramics, metals, and polymers. Thermodynamically, nucleation and growth and spinodal decomposition are distinguished by whether or not the homogeneous material is stable against small random fluctuations in composition. The distinction can be visualized by referring to the free energy curves for a binary system containing a miscibility gap (Fig. 5.79). At temperatures below the consolute (peak) temperature, an upwards inflection of the free energy curve develops, from which we can determine the phase boundary using the common tangent construction (discussed in Chapter 4). A homogeneous composition within the phase boundaries can lower its free energy by separating into the two phases defined by the common tangent. However, defining the points of inflection in the free energy curves as the *spinodes* (where $\partial^2 G/\partial C^2 = 0$), we find that the path to a phase-separated mixture involves an energy barrier if the composition lies outside the spinodes, but not if it lies within. Outside the spinodal, a small fluctuation in composition about the mean value involves an *increase* in free energy since the curve is concave upwards (Fig. 5.80*a*), thus stabilizing the solution against small fluctuations in composition. With a large enough fluctuation the free energy can be decreased: This corresponds to the nucleation of a composition of lower bulk free energy. On the other hand, within the spinodal any small fluctuation in composition immediately *decreases* the free energy since the free energy curve is concave downwards (Fig. 5.80*b*). Fluctuations in composition are stable and can grow in amplitude. It is important to point out that nucleation and growth and spinodal decomposition are competitive processes within the spinodal region. Outside the spinodal, however, only nucleation and growth can take place.

For spinodal decomposition to occur it is necessary, but not sufficient, for the free energy curvature to be negative ($\partial^2 G/\partial C^2 < 0$). This is because phase separation by the spinodal mechanism introduces new interfaces, the positive free energy of which must be overcome by the decrease in bulk free energy. This is

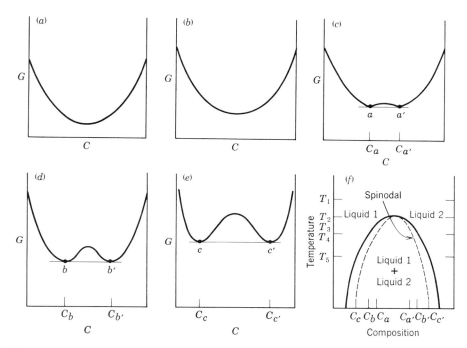

Fig. 5.79 Two-liquid immiscibility. (*a*) to (*e*) show a sequence of Gibbs free energy curves corresponding to the phase diagram shown in (f). (*a*) T = T_1; (*b*) T = T_2; (*c*) T = T_3; (*d*) T = T_4; (*e*) T = T_5. (From T.P. Seward, in *Phase Diagrams*, Vol. 1, Academic Press, Inc., New York, 1970.)

analogous to the tradeoff between the decrease in bulk free energy and the increase in surface energy which occurs in nucleation and growth. In earlier discussion we treated interfaces as atomically sharp entities with a certain excess energy per unit area (Fig. 5.2). However, in general, we expect that the interface is diffuse to a certain extent such as is schematically illustrated in Fig. 5.81. A means of characterizing the energy of a diffuse interface becomes the central issue in determining the free energy change associated with gradual compositional fluctuations.

R. Becker [*Ann. Phys.*, 32, 128 (1938)] suggested a simplied approach shown in Fig. 5.81: If we consider a planar structure containing atoms A and B, with a bond energy for AA bonds being ε_{AA}, for BB bonds being ε_{BB}, and for A-B bonds being ε_{AB}, then the pure phases are stable when $2\varepsilon_{AB} > \varepsilon_{AA} + \varepsilon_{BB}$. Defining atomic planes parallel to the interface located at p as having compositions (in atom fraction of A) of x_p, x_{p+1}, and so on, we first calculate the energy of one atomic plane bounded by neighboring planes of the same composition,

$$E_{x_p} = z\left(\frac{m}{2}\right)\left[x_p x_p \varepsilon_{AA} + (1-x_p)(1-x_p)\varepsilon_{BB} + 2x_p(1-x_p)\varepsilon_{AB}\right] \quad (5.85)$$

454 Chapter 5 / Microstructure

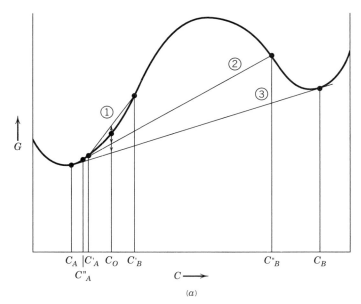

Fig. 5.80 (a) For a composition C_o which lies outside the spinodal but within the miscibility gap, a small fluctuation in composition about the mean value to the limits C_A' and C_B' raises the free energy of the system (path 1). A larger fluctuation, to the compositions C_A'' and C_B'', is able to lower the free energy (path 2), leading to a lowering of free energy when the equilibrium compositions C_A and C_B are reached (path 3).

where z is the number of bonds per atom between planes and m is the density per unit area. A similar expression holds for the x_{p+1} plane. The total bond energy between planes of differing composition, x_p and x_{p+1}, is

$$E_{x_p - x_{p+1}} = z\left(\frac{m}{2}\right)\{x_p x_{p+1} \varepsilon_{AA} + (1-x_p)(1-x_p)\varepsilon_{BB} \\ + [x_p(1-x_{p+1}) + x_{p+1}(1-x_p)]\varepsilon_{AB}\}$$
(5.86)

The extra energy resulting from the compositional gradient relative to the average uniform composition is given by

$$E_s = 2E_{x_p - x_{p+1}} - (E_{x_p} + E_{x_{p+1}})$$
$$= z\left(\frac{m}{2}\right)(2\varepsilon_{AB} - \varepsilon_{AA} - \varepsilon_{BB})(x_p - x_{p+1})^2$$
(5.87)

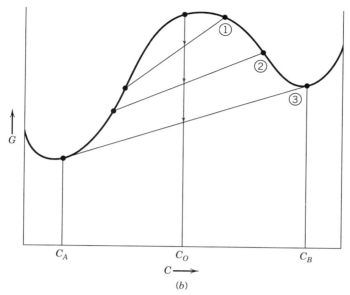

Fig. 5.80 (b) A composition C_o which lies within the spinodal will immediately undergo a decrease in free energy with a small fluctuation in composition about the mean value. The free energy decreases continuosly with further compositional separation until the equilibrium values C_A and C_B are reached. This composition is *unstable* against fluctuations in composition, whereas the composition in (a) is only *metastable*.

If $(x_p - x_{p+1})/a_o$ represents the compositional gradient $\partial C/\partial y$ and $\upsilon = (2\varepsilon_{AB} - \varepsilon_{AA} - \varepsilon_{BB})$ is the interaction energy, the excess surface energy is

$$E_s = z\left(\frac{m}{2}\right) a_o^2 \upsilon \left(\frac{\partial C}{\partial y}\right)^2 \tag{5.88}$$

where a_o is the interplanar spacing. The important result is that the energy contribution of a diffuse interface increases with the *square* of the magnitude of the concentration gradient. A more rigorous and general derivation [J.W. Cahn and J.E. Hilliard, *J. Chem. Phys.*, 28, 258 (1958)] leads to the same result.

With this in hand, the thermodynamics of spinodal decomposition can be treated as follows. Writing the diffuse interface energy as $\kappa(\nabla C)^2$ where κ is termed the gradient energy coefficient, the free energy of a solution with compositional fluctuations is given to first order by

$$G = N_v \int_V \left[g(C) + \kappa(\nabla C)^2 \right] dV \tag{5.89}$$

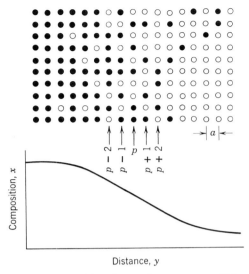

Fig. 5.81 Composition fluctuations across a diffuse interface.

in which N_v is the number of molecules per unit volume and $g(C)$ is the free energy per molecule of a solution of composition C. $g(C)$ is given by a free energy curve, such as that illustrated in Fig. 5.79, and has a dependence on composition and temperature that can be approximated using various solution models for simple systems, but is not *a priori* known for more complex ones. This free energy function $g(C)$ can also be written as a Taylor series expansion about the average composition C_0:

$$g(C) = g(C_o) + \left.\frac{\partial g}{\partial C}\right|_{C_o}(C - C_o) + \frac{1}{2!}\left.\frac{\partial^2 g}{\partial C^2}\right|_{C_o}(C - C_o)^2 + \frac{1}{3!}\left.\frac{\partial^3 g}{\partial C^3}\right|_{C_o}(C - C_o)^3 + \ldots \quad (5.90)$$

When we integrate over a volume of material in which the composition fluctuates about the average composition C_0, all odd terms in this expansion conveniently vanish because the term $(C-C_0)$ changes sign symmetrically about C_0. If we neglect the higher order terms and subtract from Eq. 5.90 the free energy of a uniform solution of composition C_0, the *difference* in free energy between the uniform and nonuniform solution is simplified to

$$\Delta G = N_v \int_v \left[\frac{1}{2}\left(\frac{\partial^2 g}{\partial C^2}\right)_{C_o}(C - C_o)^2 + \kappa(\nabla C)^2 \right] dV \quad (5.91)$$

5.6 Glasses and Glass-Ceramics

From this expression, we see that in order for ΔG to be negative, it is not only necessary for $\partial^2 G/\partial C^2$ to be negative, but the condition

$$\frac{1}{2}\left|\left(\frac{\partial^2 G}{\partial C^2}\right)_{C_o}\right|(C-C_o)^2 > \kappa(\nabla C)^2 \tag{5.92}$$

must also be satisfied. Equation 5.92 therefore expresses the need for the bulk free energy decrease to be greater than the diffuse-interfacial energy increase. Outside the spinodal, where $\partial^2 G/\partial C^2$ is positive, ΔG is positive for all small fluctuations in composition, and phase separation can only proceed by nucleation and growth.

The balance between bulk and interfacial energy is manifested in the existence of a critical length scale of spinodal decomposition below which ΔG becomes positive. According to the linearized theory of spinodal decomposition, if the compositional fluctuation is spatially sinusoidal

$$(C-C_o) = A\cos\beta x \tag{5.92a}$$

where $\beta = 2\pi/\lambda$ is the wave number, by substituting this into Eq. 5.91 the free energy becomes

$$\frac{\Delta G}{V} = \frac{A^2 \pi}{4}\left[\left(\frac{\partial^2 g}{\partial C^2}\right)_{C_o} + 2\kappa\beta^2\right] \tag{5.93}$$

from which a critical wavelength is defined by the condition $\Delta G = 0$:

$$\lambda_c = \left[\frac{-8\pi^2 \kappa}{\left(\partial^2 g/\partial C^2\right)_{C_o}}\right]^{1/2} \tag{5.94}$$

Fluctuations of a periodicity smaller than λ_c are unstable and decay away, while those larger than λ_c are stable and can grow. However, not all periodicities grow at the same rate. It has been shown that for isothermal decomposition, the fastest growing wavelength is given by

$$\lambda_m = \sqrt{2}\,\lambda_c \tag{5.95}$$

This becomes the characteristic length scale of the microstructure in the early stages of decomposition.

458 Chapter 5 / Microstructure

When there is a large fraction of both phases, a spinodally decomposed microstructure tends to become interconnected as shown by the calculated microstructure in Fig. 5.82. Such mutually interconnected microstructures are often observed in phase separated glasses and polymers. The periodicity actually observed may be larger than λ_m, since a coarsening of the microstructure occurs in later stages of decomposition.

Generally speaking, it is difficult to tell from a casual observation whether phase separation in a system with large amounts of each phase is due to nucleation and growth or to spinodal decomposition. Clearly, when a small fraction of one phase is present, it is likely that the overall composition lies outside the spinodal and nucleation and growth has occurred. At larger fractions of both phases, but away from the center of the miscibility gap, microstructures are often not co-connected after decomposition regardless of the phase separation mechanism since a minor phase tends to spheroidize. Conversely, it has been argued that interconnected structures can result from nucleation and growth at high-phase fractions,

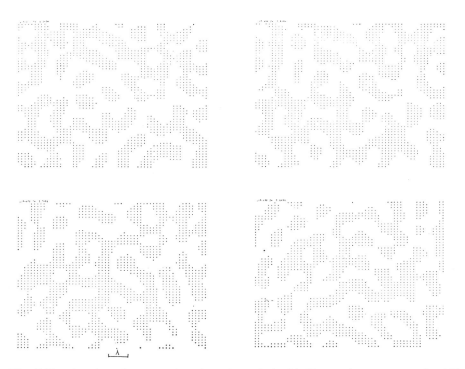

Fig. 5.82 A series of computed sections through the 50:50 two-phase structure for 100 random sine waves. Note that all particles are interconnected. The spacing between sections is $1.25/\beta_{min}$. [From J.W. Cahn, *J. Chem. Phys.*, **42**, 93 (1965).]

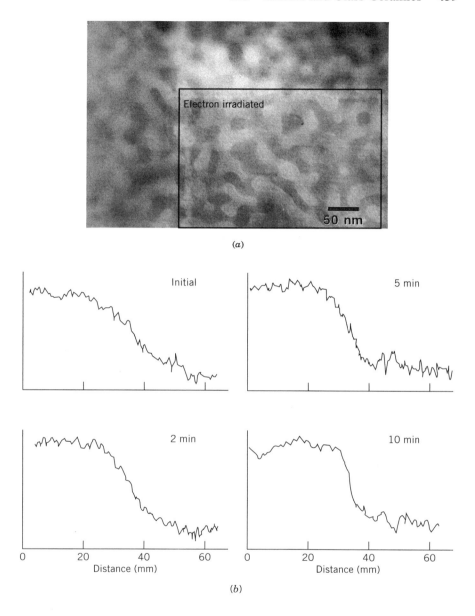

Fig. 5.83 (a) Microstructure of a phase-separated glass with diffuse interfaces, which sharpen in the region under electron irradiation due to accelerated diffusion. (b) Microdensitometer traces across a single phase boundary photographed at successive intervals as the concentration gradient becomes more abrupt. Ordinates represent normalized intensity; abscissas indicate distance along trace. [From Y.-M. Chiang and W.D. Kingery, *J. Am. Ceram. Soc.*, **66**[9], C171-72 (1983).]

due to the coalescence of particles. The existence of spinodal decomposition has generally been confirmed by kinetic studies, which use small angle x-ray and light scattering to observe the selective amplification of a characteristic wavelength as predicted by the theory. Another distinguishing characteristic of spinodal decomposition is the existence of a diffuse interface. Figure 5.83 shows the interconnected microstructure of a K_2O-Al_2O_3-CaO-SiO_2 glass which has phase-separated into high silica and high calcia phases. At high magnification in the electron microscope, the interfaces between the two phases are observed to sharpen with time due to accelerated transport under electron irradiation, thereby demonstrating the diffuseness of the quenched interfaces, and identifying the mechanism as spinodal decomposition.

Spinodal decomposition also occurs in crystalline materials, where its characteristics are modified by the elastic strain energy accompanying compositional separation. Due to the anisotropy of crystal elastic constants, composition waves tend to propagate with the interfaces parallel to elastically soft directions of the crystal, often resulting in lamellar structures. A widely studied ceramic system is the rutile system TiO_2-SnO_2, which possesses a nearly symmetric miscibility gap. Upon cooling a solid solution from high temperature into the miscibility gap, finely divided lamellae alternatively rich in Ti and Sn are formed within each polycrystalline grain (Fig. 5.84). Since TiO_2 and SnO_2 share a common structure and thus a common oxygen sublattice, decomposition in this system requires only the counterdiffusion of cations. It is found that the rate of spinodal decomposition varies tremendously as aliovalent solutes that increase or decrease cation diffusion are added; acceptor doping increases the rate to where it is difficult to quench a solid solution without decomposition, whereas slight donor doping decreases the rate to where a sample can be annealed for many hours at 1000°C without decomposition.

Multiple-Phase Separation

Since the heat treatment of glasses often involves continuous cooling, systems undergoing phase separation experience a continuous change in the composition boundaries with time. Multiple-phase separation is the phenomenon whereby a phase-separated material undergoes additional phase separation within each primary phase. This process is illustrated in Fig. 5.85a and by the microstructure in Fig. 5.85b. Initial phase separation forms two phases, which then become isolated subsystems within which further immiscibility develops with cooling. Secondary-phase separation leads to a microstructure of phases within phases, as shown in Fig. 5.85b. While the compositions of the various phases tend toward the two extremes given by the phase boundaries, as many as six microstructurally distinct phases have been observed in a barium borosilicate glass, indicating tertiary phase separation.

5.6 Glasses and Glass-Ceramics 461

Fig. 5.84 Spinodal phase separaton in an equimolar TiO_2-SnO_2 crystal homogenized at 1600°C and annealed 5 min at 1000°C. Electron diffraction pattern in lower right; optical diffraction pattern in upper left. (Courtesy M. Park and A. H. Heuer.)

A widely used glass that exhibits phase seperation is sodium-borosilicate, a ternary system which forms the basis for low-thermal-expansion glasses such as Pyrex™ and Kimax™ as well as the Vycor™ process. These glasses are formed in the region of the large immiscibility "dome" in Fig. 5.86, within which the tie lines run roughly parallel to the SiO_2-B_2O_3 join. As discussed in Chapter 1 the thermal expansion coefficient of alkaline borates and borosilicates reaches a minimum at an alkali/boron ratio of about 0.16 due to changes in structural coordination of boron. Low-thermal-expansion sodium borosilicates are prepared using about 80% SiO_2 and 20% Na_2O and B_2O_3, with the latter in approximately the 0.16 ratio. The range of compositions lies toward the silica-rich end of the immiscibility region (Fig. 5.86). (Notice from the isothermal contours that this immiscibility dome is asymmetric, as a result of which the high silica phase is slow to undergo secondary-phase separation, being on the side that is steep in temperature. On the other hand, the borate-rich phase is likely to exhibit secondary-phase separation since the phase boundary is more sloped in temperature, and because atom diffusion is much faster in the low-viscosity borate phase.) After normal

462   Chapter 5 / Microstructure

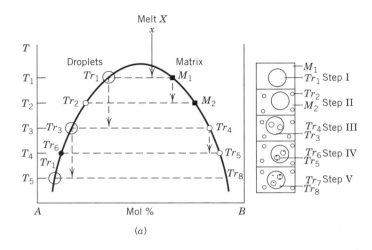

(a)

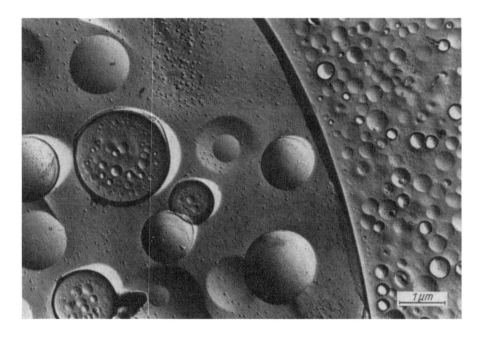

Fig. 5.85 Schematic of multiple phase separation process in which demixed phases undergo further phase separation in a stepwise manner. Schematic of resulting microstructures at each stage is shown at right. (From W. Vogel, *Chemistry of Glass*, The American Ceramic Society, Columbus, Ohio, 1985.)

5.6 Glasses and Glass-Ceramics   463

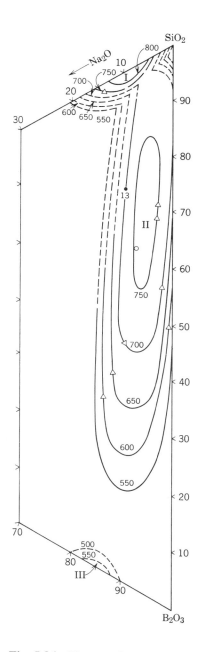
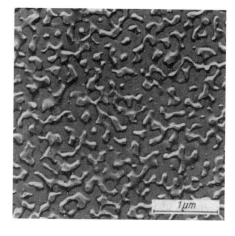

Fig. 5.86 Three regions of liquid-liquid immiscibility in the Na_2O-B_2O_3-SiO_2 system (wt %) [From W. Haller, D.H. Blackburn, F.E. Wagstaff and R. J. Charles, *J. Am. Ceram. Soc.*, **53**, 34 (1970).], and microstructure of phase-seprated glass of composition near center of immiscibility field labled II (from W. Vogel, *Chemistry of Glass*, The American Ceramic Society, Columbus, Ohio, 1985).

processing, Pyrex™-type glasses contain borate-rich regions of ~5 nm scale in a high-silica matrix. Vycor™-type glasses are prepared toward the center of the immiscibility region in order to obtain, after an annealing treatment, a phase-separated microstructure in which both borate and silica phases are of equivalent volume fraction and are interconnected. Upon leaching the borate phase with an acid, a porous high silica glass of extremely fine pore size (2–5 nm) and high surface area (>100 m^2/g) is obtained. This "thirsty" glass, so-called because of the enormous capillary action enacted by so fine a porosity, is useful in its own right as filters, membranes, and catalyst supports. There is also a high sintering pressure associated with such fine-scale porosity, however, and upon firing at about 1100°C the glass densifies by viscous sintering to form a glass that is greater than 95% silica. This process offers a great temperature advantage in comparison to the melt-processing of pure silica, which requires temperatures well in excess of the melting point of 1713°C.

SPECIAL TOPIC 5.3

THERMAL SHOCK RESISTANCE

Most of us have experienced the accidental breakage of a hot glass object suddenly thrust into cold water. Because ceramics are much weaker in tension than in compression, and because surfaces usually contain flaws that act as stress concentrators, thermal shock failure usually occurs upon cooling since the more rapidly cooled surface develops a tensile stress relative to the interior, as shown in Fig. ST62. Thermal shock resistance is an important engineering property of high temperature ceramics; those which have a low thermal expansion coefficient (giving lower differential expansion) and high thermal conductivity (smoothing out temperature gradients) tend to be more thermal shock resistant, all else being equal. However, ceramics of both high and low strength can be thermal shock resistant. Firebricks used to line furnaces are weak, porous, full of microscopic flaws, and certainly not of high thermal conductivity, but are thermal shock resistant because the copious numbers of microcracks allows thermal stresses to be locally accommodated without propagating existing flaws or generating new ones. In contrast, high-performance silicon nitride ceramics have better thermal shock resistance than many competing materials due to a high fracture strength and toughness, as well as lower elastic modulus, lower thermal expansion coefficient, and good thermal conductivity.

The arena in which we most frequently encounter thermal shock resistant ceramics is the kitchen. We may ask, how severe a thermal shock can we expect commercial glass and glass-ceramic products to endure? The maximum surface stress generated in a thick body quenched into a medium is

5.6 Glasses and Glass-Ceramics

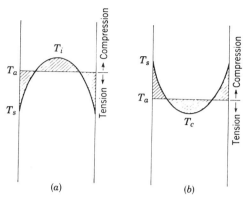

Fig. ST62 Temperature and stress distributions for plate that is (a) cooled from the surface and (b) heated from the surface.

$$\sigma = \frac{E\alpha}{1-\mu}(T_a - T_s)$$ (Eq. 1)

where E is the Young's modulus (typically ~ 10^7 psi (69 GPa) for glasses and glass ceramics), α is the thermal expansion coefficient, μ is Poisson's ratio (typically ~0.2), T_a is the average temperature, and T_s the surface temperature. For the worst-case scenario of a body thick enough and heat transfer fast enough for the average temperature to remain at the initial temperature and the surface temperature to reach that of the quenching medium, we may take $(T_a - T_s) \approx \Delta T$, where ΔT is the magnitude of the quench. Rearranging Eq. 1, the thermal shock which can be withstood without failure is

$$\Delta T = \frac{(1-\mu)\sigma_f}{E\alpha}$$ (Eq. 2)

where σ_f is the failure strength of the material. For most glasses and lithia-alumina-silica glass-ceramics which have not been strengthened, σ_f is on the order of 10^4 psi (69 MPa). Table ST1 shows the termal expansion coefficient of various materials and the thermal shock which can be expected. We see that a soda-lime silicate water glass or glass jar should not be expected to survive the 100°C quench from boiling water into ice water. This is consistent with common everyday experience. Sodium borosilicate glassware, on the other hand, may survive a broiler-to-ice water quench of ~550°F (288°C) due to its low thermal expansion coefficient of ~ 3×10^{-6}/°C. (Do not test this at home!) Fused silica glass has an extremely low thermal expansion coefficient of 0.5×10^{-6}/°C and can survive a quench from 1600°C,

Table ST1 Thermal Shock Resistance of Common Glasses and Glass Ceramics

Material	Thermal Expansion Coefficient (°C^{-1})	Thermal Shock to Failure[1*]
Soda-lime-silicate	10^{-5}	80°C
Sodium borosilicate (Pyrex™ type)	3×10^{-6}	270°C
Fused silica	0.5×10^{-6}	1600°C
Lithia-alumina-silicate glass ceramic (Pyroceram™ type)	1.2×10^{-6}	670°C
Transparent lithia-alumina -silicate glass ceramic (Visions™ type)	0.6×10^{-6}	1330°C

[1*] Assuming failure strength in tension at a stress of 69 MPa.

which is above its glass transition temperature (of ~ 1200°C)! The utility of lithia-alumina-silica glass ceramics is clear, since with thermal expansion coefficients of 0.6-1.2 × 10^{-6}/°C, they can easily survive the thermal shocks encountered in most kitchens *if* the object has not been significantly weakened by the introduction of flaws.

5.7 COMPOSITE PROPERTIES

We shall use the term *composite* in its most general sense to refer to any polyphase structure; the resultant ceramic properties and processing requirements depend on the relative amounts, shapes, distributions, and preparations of the two or more constituents. In sections 5.1 and 5.5 we saw that the physical distribution and interconnectivity of the separate phases in a microstructure are often determined by capillarity. This occurs when there is sufficient atom transport to allow surface tensions to dictate microstructure. However, a particular distribution of phases in a composite microstructure can also be artificially constructed, for instance by fabricating a continuous-fiber reinforced composite (Fig. 5.87), a whisker-reinforced composite (Fig. 5.88), or by using a removable template around which a phase is deposited. In Fig. 5.89, a ceramic slurry has been infiltrated into a polymer fiber network, and the fibers have been subsequently burned to leave a replicate pore structure, which is then infiltrated with a metal. The resulting microstructure in synthesized composites often is neither what capillarity would dictate nor what high temperature phase equilibria would allow.

5.7 Composite Properties 467

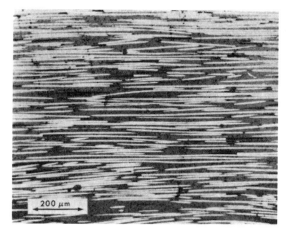

Fig. 5.87 Cross-section parallel to the fibers in a carbon fiber reinforced Pyrex glass matrix composite. [From R.A.J. Sambell, A. Briggs, D.C. Phillips, and D.H. Bowen, *J. Mat. Sci.*, **17**, 676 (1982).]

Fig. 5.88 (*a*) Polished cross-section of SiC whisker-reinforced Al_2O_3 composite, and (b) fracture surface revealing SiC whiskers. [From P.F. Becher and G.C. Wei, *J. Am. Ceram. Soc.*, **67**[12], C267 (1984).]

468 Chapter 5 / Microstructure

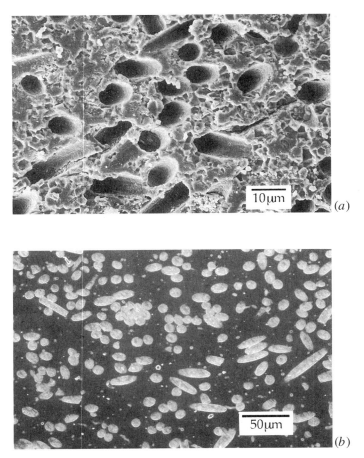

Fig. 5.89 (*a*) Ceramic preform where organic fibers have been removed by pyrolysis, and (*b*) composite after metal infiltration. (Courtesy of F. F. Lange.)

Rules of Mixing

Regardless of how a composite structure is arrived at, its physical properties are generally determined by the properties of the constituent phases and the way in which they are joined. *Rules of mixing* tell us how the properties of the composite can be obtained from those of the individual phases. Different types of mixture rules apply to different properties, and to different physical arrangements of the phases.

One of the simplest composites is a mixture of solid with porosity. In previous sections we have discussed the elimination of porosity during sintering. For some properties, such as the thermal expansion coefficient, the addition of porosity has no effect. The thermal expansion coefficient of a "two-phase" porous material is the same as that of the fully dense material, or of any individual grain. On the other hand the density of a porous material is reduced in direct proportion to the volume fraction of porosity ($\rho = (1-V_p)\rho_c$, where V_p is the pore volume fraction and ρ_c is the crystal density). The permeability of a porous solid is more sensitive still to the volume fraction of porosity and in a highly nonlinear way, since as porosity is eliminated, there is a change from a situation where the pores form an interconnected network and allow permeability of gas of liquid toward a configuration where the permeability drops rapidly to zero, at 5–8 volume percent porosity (Fig. 5.90). The *percolation limit* at which a minority phase (here, porosity) forms a continuous network is important for permeability, conductivity, and other properties, and is discussed later.

When two phases are mixed, most properties take on a value intermediate between those of the two. The volume-averaged value may be simply a linear combination of the two (linear rule of mixing), but more often will exhibit a negative deviation from linearity. For processes that involve physical deformation of the individual phases this is easy to understand—each phase places a constraint on the other. Positive deviations from ideality on a volume basis are less common, and indicate a synergistic effect since the whole is greater than the sum of its parts.

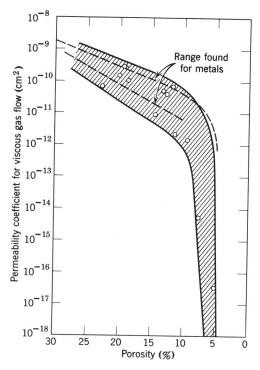

Fig. 5.90 Permeability coefficient for viscous gas flow in beryllia ware of differing porosity. Permeability drops rapidly when the porosity is below 5 to 8%. (After J.S. O'Neill, A.W. Hey, and D.T. Livey, U.K. AERE-R3007.)

470 Chapter 5 / Microstructure

For many properties the mixture rules can be expressed in the general form

$$K_{composite}^n = v_1 K_1^n + v_2 K_2^n \qquad (5.96)$$

where v_1 and v_2 are the volume fractions of phases 1 and 2, and K_1 and K_2 are the respective values of the property of interest, and n is an exponent, which may be +1, -1, or 0. Consider the simple composite arrangement where parallel slabs of two phases of widely different properties are assembled (Fig. 5.91). For electrical (σ) or thermal conductivity (k), the properties are additive when the flow is parallel to the slabs (Fig. 5.91b), and we have the linear rule of mixture ($n = 1$):

$$\sigma_{composite} = v_1 \sigma_1 + v_2 \sigma_2 \qquad (5.97a)$$

$$k_{composite} = v_1 k_1 + v_2 k_2 \qquad (5.97b)$$

The dielectric constant κ for an electric field applied parallel to the slabs (Fig. 5.91a) also adds linearly, since the field is the same for each element (i.e., the capacitors are in parallel):

$$\kappa_{composite} = v_1 \kappa_1 + v_2 \kappa_2 \qquad (5.98)$$

The elastic moduli similarly follow linear mixing if it is assumed that the strain in each slab is the same (the direction of loading being parallel to the slabs). This is known as the Voigt bound (Fig. 5.92), under which the Young's modulus is

$$E_{composite} = v_1 E_1 + v_2 E_2 \qquad (5.99)$$

Similar relations apply to the bulk and shear moduli.

If the direction of flow, electric field, or applied load is normal to the slabs, the inverse of the thermal and electrical conductivities (i.e., the resistivities), inverse capacitances, and inverse moduli are additive. Taking $n = -1$ in Eq. 5.96 results in the mixing rules:

$$\frac{1}{\sigma_{composite}} = \frac{v_1}{\sigma_1} + \frac{v_2}{\sigma_2}$$

$$\frac{1}{k_{composite}} = \frac{v_1}{k_1} + \frac{v_2}{k_2}$$

$$\frac{1}{\kappa_{composite}} = \frac{v_1}{\kappa_1} + \frac{v_2}{\kappa_2} \qquad (5.100)$$

$$\frac{1}{E_{composite}} = \frac{v_1}{E_1} + \frac{v_2}{E_2}$$

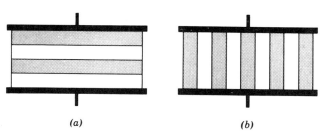

Fig. 5.91 Possible arrangements of layers having different characteristics in a dielectric.

These show large negative deviations from the linear relationship, as seen for the Young's modulus in Fig. 5.92 (known as the Reuss bound, which assumes identical stress in each phase), and for the dielectric constant in Fig. 5.93.

The linear and inversely-linear limiting cases are generally too restrictive since such simple physical arrangements are rarely encountered. They result in bounds that are often too far apart to be of much use in predicting the properties of more general microstructures. Thus much effort has been devoted towards finding mixture rules which give narrower bounds than Eqs. 5.98 and 5.100 for more realistic and random configurations, such as when one phase is dispersed and the other continuous. These mixture rules have often been developed independently for different physical properties. Maxwell derived a relationship for the composite

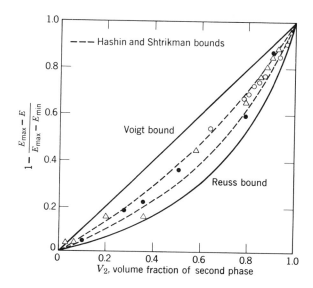

Fig. 5.92 Comparison of observed with predicted variations of Young's modulus with volume fraction of second phase material. [From R.R. Shaw and D.R. Uhlmann, *J. Non-Cryst. Solids*, **5**, 237 (1971).]

dielectric constant of a dispersion of spherical particles of dielectric constant κ_d in a matrix of dielectric constant κ_m:

$$\kappa_{composite} = \frac{v_m \kappa_m \left(\frac{2}{3} + \frac{\kappa_d}{3\kappa_m}\right) + v_d \kappa_d}{v_m \left(\frac{2}{3} + \frac{\kappa_d}{3\kappa_m}\right) + v_d} \qquad (5.101)$$

which give bounds (for $\kappa_d > \kappa_m$ and $\kappa_d < \kappa_m$), which lie within those given by Eq. 5.96 for $n = -1$ and $+1$, as shown in Fig. 5.93. Adding a phase of high dielectric constant as a continuous matrix phase ($\kappa_d < \kappa_m$) causes a greater increase in the composite dielectric constant than does adding it as a dispersed phase ($\kappa_d < \kappa_m$). Hashin and Shtrikman (*J. Mech. Phys. Solids*, 11,127, 1963) derived upper and lower bounds for the elastic moduli of a two-phase mixture, without requiring specific assumptions about the phase geometry, that are considerably narrower than the Voigt and Reuss bounds (Fig. 5.92). Their results

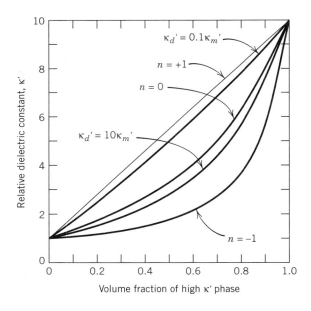

Fig. 5.93 Expressions for the dielectric constant of various mixtures of two dielectrics, Eqs. 5.96 and 5.101.

for the upper and lower bound of the bulk and shear moduli (K and G) are, for $K_2>K_1$ and $G_2>G_1$,

$$K_{lower} = K_1 + \frac{V_2}{\frac{1}{(K_2-K_1)}+\frac{3(1-V_2)}{(3K_1+4G_1)}}$$

$$K_{upper} = K_2 + \frac{1-V_2}{\frac{1}{(K_1-K_2)}+\frac{3V_2}{(3K_2+4G_2)}}$$

$$G_{lower} = G_1 + \frac{V_2}{\frac{1}{(G_2-G_1)}+\frac{6(K_1+2G_1)(1-V_2)}{5G_1(3K_1+4G_1)}} \quad (5.102)$$

$$G_{upper} = G_2 + \frac{1-V_2}{\frac{1}{(G_1-G_2)}+\frac{6(K_2+2G_{21})V_2}{5G_2(3K_2+4G_2)}}$$

The curves for Young's modulus in Fig. 5.92 are obtained from the respective upper and lower bound values of K and G using the relation

$$E = \frac{4KG}{3K+G} \quad (5.103)$$

The lower bounds in Eq. 5.102 also correspond to exact solutions obtained by Hashin and by Kerner for the special case of spherical second-phase particles of modulus K_2 and G_2 in a matrix of lower moduli K_1 and G_1. The upper bounds can be obtained by simply reversing the role of matrix and inclusion in the analysis. As in the case of the dielectric constant (Fig. 5.93), adding the higher modulus phase as a continuous matrix (i.e., the upper bound) leads to a higher composite modulus at a given volume fraction (Fig. 5.92). The thermal expansion coefficient of two-phase composites follows similar mixture rules.

It is often found that a logarithmic mixing rule:

$$\log K_{composite} = v_1 \log K_1 + v_2 \log K_2 \quad (5.104)$$

describes the properties of a random composite fairly well. Figure 5.93 for the dielectric constant shows that this curve lies in between Maxwell's bounds. We may think of the logarithmic mixing rule as corresponding to Eq. 5.96 with an n value approaching zero for the random mixture (since as n approaches zero, K^n equals $1 - n \log K$), exactly in between the limiting values of $+1$ and -1 for parallel and perpendicular layers.

Percolation

Transport properties in a composite such as electrical conductivity and thermal conductivity are especially sensitive to not only the amount of each phase but also the degree to which phases are connected. The greater the difference in specific properties of the end members, the wider the separation between the upper and lower bounds given by mixing rules; and after a point these rules cannot predict the composite properties with accuracy. This is illustrated in Fig. 5.94, where the electrical conductivity of a composite containing a conductive phase (1 Ω cm resistivity) and an insulating phase (10^6 Ω cm resistivity) is shown for different models. Notice that the resistivity is shown on a logarithmic scale, and all curves lie below that for linear mixing. We see first of all that the upper and lower Hashin-Shtrikman bounds, curves (a) and (e), are so widely separated as to be virtually useless for predicting the composite conductivity. Curve (c) shows the composite conductivity for a microstructure in which the high-conductivity phase becomes continuous, or *percolates*, at a critical volume fraction of 1/2. A dramatic drop in resistivity occurs at the percolation limit. Curve (d) shows the result when the percolation limit is at zero volume fraction, and curve (b), when it is at unity. An actual measurement of conductivity versus volume fraction in a

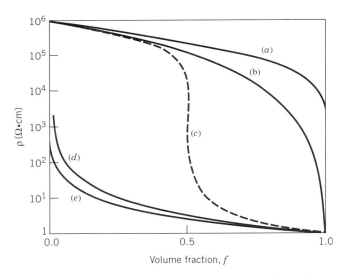

Fig. 5.94 Comparison of resistivity (on a logarithmic scale) predicted for a composite of conducting particles (1 Ω–cm resistivity) in an insulating matrix (10^6 Ω–cm resistivity) according to different models. (*a*) and (*e*) are the upper and lower Hashin-Shtrikman bounds, respectively, (*b*) is the result from "generalized effective medium" theory with a percolation threshold of unity, (*c*) represents Bruggeman and generalized effective medium relationships with a percolation threshold of one-half, and (*d*) is generalized effective medium theory with a percolation threshold of zero. [See D.R. Clarke, *J. Am. Ceram. Soc.*, **75**[4], 739-59 (1992) for details.]

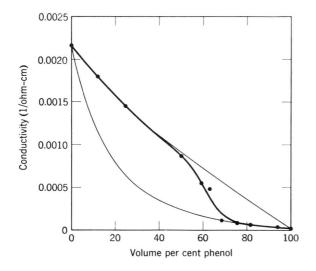

Fig. 5.95 The electrical conductivity of a phenol - KI emulsion increases sharply at a percolation limit of about 30%. [From P. Grootenhuis et al., *Proc. Phys. Soc. (London)*, **B65**, 502 (1952).]

phenol-KI emulsion is shown in Fig. 5.95, indicating percolation at about 30% KI. The thermal conductivity of two phase mixtures of MgO and $MgSiO_4$, shown in Fig. 5.96, also exhibit S-shaped variations due to percolation. The curves follow a lower bound at high volume fractions corresponding to $MgSiO_4$ being the continuous phase, and an upper bound at low volume fractions when MgO is the continuous phase. Below about 40% forsterite, the high-conductivity MgO phase forms a percolating network. Above 40%, the MgO phase forms dispersed particles within a forsterite matrix. The gas permeability in a porous ceramic

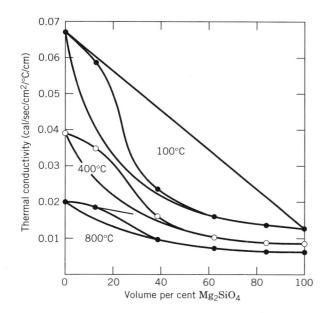

Fig. 5.96 Thermal conductivity in the two-phase system $MgO-MgSiO_4$.

(Fig. 5.90) is another example, where sharp increases are seen once the porosity is percolating.

If spheres of a good conductor are mixed at random with like-sized spheres of an insulator (Fig. 5.97), percolation occurs when 27% of the spheres are metal. Since random close-packing of like-sized spheres is only 60% dense, at the percolation limit the metal spheres occupy ~16% of the total space. If ordered arrays of spheres (fcc, bcc, hcp, etc.) are constructed, the critical number fraction of metal spheres necessary for percolation decreases, but at the same time the packing density of the array increases compared to random close-packing. It is found that the critical volume fraction of spheres for percolation remains at ~16%, regardless of the packing geometry (In two dimensions, the critical area fraction is ~45%).

However, packed spheres are obviously an oversimplified representation of polycrystalline microstructures. In composite structures the percolation limit depends greatly on the physical configuration of the two or more phases. As discussed earlier the configuration of a minor phase in a polycrystalline microstructure is determined by relative surface tensions. Thus, a percolation limit of essentially zero may result when the dihedral angle is such that the conductive phase wets along three-grain junctions of grain boundaries (see Fig. 5.15a). One example of such a composite is a cemented carbide (such as used for cutting tools and wear parts), where a small amount of a metal phase such as Co or Ni used as a liquid phase sintering aid remains as a continuous grain boundary film between the WC grains. Conversely, a percolation limit of unity may occur if the wetting phase is insulating and the grains conductive; this mimics the structure of some

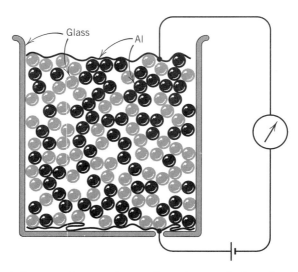

Fig. 5.97 Percolation experiment using random close-packed conducting (aluminum) and insulating (glass) spheres. [After R. Zallen, *Physics of Amorphous Solids*, John Wiley and Sons, New york, 1983. See also J.P. Fitzpatrick, R.B. Malt, and F. Spaepen, *Phys. Lett.*, **A47**, 207, 1974).]

grain boundary barrier-layer electrical ceramics. From Fig. 5.15, one can see that intermediate values of the percolation limit may easily occur and are sensitive to the dihedral angle between phases. In assembled composites, such as the fiber and whisker reinforced structures in Figs. 5.87–5.88, the percolation limit depends upon the degree of alignment and aspect ratio of the second phase and sometimes on dispersion interactions introduced during processing.

Immediately above the percolation limit, properties related to percolation increase sharply with volume fraction and are found to obey a universal power law of the form:

$$\sigma = (v - v_c)^n \tag{5.105}$$

where n is a critical exponent (so-called because it describes the behavior near the critical point of percolation), which has various values depending on the property of interest and the dimensionality. For transport (electrical, thermal, diffusive) in three dimensions, n is equal to 1.65.

Properties that are the product of two or more elementary properties are often not so simply described by either mixture rules or percolation models. For example, the acoustic velocity of a single phase is given by $V = (E\rho)^{1/2}$ where E is the elastic modulus and ρ is the density. For composite structures of oriented high-velocity wires in a low-velocity matrix (e.g., metal wires in a polymer matrix), the mixing rules for E and ρ are different. While the longitudinal wave velocity is intermediate between the individual constituent values, the transverse wave velocity turns out to be smaller than either metal or polymer. Composites that exhibit synergistic effects are of particular interest to materials engineers. A dense-oriented composite of ferroelectric barium titanate and ferromagnetic cobalt titanium ferrite can be electrically poled to make the $BaTiO_3$ phase piezoelectric [A. M. J. G. van Run, D. R. Terrell, J. H. Scholing, *J. Mat. Sci.*, 9, 1710, (1974)]. When a magnetic field is applied, the resulting magnetostriction makes the ferrite grains change shape, which applies a strain to the piezoelectric grains, which in turn causes an electrical polarization that results in very large magnetoelectric effects. As discussed in the next section, interactive material properties are particularly important in developing mechanical toughness in normally brittle single-phase ceramics.

5.8   STRENGTH AND TOUGHNESS

Until 1980 or so, it was accepted that ceramics were brittle and attention was focused on increasing the fracture stress to values closer to the theoretical limit. With the finding of increased toughness for partially stabilized zirconia and the development of fiber-reinforced composites, fracture toughness has increasingly occupied the spotlight. In fact, great improvements have been made both in fracture stress and in toughness through the control of microstructure characteristics. Many of these developments have been based on an improved understanding of fracture mechanics.

Brittle Fracture

Taking first the question of fracture stress, we may estimate the theoretical strength of a body, σ_{th}, as the stress necessary to separate it into two parts simultaneously across the cross section. If we pull a cylinder bar of unit cross section, the force of cohesion between two planes of atoms varies with their separation λ as shown in Fig. 5.98. The resistance to separation first increases as atoms are separated, reaches a peak at the theoretical strength, and then decays with further separation. A portion of this relationship can be approximated by

$$\sigma = \sigma_{th} \sin\left(\frac{2\pi X}{\lambda}\right) \tag{5.106}$$

The work per unit area required to separate the two planes of atoms is given by

$$\int_0^{\lambda/2} \sigma_{th} \sin\left(\frac{2\pi X}{\lambda}\right) dx = \frac{\lambda \sigma_{th}}{\pi} \tag{5.107}$$

which may be equated with the surface energy 2γ of the new surfaces that are formed,

$$\sigma_{th} = \frac{2\pi\gamma}{\lambda}. \tag{5.108}$$

For the first part of the curve near the equilibrium spacing, a_0, the relationship between force and displacement can be approximated as linear (Hooke's law) and is given by

$$\sigma = E\frac{X}{a_0} \tag{5.109}$$

where E is the modulus of elasticity. For this part of the curve, from Eq. 5.106 the slope is

$$\frac{d\sigma}{dx} = \frac{2\pi\sigma_{th}}{\lambda}\cos\left(\frac{2\pi x}{\lambda}\right) \approx \frac{2\pi\sigma_{th}}{\lambda} \quad \text{(at small } x\text{)} \tag{5.110}$$

Equating this with $d\sigma/dx$ from Eq. 5.109, we have

$$\frac{d\sigma}{dx} \approx \frac{2\pi\sigma_{th}}{\lambda} \approx \frac{E}{a_0} \tag{5.111}$$

5.8 Strength and Toughness 479

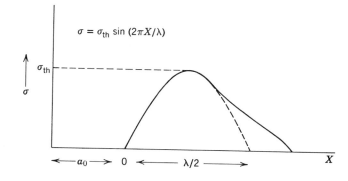

Fig. 5.98 Force versus separation relation (schematic).

Combining Eqs. 5.108 and 5.111, we have for the theoretical strength

$$\sigma_{th} \approx \left[\frac{E\gamma}{a_o}\right]^{1/2} \tag{5.112}$$

However, the observed strengths of brittle materials are often much less than that predicted by Eq. 5.112. For silicate glasses, a typical value of the Young's modulus is 70 GPa. (Table 5.7 lists values for other materials). The solid surface has an energy of ~1 J/m² and the atomic separation is ~3 × 10⁻¹⁰ m. Thus from Eq.

Table 5.7 Young's Modulus of Various Ceramics (at room temperature)

	Young's Modulus (GPa)
Fused silica	72
Soda-lime-silica glass	73
Pyrex borosilicate	69
Aluminosilicate glass	89
Single crystal Al_2O_3	380
Dense sintered Al_2O_3	366
Dense sintered Si_3N_4	304
MgO	207
ZrO_2	138
SiC	414
TiC	430
B_4C	290
Al_3O_3N spinel	330
Diamond	1035

5.112 the predicted theoretical strength is about 15 GPa, or roughly a fifth of the modulus of elasticity. While these strengths have been nearly obtained with thin pristine fibers of fused silica, more typical strengths of silicate glasses are in the range of $E/1000$ (i.e., ~70 Mpa). The fracture stress problem is essentially to relate the actual strengths to this theoretical limit.

The fundamental reason that actual strengths do not reach the intrinsic values is that fracture of brittle materials does not occur by the simultaneous separation of many bonds as assumed above, but instead by the initiation and propagation of a flaw. Griffith (*Philosophical Transactions Royal Society*, A221, 163, 1920) suggested that flaws in the materials act as stress concentrators to explain the substantial discrepancy between theoretical and observed strength. Griffith proposed that a crack propagates when the decrease in stored elastic energy associated with its extension exceeds the increase in surface energy created. Taking the example of an elliptical crack of major axis $2c$ in a thin plate of thickness l under biaxial stress σ, the energy U is the sum of strain energy and surface energy terms:

$$U = -\frac{\pi c^2 \sigma^2 l}{E} + 4\gamma c l \tag{5.113}$$

Since the surface energy increases in proportion to the crack size c, whereas the elastic energy decreases as the crack size squared, above a certain crack size it is energetically downhill for the crack to propagate ($dU/dc < 0$), and catastrophic failure occurs. The critical condition is obtained by taking the derivatives and equating,

$$\frac{d}{dc}\left(\frac{\pi c^2 \sigma^2}{E}\right) \geq \frac{d}{dc}(4\gamma c)$$

$$\frac{\pi \sigma^2 c}{E} \geq 2\gamma \tag{5.114}$$

The left-hand side of Eq. 5.114 represents the elastic energy per unit crack surface area (J/m²) that is available upon infinitesimal crack extension and is called the *energy release rate*, G. The right-hand side of Eq. 5.114 is the surface energy increase and is referred to as the *crack resistance*, R. Unstable crack growth thus occurs when G is equal to or greater than R. If R is a constant, as in an ideally brittle material, fracture occurs when a critical value G_c is reached, corresponding to a critical stress level σ_c,

$$\frac{\pi \sigma^2 c}{E} \geq \frac{\pi \sigma_c^2 c}{E} = G_c = R \tag{5.115}$$

Rearranging, an important result of the Griffith theory is that for a particular value of the crack resistance, the failure stress varies inversely with the square root of the crack size:

$$\sigma_c = \left(\frac{EG_c}{\pi c}\right)^{1/2} \tag{5.116}$$

where in the ideally brittle case $G_c = R = 2\gamma$. Orowan, in 1948 (*Rep. Prog. Phys.*, 12, 185, 1948), suggested that the Griffith energy balance approach could be applied to materials having some plasticity, by defining an "effective" surface energy for the plastic strain work, γ_p. Then, Eq. 5.114 becomes

$$\frac{\pi\sigma^2 c}{E} = 2(\gamma + \gamma_p) \tag{5.117}$$

At failure, the result corresponding to Eq. 5.116 is

$$\sigma_c = \left(\frac{EG_c}{\pi c}\right)^{1/2} \approx \left(\frac{E(\gamma + \gamma_p)}{c}\right)^{1/2} \tag{5.118}$$

Stress Intensity Factor and Fracture Toughness

Griffith had based his stress concentration approach on earlier results by Inglis [*Trans. Inst. Naval Archit.*, 55, 219 1913)] showing that stresses in a plate with a flaw are maximized where the flaw has its sharpest curvature. A subsequent major advance by Irwin [*Welding J.*, 32, 450 (1952)] determined the local stress field near the tip of a sharp crack subject to a macroscopic stress. From linear elastic theory Irwin showed that the resulting stresses vary with distance r and angle θ (in polar coordinates) away from the crack tip, and take on the form

$$\sigma_{ij} = \frac{K}{\sqrt{2\pi r}} f_{ij}(\theta) + \ldots \tag{5.119}$$

K is the *stress intensity factor*, which is a scaling factor giving the extent to which an applied stress is amplified near the crack tip. Dimensional analysis requires K to be linearly related to stress and to the square root of a characteristic length; it is commonly given in units of MPa·m$^{1/2}$ or Ksi·in$^{1/2}$. While Eq. 5.119 is not applicable precisely at the crack tip ($r = 0$) or at very large r, the concept of a stress intensity factor is highly useful. Irwin showed that as the externally applied stress is increased, the achievement of a *critical stress intensity factor*, K_c, is equivalent to reaching a critical energy release rate G_c causing unstable crack propagation as described by Griffith. While the stress intensity factor K depends only on the shape of the crack and the manner in which it is loaded,[4] and not on the specific material, the *critical* stress intensity factor causing crack propagation is highly material dependent, and is termed the *fracture toughness* of the material. Due to the propensity for brittle materials to fail in opening mode, the fracture toughness quoted for ceramics is usually the value of K_{IC}.

Referring to the earlier results for fracture stress given in Eqs. 5.117 and 5.118, notice that all materials constants appear in the numerator; the results can be expressed in the form

$$\sigma_c = \frac{K_c}{\sqrt{\pi c}}$$

More generally we can rewrite Griffith's result in terms of the stress intensity factor as

$$K = \sigma \sqrt{\pi c} \cdot f\left(\frac{c}{w}\right) \tag{5.120}$$

where $f(c/w)$ is a dimensionless parameter of order unity which depends on the geometry of the crack (of length c) and the sample (of width w). Figure 5.99 illustrates different geometries of a small crack in a large sample ($c \ll w$) and the corresponding stress intensity factors for mode I loading. For tensile loading with a plane stress as shown in Fig. 5.98a, $f(c/w)$ is unity, and combining Eqs. 5.117 and 5.120,

$$G_c = \frac{K_c^{\,2}}{E} \tag{5.121}$$

---

[4] The notations K_I, K_{II}, and K_{III}, are frequently used. The subscripts I, II and III refer to the way the crack is loaded. Mode I, or "opening mode" loading, is the application of tensile stress normal to the plane of the crack, as illustrated for the cracks in Fig. 5.99. Mode II refers to a shearing load applied parallel to the plane of the crack along its direction of propagation, while Mode III is a torsional (tearing) load with the applied stresses again parallel to the crack plane but normal to the direction of propagation. Because brittle materials are weakest in tension, and a propagating crack often turns so that it is normal to the direction of maximum tension regardless of the applied stress configuration (i.e., it chooses mode I loading) K_I is usually the stress intensity factor of greatest interest. The value of K_I which causes crack propagation in a given material is its critical stress intensity factor, or fracture toughness, K_{IC}.

5.8 Strength and Toughness

Thus combining the Griffith–Orowan–Irwin approaches, we have for a through-cracked plate

$$K = \sigma\sqrt{\pi c} = \left[2E(\gamma + \gamma_p)\right]^{1/2} = (EG)^{1/2} \tag{5.122}$$

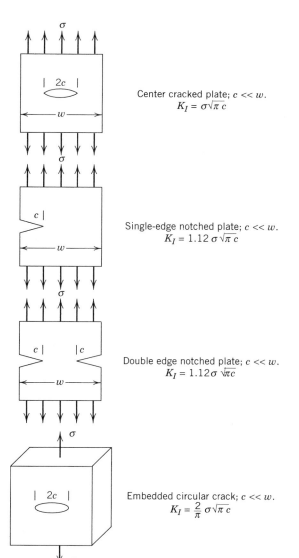

Center cracked plate; $c \ll w$.
$K_I = \sigma\sqrt{\pi}\,c$

Single-edge notched plate; $c \ll w$.
$K_I = 1.12\,\sigma\sqrt{\pi}\,c$

Double edge notched plate; $c \ll w$.
$K_I = 1.12\sigma\sqrt{\pi c}$

Embedded circular crack; $c \ll w$.
$K_I = \frac{2}{\pi}\sigma\sqrt{\pi}\,c$

Fig. 5.99 Stress intensity factor for mode I loading (K_I) with various crack configurations.

484 Chapter 5 / Microstructure

At the failure stress σ_c, the stress intensity factor has reached its critical value

$$K_c = \sigma_c \sqrt{\pi c} \qquad (5.123)$$

Equation 5.123 states that for any crack size c, there is a critical stress σ_c at which the stress intensity at the crack tip has reached the critical value, K_c, and crack propagation occurs. That is, fracture toughness can be determined experimentally by measuring the stress necessary for crack extension for a particular specimen/crack geometry.

A lower limit to the fracture toughness is $K_{IC} = \sqrt{2\gamma E}$, which assumes that surface energy presents the only resistance to crack propagation ($R=2\gamma$). When fracture is not purely brittle, the actual energy released during crack propagation exceeds that given by surface energy alone. During the past decade major ad-

Table 5.8 Fracture Toughness of Ceramics and Composites at Room Temperature

	K_{IC} (MPa·m$^{1/2}$)
Silicate glasses	0.7–0.9
Glass ceramics	~2.5
Single crystal NaCl	~0.3
Single crystal Si	~0.6
Single crystal MgO	~1
Single crystal SiC	1.5
Sintered, hot-pressed SiC	4–6
Single crystal Al$_2$O$_3$	
$\quad$ (0001)	4.5
$\quad$ ($10\bar{1}0$)	3.1
$\quad$ ($10\bar{1}2$)	2.4
$\quad$ ($11\bar{2}0$)	2.4
Polycrystalline Al$_2$O$_3$	3.5–4
Al$_2$O$_3$-Al composites	6–11
Reaction-bonded Si$_3$N$_4$	2.5–3.5
Sintered, hot-pressed, and gas-pressure sintered Si$_3$N$_4$	6–11
Cubic stabilized ZrO$_2$	~2.8
MgO-partially stabilized zirconia (PSZ)	9–12
Tetragonal zirconia polycrystals (Y-TZP, Ce-TZP)	6–12
Al$_2$O$_3$-ZrO$_2$ composites	6.5–13
Single crystal WC	~2
Metal (Ni, Co) bonded WC	5–18
Aluminum alloys	35–45
Cast iron	37–45
Steels	40–60

5.8 Strength and Toughness

vances have been made in developing ceramic microstructures which increase the energy absorption during crack propagation. The crack resistance R (having units of J/m^2) is increased to include additional work terms thereby increasing the stress intensity factor necessary to extend a crack, $K = \sqrt{ER}$. Typical values of the fracture toughness for a variety of ceramic materials are listed in Table 5.8. Glasses, and single crystals with easy cleavage planes, have the lowest fracture toughnesses, 0.5–2 MPa·m$^{1/2}$. Crystals without easy cleavage (such as sapphire) and typical polycrystals have higher values, 2–5 MPa·m$^{1/2}$, owing to a more tortuous crack path with greater fracture surface energy. Toughened ceramics have the highest fracture toughnesses, due to energy absorbing microstructures discussed later.

Variability in Strength

An important consequence of the Griffith theory is that brittle fracture is initiated at the most serious flaw, the weakest link in a chain. This is evident in the inverse square root dependence of failure stress on crack size, Eqs. 5.116 and 5.118. A corollary is that the fracture strength of a brittle solid is statistical in nature, depending on the probability that a flaw capable of initiating fracture is present when a particular stress is applied. That is, the observed strength should be related in some way to the surface area or the volume of material under stress. This expectation that observed strengths should vary with the specimen size, and the manner in which tests are conducted, has been confirmed in practice. Various efforts have been made to develop a statistical theory for the strength of brittle solids involving assumptions about the number of dangerous flaws associated with the volume or surface area. Direct observations suggest that this relationship changes from one material to another, depending on the fabrication method and surface treatments; no generalized statistical strength theory can be expected to hold for all materials. The most widely used theory is that developed by Weibull [*Ing. Vetensk. Akad.*, Proc. 151, No. 153, 1939)]. He assumed that the risk of rupture is proportional to a function of the stress in the volume of body, $f(\sigma)$, and should be integrated over the volume under applied tensile stress,

$$R = \int_v f(\sigma)\,dv \qquad (5.124)$$

Weibull proposed a particular form for $f(\sigma)$,

$$f(\sigma) = \left(\frac{\sigma}{\sigma_0}\right)^m \qquad (5.125)$$

where σ_0 is a characteristic strength depending on the distribution function and m is a constant related to material homogeneity, termed the Weibull modulus. The "average" strength value, taken to be that where the probability of failure is 1/2, is given by

486 Chapter 5 / Microstructure

$$\sigma_{R=1/2} = \left(\frac{1}{2}\right)^{1/m} \sigma_o \qquad (5.126)$$

The larger the value of m, the more homogeneous the strength. As m becomes infinite, fracture occurs only at the characteristic strength σ_0; that is, the strength is absolutely predictable. With current ceramics technology, a value of $m = 25$ is considered to be high, whereas values of 5 to 10 are more typical.

Surface Flaws

Experience has shown that the most dangerous flaws in most ceramic bodies occur in association with the surface. If a flame-polished surface of a SiO_2 glass is tested immediately after preparation, it has a strength nearly equivalent to the theoretical value. After standing in the atmosphere or being waved about through the air and being tested again, its strength is remarkably decreased. Similarly, chemical polishing leads to substantial improvement of the strengths of brittle materials while surface abrasion substantially decreases strength. As a result, the most successful efforts to increase practical fracture strength values of ceramic materials have been to avoid the development of surface flaws or to develop surface compressive stresses which will counteract the effects of flaws.

One method of developing residual surface compressive stresses is by thermal tempering. A plate of glass which is rapidly cooled, typically with air jets, will have a surface layer which has been more rapidly cooled through the glass transition temperature than the interior. Since the specific volume of glass increases as the cooling rate increases (see Fig. 1.45), the surface layer has a greater specific volume than the interior and retains a compressive stress while the interior is in tension (Fig. 5.100). An equivalent result can be obtained by chemical treatment of the glass in order to develop a surface compressive layer. If a sodium lime silica glass is immersed in a potassium salt (e.g., potassium nitrate), there is an exchange of large potassium ions for the smaller sodium ions and the development of a compressive layer in the surface, which acts to negate the effects of Griffith surface flaws. Another approach to this is to use a laminated structure in which a low thermal expansion surface layer is applied on an underlying glass of greater thermal expansion, such that, on cooling, a high compressive stress develops in the surface. This compressive stress negates the influence of Griffith flaws and leads to a high resistance to brittle fracture. Still another approach is to diffusively change the surface composition of the glass or glass ceramic (Corning's Chemcor™ process) to one of low thermal expansion, such as a lithium alumina silicate. Upon cooling a compositionally graded structure of this type, the interior undergoes greater thermal contraction than the surface layers and creates residual surface compression.

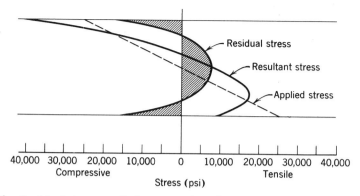

Fig. 5.100 Residual stress, applied stress, and resultant stress distribution for transverse loading of a tempered glass plate.

Microstructural Toughening

Strengthened ceramics are not necessarily *tougher*; the "dicing" fracture of a high-strength, thermally tempered glass shows a dramatic lack of resistance to crack propagation once the failure stress is exceeded. The values for fracture toughness of glass and single crystal ceramics listed in Table 5.8 are low compared to the values for common structural metals (which exceed 30 MPa·m$^{1/2}$ due to extensive plastic work and crack blunting during fracture). However, over the past two decades substantial increases in the fracture toughness and strength of polycrystalline and polyphase ceramics have been achieved through the control of microstructure. High-strength (~1 GPa) silicon nitride can be produced with fracture toughness values in the range of 6–10 MPa·m$^{1/2}$. Partially stabilized zirconias (PSZ's) and tetragonal zirconia polycrystals (TZP's) have fracture toughnesses in the range of 6–10 MPa·m$^{1/2}$ and strengths ranging from 0.6 GPa to more than 1 GPa. Ductile metal-toughened ceramics (which we may somewhat arbitrarily define as those containing less than ~30 volume % metal, in order to distinguish them from ceramic-reinforced metal matrix composites) can have still higher fracture toughnesses (10–15 MPa·m$^{1/2}$), and certain fiber-reinforced ceramic composites do not fail in a catastrophic manner at all. Well-engineered, fiber-reinforced composites fail by progressive delamination and pull-out of fibers from the surrounding matrix and can absorb large amounts of fracture work over strains of several percent before finally separating. Standardized measurements yield apparent fracture toughness values of 20–25 MPa·m$^{1/2}$ for these composites, but the failure process can be quite different from the propagation of a sharp crack as described by Griffith theory. These improvements in toughness have enabled a number of new structural applications of ceramics, including silicon nitride for automotive components (e.g., turbocharger rotors) and high temperature gas turbines, tranformation toughened zirconia polycrystals and composites for a broad range of low temperature applications, and fiber- and whisker-reinforced glasses, glass ceramics, and polycrystalline ceramics for jet engine components, cutting tools, bearings, and other demanding applications.

As we have discussed, one approach to increasing both strength and reliability in brittle materials is to decrease the characteristic flaw size, and to narrow the distribution in flaw sizes. In pursuit of this ideal, much effort has been devoted to improving the flaw population in structural ceramics through careful processing. Microstructural toughening is a complementary approach, which emphasizes *flaw tolerance*. In the following examples, we will see that many tough ceramics are flaw-tolerant due to the fact that the resistance to crack growth is not a constant, but increases as a flaw propagates. Reliability increases since smaller flaws can often grow to a limiting size before failure; the initial distribution of flaw sizes becomes less important.

It is almost always desirable to increase both strength and fracture toughness. All else being equal, an increase in fracture toughness will increase strength due to the proportionality between σ and K (Eq. 5.123). However, if the microstructure necessary to achieve high toughness requires larger characteristic flaws, then increases in fracture toughness can come at the expense of strength. Ceramics that derive thermal shock resistance from extensive microcracking, such as coarse-grained porous firebrick and polycrystalline aluminum titanate (which has unusually high crystalline thermal expansion anisotropy), are examples of relatively tough, but also relatively weak ceramics.

Transformation-Toughened Zirconia

The exemplar of microstructural toughening is illustrated with the partially stabilized zirconia microstructure containing fine platelike precipitates of the metastable tetragonal phase (Fig. 5.101). This microstructure leads to three different sorts of toughening mechanisms. First, as a crack advances, the intensified stress forward of the crack causes transformation of the tetragonal precipitates to the monoclinic form with absorption of energy and a volume expansive strain resulting from the volume change during transformation (see also the discussion of the zirconia polymorphs in Chapter 1). In fig. 5.101*b*, particles which have transformed in the vicinity of the crack appear bright due to a difference in diffraction contrast from those which have not. Additional energy absorption comes from the local stress fields around the transformed particles, which cause microcracks to develop in the intensified stress field as the crack advances. These microstructural changes have the effect of reducing the effective stress intensity at the crack tip. Third is the crack deflection caused by localized residual stress fields resulting from the transformation and from the fracture resistance of the individual particles. Crack deflection adds to the toughening by increasing the crack surface area and effective surface energy, and is also employed in particle- and short fiber-reinforced composites.

5.8 Strength and Toughness 489

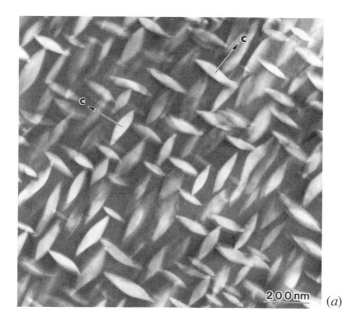

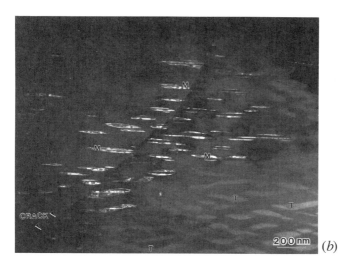

Fig. 5.101 Transformation-toughened zirconia. (*a*) Microstructure showing bright particles of tetragonal zirconia of oblate spheroid shape in a matrix of cubic zirconia, in a magnesia partially-stabilized-zirconia. (*b*) Transformation zone in a thin foil in the vicinity of a crack. Dark field electron micrograph taken with m-ZrO_2 reflection reveals bright particles of transformed monoclinic zirconia. [Courtesy of A.H. Heuer.]

490 Chapter 5 / Microstructure

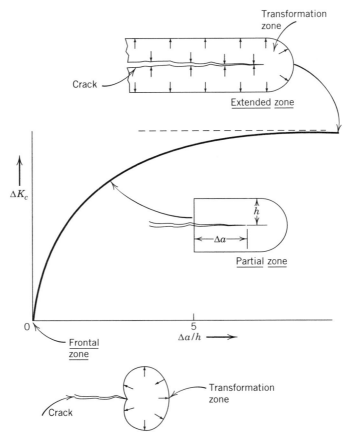

Fig. 5.102 Predicted crack resistance R-curve for transformation toughening showing asymptotic approach to a limiting K_c as the crack extension (a reaches $5h$, where h is the width of the transformation zone. (Courtesy of A.G. Evans.)

Each of these energy absorption processes occurs as the *result* of crack propagation. As the crack begins to propagate, at first only a frontal zone is affected. As the process continues, a larger zone in the wake of the crack is affected, and after a certain amount of crack growth a steady state is reached in which the transformed zone around the crack, the *process zone*, is fully developed, and the crack resistance is increased by an amount ΔK_R, exemplified by the so-called *R*-curve illustrated in Fig. 5.102. The *R*-curve describes the crack resistance as a function of crack extension, $K_R (\Delta c)$, and increases approximately as $tan^{-1}(\Delta c/h)$,[5] where h is the width of the transformation zone.

---

[5]A.G. Evans, pp. 193-212 in *Advances in Ceramics*, Vol. 12, *Science and Technology of Zirconia II*, edited by N. Claussen, M. Ruhle, and A.H. Heuer, American Ceramic Society, Columbus, Ohio, 1984.

5.8 Strength and Toughness

The influence of the R-curve on fracture is illustrated in Fig. 5.103 as plots of K versus c. The applied stress intensity varies parabolically with crack length as $K_I \propto \sigma c^{1/2}$, and for increasing applied stress, a family of K_I curves may be drawn as shown in Fig. 5.103a. For conventional brittle materials where the fracture toughness is a constant, Griffith fracture occurs when the product of stress and crack length reaches K_{IC}. In a material with a rising resistance curve, the crack resistance K_R increases along the heavy line shown in Fig. 5.103b as c increases. If the

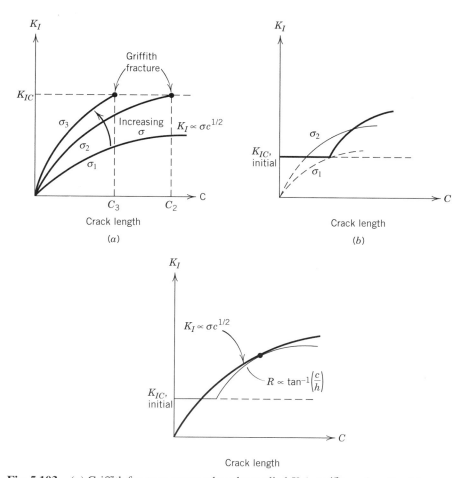

Fig. 5.103 (a) Griffith fracture occurs when the applied K_I ($\propto \sigma c^{1/2}$) reaches K_{IC}; this corresponds to reaching a critical stress for the flaw size present. Curves represent K_I for increasing values of stress σ. (b) With an R-curve being present, the crack resistance increases beyond its initial value as the crack grows. If the crack resistance rises faster with crack extension than does the applied K_I, the crack can extend stably without failure occuring. (c) Eventually the R-curve flattens, and crack resistance cannot keep up with the applied K_I. The point of tangency between the two curves shown defines the failure criterion.

crack resistance grows faster than does the applied K_I, the crack remains subcritical; this condition is

$$\frac{dK_R}{dc} > \frac{dK_I}{dc}$$

As long as the crack resistance curve rises faster with c than does the applied K_I, the crack can continue to grow, *stably*, without fracture occurring. However, this condition cannot continue indefinitely since the R-curve eventually flattens. The critical condition for fracture occurs when tangency between the two curves is reached, as shown in Fig. 5.103c. This tangency defines the strength and fracture toughness of the material.

The R-curve can be experimentally measured by growing flaws of controlled size, at subcritical stresses, and measuring the stress necessary to propagate them. In partially stabilized zirconias, continuing increases in R up to crack lengths of several millimeters, and plateau values of R approaching 15 MPa·m$^{1/2}$, have been observed. However, it is important to recognize that the plateau value is not generally the fracture toughness of the material, due to the tangent criterion. In transformation-toughened zirconias, failure may occur at stresses where the corresponding fracture toughness is only about one-half of its plateau value. Improving the fracture toughness at small flaw sizes, and obtaining a rapidly rising R-curve, is therefore an important objective of microstructure engineering.

Other Flaw-Tolerant Microstructures

The R-curve concept, while first developed for transformation-toughened zirconias, may be generalized to any material in which the microstructure causes the fracture resistance to increase as a flaw propagates. In addition to microcracking and phase transformations, it has become recognized that ligaments which continue to bridge the crack after the crack front has passed have a toughening effect by taking up some of the stress tending to open the crack, and thereby lessen the applied stress intensity at the crack tip. This effect is perhaps obvious in fiber-reinforced ceramics, Fig. 5.104a, where fibers which are not too tightly bonded to the matrix can debond along some length, slip, and continue to bridge the crack after the crack front has passed. The critical stress intensity factor which results from frictional bridging by strong debonded fibers has been analyzed by A. G. Evans and R. M. McMeeking,[6] and is given approximately by

$$K_c \approx \left(\frac{\sigma_f^3 G}{3 E_f \tau}\right)^{1/2} \sqrt{r_f A_f} \qquad (5.127)$$

where σ_f, E_f, and r_f are, respectively, the failure strength, Young's modulus, and radius of the fiber, G is the shear modulus of the matrix, τ is the shear strength

5.8 Strength and Toughness

of the fiber-matrix interface, and A_f is the area fraction of fibers intersecting the crack plane (a function of fiber volume fraction and orientation). Notice that the fracture toughness increases with the area (volume) fraction of fibers and with fiber size. A strong fiber of lower modulus and a weak interface are also important, as this increases the length of debonded interface. A fiber that is very strongly bonded to the matrix fractures close to the crack plane and provides little advantage over a monolithic material.

Another example of crack bridging occurs in ductile metal-phase-toughened ceramics, depicted in Fig. 5.104b. Ductile ligaments which remain well-bonded to the crack faces have a similar effect to fibers, remaining bridged behind the crack and undergoing plastic deformation, necking, and fracture. The fracture toughness has the theoretical form:[6]

$$K_c \approx \sqrt{\theta C \sigma_y G r_p A_p \left(0.5 + \exp(\varepsilon_f)\right)} \quad (5.128)$$

where θ and C are empirical constants, r_p and A_p are the radius and crack-plane area fraction of the particles, and σ_y and ε_f are, respectively, the yield strength and strain-to-failure of the metal particles. Since in both fiber and ductile-particle reinforced materials the crack must first propagate in order to develop bridging ligaments, R-curve behavior is a natural consequence of crack bridging.

What was more recently appreciated[7] is that crack bridging also plays an important role in the toughening of conventional polycrystalline ceramics, particularly those with large and anisotropic grains. In the wake of the crack front, frictional forces from a small fraction of grains which remain interlocked have a similar effect to that of fibers or ductile ligaments. Figure 5.105 shows a photographic sequence of a grain which remains bridging in the wake of a propagating crack for some extent before finally separating. The extent of crack bridging is expected to increase with grain size and grain shape anisotropy. The fracture toughness due to bridging by a strong reinforcement of this type is given by Evans and McMeeking as:[6]

$$K_c \approx 1.1 \sigma_p \sqrt{r_p A_p \left(1 - \sqrt{A_p}\right)\left(1 - A_p\right)} \quad (5.129)$$

where r_p is a particle radius and σ_p the failure strength of the bridge. The greater toughness of most polycrystalline ceramics in comparison to their single crystal counterparts, and of whisker- and particle-reinforced ceramics, results partly from this kind of bridging, and partly from the effects of crack deflection along grain

---

[6] A.G. Evans and R.M. McMeeking, *Acta Met.*, 34[12], 2435, 1986.

[7] R. Knehans and R. Steinbrech, *Sci. Ceram.*, 12, 613, 1983, P.L. Swanson, C.J. Fairbanks, B.R. Lawn, Y.-W. Mai, and B.J. Hockey, *J. Am. Ceram. Soc.*, 70[4], 279, 1987.

494 Chapter 5 / Microstructure

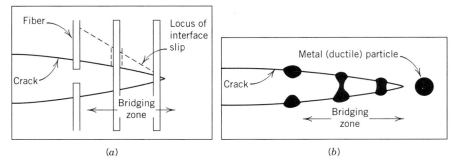

Fig. 5.104 Reduction of stress intensity at crack tip due to bridging by (*a*) fibers; and (*b*) ductile particles.

boundaries or the interfaces between reinforcements and the matrix, as illustrated in Fig. 5.106. An analysis by K.T. Faber and A.G. Evans[8] shows that due to a reduction in the stress intensity factor at the tip of a deflected (tilted and twisted) crack compared to a planar crack, there is an increase in fracture toughness above and beyond that which results from the increased fracture surface area alone. It is predicted that the relative increase in toughness increases with the volume fraction of deflection material and its aspect ratio, but not its size. Increases in fracture toughness of a factor of 4 or so are expected for 50% loading by a rodlike reinforcement with an aspect ratio of 10:1.

Silicon Nitride

Silicon nitride which is sintered or hot-pressed using a small quantity of oxide additive (including various combinations of the additives MgO, Al_2O_3, CaO, Y_2O_3), exhibits acicular, interlocking grains that yield a high-strength, high-toughness microstructure sometimes referred to as an in-situ composite, shown in Fig. 5.32. The elongated grains, which are of the β-phase of Si_3N_4, form from equiaxed grains of the α phase during sintering and grain growth by a solution and reprecipitation process facilitated by the liquid. This liquid is a silicon oxynitride, formed during sintering from the oxide additive and the surface oxygen impurity, which is almost always present in silicon nitride powders. The needle-like morphology of the β-Si_3N_4 grains is shown particularly clearly in Fig. 5.107, where the grains of a sintered polycrystal have been separated by chemical dissolution of the intervening glass. The polycrystalline microstructure is virtually a composite of short fibers more or less randomly oriented (although hot-pressing can introduce some texture). Crack propagation accompanied by copious grain bridging yields *R*-curve behavior, with fracture toughnesses which are typically ~7 MPa·m$^{1/2}$, and in some materials, as high as 11 MPa·m$^{1/2}$. In contrast, reaction-bonded silicon

---

[8]K.T. Faber and A.G. Evans, *Acta Metall.*, 31[4], 565 (1983).

5.8 Strength and Toughness   495

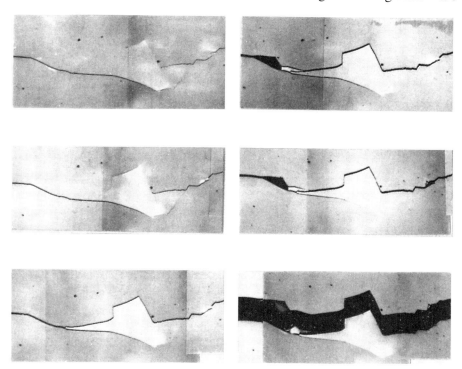

Fig. 5.105 Crack bridging by a grain in the wake of a crack in coarse-grained alumina. Continued crack opening eventually causes the bridge to fail. [From P. Swanson, C.J. Fairbanks, B.R. Lawn, Y.-W. Mai, and B.J. Hockey, *J. Am. Ceram. Soc.*, **70**[4] 279 (1987).]

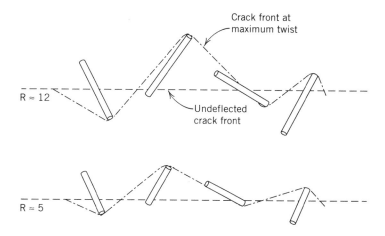

Fig. 5.106 Deflection of cracks by rod-shaped reinforcements of two aspect ratios R. Higer aspect ratio and reinforcement volume fraction lead to greater increases in fracture toughness. [From K.T. Faber and A.G. Evans, *Acta Metall.*, **31**[4], 565 (1983).]

496 Chapter 5 / Microstructure

Fig. 5.107 The grain morphology of sintered silicon nitride, an *in-situ* formed composite, is revealed to be rodlike by chemically dissolving the intergranular glass phase and separating the grains. Cf. Fig. 5.31. (Courtesy J. Wallace, National Institute for Standards and Technology.)

nitride, which is formed by the direct nitridation of silicon powder compacts using nitrogen gas, is equiaxed in grain structure and primarily of the α phase. Thus it has only about one-half the fracture toughness of sintered and hot-pressed silicon nitride, Table 5.8. However, this material retains strength and creep resistance to higher temperatures, owing to the absence of the residual glass phase, which otherwise limits use temperatures in sintered silicon nitride to about 1200°C.

Fiber- and Whisker-Reinforced Ceramics

While the basic concept of reinforcing a brittle material with fibers is an ancient one, perhaps first exploited in the strengthening of clay building materials with straw, and has been for decades employed to great effect in reinforced concrete (using steel bars or mesh), a new generation of fiber-reinforced ceramics began in the 1970s with studies of carbon fiber-reinforced glasses [R. A. J. Sambell, A. Briggs, D. C. Phillips, and D. H. Bowen, *J. Mater. Sci.*, 7, 663 (1972), and D.C. Phillips, *J. Mater. Sci.*, 7, 1175 (1972)]. Subsequently, continuous silicon carbide fibers, with better high temperature oxidation resistance, were also used to reinforced glasses and glass-ceramics (K. M. Prewo and J. J. Brennan, *J. Mater. Sci.*, 15[2], 463 (1980), and J. J. Brennan and K. M. Prewo, *J. Mater. Sci.*, 17, 7371 (1982)]. These composites are typically prepared by impregnating continuous fiber "yarns" with a glass powder slurry, layering the impregnated fibers, often with alternating fiber directions in a configuration not unlike that of plywood, and hot-pressing in vacuum to densify the composite by viscous flow of the glass matrix. Sodium borosilicate glasses and lithia-alumina-silica glass ceramics similar to the Pyrex™ and Pyroceram™ compositions, respectively, have been widely used. Since these matrices are lower in thermal expansion coefficient than carbon or silicon carbide fibers, residual compressive stresses develop in the matrix after cooling. In analogy to thermally tempered materials, the residual compressive stress must be overcome by the applied stress before initial cracking of the matrix glass occurs. A number of fiber/matrix combinations, including crystalline oxide and covalent ceramic matrices, have also been studied.

Continuous fiber reinforcements provide both strengthening and toughening. The fibers are much stronger (failure strengths usually ≥ 2 GPa) and of higher modulus (≥ 200 GPa) than the matrices used, and they support most of the applied load. Ultimate composite fracture strengths exceeding 1 GPa have been achieved. However, well before this limit is reached, the matrix has usually cracked. Neglecting the effects of residual thermal stresses, the stress in the matrix of a unidirectionally reinforced composite containing a volume fraction of fibers V_f is obtained by equating the strains in the fibers and the matrix (i.e., by assuming no fiber slipping):

$$\sigma_m = \frac{\sigma_c}{(1-V_f) + V_f \left(\dfrac{E_f}{E_m} \right)} \tag{5.130}$$

where σ_c is the stress applied to the composite, and E_f and E_m are the moduli of the fiber and matrix, respectively. Matrix cracking occurs when this stress exceeds the failure strength of the glass or glass-ceramic; the matrix stress is reduced by increasing the volume fraction of fibers and by having a higher fiber modulus relative to the matrix modulus.

The failure behavior of fiber-reinforced ceramic composites is unusual in that it is not suddenly catastrophic. A large fraction of the applied load can continue to be supported by the fibers even after the matrix has fractured. The stress-strain curve for a typical composite, Fig. 5.108, shows an initial change in slope (i.e., modulus) when the matrix fractures. Matrix cracking can define the limiting stress for high

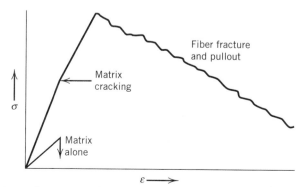

Fig. 5.108 Schematic stress-strain curve for a ceramic composite reinforced with continuous fibers of higher elastic modulus.

temperature use in oxidizing ambients, since both carbon and silicon carbide fibers weaken upon oxidation (the latter are more stable). After the onset of matrix cracking, the stress is increasingly born by the fibers, reaching a peak value before the fibers themselves begin to fracture and pull out from the matrix. The stress-strain curve shows a great deal of nonlinearity as the fibers break and pull out in a step-like fashion. It is necessary to have a relatively weak interfacial bond between the fiber and the matrix in order to develop this type of noncatastrophic failure. Figures 5.109 and 5.110 compare the fibrous fracture surface of this type of composite with one in which a strong interfacial bond has caused brittle fracture through the fibers as well as the matrix. As discussed above, the fracture toughness is increased by crack bridging as the crack propagates. The total work of fracture due to fiber pullout is given by

$$W = \frac{V_f \sigma_f^2 r_f}{12 \tau} \tag{5.131}$$

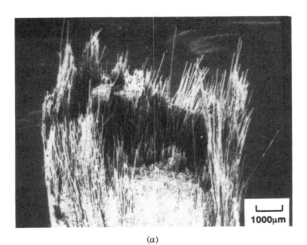

(a)

(b)

Fig. 5.109 Fibrous fracture with extensive fiber pullout characteristic of a high toughness composite; silicon carbide fiber reinforced lithia-aluminosilicate glass-ceramic matrix. [From J.J. Brennan and K.M. Prewo, *J. Mat. Sci.*, **17**, 2371 (1982).]

5.8 Strength and Toughness   499

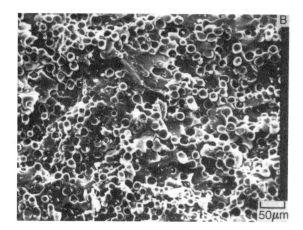

Fig. 5.110 Brittle fracture with no discernible fiber pullout which results when fiber-matrix interface bond is excessively strong. [From J.J. Brennan, pp. 387-399 in *Ceramic Microstructures '86*, J.A. Pask and A.G. Evans, Plenum Press, New York, 1987.]

where σ_f is the fiber strength and τ is the shear strength of the fiber–matrix interface. It is perhaps not intuitive that a strong fiber, a large fiber radius, and a low interfacial strength all tend to increase the work of fracture. This results from the fact that the length of fiber over which debonding from the matrix and pull-out occurs is increased, Fig. 5.104a. A strong interfacial bond causes fracture of the fiber closer to the plane of the crack and lowers the work of fracture, Fig. 5.110. Often the weak interface is achieved by deliberate fiber coatings of a weak material such as carbon or by a chemical reaction between the silicon carbide fiber and the oxide matrix, which produces a thin interfacial layer of carbon (Fig. 5.111). A

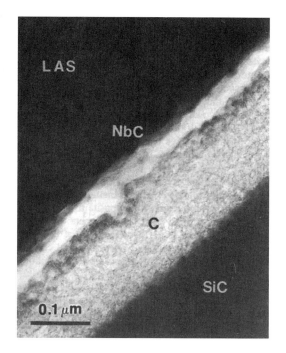

Fig. 5.111 Carbon interfacial layer resulting from chemical reaction (oxidation of SiC) in silicon carbide fiber reinforced lithia-aluminosilicate composite with niobium additions. [From, *J. Am. Ceram. Soc.*, **72**[5], 741 (1989).]

drawback of a weak interfacial bond that while the strength and toughness are increased in the direction of the fibers, the transverse strength suffers. The anisotropy in properties of continuous fiber-reinforced composites can be adjusted to some extent by using multidirectional fiber architectures.

ADDITIONAL READING

Interfaces

L. E. Murr, *Interfacial Phenomena in Metals and Alloys*, Addison-Wesley, Reading, Mass., 1975.

J. Israelachvili, *Intermolecular and Surface Forces (*2nd Edition*)*, Academic Press, London, 1992.

J. M. Blakely, *Introduction to the Properties of Crystal Surfaces*, Pergamon Press, Oxford, 1973.

R. W. Balluffi and A. P. Sutton, *Interfaces in Crystalline Materials,* Oxford Univ. Press, 1995.

D. Wolf and S. Yip, Editors, *Materials Interfaces: Atomic Level Structure and Properties*, Chapman and Hall, London, 1992.

Microstructure Development and Sintering

C. S. Smith, "Grains, Phases, and Interfaces: An Interpretation of Microstructure," *Trans. A.I.M.E.*, 175[1], 15-51 (1948).

E. E. Underwood, *Quantitative Stereology*, Addison-Wesley, Reading, Mass., 1970.

R. M. Cannon and R. L. Coble, "Paradigms for Ceramic Powder Processing," pp. 151–170 in *Processing of Crystalline Ceramics*, Mat. Sci. Res. Vol. 11, H. Palmour III, R. F. Davis, and T.M. Hare, Eds., Plenum Press, New York, 1978.

M. F. Yan, R. M. Cannon, and H. K. Bowen, "Grain Boundary Migration in Ceramics," pp. 276-307 in *Ceramic Microstructures '76*, J. A. Pask and R. A. Fulrath, Ed., Westview Press, Boulder, Colorado, 1977.

S. J. Bennison and M. P. Harmer, "A History of the Role of Magnesia in the Sintering of Alumina," *Ceramic Transactions*, Vol. 7, p. 13, The American Ceramic Society,Columbus, Ohio, 1990.

R. M. German, *Liquid Phase Sintering*, Plenum Press, New York, 1985.

Processing

J. S. Reed, *Introduction to the Principles of Ceramic Processing*, John Wiley and Sons, New York, 1988.

J. W. Evans and L. C. De Jonghe, *The Production of Inorganic Materials*, MacMillan, New York, 1991.

J. D. Mackenzie and D. R. Ulrich, *Ultrastructure Processing of Advanced Ceramics*, John Wiley and Sons, New York, 1988.

C. J. Brinker and G. W. Scherer, *Sol-Gel Science*, Academic Press, San Diego, 1990.

Glasses and Glass-Ceramics

W. Vogel, *Chemistry of Glass*, The American Ceramic Society, Columbus, Ohio, 1985.

D. R. Uhlmann and N.J. Kreidl, editors, *Glass: Science and Technology*, Vols. 1–5, Academic Press, 1980-1990 (Vol. 1, 1983; Vol. 2, 1984; Vol. 3, 1986; Vol. 4A and 4B, 1990, Vol. 5, 1980).

S. D. Stookey, *Journey to the Center of the Crystal Ball*, The American Ceramic Society, Columbus, Ohio, 1985.

G. H. Beall, "Design and Properties of Glass-Ceramics," *Ann. Rev. Mater. Sci.*, 22, 91-119 (1992).

Mechanical Properties of Ceramics

J.E. Gordon, *The New Science of Strong Materials*, Princeton University Press, Princeton, New Jersey, 1976.

D.W. Richerson, *Modern Ceramic Engineering*, 2nd Ed., Marcel Dekker, New York, 1992.

R.W. Davidge, *Mechanical Behavior of Ceramics*, Cambridge University Press, U.K., 1979.

B. Lawn, *Fracture of Brittle Solids*, 2nd Ed., Cambridge University Press, U.K., 1993.

A.G. Evans and T.G. Langdon, *Structural Ceramics*, *Prog. in Mater. Sci.*, Vol 21, 3/4, Pergamon Press, 1976.

Transformation Toughened Ceramics

D.J. Green, R.H.J. Hannink, M.V. Swain, *Transformation Toughening of Ceramics*, CRC Press, Boca Raton, Florida, 1989.

A.H. Heuer, "Transformation Toughening in ZrO_2-Containing Ceramics," *J. Am. Ceram. Soc.*, 70[10], 689, 1987.

A.H. Heuer and L.W. Hobbs, Editors, *Science and Technology of Zirconia*, Advances in Ceramics, Vol. 3, The American Ceramic Society, Columbus, Ohio, 1981.

N. Claussen, M. Ruehle, and A.H. Heuer, Editors, *Science and Technology of Zirconia II*, Advances in Ceramics, Vol. 12, The American Ceramic Society, Columbus, Ohio, 1984.

S. Somiya, N. Yamamoto, and H. Yanagida, Editors, *Science and Technology of Zirconia III*, Advances in Ceramics, Vol. 24B, The American Ceramic Society, Westerville, Ohio, 1988.

S.P.S. Badwal, M.J. Bannister, and R.H.J. Hannink, *Science and Technology of Zirconia V*, Technomic Publishing, Lancaster, Penn., 1993.

Composites

D. Hull, *An Introduction to Composite Materials*, Cambridge University Press, U.K., 1981.

M.R. Piggott, *Load Bearing Fiber Composites*, Pergamon Press, New York, 1980.

D.R. Clarke, "Interpenetrating Phase Composites," *J. Am. Ceram. Soc.*, 75[4] 739 (1992).

PROBLEMS

1. At 0°C the solid–liquid interfacial energy in the ice-water system is 70 ergs/cm^2, the grain boundary energy is 28 ergs/cm^2, the liquid surface energy is

502   Chapter 5 / Microstructure

76 ergs/cm^2, and liquid water completely wets (spreads across) a free ice surface.

(i) What will be the angle of grain boundary etching for polycrystalline ice submerged in water?

(ii) Ethanol added to water is observed to cause complete penetration along the grain boundaries and separation of the grains. How much must alcohol lower the ice-water interfacial energy for this to occur?

2. In the processing of thin films by various methods such as sputtering, evaporation, and wet chemical methods, pinholes can often occur. Let's assume that we have deposited a film on a substrate, that a number of pinholes are present, and we will now be annealing the film at a temperature high enough for mass transport to take place. Identify the conditions under which the pinhole will grow or shrink, depending on the relative values of interfacial energies. To skirt the issues raised by dihedral angles where grain boundaries meet the surface and by anisotropic surface energies, assume the film to be amorphous and all interfacial energies to be isotropic.

3. Most oxides do not vaporize congruently; that is, the vapor phase species have a different metal/oxygen stoichiometry than the solid. BaO does vaporize congruently under oxidizing conditions, and data for the partial pres-

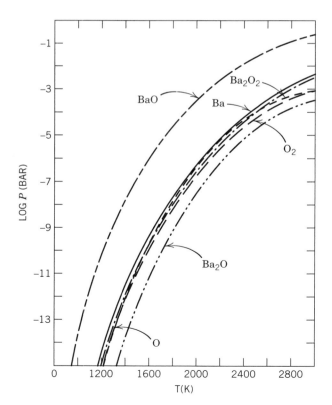

sure of BaO (vapor) as well as minor gas phase species as a function of temperature is given in the figure.

Consider a flat BaO crystal surrounded by a bed of fine powder of particle radius r, all of which is enclosed by a crucible. We heat this crucible up to temperature T = 2000 K and allow the gas phase to come to equilibrium over the solid.

(i) Will the crystal grow or shrink? Why?

(ii) Calculate the rate of evaporation-condensation growth/shrinkage for this crystal, if the surrounding powder is of $r = 0.03$ µm and 1 µm respectively, assuming that the rate of Langmuir condensation is rate limiting rather than gas phase diffusion, and that the sticking coefficient α is 0.5. Ignore the fact that the particles will coarsen or densify among themselves (i.e., calculate for the early stage.). The molecular weight of BaO is 153.34 g/mole and the density is 5.72 g/cm³.

4. Air is roughly 80% nitrogen, a gas that is insoluble in most oxides. Oxygen, however, is soluble. If we fire a powder compact in air (at 1 atm. pressure) and the pores close off when they are 2 µm in diameter, what will the nitrogen pressure in the pores be when equilibrium is reached? Assume a reasonable value for the surface tension of the oxide.

5. Calculate the sintering pressure as a function of shrinkage, $\Delta L/L_o$, for particles of radius $r = 0.01$, -0.1, $-$ and 1.0-µm. Only consider neck size/particle size ratios that are in the initial stage of sintering. For simplicity, you may assume simple cubic packing of the particles. Use the geometric relation between neck radius of curvature and shrinkage given in Eq. 5.45 and assume a reasonable value for the surface tension.

6. Calculate the initial rate of sintering of 1-µm spheres of Al_2O_3 at 1400°C, given the following diffusion coefficients:

$$D_{gb}^{Al} = 5.6 \times 10^{-17} \text{ cm}^3/\text{sec}$$
$$D_{gb}^{O} = 7 \times 10^{-14} \text{ cm}^3/\text{sec}$$
$$D_{lattice}^{Al} = 4 \times 10^{-14} \text{ cm}^3/\text{sec}$$
$$D_{lattice}^{O} = 1 \times 10^{-17} \text{ cm}^3/\text{sec}$$

The molecular weight of Al_2O_3 is 101.96 g/mole, and the density is 3.97 g/cm³.

7. It has been found that a moving pore adopts the cross-sectional shape shown in Fig. 5.47.

(a) Is there a gradient in vapor pressure between the leading and trailing surfaces of the pore? If so, which way is the gradient? Explain your answer, using an equation that identifies the important variables.

(b) Should diffusional transport occur from one side to the other? Explain fully. Identify the paths along which this transport could take place.

(c) Sometimes the addition of a solute that segregates to the surface can change the surface energy relative to the grain boundary energy. Write a relationship to show how the dihedral angle, ϕ, is related to the relative values of these two energies.

(d) For a fixed pore volume, explain qualitatively (using a sketch) how an increase or decrease in the dihedral angle will change the magnitude of the gradient. What influence does dihedral angle have on the drag force of the pore on the grain boundary?

(e) Draw a grain size-density plot showing the grain size trajectory followed during later-stage sintering for: (i) a material in which pores become entrapped; and (ii) where sintering to theoretical density is achieved. Based on your answers to parts (a)–(c), should a decrease in the dihedral angle assist in sintering to complete density? Explain your reasoning.

8. We find that upon sintering Al_2O_3 (of 0.2 µm starting particle size) doped with ~0.25 wt % MgO in an oxygen atmosphere at 1800°C for a few hours we end up with a fully dense polycrystalline material of about 15-micron average grain size.

If we fire in argon (1-atm pressure), we get a 95% dense final product of grain size G. By examining the time evolution of microstructure during firing, through a series of quenching experiments (fire at 1800°C for a specified time, then quench) we find that when the intermediate stage ends (that is, the pore channels are pinched off and "closed porosity" occurs) the average grain size is about 2 µm and the average pore size is 0.25 µm. Predict the grain size G at which the sample fired in argon will end up. State any notable assumptions you make. Where will the pores mostly be located?

If the pores coalesce during firing to where the average size is now 2 µm, will the observed volume fraction porosity be different? Why or why not, and if it is different, how different?

If we fire the same powder without MgO, what microstructure should we end up with and why? Do you expect much difference in the results from firing in O_2 versus argon?

9. In a fully dense and ideally pure polycrystalline MgO that exhibits normal (parabolic) grain growth, what would one expect the final grain size to be after 1/2 hour at 1900°C if the initial grain size is 0.5µm? After 1 hour and 2 hours? You will need to calculate a grain boundary mobility: use a grain boundary diffusion coefficient of $D_{gb}=10^{-7}$ cm²/sec. Assume the grain boundary energy is 10^3 ergs/cm², and a grain boundary width of $\delta=0.5$ nm. Other useful data:

$R = 8.314 \times 10^7$ ergs/mole K
Molecular weight (MgO) = 40.31 g/mole
Density (MgO) = 3.58 g/cm³

If some SiO_2 is present, we see from the phase diagram in Fig. 4.24 that at this temperature a liquid phase of roughly forsterite composition (Mg_2SiO_4) forms. If a wetting grain boundary film of 50 nm width forms, how will the grain boundary mobility vary if:

(i) The liquid diffusivity D_l is assumed equal to D_{gb}?
(ii) The liquid diffusivity is 10^2 faster?

10. During sintering of polycrystalline materials, we often end up with pores of an assortment of sizes, located both within the grains and at grain junctions, as shown schematically here:

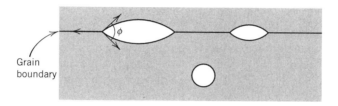

(a) Which pore will disappear first, and why? Which pore disappears last, and why?

(b) If the material in question is spinel, $MgAl_2O_4$, identify the possible rate-limiting mechanisms for the shrinkage of the pores on the grain boundary. (A mechanism is here defined as a combination of an atomic species and a transport path.) Do the same for the pore within the grain. If numerical transport data were available for all of the mechanisms you have identified, how would you go about deducing which is likely to be the rate-limiting one?

(c) If the three pores shown above are close enough for atom diffusion to take place readily, will there be any mass transported between the pores? (Aside from the tendency of each to shrink.) Show the direction of mass transport, and explain.

11. The term "calcining" in ceramics processing refers to the decomposition of a precursor, such as a metal salt or combination of metal salts, to form an oxide powder. The phrase is derived from the "calcination" of calcium carbonate:

$$CaCO_3(s) \rightarrow CaO(s) + CO_2(g),$$

a process which has a standard free energy given by

$$\Delta G° = 182.50 - 0.16(T) \quad \text{(kJ/ mole)}$$

where T is the temperature in Kelvin. The partial pressure of CO_2 in air is about 2×10^{-3} atm.

(a) If we heat $CaCO_3$ in an open crucible in air with plenty of ventilation, at what temperature will it begin to decompose to CaO?

(b) If we calcine in a tightly closed box furnace, what will happen (qualitatively)?

(c) Will it help to calcine in a tightly sealed box furnace back-filled with an inert gas such as helium or argon, at one atmosphere pressure?

(d) At what temperature will it decompose if we calcine in a vacuum furnace with a base pressure of 10^{-5} Torr (1 Torr = 1 mm of mercury). What drawbacks, if any, do you see in vacuum calcination?

12. Zinc oxide has a fairly high vapor pressure at temperatures above half the melting point, sufficiently high to cause problems in densification without some assistance. Usually a liquid-forming additive is used. Instead, we will try to densify it by hot-pressing (in air).

    (a) We know that applied pressure can increase the rate of densification. By how much does the driving force for densification increase over that which is due to capillarity alone, when a pressure of 14,500 psi (100 MPa) is applied to a powder compact in which the average particle diameter is 3 micrometers? (1 Pa = 1 N/m²)

    (b) Does the applied pressure have a direct influence on the rate of coarsening by vapor phase transport? Explain.

13. One can purchase cutlery such as scissors and knives made of "tetragonal zirconia polycrystals" (TZP), a transformation-toughened material. These cutting blades come in two varieties - a cream-colored zirconia that is sintered to high density in air, and a grayish-black version which is made by first sintering the zirconia to the closed pore stage in oxygen, and then hot-isostatically pressing with an inert gas at high temperature until full density is reached. The dark color comes from the fact that the zirconia is slightly reduced; this version of the material is more expensive and of higher strength since fewer pore clusters are left to act as internal "critical flaws." (In Japan certain high-priced sushi knives are made using HIP'd zirconia. Aside from superior edge-holding, it is held that sushi prepared thusly avoids a metallic aftertaste detectable by true aficionados when steel knives are used!)

    Lattice diffusion of oxygen is extremely fast in stabilized zirconias; this is why zirconia is also used in oxygen sensors as a "fast-ion conductor" or "solid electrolyte." Some typical data are shown in Figs. 3.13 and 3.14. The grain boundary diffusion coefficient of oxygen has not been specifically measured but is of the same order of magnitude. The lattice and grain boundary diffusion coefficients of Zr are at least 10^6 lower over the sintering temperature range of 1400–1600°C, but we do not know which is the greater.

The vapor pressure of zirconia is negligible in this temperature range.

(a) With the information given, can you conclusively state what the sintering mechanism of stabilized zirconia is? Explain. If not, can you narrow the choices down? What additional information would you need in order to form a more conclusive opinion?

(b) During the hot-isostatic pressing of zirconia which has first been sintered to the closed porosity stage, residual oxygen-filled pores are removed. Consider the two types of pores shown in problem 10. Can you conclude, on the basis of the information given, which diffusion coefficient will be rate-limiting in the pore removal process?

14. It is possible to vitrify (render glassy) liquid water if organic additives are used that change its physical properties. (This is an approach used in cryopreservation, the freezing of animal tissue without cell death.) However, most forms of ice which we encounter (snow, ice cubes) are crystalline, suggesting that the crystallization rate of undercooled liquid water is fast, and that special measures are necessary to obtain glassy ice. (For example, amorphous ice can be made by condensing water vapor onto a chilled substrate.) In this problem you are asked to examine why this is so.

Some additional information is necessary. The figure shows an educated guess at the temperature dependence of the viscosity of undercooled liquid water. The crystal-liquid interfacial energy has been determined to be 0.031

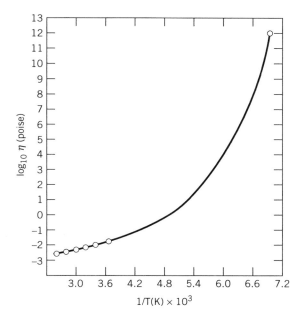

[From D.R. Uhlmann, *J. Non-Cryst. Solids*, 7, 337 (1972).]

Chapter 5 / Microstructure

J/m², based on the temperature at which homogeneous nucleation takes place when water is undercooled (this temperature is 233 K). For this problem you may use Turnbull's approximation for the free energy of crystallization. Using these results and the data below, answer the following questions. State any significant assumptions you make along the way.

(a) What linear cooling rate is necessary to form glassy ice from pure liquid water?

(b) How might a quench of this magnitude be accomplished, experimentally?

(c) If you were in fact able to quench the water without crystallization, at what temperature (approximately) would it then undergo the glass transition?

Data:

Melting point of water	$T_m = 273.2$ K
Heat of fusion	$\Delta H_m = 5900$ J/mole
Density	$\rho = 1$ g/cm³
Molecular weight	M.W. $= 18$ g/mole

Crystal nucleation frequency:

$$I_v = \frac{10^{30} \; P/cm^3 \; \sec}{\eta(P)} \exp\left[\frac{-16\pi}{3kT} \frac{V_m^2 \gamma^3}{\Delta G_m^2}\right]$$

Crystal growth rate:

$$u = f v a_o \left[1 - \exp\left(\frac{-\Delta G_m}{RT}\right)\right]$$

Molecular vibrational frequency:

$$v = \frac{kT}{3\pi a_o^3 \eta}$$

15. It is deduced from the entropy of fusion of SiO$_2$ that crystals growing from the pure melt should undergo so-called "normal" growth in which the interface is atomically rough, rather than faceted. In fact, SiO$_2$ has the lowest entropy of fusion of all known pure oxides. (The enthalpy of fusion is 7.68 kJ/mole; the melting point is 2007 K.)

5.8 Strength and Toughness

Quartz crystals can be found in nature with a clearly faceted growth habit - they grow in the shape of a hexagonal prism. It is known that these crystals are grown "hydrothermally," meaning that they grow from dilute solutions of silica, dissolved in water at high temperatures and pressures in the earth's crust.

This observation appears to contradict our predicted crystal growth morphology. Do you agree? Discuss why the observed growth morphology might differ from what we expect based on the entropy of fusion.

16. Turnbull's approximation for the molar free energy change of crystallization of an undercooled liquid is

$$\Delta G_m = \frac{\Delta H_m \Delta T}{T_m}$$

where ΔH_m is the heat of fusion and T_m is the melting point. Derive this relationship and state any assumptions implicit in the result.

17. In order for a glass-ceramic composition based on Li_2O-Al_2O_3-SiO_2 to be transparent, the crystallites need to be less than about $0.2 \mu m$ in diameter. What density of nuclei at the end of the nucleation stage of processing is necessary in order to keep the average crystal size at this limit? Assume that during the growth stage of processing, the crystals grow to impingement and do not coarsen thereafter. What final crystal size would you expect if the nuclei density is 10^{10} cm^{-3}?

Transparent lithia-alumina-silica glass ceramics achieve these fine grain sizes using TiO_2 and ZrO_2 as nucleation aids. Draw typical processing cycle for these glass ceramics. Do you think it is possible to achieve a highly crystalline, transparent product without the use of nucleation aids by carrying out *homogeneous* nucleation at a low temperature? Do you think this would be practical? Explain. (Hint: Does the nucleation frequency increase continuously without limit as temperature is lowered?)

18. Based on its chemical and structural similarity to silica, which we know is a good glassformer, it seems that germania (GeO_2) should also be a good glass former. See if this is so, by estimating the critical cooling rate necessary to form glassy germania from the melt. Potentially useful data and equations are listed below. The viscosity of GeO_2 as a function of temperature is shown in the plot below.

Melting point:	$T_m = 1389$ K
Heat of fusion	$\Delta H_m = 14.63$ kJ/mole
Edge dimension of the GeO_4^{4-} tetrahedra	2.5 angstroms
Molecular weight	M.W. = 111.59 g/mole
Liquid-solid interfacial energy	$\gamma = 0.05$ J/m^2
Density	$\rho = 3.6$ g/cm^3

Chapter 5 / Microstructure

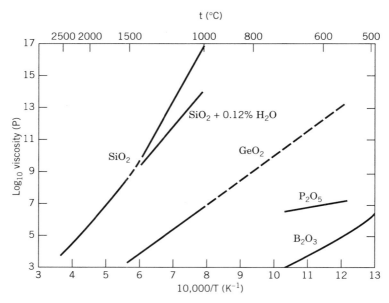

Viscosities of several glass-forming oxides as a function of reciprocal temperature. (From *Handbook of Glass Properties*, N.P. Bansal and R.H. Doremus, Academic Press, 1986.)

19. TiO_2 and SnO_2 both crystallize in the rutile structure, and form a complete solid solution above 1400°C. However, at lower temperatures there is a nearly symmetric immiscibility gap centered at ~50% of each component. As a result, spinodal decomposition can occur upon cooling TiO_2-SnO_2 solid solutions, the early stages of which are characterized by separation into Ti-rich and Sn-rich, lathe-like regions oriented in specific crystallographic directions. No charge separation occurs during this diffusional process; we can think of the oxygen sublattice as remaining fixed while the cations demix. Explain, using defect incorporation reactions, what influence you expect the additives Nb_2O_5, Ta_2O_5, Al_2O_3, and Fe_2O_3 to have on the rate of spinodal decomposition.

20. For SiO_2, $Na_2O \cdot 2SiO_2$, and pure Fe, estimate the critical cooling rate (R_c) necessary to form a glass from the melt, using the data given below. Describe the expected precipitation morphology in each system. For facetted precipitate growth, assume a screw dislocation mechanism where the fraction of preferred growth sites at the interface is $f = 0.05$. For all cases, consider the undercooling sufficiently large for the Hoffman approximation to ΔG_m to be applicable. What is the relative importance of avoiding nucleation vs. avoiding growth for each system?

	SiO$_2$	Na$_2$O·SiO$_2$	Fe
T_m (K)	2007	1147	1809
ΔH_m (kJ/mole)	7.68	33.76	15
γ_{S-L} (ergs/cm^2)	50	55	200
MW (g/mole)	60.08	182.15	55.85
ρ (g/cm^3)	2.3	2.5	7.86

In the case of Na$_2$O·SiO$_2$, an experimental K (Eq. 5.66) of 10^{26} cm^{-3}sec^{-1}P has been measured. The viscosity of liquid Fe may be taken to be 0.01P.

21. Most glasses have a fracture toughness K_{IC} of about 0.7 MPa·m$^{1/2}$. An abraded glass rod under uniform tensile loading fails at 7250 psi load (50 MPa). What, approximately, is the size of the surface flaw that caused failure?

22. If a tempered glass window is to survive stresses of 300 MPa in use, even after prolonged exposure to an abrasive environment which introduces surface flaws ranging in size from 0.01 to 20 μm, with an average size of 5 μm, how much surface compressive stress would you aim to achieve in the thermal tempering process?

23. Using the values for surface energy in Table 5.2, and Young's modulus in Table 5.7, rank in decreasing order the fracture toughness of single crystal MgO, Al$_2$O$_3$, SiC, and ZrO$_2$ (cubic). Estimate a value for K_{IC} in each case, stating clearly any assumptions.

24. Which of the following polyphase ceramic microstructures might exhibit R-curve behavior, and which are likely to show conventional brittle fracture? For each, give an educated guess as to the approximate value of K_{IC}.

    (a) Alumina with an elongated grain microstructure

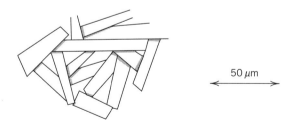

    (b) Phase-separated glass

(c) Single crystal of MnZn ferrite

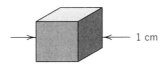

(d) Laminated alumina/cubic zirconia composite

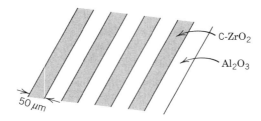

(e) Particulate alumina-tetragonal zirconia composite

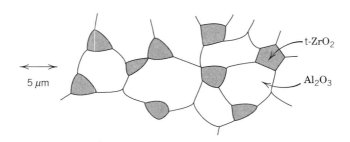

(f) SiC whisker-reinforced alumina

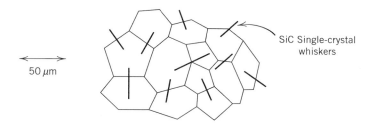

(g) SiC continuous fiber reinforced borosilicate glass

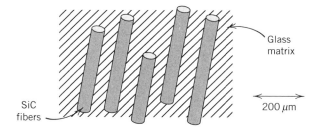

Index

Acceptors, 126–129
Al_2O_3
    color in, 133–135
    crack bridging in, 495
    crystal structure of, 32–34
    diffusional creep in, 249, 251
    dislocations in, 169
    microstructures, 385, 415, 418–420
    sintering, 413–421
Al_3O_3N, 71
Aliovalent solute, 103
Alkemade line, definition of, 334
AlN, secondary phases in, 367
Alumina, *see* Al_2O_3
Ambipolar diffusion
    coefficients, 241, 242, 244, 250
    in diffusional creep, 245–251
    in oxidation, 238–242
    in sintering, 242–245
Amorphous solids, *see* Glass
Amphiboles, 73, 74
Annealing point of glass, 82
Antiferroelectricity, 43
Antiferromagnetism, 53, 54

Antifluorite structure, 24–27
Asbestos, 73

B-H loop, 57, 58
$Ba_{1-x}K_xBiO_3$, 59
Bandgap, 119
    table of, 120
    temperature dependence of, 125
$BaPb_{1-x}Bi_xO_3$, 59
Barium titanate, *see* $BaTiO_3$
$BaTiO_3$,
    defects in, 128
    ferroelectric domains, 43, 44
    grain growth in, 385
    PTC resistor, 225, 230–233
    structure, 38–41
$Be_3Al_2Si_6O_{18}$, 73
Bi_2O_3, 228, 236
$Bi_2Sr_2Ca_2Cu_3O_{10}$, 65
$Bi_2Sr_2CaCu_2O_8$, 63
Binary phase diagram, 271ff
    eutectic, 273–276
    peritectic, 283–285, 320
    solid solution, 272

515

516   Index

Bixbyite, 27
Bloch wall, 56
BN, 27, 68
Bohr magneton, 52
Bonding
　covalent, 2, 68
　ionic, 2, 10–13, 68
Borate glass, 91–93
Boron anomaly, 92
Borosilicate glass
　fiber-reinforced composites, 467, 496
　phase separation, 461, 463
　structure and composition, 91–93
　thermal shock, 465–466
Boundary curves, see ternary phase diagram
Brouwer diagram, 136ff
Burgers vector, 166–168

CaF_2, 25, 26
Capillary force, 351ff
Capillary pressure, 354, 355
$CaSiO_3$, 73
CdS, 27
CeO_2, 35
Chemcor™, 448, 486
Chemical potential
　at curved surfaces, 356, 357
　see also Electrochemical potential
　and mobility, 212, 213
　thermodynamic equilibrium, 264, 268
Clay minerals, 73–79
Close-packing, 3–6
Coarsening
　of particles, see Ostwald ripening
　of powder compacts, 399, 400
Coercive field,
　ferroelectric, 45
　magnetic, 58
Coincident site lattice, 174, 175
Color, in sapphire, 133–135
Composites, 466–477, 492–500
Congruent melting, 291, 297, 298
Consolute temperature, 280
Continuous random network, 80, 83, 84
CoO
　electrical conductivity in, 217, 218
　nonstoichiometry, 117

Coordination polyhedron, 14, 17
Corelle™, 431
Correlation factor, for diffusion, 198–201
Corundum structure, 32–34
　derivative structures, 34–36
Covalent bond, see bonding
Covalent ceramics, 27–29, 68, 69
Crack bridging, 492, 493, 495
Crack deflection, 493, 494
Cristobalite structure, 75
Crystal morphology, 353–355
Crystal growth methods, 265, 291–293
Crystal structures,
　FCC based, 23ff
　HCP based, 31ff
　table of, 20–22
Crystallization, 431ff
　growth rate, 437–442
　in glass ceramics, 446–449
　nucleation, 433–437
　time-temperature-transformation diagrams, 432, 433, 441–446
Cuprate superconductors, 59ff, 389
Curie point, ferroelectric, 39, 40, 232, 233
Czochralski method, 291

Debye-Hückel theory, 152
Defect chemistry, 102
Defects, 101ff
　antisite, 154
　association, 146ff
　chemical reactions, 111–117
　concentrations, table of, 108
　crystal density, 131
　electronic, 118ff
　entropy of, 106, 107
　formation energies, table of, 109
　Frenkel, 105–107
　in $LiNbO_3$, 152–155
　in MgO, 125, 126, 146–148
　in NaCl, 125, 126
　in Si, 127
　in TiO_2, 129, 130
　in ZrO_2, 131, 133
　ionization, 115
　nonstoichiometry, 117
　Schottky, 105, 106
　self-compensating, 114

Densification, of powder compacts
   by viscous flow, 394–398
   of crystalline ceramics, 400ff
Diamagnetism, 52
Diamond
   structure of, 28
   synthesis, 265–267
Dielectric behavior, 42, 47–49
   dielectric constant of composites, 470–472
   loss, 48
   susceptibility, 42
Diffusion, 186ff
   ambipolar, 236–251
   diffusion coefficients for ceramics, 187
   interstitial mechanism, 196
   in MgO, 202–205
   in NaCl, 201, 202
   random walk, 192, 193
   self, 196
   thermally activated, 193–195
   tracer, 197
   in transition metal monoxides, 205–208
   vacancy mechanism, 196
   in zirconia, 208–211
Dihedral angle, 360–363, 365
   and sintering, 417, 422, 425,
   in Al_2O_3, 363, 417
   in $AlN-Y_2O_3$, 367
   in MgO, 363
   in $ZnO-Bi_2O_3$, 423, 428
   Rayleigh instability, 370, 371
   second-phase distributions, 365, 368
Dislocations, 166–171
   Burgers vector, 166–168
   climb, 170
   glide, 169–170
   in alumina, 169
Disorder, see Defects
Donors, 126–129

Elastic modulus
   of composites, 470–473
   typical values, 479
Electrical conductivity, 211ff
   in CoO, 217–219
   in MgO, 219–221
   mixed, 219
   in NiO, 217–219
   nonlinear, see Varistors
   partial, 216, 217
   in $SrTiO_3$, 222–225
   in ZrO_2, 221, 222
   of composites, 474–476
Electrochemical potential, 233ff
Electronic structure, 118
Equilibrium, thermodynamic, 263–265
Equivalence diagram, see Reciprocal salt systems
Eutectoid, 286
Exchange interaction, in magnetism, 53

Face-centered cubic lattice (FCC), 3–6
Fast firing, 403
Fast-ion conductors, table of, 236
Fe_2SiO_4, 73
Fe_3O_4, 54
FeO
   nonstoichiometry, 117, 288, 289
   diffusion, 207, 208
Fermi level, 123, 124
Fermi-Dirac function, 121, 122
Ferrimagnetism, 53
Ferrites, 52–58
Ferroelectrics, 42ff
   domain walls, 43
   domains, 43–44, 232
   hysteresis, 44–47
   polarization, 45
   strain, 46
Ferromagnetism, 52–53
Fiber-reinforced ceramics, 466, 467, 492, 493, 494, 496–500
Fick's laws, 186–190
Fluorite structure, 24–27, 29
Fracture toughness of ceramics, 477ff
   microstructural toughening, 487ff
   transformation toughening, 488–492
   typical values, 484
Fracture, brittle, 478–481
Free energy curves, 272, 277–281

GaAs, 27
$Gd_2Ti_2O_7$ - $Gd_2Zr_2O_7$, 27, 236
GeO_2, crystal growth rate, 439

518   Index

Gibbs phase rule, 267–271
Gibbs adsorption isotherm, 358
Gibbs-Thompson equation, *see* Thompson-Freundlich equation
Glass
    amorphous silicon, 84
    borosilicate, 91–93, 461, 463, 465–467, 496
    critical cooling rate for glass formation, table of, 446
    crystallization, 431–433, *see also* Glass-ceramics
    glass transition, 81, 82
    phase separation, 449–464
    processing, 430ff
    silica, 83, 84, 86, 466
    soda-lime-silica, 89, 466
    sodium silicate, 89
    structure of, 80ff
    thermal expansion coefficients, 92, 93, 465, 466
    thermal shock, 464–466
    viscosity, 82
Glass-ceramics, 431, 446–449, 464–466
    fiber-reinforced composites, 496
    machinable, 431, 449
    nucleating agents, 447
    processing 447, 448
    thermal expansion, 447, 466
    transparent, 447
Grain boundary
    enthalpy, 372–374
    entropy, 374
    films at, 176–181
    in NiO, 178
    in Si_3N_4, 180
    in thick-film resistors, 181
    in TiO_2, 162
    in ZnO, 228
    mobility, 375–383
    pinning, 379–383
    segregation, 156–165
    structure of, 171–176
Grain growth, 371ff
    abnormal, *see* discontinuous
    anisotropy, 386, 388, 389, 417, 418, 420
    discontinuous, 372, 383–386
    during sintering, 405–408
    heat evolution during, 373
    in Al_2O_3, 385, 415–419
    in $BaTiO_3$, 385
    in MgO with liquid, 391
    in oxide superconductor, 389
    in Si_3N_4, 388, 484, 496
    in spinel, 386
    in TiO_2, 373
    kinetics, 375, 376
    normal, 372, 375
Grain shape, 364–368
Griffith theory, 480, 481, 485, 487

Hard magnet, 58
Herring's scaling law, 403
Hexagonal-close-packed lattice (HCP), 3–6
Heywang, W., 231
Homologous temperature, 102
Hot-isostatic pressing, 371
Hot-pressing, 429, 430

Ilmenite
    kinetic demixing in, 253, 254
    structure of, 34–36
Immiscibility, *see* phase separation
Incongruent melting, 283, 291, 297, 298
Interfaces
    energy, table of, 360
    in fiber-reinforced composites, 499, 500
Intergranular films, 176–181
    and grain boundary mobility, 376, 377
International Tables for Crystallography, 2
Interstitial sites, 7–10
Ionic bond, *see bonding*
Ionic radii, 14
    table of, 15–16
Isothermal sections, in ternary phase diagrams, 310–314

Kaolinite, structure of, 78, 79
Kimax', 93, 461
Kinetic demixing
    oxygen potential gradient, 251–254
    stress-induced, 254–255
Kröger-Vink

diagram, *see* Brouwer diagram
notation, 110, 111

$La_{2-x}Sr_xCuO_4$, 62, 66
La_2CuO_4, 60, 65
$LaAlO_3$, 39
Laser, ruby, 135
Lead zirconate-titanate, *see* PZT
Lead lanthanum zirconium titanate, *see* PLZT
Lever rule
  binary, 290
  ternary, 315–317
Li_2O, 25
Li_2O-Al_2O_3-SiO_2 glass ceramics, 431, 446–448, 449, 450, 466
$LiNbO_3$ structure, 34–36
  Curie point, 45
  defects in, 152–155
  phase diagram, 292
Liquid-liquid immiscibility, 302, 309, 314, 315, 463
Liquidus, 273, 293, 297
Lithium niobate, *see* $LiNbO_3$
Low thermal expansion glass, 465–466
Lucalox', 413

Madelung constant, 9–13
  table of, 12
Magnetic
  ceramics, *see* ferrites
  hysteresis, 56
  permeability, 56
Metastable phase diagram, 281, 282
Mg_2SiO_4, 73
$MgAl_2O_4$
  discontinuous grain growth in, 386
  nonstoichiometry, 153
  structure, 49–50
MgO
  ambipolar diffusion, 244
  defects in, 125, 126, 146–148
  diffusion, 202–205
  electrical conductivity, 219–221, 237
  grain boundaries, 160, 179
  grain growth with liquid, 391
  precipitation in, 148–151
$MgSiO_3$, 73

Mica, structure of, 79
Miller indices, 2
$Mn_{1-x}Zn_xFe_2O_4$, 55, 387
Mobility, of atoms and defects,
  absolute, 213
  and diffusivity, 214, 215
  carrier, table of, 216
  drift, 215
  electrical, 214
Montmorillonite, structure of, 79
Mullite, 282, 296
Multiple phase separation, 460–462

Na_2O-B_2O_3-SiO_2 glass, 91–93, 465, 466, 463
NaCl
  defects in, 125, 126
  diffusion, 201–202
  structure of, *see* rocksalt structure
Nernst equation, 234, 236
Nernst-Einstein equation, 214–215
NiO
  electrical conductivity, 217–219
  nonstoichiometry, 117
$NiTiO_3$, 254
Nucleation, 433–437
  heterogeneous, 443–445
  temperatures, table of, 437

O/Si ratio, in silicate structures, 73, 74
Orthosilicates, 74
Ostwald ripening, 371, 388–391
Oxidation, defect formed during, 116
Oxynitrides, 69–72

p-n junction, 124
Paraelectric, 42
Paramagnetism, 52
Pauli exclusion principle, 10
Pauling's rules, 13, 14, 17–19
$Pb(Mg_{1/3}Nb_{2/3})O_3$, 39, 48
$Pb(Mg_{1/3}Nb_{2/3})O_3$, 48
$Pb(Sc_{1/2}Ta_{1/2})O_3$, 39
$Pb_2Ru_2O_7$, 27
$PbZrO_3$, 38, 43
Percolation, 469, 474–477
Peritectoid, definition of, 287
Permeability of porous ceramics, 469

Perovskite, 38–41
Phase diagram,
    $BaO-TiO_2$, 285
    $BaTiO_3$, 155
    Be-Si-Al-O-N, 306
    $CaO-Al_2O_3-SiO_2$, 332
    Fe-O, 289
    $FeO-SiO_2$, 309
    H_2O, 269
    $KNbO_3-K_2CO_3$, 292
    $Li_2O-Nb_2O_5$, 292
    $MgO-Al_2O_3$, 276
    $MgO-Al_2O_3-SiO_2$, 322, 323
    MgO-CaO, 274
    MgO-NiO, 272
    $MgO-ZrO_2$, 30
    $Na_2O-B_2O_3-SiO_2$, 463
    $NaF-KF-MgF_2$, 346
    quartz-feldspar-clay, 343
    Si-Al-O-N, 304, 305
    Si-Y-O-N, 350
    $SiO_2-Al_2O_3$, 282
    $SiO_2-Al_2O_3$-FeO, 296, 311, 313–315
    $SiO_2-Al_2O_3-Y_2O_3$, 348
    SiO_2-TiO_2-CaO, 349
    ZrO_2-CaO, 286
Phase rule, *see* Gibbs phase rule
Phase separation, 449–464
    immiscibility gap, 280
    see also Liquid-liquid immiscibility
Phase transformations,
displacive, 29, 73
reconstructive, 29
Piezoelectric ceramics, 40, 42, 46
electromechanical coupling factor, 46
strain, 46
PLZT, 41, 46
PMN, 48
Polarization, in ferroelectrics, 45
Polymorphism, 12, 29–31
    of SiO_2, 75
    of ZrO_2, 29
    of $BaTiO_3$, 41
Polytypism, 29–31
Polytypoids, 71
Porcelain, 342–345
Pore
    agglomeration, 384, 409, 410

drag, 381–383, 408
mobility, *see* pore drag
pore-boundary separation, 382, 383, 405–408, 416, 417
removal during sintering, 406–409
stability, 409
Positive-temperature-coefficient (PTC) resistor, 225, 230–233
Precipitation, 148–151, *see also* Phase separation
Primary phase field, definition of, 297
Process zone, 490
Pyrex™, 93, 461, 464, 466
Pyroceram™, 431, 447, 466
Pyrochlore, 27
Pyrophyllite, structure of, 78
Pyrosilicates, 74
Pyroxenes, 73, 74
PZT, 40, 46

Quartz, structure of, 75

R-curve, 490–493
Radial distribution function, 85–87
Radius ratio, of ions, 14, 17
Ramsdell notation, 31, 71
Random close-packed structure, 80, 85
Rayleigh instability, 368–371
    and sintering, 404
Reciprocal salt systems, 302–307, 350
Reduction, defects formed during, 116
Relaxor ferroelectric, 48
Residual stress, 486, 487
Resorption of phases, 335–337, 340
Rocksalt structure, 23–24
Ruby, 133
Rules of mixing, 468–473
Rutile structure, 37–38

Sapphire, *see* Al_2O_3, Corundum
    blue, 135
Saturation magnetization, in spinel ferrites, 54, 55
Segregation
    grain boundary, 156–165
    in two-component system, 360
Sensors
    oxygen, 234–235

temperature, *see* Thermistors
Si$_2$N$_2$O, 71
Si$_3$N$_4$ crystal structure, 68–69, 70
    polycrystalline microstructure, 388, 496
    toughened, 494, 496
Sialons, 69, 71, 72
SiC structure, 30–31
Silica, amorphous, 83, 84
Silicates, crystalline, 72ff
Silicon nitride, *see* Si$_3$N$_4$
Sintering, 392ff
    coarsening, 399, 400
    final stage, 404–409
    grain size-density map, 407
    initial stage, 394, 399–403
    intermediate stage, 404
    liquid phase, 393, 421–426
    magnesia-doped alumina, 413–421
    particle size dependence, 403
    rate-limiting mechanisms, 402
    reactive, 421–427
    sintering pressure, 392
    solid state, 393, 399ff
    table, methods for various ceramics, 393
    viscous, 392–398
    ZnO-Bi$_2$O$_3$, 421, 422
Soda lime silicate glass, 89
Sodium borosilicate glass, *see* Borosilicate glass
Sodium vapor lamp, 413–414
Soft magnet, 58
Solid state amorphization, 81
Solidus, 273, 301
Solubility, surface effects on, 357
Solute drag, 377–379
Space-charge, 156ff
    solute segregation, 158–165
Spinel structure, 49–51, 54, 55
Spinodal decomposition, 450–461
Spreading coefficient, 361
Square loop magnet, 58
SrTiO$_3$ structure, 38
    electrical conductivity in, 222–225
SrTiO$_3$, secondary phase in, 366
Stability diagram, 263
Stokes-Einstein relation, 434

Strain point of glass, 82
Strength of ceramics, 477ff
    high strength ceramics, 487
    theoretical strength, 478–480
    variability, 485–486
Stress intensity factor, 481–485
Subsolidus phase equilibria, 285
Superconductivity, 52, 65ff
Superconductors, structure of, 59–65
    microstructure, 389
Superplastic ceramics, 248
Surface compression strengthening, 431, 486, 487
Surface energy,
    and sintering, 392
    anisotropy in, 354
    definition of, 352
    instability, *see* Rayleigh instability
    table of, 359
Surface flaws, 486
Surface stress, definition of, 353
Surface tension, definition of, 352

Ternary phase diagram, 293ff
    boundary curve, 299
    compatibility triangle, 301, 314, 323
    eutectic reaction, 318, 323, 332–334
    eutectic systems, 294, 298, 319
    invariant point, 300, *see also* Gibbs phase rule
    nonequilibrium crystallization, 340–342
    peritectic reaction, 319, 320, 328, 329, 339
    peritectic systems, 298, 321
    solid solution, 297
    temperature contours, 299
Tetrakaidecahedron, 364
Thermal conductivity, of two-phase mixture, 475
Thermal tempering of glass, 486, 487
Thermal shock, 464–466
Thermistors,
    NTC, 225
    PTC barium titanate, 230
Thompson-Freundlich equation, 357
    and particle coarsening, 388, 389
Time-temperature-transformation

522　Index

diagram, 432, 442–445
TiN, 68
TiO_2
　defects in, 118, 129, 130, 144–146, 164
　electrical conductivity of, 144
　grain boundaries, 161–165
　nonstoichiometry, 118
　oxidation of, 238–239
　oxygen sensor, 142–146
　-SnO_2, spinodal decomposition in, 460, 461
　structure of, 37–38
$Tl_2Ba_2Ca_2Cu_3O_{10}$, 64, 65
Toughness, see Fracture toughness
Transducers, piezoelectric, 40
Transference number, 216
Transformation toughening, 30, 488–492
Tridymite, 75

UO_2, 25, 27, 354

Vapor pressure, over curved surfaces, 356–358
Varistors, 225–230
Viscosity
　and diffusion, 434
　of glass, 82, 510
　of water, 507
Visions™, 431, 447, 466
Vycor™, 93, 431, 461, 464

Weibull theory, 485–486

Weiss domains, 56
Wetting angle, 360, 361
Wollastonite, 73
Working point of glass, 82
Wulff plot, 353, 354
Wurzite structure, 31–32

XR7 capacitor, 48

Y_2O_3, 27
$YBa_2Cu_3O_{7-x}$, 59–61, 66
Young's modulus, table of, 479
　of composites, 471, 473

Zachariesen, W.H., 83, 87
Zener, C., analysis of grain growth, 379–381
Zincblende structure, 27–29
$ZnFe_2O_4$, 54, 56
ZnO
　grain boundary segregation, 228
　varistors, 225–230
ZnO, Bi_2O_3-rich secondary phase, 423, 428
ZrO_2
　defects in, 131–133
　electrical conductivity, 221, 222
　oxygen sensor, 234, 235
　stabilized, 29–30, 489
　structure of, see fluorite
　transformation toughened, 488–492
$ZrSiO_4$, 73
Zirconia, see ZrO_2

UNIT CONVERSION FACTORS

Length

1 Å = 10^{-10} m
1 nm = 10^{-9} m
1 μm = 10^{-6} m
1 mm = 10^{-3} m
1 cm = 10^{-2} m
1 in. = 25.4 mm
1 in. = 2.54 cm

Volume

1 m^3 = 10^6 cm^3
1 m^3 = 35.32 ft^3
1 $in.^3$ = 16.39 cm^3

Mass

1 kg = 10^3 g
1 lb_m = 453.6 g

Density

1 g/cm^3 = 10^3 kg/m^3
1 g/cm^3 = 62.4 lb_m/ft^3
1 g/cm^3 = 0.0361 $lb_m/in.^3$

Force

1 N = 10^5 dynes
1 lb_f = 4.448 N

Stress/Pressure

1 MPa = 145 psi = 10 bar
1 Pa = 1 N/m^2 = 10 $dynes/cm^2$

Fracture Toughness

1 Mpa $\sqrt{m}$ = 910 psi $\sqrt{in.}$

Energy

1 J = 10^7 ergs
1 J = 6.24 × 10^{18} eV
1 J = 1 N·m = 1 kg m^2/s^2
1 cal = 4.184 J
1 Btu = 1054 J
1 ft-lb_f = 1.356 J
1 J = 1 V·c

Power

1 W = 1 J/sec
1 W = 0.239 cal/s
1 W = 3.414 Btu/h

Viscosity

1 P = 0.1 Pa-s = 1 g/cm·s

Temperature, T

$T(K) = 273 + T(°C)$
$T(°C) = 5/9[T(°F) - 32]$

Specific Heat

1 J/kg-K = 2.39 × 10^{-4} cal/g-K
1 J/kg-K = 2.39 × 10^{-4} Btu/lb_m-°F

Thermal Conductivity

1 W/m-K = 2.39 × 10^{-3} cal/cm-s-K
1 W/m-K = 0.578 Btu/ft-h-°F

Values of Selected Physical Constants

Quantity	Symbol	SI Units	cgs Units
Avogadro's number	N_A	6.023×10^{23} molecules/mol	6.023×10^{23} molecules/mol
Boltzmann's constant	k	1.38×10^{-23} J/atom-K	1.38×10^{-16} erg/atom-K 8.62×10^{-5} eV/atom-K
Bohr magneton	μ_B	9.27×10^{-24} A-m²	9.27×10^{-21} erg/gauss[a]
Electron charge	e	1.602×10^{-19} C	4.8×10^{-10} statcoul[b]
Electron mass	—	9.11×10^{-31} kg	9.11×10^{-28} g
Faraday's constant	F	96,500 C/mol	—
Gas constant	R	8.31 J/mol-K	1.987 cal/mol-K
Permeability of a vacuum	μ_0	1.257×10^{-6} henry/m	unity[a]
Permittivity of a vacuum	ϵ_0	8.85×10^{-12} farad/m	unity[b]
Planck's constant	h	6.63×10^{-34} J-s	6.63×10^{-27} erg-s 4.13×10^{-15} eV-s
Velocity of light in a vacuum	c	3×10^8 m/s	3×10^{10} cm/s

[a] In cgs-emu units.
[b] In cgs-esu units.

Unit Abbreviations

A = ampere	in. = inch	N = newton
Å = angstrom	J = joule	nm = nanometer
Btu = British thermal unit	K = degrees Kelvin	P = poise
C = Coulomb	kg = kilogram	Pa = pascal
°C = degrees Celsius	lb_f = pound force	s = second
cal = calorie (gram)	lb_m = pound mass	T = temperature
cm = centimeter	m = meter	μm = micrometer
eV = electron volt	Mg = megagram	(micron)
°F = degrees Fahrenheit	mm = millimeter	W = watt
ft = foot	mol = mole	psi = pounds per square inch
g = gram	MPa = megapascal	

SI Multiple and Submultiple Prefixes

Factor by Which Multiplied	Prefix	Symbol
10^9	giga	G
10^6	mega	M
10^3	kilo	k
10^{-2}	centi[a]	c
10^{-3}	milli	m
10^{-6}	micro	μ
10^{-9}	nano	n
10^{-12}	pico	p

Limited Use License Agreement

This is the John Wiley & Sons, Inc. (Wiley) limited use License Agreement, which governs your use of any Wiley proprietary software products (Licensed Program) and User Manual(s) delivered with it.

Your use of the Licensed Program indicates your acceptance of the terms and conditions of this Agreement. If you do not accept or agree with them, you must return the Licensed Program unused within 30 days of receipt or, if purchased, within 30 days, as evidenced by a copy of your receipt, in which case, the purchase price will be fully refunded.

License: Wiley hereby grants you, and you accept, a non-exclusive and non-transferrable license, to use the Licensed Program and User Manual(s) on the following terms and conditions:

- The Licensed Program and User Manual(s) are for your personal use only.
- You may use the Licensed Program on a single computer, or on its temporary replacement, or on a subsequent computer only.
- You may modify the Licensed Program for your use only, but any such modifications void all warranties expressed or implied. In all respects, the modified programs will continue to be subject to the terms and conditions of this Agreement.
- A backup copy or copies may be made only as provided by the User Manual(s), but all such backup copies are subject to the terms and conditions of this Agreement.
- You may not use the Licensed Program on more than one computer system, make or distribute unauthorized copies of the Licensed Program or User Manual(s), create by decompilation or otherwise the source code of the Licensed Program or use, copy, modify, or transfer the Licensed Program, in whole or in part, or User Manual(s), except as expressly permitted by this Agreement.
- If you transfer possession of any copy or modification of the Licensed Program to any third party, your license is automatically terminated. Such termination shall be in addition to and not in lieu of any equitable, civil, or other remedies available to Wiley.

Term: This License Agreement is effective until terminated. You may terminate it at any time by destroying the Licensed Program and User Manual together with all copies made (with or without authorization).

This Agreement will also terminate upon the conditions discussed elsewhere in this Agreement, or if you fail to comply with any term or condition of this Agreement. Upon such termination, you agree to destroy the Licensed Program, User Manual(s), and any copies made (with or without authorization) of either.

Wiley's Rights: You acknowledge that the Licensed Program and User Manual(s) are the sole and exclusive property of Wiley. By accepting this Agreement, you do not become the owner of the Licensed Program or User Manual(s), but you do have the right to use them in accordance with the provisions of this Agreement. You agree to protect the Licensed Program and User Manual(s) from unauthorized use, reproduction or distribution.

Warranty: To the original licensee only, Wiley warrants that the diskettes on which the Licensed Program is furnished are free from defects in the materials and workmanship under normal use for a period of ninety (90) days from the date of purchase or receipt as evidenced by a copy of your receipt. If during the ninety day period, a defect in any diskette occurs, you may return it. Wiley will replace the defective diskette(s) without charge to you. Your sole and exclusive remedy in the event of a defect is expressly limited to replacement of the defective diskette(s) at no additional charge. This warranty does not apply to damage or defects due to improper use or negligence.

This limited warranty is in lieu of all other warranties, expressed or implied, including, without limitation, any warranties of merchantability or fitness for a particular purpose.

Except as specified above, the Licensed Program and User Manual(s) are furnished by Wiley on an "as is" basis and without warranty as to the performance or results you may obtain by using the Licensed Program and User Manual(s). The entire risk as to the results or performance, and the cost of all necessary servicing, repair, or correction of the Licensed Program and User Manual(s) is assumed by you.

In no event will Wiley be liable to you for any damages, including lost profits, lost savings, or other incidental or consequential damages arising out of the use or inability to use the Licensed Program or User Manual(s), even if Wiley or an authorized Wiley dealer has been advised of the possibility of such damages.

General: This Limited Warranty gives you specific legal rights. You may have others by operation of law which varies from state to state. If any of the provisions of this Agreement are invalid under any applicable statute or rule of law, they are to that extent deemed omitted.

This Agreement represents the entire agreement between us and supercedes any proposals or prior Agreements, oral or written, and any other communication between us relating to the subject matter of this Agreement.

This Agreement will be governed and construed as if wholly entered into and performed within the State of New York. You acknowledge that you have read this Agreement, and agree to be bound by its terms and conditions.

KERAMOS

WELCOME

Thank you for trying KERAMOS. We hope that it will help you in your efforts to learn ceramic science and engineering.

REQUIREMENTS

KERAMOS is a DOS program that requires a 286 or better IBM compatible personal computer, VGA graphics or better, and at least 507 KB free conventional memory (out of a total of 640 KB) to run. It is most effective when used with a mouse, in which case a DOS mouse driver must be running (as opposed to a WINDOWS mouse driver).

INSTALLATION

There are two ways to install the program. (1) Run the INSTALL.EXE program provided on the floppy disk or (2) manually copy all of the files from the floppy disk to a single empty subdirectory on your hard disk. KERAMOS needs to have all of its files in the single subdirectory. If the INSTALL.EXE program fails to work for any reason, use the manual method.

The INSTALL.EXE program will ask you a few questions before copying all of the files from the floppy diskette to a subdirectory on your hard disk.

RUNNING KERAMOS

To run the program switch to its subdirectory and type "KERAMOS". (If you receive an "insufficient memory" message, then it is necessary to quit one or more applications to free up conventional memory.)

KERAMOS can be run from the floppy disk, but some of the functions will be slower.

DEINSTALLATION

To remove KERAMOS from your machine delete all of the program files in the \KERAMOS directory (or whichever subdirectory it has been installed in). KERAMOS does not install files into any other subdirectory, such as \DOS or \WINDOWS. A complete list of file names is provided in FILELIST.TXT.

Copyright © 1996 by John Wiley & Sons, Inc.